13
Topics in Organometallic Chemistry

Topics in Organometallic Chemistry

Recently Published and Forthcoming Volumes

Metal Carbenes in Organic Synthesis
Volume Editor: K.H. Dötz
Vol. 13, 2004

Theoretical Aspects of Transition Metal Catalysis
Volume Editor: G. Frenking
Vol. 12, 2005

Ruthenium Catalysts and Fine Chemistry
Volume Editors:
C. Bruneau and P.H. Dixneuf
Vol. 11, 2004

New Aspects of Zirconium Containing Organic Compounds
Volume Editor: I. Marek
Vol. 10, 2004

CVD Precursors
Volume Editor: R. Fischer
Vol. 9, 2005

Metallocenes in Stereoselective Synthesis
Volume Editor: T. Takahashi
Vol. 8, 2004

Transition Metal Arene π-Complexes in Organic Synthesis and Catalysis
Volume Editor: E.P. Kündig
Vol. 7, 2004

Organometallics in Process Chemistry
Volume Editor: R.D. Larsen
Vol. 6, 2004

Organolithiums in Enantioselective Synthesis
Volume Editor: D.M. Hodgson
Vol. 5, 2003

Organometallic Bonding and Reactivity: Fundamental Studies
Volume Editor: J.M. Brown, P. Hofmann
Vol. 4, 1999

Activation of Unreactive Bonds and Organic Synthesis
Volume Editor: S. Murai
Vol. 3, 1999

Lanthanides: Chemistry and Use in Organic Synthesis
Volume Editor: S. Kobayashi
Vol. 2, 1999

Alkene Metathesis in Organic Synthesis
Volume Editor: A. Fürstner
Vol. 1, 1998

Metal Carbenes in Organic Synthesis

Volume Editor: K. H. Dötz

With contributions by
J. Barluenga · K. H. Dötz · M. P. Doyle · F. J. Fañanás · J. Flórez ·
L. S. Hegedus · J. Hermanns · A. de Meijere · A. Minatti · J. Mulzer ·
E. Öhler · F. Rodríguez · B. Schmidt · T. Strassner · Y.-T. Wu

Springer

The series *Topics in Organometallic Chemistry* presents critical overviews of research results in organometallic chemistry, where new developments are having a significant influence on such diverse areas as organic synthesis, pharmaceutical research, biology, polymer research and materials science. Thus the scope of coverage includes a broad range of topics of pure and applied organometallic chemistry. Coverage is designed for a broad academic and industrial scientific readership starting at the graduate level, who want to be informed about new developments of progress and trends in this increasinly interdisciplinary field. Where appropriate, theoretical and mechanistic aspects are included in order to help the reader understand the underlying principles involved.
The individual volumes are thematic and the contributions are invited by the volumes editors.
In references Topics in Organometallic Chemistry is abbreviated
Top. Organomet. Chem. and is cited as a journal

Springer WWW home page: springeronline.com
Visit the TOMC contents at springerlink.com

ISSN 1436-6002

DOI 10.1007/978-3-540-40910-6

Bibliographic information published by Die Deutsche Bibliothek
Die Deutsche Bibliothek lists this publication in the Deutsche Nationalbibliographie; detailed bibliographic data is available in the Internet at <http://dnb.ddb.de>.

springeronline.com

Originally published by Springer-Verlag Berlin Heidelberg New York in 2004

MyCopy version of the original edition 2004

Typesetting: Fotosatz-Service Köhler GmbH, Würzburg
Production editor: Christiane Messerschmidt, Rheinau
Cover: design & production GmbH, Heidelberg

Printed on acid-free paper 02/3020 – 5 4 3 2 1 0
www.springer.com/mycopy

Volume Editors

Professor Dr. K. H. Dötz
Kekulé-Institut für Organische Chemie
und Biochemie
Rheinische Friedrich-Wilhelms-Universität
Gerhard-Domagk-Strasse 1
53121 Bonn, Germany
Doetz@uni-bonn.de

Editorial Board

Topics in Organometallic Chemistry is also Available Electronically

For all customers who have a standing order to Topics in Organometallic Chemistry, we offer the electronic version via SpringerLink free of charge. Please contact your librarian who can receive a password for free access to the full articles by registration at:

springerlink.com

If you do not have a subscription, you can still view the tables of contents of the volumes and the abstract of each article by going to the SpringerLink Homepage, clicking on "Browse by Online Libraries", then "Chemical Sciences", and finally choose Topics in Organometallic Chemistry.

You will find information about the

- Editorial Board
- Aims and Scope
- Instructions for Authors
- Sample Contribution

at springeronline.com using the search function.

Preface

In 1915 a paper submitted to the Russian Physical and Chemical Society by L. Tschugajeff, professor at the Inorganic Division of the Chemical Institute of the University of St. Petersburg, stated that the reaction of a potassium chloroplatinum complex with methylisocyanide and hydrazine hydrate affords red shiny crystals; a careful and correct elemental analysis encouraged him to suggest the structure of a hydrazide-bridged platinum dimer. In 1968 – after E. O. Fischer's pioneering rational synthesis and complete analytical characterization of carbonyl carbene complexes of chromium and tungsten – Tschugajeff's reaction was reinvestigated, and the complex was identified as a cyclic diaminocarbene coordinated to platinum. It revealed that by serendipity Tschugajeff had the first metal carbene complex in his hands, an idea which was beyond imagination in the early 1900's.

Indeed, metal carbene chemistry started in 1964 with the seminal work of E. O. Fischer. He demonstrated that the sequential addition of an organolithium nucleophile and an *O*-alkylating or acylating electrophile across the C=O bond – a well-known protocol for aldehydes and ketones – can be extended to CO ligands in metal carbonyls. Subsequent studies in the Munich laboratories on synthesis, strucure and reactivity have characterized carbonyl carbene complexes as an electrophilic metal-substituted carbenium species which laid the basis for both organometallic coordination chemistry and organic synthesis. When R. R. Schrock discovered a nucleophilic metal carbene counterpart in 1974 the diversity of the field and its scope became obvious. It revealed that the reactivity of carbene ligands may be tuned by the carbene substitution pattern as well as by an appropriate choice and combination of the metal center and the coligand sphere. Up to now carbene complexes are known for most of the transition metals, and some of those have been developed to useful reagents and catalysts in organic synthesis.

The concept that the electronic properties of the carbene carbon atom can be tuned by the metal coligand fragment, which serves as an organometallic functional group, has led to an impressive variety of unprecedented carbon carbon bond forming reactions as demonstrated by the contributions of A. de Meijere and J. Barluenga. The chapter by Th. Strassner illustrates how the rationalization of experimental results is supported by the rapid progress in theoretical methodology which now also provides a guideline for the design of novel reactions. Beyond its role as a functional group the transition metal may serve as a template which allows for a preorganization of the relevant substrates required for a successful subsequent coupling process. This principle is illustrated by the chromium-templated benzannulation to give fused arenes presented by our group as well as by the photo-induced generation of chromium ketene intermediates applied by L. Hegedus to cycloaddition and nucleophilic addition reactions.

Apart from complexes which are stable under standard conditions metal carbenes have a tradition as catalysts formed in situ. The methodology of copper-catalyzed reactions of diazo compounds has been extended to binuclear rhodium systems that provide selective catalysts for domino-type addition, insertion and cyclization reactions as illustrated by M. Doyle. Perhaps the most spectacular recent development in organic synthetic methodology refers to olefin metathesis which was discovered in the mid 1960's and subsequently commercially applied in a heterogenous process. Based on the increasing knowledge of metal carbene chemistry Chauvin proposed a non-pairwise alkylidene exchange mechanism which fostered the development of improved catalysts. Low-coordinate carbene complexes of molybdenum and tungsten have been designed by Schrock, and more recently, Grubbs and others have developed ruthenium carbene catalysts for the ring-closing variant (RCM) to the most efficient methodology of macrocyclization: The principles of this type of reaction are presented by B. Schmidt while its scope and versatility are highlighted by J. Mulzer who describes elegant approaches to complex natural products.

The aim of this volume is to convince the reader that metal carbene complexes have made their way from organometallic curiosities to valuable – and in part unique – reagents for application in synthesis and catalysis. But it is for sure that this development over 4 decades is not the end of the story ; there is both a need and considerable potential for functional organometallics such as metal carbon multiple bond species which further offer exciting perspectives in selective synthesis and catalysis as well as in reactions applied to natural products and complex molecules required for chemical architectures and material science.

Bonn, April 2004 Karl Heinz Dötz

Contents

Topics Organomet Chem (2004) 13: 1–20
DOI 10.1007/978-3-540-40910-6

Electronic Structure and Reactivity of Metal Carbenes

Thomas Strassner (✉)

Institut für Physikalische Organische Chemie, Technische Universität Dresden,
Mommsenstr. 13, 01062 Dresden, Germany
thomas.strassner@chemie.tu-dresden.de

Abstract Metal carbenes have for a long time been classified as Fischer or Schrock carbenes depending on the oxidation state of the metal. Since the introduction of *N*-heterocyclic carbene complexes this classification needs to be extended because of the very different electronic character of these ligands. The electronic structure of these different kinds of carbene complexes is analysed and compared to analogous silylenes and germylenes. The relationship between the electronic structure and the reactivity towards different substrates is discussed.

Keywords Reactivity · Theory · Density functional theory (DFT) calculations · Carbenes

Abbreviations

BDE	Bond dissociation energy
CDA	Charge decomposition analysis
Cp	Cyclopentadienyl
Cy	Cyclohexyl
DFT	Density functional theory
EDA	Energy decomposition analysis
Hal	Halogen
HF	Hartree–Fock
Me	Methyl
Ph	Phenyl
PPh_3	Triphenylphosphine
post-HF	post-Hartree–Fock
TM	Transition metal

1 Introduction

Carbenes – molecules with a neutral dicoordinate carbon atom – play an important role in all fields of chemistry today. They were introduced to organic chemists by Doering and Hoffmann in the 1950s [1] and to organometallic chemists by Fischer and Maasböl about 10 years later [2, 3]. But it took another 25 years until the first carbenes could be isolated [4–8]; examples are given in Scheme 1.

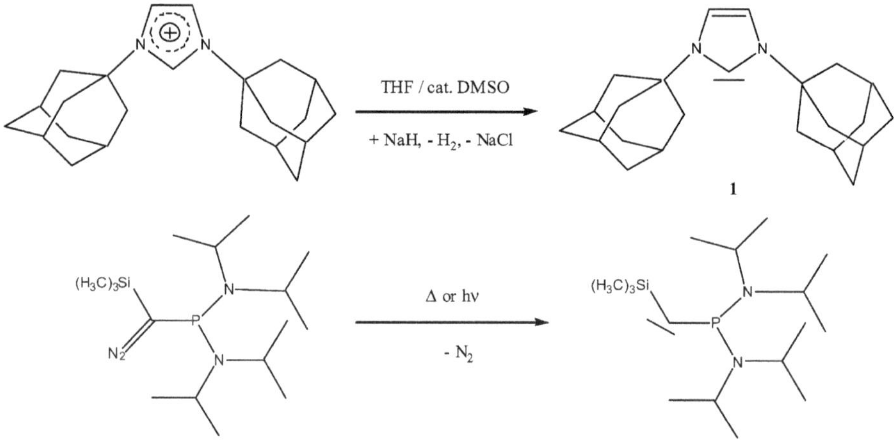

Scheme 1 Examples of isolated carbenes

The surprising stability of *N*-heterocyclic carbenes was of interest to organometallic chemists who started to explore the metal complexes of these new ligands. The first examples of this class had been synthesized as early as 1968 by Wanzlick [9] and Öfele [10], only 4 years after the first Fischer-type carbene complex was synthesized [2, 3] and 6 years before the first report of a Schrock-type carbene complex [11]. Once the *N*-heterocyclic ligands are attached to a metal they show a completely different reaction pattern compared to the electrophilic Fischer- and nucleophilic Schrock-type carbene complexes.

Wanzlick showed that the stability of carbenes is increased by a special substitution pattern of the disubstituted carbon atom [12–16]. Substituents in the vicinal position, which provide π-donor/σ-acceptor character (Scheme 2, X), stabilize the lone pair by filling the p-orbital of the carbene carbon. The negative inductive effect reduces the electrophilicity and therefore also the reactivity of the singlet carbene.

Based on these assumptions many different heteroatom-substituted carbenes have been synthesized. They are not limited to unsaturated cyclic diaminocarbenes (imidazolin-2-ylidenes; Scheme 3, A) [17–22] with steric bulk to avoid dimerization like **1**; 1,2,4-triazolin-5-ylidenes (Scheme 3, B), saturated

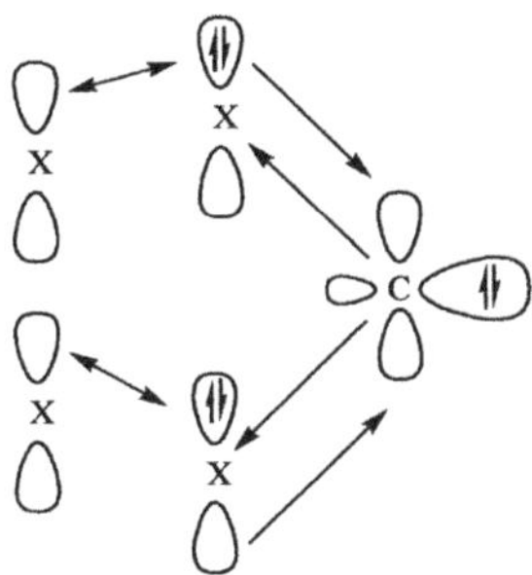

Scheme 2 Stabilization by vicinal substituents with π-donor/σ-acceptor character

imidazolidin-2-ylidenes [6,7,23] (Scheme 3, C), tetrahydropyrimid-2-ylidenes [24, 25] (Scheme 3, D), acyclic structures [26, 27] (Scheme 3, E), or systems where one nitrogen was replaced by an oxygen (Scheme 3, F) or sulphur atom (Scheme 3, G and H) have also been synthesized [28]. Several synthetic routes from different precursors can be found in the literature [29–31].

During the last decade *N*-heterocyclic carbene complexes of transition metals have been developed for catalytic applications for many different or-

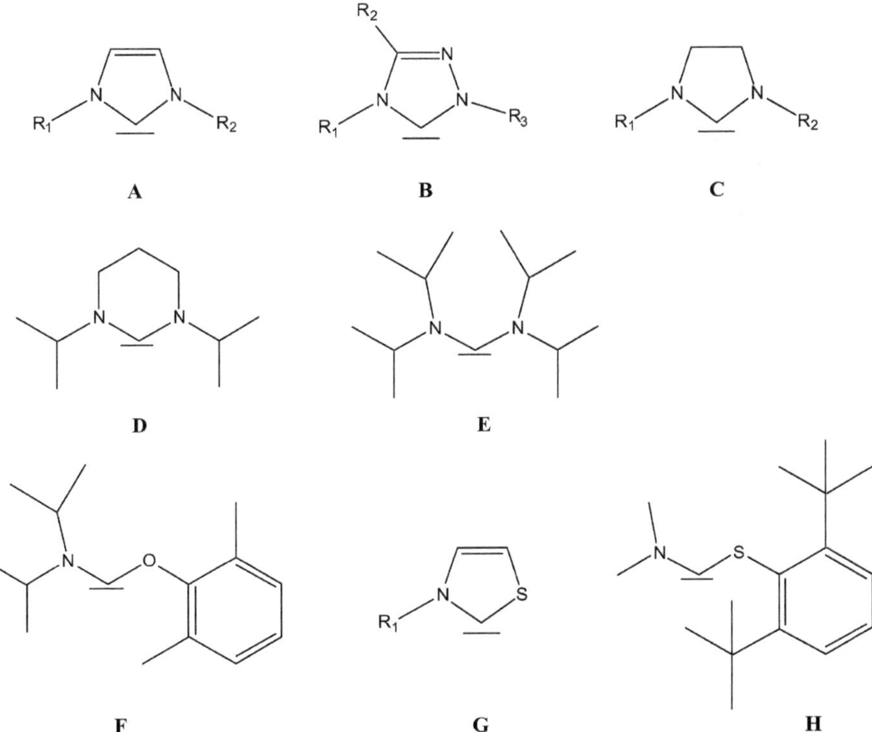

Scheme 3 Different classes of synthesized (*N*-heterocyclic) carbenes

ganic transformations. The most prominent examples are probably the olefin metathesis reaction by the Herrmann/Grubbs catalyst or the methane functionalization, which are described later in more detail.

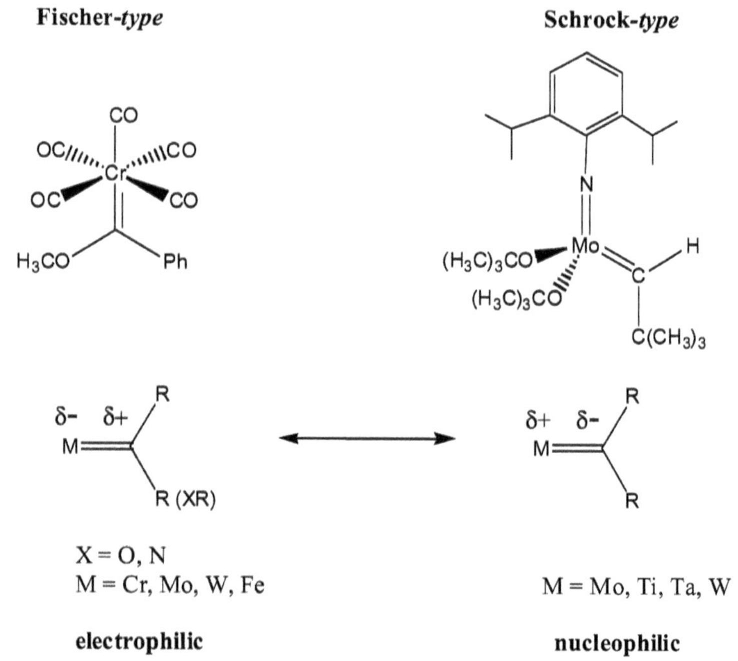

Fischer-*type* **Schrock-*type***

X = O, N
M = Cr, Mo, W, Fe

M = Mo, Ti, Ta, W

electrophilic **nucleophilic**

Scheme 4 Schrock-type and Fischer-type carbene complexes

Fischer-type carbene complexes (Scheme 4) are electrophilic heteroatom-stabilized carbenes coordinated to metals in low oxidation states. They can be prepared from $M(CO)_6$ (M=Cr, Mo, W) by reaction of an organolithium compound with one of the carbonyl ligands to form an anionic lithium acyl "ate" complex. This is possible because of the anion-stabilizing and delocalizing effect of the remaining five π-accepting electron-withdrawing CO ligands. The first synthesis of a Fischer-type carbene complex is shown in Scheme 5.

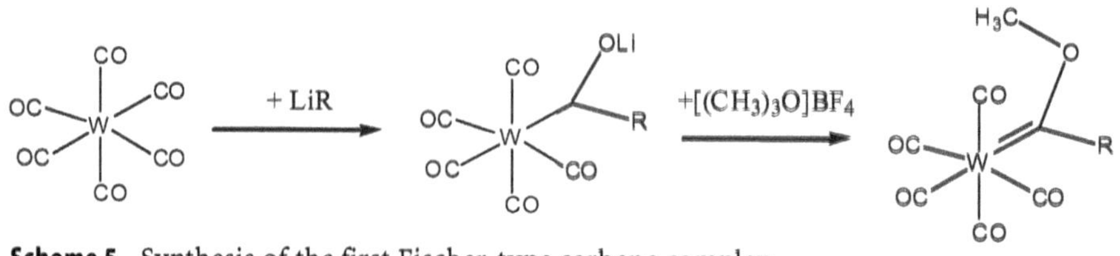

+ LiR
$+[(CH_3)_3O]BF_4$

Scheme 5 Synthesis of the first Fischer-type carbene complex

The reactivity of these carbene complexes can be understood as an electron-deficient carbene carbon atom due to the electron-attracting CO groups, while the alkoxy group stabilizes the carbene. They are therefore strongly electrophilic and can easily be attacked by nucleophiles. Derivatives can be synthesized by replacing the alkoxy group by amines via an addition-elimination mechanism [32–34]. Additionally, the hydrogens at the α-carbon are acidic and can be deprotonated with a base. Electrophiles therefore would attack at the α-carbon.

Because of the strongly electron-withdrawing character of the $Cr(CO)_5$ unit, the reaction with alkynes to hydroquinone and phenol derivatives [35–37] (Dötz reaction) is possible according to Scheme 6 (see also Chap. 4 "Chromium-templated Benzannulation Reactions").

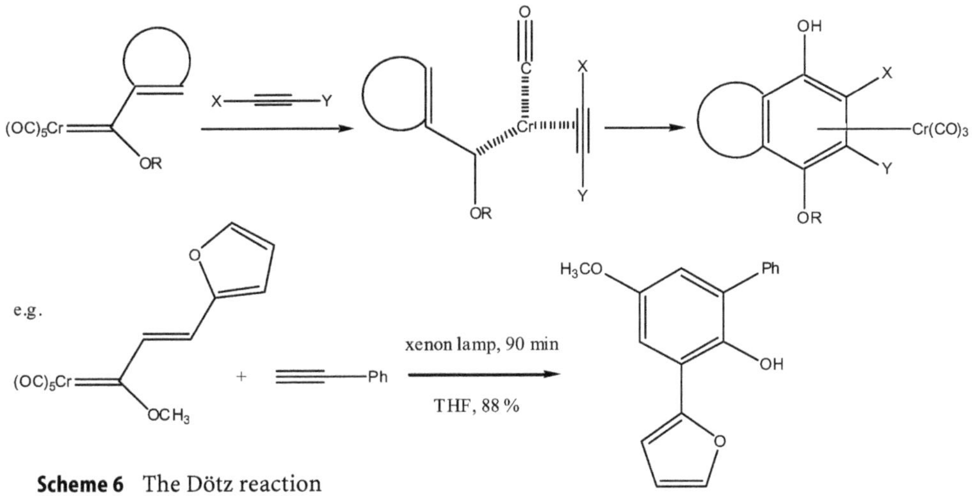

(OC)5Cr, OR, X—≡—Y, O, C, Cr, X, Y, OR, OH, X, Cr(CO)3, Y, OR

e.g. (OC)5Cr, O, OCH3, + ≡—Ph, xenon lamp, 90 min, THF, 88 %, H3CO, Ph, OH, O

Scheme 6 The Dötz reaction

Schrock-type carbenes are nucleophilic alkylidene complexes formed by coordination of strong donor ligands such as alkyl or cyclopentadienyl with no π-acceptor ligand to metals in high oxidation states. The nucleophilic carbene complexes show Wittig's ylide-type reactivity and it has been discussed whether the structures may be considered as ylides. A tantalum Schrock-type carbene complex was synthesized by deprotonation of a metal alkyl group [38] (Scheme 7).

+, Ta, CH3, CH3, H, C, H, Ta, CH3

Scheme 7 Synthesis of the first Schrock-type carbene complex

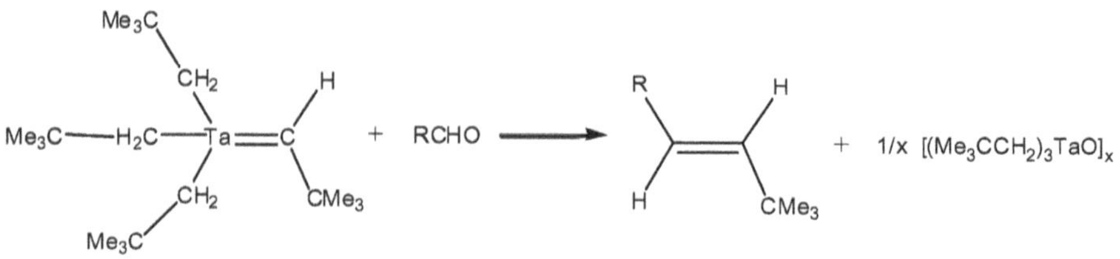

Scheme 8 Typical reaction of alkylidene complexes

These alkylidene complexes are reactive and add electrophiles to the alkylidene carbon atom according to Scheme 8. Wittig-type alkenation of the carbonyl group is possible with Ti carbene compounds, easily prepared in situ by the reaction of CH_2Br_2 with a low-valent titanium species generated by treatment of $TiCl_4$ with Zn, where the presence of a small amount of Pb in Zn was found to be crucial [39, 40]. It is synthetically equivalent to $Cl_2Ti{=}CH_2$. Replacement of the chlorine by cyclopentadienyl ligands leads to the so-called Tebbe reagent [41–44]. It is formed by the reaction of Cp_2TiCl_2 with $AlMe_3$. Due to the high oxophilicity it reacts smoothly with ketones, esters and lactones to form oxometallacycles.

These carbene (or alkylidene) complexes are used for various transformations. Known reactions of these complexes are (a) alkene metathesis, (b) alkene cyclopropanation, (c) carbonyl alkenation, (d) insertion into C–H, N–H and O–H bonds, (e) ylide formation and (f) dimerization. The reactivity of these complexes can be tuned by varying the metal, oxidation state or ligands. Nowadays carbene complexes with cumulated double bonds have also been synthesized and investigated [45–49] as well as carbene cluster compounds, which will not be discussed here [50].

2 Fischer-Type Complexes

Fischer-type carbene complexes, generally characterized by the formula $(CO)_5M{=}C(X)R$ (M=Cr, Mo, W; X=π-donor substitutent, R=alkyl, aryl or unsaturated alkenyl and alkynyl), have been known now for about 40 years. They have been widely used in synthetic reactions [37, 51–58] and show a very good reactivity especially in cycloaddition reactions [59–64]. As described above, Fischer-type carbene complexes are characterized by a formal metal-carbon double bond to a low-valent transition metal which is usually stabilized by π-acceptor substituents such as CO, PPh_3 or Cp. The electronic structure of the metal–carbene bond is of great interest because it determines the reactivity of the complex [65–68]. Several theoretical studies have addressed this problem by means of semiempirical [69–73], Hartree–Fock (HF) [74–79] and post-HF [80–83] calculations and lately also by density functional theory (DFT) calculations [67, 84–94]. Often these studies also compared Fischer-type and

Schrock-type carbenes [67, 74, 75, 93] and the general agreement is that Schrock-type carbenes can be characterized by the interaction of a triplet carbene ligand with a transition metal fragment in the triplet state (Fig. 1B). This leads to a balanced electronic interaction and nearly covalent σ and π bonds. On the other hand, Fischer-type carbene complexes are formed by coordination of a singlet carbene ligand to a transition metal fragment in the singlet state, with significant carbene to metal σ donation and metal to carbene π back-donation (Fig. 1A). Both interactions have been found to be important for the correct description of the bond and the electrophilic character at the carbene carbon atom [86, 88, 93, 94].

The kinetic and thermodynamic properties of Fischer-type carbene complexes have also been addressed by Bernasconi, who relates the strength of the π-donor substituent to the thermodynamic acidity [95–101] and the kinetics and mechanism of hydrolysis and reversible cyclization to differences in the ligand X [96, 102].

A recent study by Frenking [84] investigated in great detail the influence of the carbene substitutents X and R at a pentacarbonyl-chromium Fischer-type complex. The electronic characteristics of these substituents control the reac-

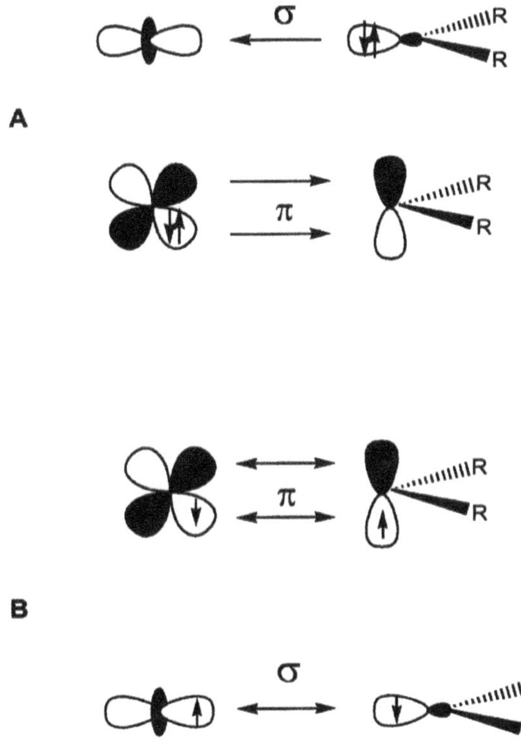

Fig. 1A,B Dominant orbital interactions in Fischer-type carbene complexes (**A**) and Schrock-type carbene complexes (**B**)

tivity of these complexes, which have been shown to be useful in many synthetic applications, most prominently the Dötz benzannulation reaction [36]. As described above (Scheme 6) this reaction, starting from aryl- or alkenyl-substituted alkoxycarbene complexes of chromium affords alkoxyphenol derivatives by insertion of the alkyne and one CO ligand in an α,β-unsaturated carbene and subsequent ring closure. In general, phenols are the main reaction product, which was investigated by a theoretical study and found to be the thermodynamically preferred product [103].

The study by Frenking investigated 25 different chromium carbene complexes, varying the σ- and π-donor strength by systematically combining different ligands X (X=H, OH, OCH_3, NH_2, $NHCH_3$) and R (R=H, CH_3, $CH=CH_2$, Ph, C≡CH). To analyse the nature of the metal–carbon bond they conducted an energy [104–108] and charge [109, 110] decomposition analysis.

The BP86 calculations together with a basis set of triple-ζ quality reproduce the geometries of experimentally known structures of that series very well, underestimating the $Cr–C_{carbene}$ bond length by only 0.048 Å with the differences for the Cr–CO and C–O bond lengths even smaller. According to Ziegler and co-workers the BP86 functional is especially well suited for $Cr(CO)_6$ and its accuracy is comparable to that of CCSD(T) calculations [111]. The shortest $Cr–C_{carbene}$ bond lengths for any given substituent R always correspond to the complex where X=H, the weakest π-electron donor. Increasing the π donation, e.g. by changing R=OH to R=NH_2, leads to a significant shortening of the $Cr–C_{carbene}$ bond length by about 0.05 Å.

This can be interpreted in terms of the Dewar–Chatt–Duncanson (DCD) model [112, 113] as a regular behaviour where larger $Cr–C_{carbene}$ bond lengths are supposed to go along with shorter $Cr–CO_{trans}$ and $C–O_{trans}$ bond distances. In line with that expectation the Fischer-type complexes with NH_2 or $NHCH_3$ show the shortest $Cr–CO_{trans}$ bond lengths (1.886–1.897 Å), those with OH or OCH_3 substituents distances of 1.899–1.915 Å and for R=H bond lengths of 1.916–1.937 Å. The calculated bond dissociation energies range from 64.5 to 97.9 kcal/mol and a direct relationship between them and the $Cr–C_{carbene}$ bond lengths is not observed, although in general a larger $Cr–C_{carbene}$ bond length relates to a smaller BDE. The π-electron-donating character does play a major role; for any substituent X the complex with R=H always shows the largest BDE and the larger π donation of the amino group reduces the back-donation to the carbene.

The CDA analysis provides the amount of electronic charge transfer in the carbene→metal donation and metal→carbene back-donation. For most investigated systems of the study [84] the carbene→metal donation is more than two times larger than the metal→carbene back-donation. Correlation of bond lengths with charge donation values is poor, while the back-donation values give a reasonable agreement. The authors explained the greater influence of the back-donation on the structural parameters of the complexes by the fact that the donation values are almost uniform for all complexes analysed, while the charge back-donation differs quite a bit over all complexes. This compares well with a previous CDA study of $M(CO)_5L$ complexes (M=Cr, Mo, W; L=CO, SiO,

CS, N_2, NO^+, CN^-, NC^-, HCCH, CCH_2, CH_2, CF_2, H_2), which showed that the metal→ligand back-donation correlates well with the change of the M–CO_{trans} bond length, while the ligand→metal donation does not [88].

The energy decomposition analysis of the chromium–carbene bond dissociation energy into a deformation (ΔE_{def}) and an interaction (ΔE_{int}) energy term proved that the interaction term is responsible for the differences between the Fischer-type carbene complexes. Pauli repulsion and electrostatic terms basically cancel out and the orbital interaction term exhibits a good correlation with the Cr–$C_{carbene}$ bond lengths. The results from the EDA are in good agreement with the conclusions from the CDA. The electrophilicity results from the difference between donation and back-donation, leading to a charge separation with a partially positive charge on the carbene carbon atom, which was quantified by the electrophilicity index ω [114]. The calculated values show a clear dependence of the electrophilicity from the π-donor substituents. Strong donors reduce the electrophilicity because the acceptor orbital of the carbene becomes occupied by π donation. For a given substituent R, back-donation increases in the order H>OH>OCH_3>NH_2>$NHCH_3$, and it becomes larger with decreasing π-donor character of the group X.

3
Schrock-Type Complexes

A decade after Fischer's synthesis of $[(CO)_5W{=}C(CH_3)(OCH_3)]$ the first example of another class of transition metal carbene complexes was introduced by Schrock, which subsequently have been named after him. His synthesis of $[((CH_3)_3CCH_2)_3Ta{=}CHC(CH_3)_3]$ [11] was described above and unlike the Fischer-type carbenes it did not have a stabilizing substituent at the carbene ligand, which leads to a completely different behaviour of these complexes compared to the Fischer-type complexes. While the reactions of Fischer-type carbenes can be described as electrophilic, Schrock-type carbene complexes (or transition metal alkylidenes) show nucleophilicity. Also the oxidation state of the metal is generally different, as Schrock-type carbene complexes usually consist of a transition metal in a high oxidation state.

The different chemical behaviour was explained by a different bonding situation in Schrock-type complexes, where more covalent double bond character from the combination of a triplet carbene with a transition metal fragment in a triplet state was attributed. The nature of this bond was the subject of several theoretical studies [77–81, 85, 87, 115–119] using different levels of theory. In a pioneering study, Hall suggested that the difference in the chemical behaviour results from changes in the electronic configuration of the transition metal [80]. In a recent paper [93], Frenking reported accurate ab initio calculations on several low-valent carbene complexes of the type $[(CO)_5WCX_2]$ and high-valent alkylidenes of the type $[(Hal)_4WCX_2]$, the bonding situation being examined by Bader [120–122], NBO [123] and CDA [109, 110] analyses. They did find that the bonding situation in the neutral low-valent and high-valent complexes is significantly different. The Schrock-type carbene complexes have a much shorter W–$C_{carbene}$ bond than the low-valent complexes, which is in agreement with experimentally known geometries [38]. This can be explained by the smaller radius of the metal atom in a higher oxidation state or a different type of metal–carbene bonding interaction, which was found to be the case in the complexes studied. Topological analysis of the electron density distribution (Bader analysis) clearly shows the differences between Fischer-type and Schrock-type carbene complexes. The Laplacian distributions show that the charge distribution around the carbene carbon atom, i.e. the lone-pair electrons of the carbene, are independent of the metal fragment in both types of complexes, while the Laplacian distribution in the π plane of the carbene ligand shows significant differences. Fischer complexes show an area of charge depletion in the direction of the p(π) orbitals, leading to holes in the electron concentration and therefore possible sites of nucleophilic attack, while the Schrock complexes are shielded by continuous areas of charge concentration. It was found that the Laplacian distribution in Fischer carbenes is similar to the situation in a singlet ($^1A_1$) methylene group, while the Laplacian distribution in Schrock complexes agrees well with a triplet ($^3B_1$) methylene group [93]. Evaluation of the calculated bond critical points of the tungsten–carbene bond shows that in the case of the Schrock complexes, the bond critical point is closer to the charge concentration of the carbene carbon atoms compared to the Fischer-type complexes. The calculated values show that the energy density at the bond critical point of the tungsten–carbene bond has much higher negative values for the Schrock complexes, indicating a larger degree of bond covalency [124]. Another measure of the double bond character is the calculated ellipticities, which demonstrate that the Schrock-type complexes show a much larger double bond character.

This is in agreement with the results of the NBO calculations, where Fischer-type complexes show a tungsten–carbene bond which is polarized towards the metal end, while the Schrock-type complexes show σ and π bonds that are both polarized towards the carbon end. The carbene ligands carry a significant negative partial charge and the population of the p(π) carbene orbital is higher in the Schrock-type complexes. The results of the NBO analysis, which focuses on the orbital structure, are in good agreement with the Bader analysis, which is based on the total electron density. The CDA results clearly show that the Schrock carbene complexes should be interpreted as an interaction between a triplet metal moiety and a ($^3B_1$) triplet carbene.

4 *N*-Heterocyclic Carbene (NHC) Complexes, Silylenes and Germylenes

The report of the successful isolation of a stable carbene by Arduengo in 1991 [6,7] (Scheme 1, **1**) and the realization of the extraordinary properties of these new ligands stimulated the research in this area, and many imidazol-2-ylidenes have been synthesized in the last 10 years [8]. The 1,3-diadamantyl derivative of the imidazol-2-ylidenes is stable at room temperature and the 1,3-dimesityl-4,5-dichloroimidazol-2-ylidene [125] is reported to be even air-stable. A variety of stable carbenes have been synthesized in between (Scheme 3), and it was shown that steric bulk is not a requirement for the stability (the 1,3-dimethylimidazolin-2-ylidene can be distilled without decomposition [126]), although it certainly influences the long-term stability by preventing dimerization. Applying the same principles which made the isolation of these carbenes possible led to the synthesis of the analogous silylenes [127, 128] and germylenes [129] (Scheme 9).

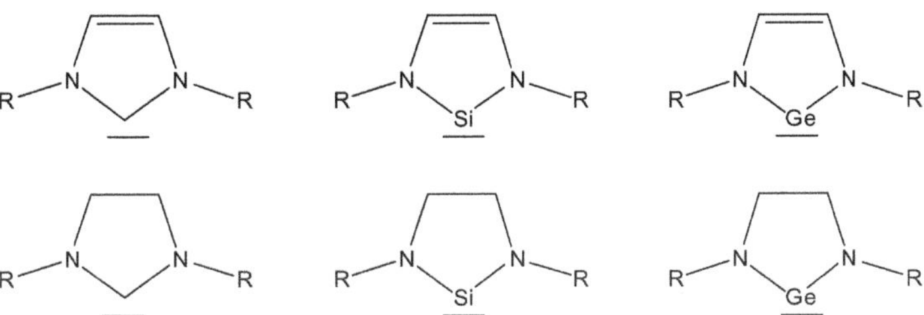

Scheme 9 Saturated and unsaturated carbenes, silylenes and germylenes

Scheme 3 shows clearly that it is absolutely not necessary to have a cyclic delocalization of π electrons in those NHC ligands to be able to isolate stable carbenes, as was believed in the beginning, although this provides additional stability [14, 130, 131]. Generally these ligands are formally neutral, two-electron donors which, contrary to Fischer-type or Schrock-type carbene complexes, are best described as pure σ-donor ligands without significant metal-ligand π back-bonding [132–135]. This might be due to a rather high occupancy of the formally empty p_π orbital of the carbene carbon atom by π delocalization [136].

Early theoretical studies [133, 135, 137–147] investigated the electronic structure of the carbenes, silylenes and germylenes shown in Scheme 9 to elucidate the reasons for the surprising stability, and came to different conclusions concerning the importance of the stabilizing effect of the π delocalization. Early studies predicted that the C–N π interaction does not play a major role [130], while others found that the p_π population at the carbene carbon atom is 30% higher for the unsaturated case, indicating that cyclic delocalization is clearly enhanced in the unsaturated carbene [147] as well as in unsaturated silylenes and germylenes [135, 146]. The electronic structure of silylenes and germylenes is thought to be qualitatively similar to that of carbenes [128, 136]. A photoelectron spectroscopy [148] study on a series of *tert*-butyl-substituted unsaturated compounds, together with an interpretation based on Kohn–Sham orbitals, gave surprising differences concerning the nature of the highest occupied (HOMO) and lowest unoccupied (LUMO) molecular orbitals compared to previous ab initio studies [146, 147]. Analysis of the chemical shielding tensors supported a non-conjugated resonance structure over a π-bonded ylidic resonance structure.

Frenking [133] showed that the higher stability of the imidazolin-2-ylidenes is caused by enhanced p_π–p_π delocalization leading to a significant electronic charge in the formally "empty" p_π orbital of the carbene carbon atom. The unsaturated imidazolin-2-ylidenes as well as the saturated imidazolidin-2-ylidenes are strongly stabilized by electron donation from the nitrogen lone pairs into the formally "empty" p_π orbital. The cyclic 6π-electron delocalization shows some aromatic character according to energetic and magnetic analysis. Silylenes and germylenes are also stabilized by p_π–p_π delocalization. The electronically less stable saturated imidazolidin-2-ylidenes need additional steric protection of the carbene carbon atom to become isolable.

N-heterocyclic carbenes show a pure donor nature. Comparing them to other monodentate ligands such as phosphines and amines on several metal-carbonyl complexes showed the significantly increased donor capacity relative to phosphines, even to trialkylphosphines, while the π-acceptor capability of the NHCs is in the order of those of nitriles and pyridine [29]. This was used to synthesize the metathesis catalysts discussed in the next section. Experimental evidence comes from the fact that it has been shown for several metals that an exchange of phosphines versus NHCs proceeds rapidly and without the need of an excess quantity of the NHC. X-ray structures of the NHC complexes show exceptionally long metal–carbon bonds indicating a different type of bond compared to the Schrock-type carbene double bond. As a result, the reactivity of these NHC complexes is also unique. They are relatively resistant towards an attack by nucleophiles and electrophiles at the divalent carbon atom.

A study [134] of the complexation of MCl (M=Cu, Ag, Au) to carbenes, silylenes and germylenes showed that metal→ligand bond dissociation energies follow the order C>Si>Ge. The strongest bond is predicted for the carbene-AuCl complex, which has a higher BDE than the classical Fischer-type complex $(CO)_5W$–CH(OH). The most important change of the ligand geometries is the shortening of the N–X (X=C, Si, Ge) bond, indicating a stronger π donation. While σ donation is still the dominant term, metal→ligand π back-donation becomes somewhat stronger for silylenes and germylenes, while it is negligible for the carbenes. The weak aromaticity of the *N*-heterocyclic ligands increases only slightly when they become bonded to the different metal chlorides.

A theoretical study of methyl-Pd heterocyclic carbene, silylene and germylene complexes revealed a very low activation barrier for the methyl migration in the silylene and germylene ligands [136]. Unlike the reaction of the carbene ligand, which experimentally occurs via concerted reductive elimination, the reaction in the silylene and germylene case is better described as an alkyl migration to the neutral ligand.

5
Grubbs/Herrmann Metathesis Catalysts

Metal-carbene complexes of the Fischer and Schrock types have been very useful for the transfer of CR_2 moieties (R=H, alkyl, aryl, alkoxy, amino) in cyclopropanation reactions and olefin metathesis. Ring-opening polymerization (ROMP), acyclic diene metathesis (ADMET) and ring-closing metathesis (RCM) are the best-known examples. Together with Schrock's molybdenum-imido complex **2**, the ruthenium-phosphine complexes **3** and **4** (Scheme 10) have been very successful olefin metathesis complexes. Excellent reviews [149] on these topics have been written and one of the chapters of this book, written by B. Schmidt, is devoted to the principles and applications of this reaction towards organic synthesis. Therefore I will only focus on the development of what are nowadays known as the Grubb's catalysts. Ruthenium became the most promising metal mostly because of its tolerance of various functional groups and mild reaction conditions.

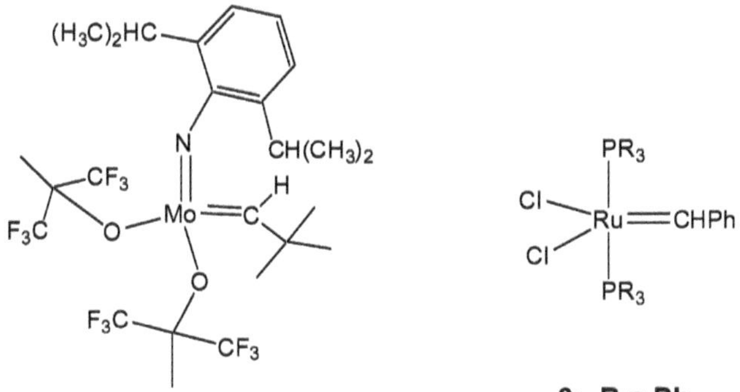

2

3: R = Ph
4: R = Cy

Scheme 10 Successful catalysts for olefin metathesis

In particular the exchange of the triphenylphosphine ligands by the more electron donating and sterically more demanding tricyclohexylphosphines was accompanied by a significantly higher stability and reactivity [150–152]. Therefore the development of complex **5** (Fig. 2) was the logical extension of that concept, keeping in mind the demonstrated excellence of NHC ligands over standard phosphane ligands.

The synthesis of these complexes can easily be accomplished by substitution of one or both PCy_3 groups of **3** by NHC ligands. The X-ray structure of **6** shows significantly different bond lengths: the "Schrock double bond" to the CHPh group is 1.821(3) Å, while the "NHC bond" to the 1,3-diisopropylimidazolin-2-ylidene is 2.107(3) Å. Complexes with imidazolidin-2-ylidenes were also synthesized and screened in an extensive study by Fürstner [153], who found that the performance of those catalysts depends strongly on the application and that

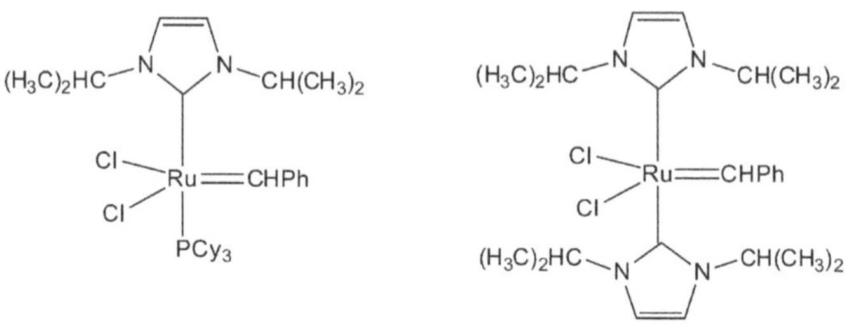

5 **6**

Fig. 2 Ruthenium-NHC complexes active in catalytic olefin metathesis

there is not just one single catalyst which outperforms all others. The mixed-ligand olefin metathesis complexes of one phosphane and one NHC ligand were first patented by Herrmann [154] and then communicated at a meeting before appearing in journals in 1999 [155]. Papers on the same topic by Nolan [156] and Grubbs [157] were published later; nevertheless these catalysts are nowadays known as "the Grubbs catalysts".

Mixed phosphane/NHC complexes have been the subject of a DFT study, where theory and experiment agree that the ligand dissociation energy for an NHC ligand is higher than for a phosphane ligand [155]. However, ligand-exchange studies revealed that the π bonding of the olefin might be the decisive factor [158, 159]. But the mechanistic discussion is still going on. Chen et al. conducted electrospray ionization tandem mass spectroscopy investigations [160–163] and concluded that the metallacyclobutane is a transition state rather than an intermediate, while calculations by Bottoni et al. found it to be an intermediate [164]. Additionally several other reaction pathways and intermediates have been proposed [118, 165–170], but there is still the need to collect additional data before a definitive answer on the mechanism of olefin metathesis catalysed by Grubbs/Herrmann catalysts can be given.

6 Platinum and Palladium NHC Complexes

Carbon–carbon bond formation reactions and the CH activation of methane are another example where NHC complexes have been used successfully in catalytic applications. Palladium-catalysed reactions include Heck-type reactions, especially the Mizoroki–Heck reaction itself [171–175], and various cross-coupling reactions [176–182]. They have also been found useful for related reactions like the Sonogashira coupling [183–185] or the Buchwald–Hartwig amination [186–189]. The reactions are similar concerning the first step of the catalytic cycle, the oxidative addition of aryl halides to palladium(0) species. This is facilitated by electron-donating substituents and therefore the development of highly active catalysts has focussed on NHC complexes.

Palladium(II) complexes provide convenient access into this class of catalysts. Some examples of complexes which have been found to be successful catalysts are shown in Scheme 11. They were able to get reasonable turnover numbers in the Heck reaction of aryl bromides and even aryl chlorides [22, 190–195]. Mechanistic studies concentrated on the Heck reaction [195] or separated steps like the oxidative addition and reductive elimination [196–199]. Computational studies by DFT calculations indicated that the mechanism for NHC complexes is most likely the same as that for phosphine ligands [169], but also in this case there is a need for more data before a definitive answer can be given on the mechanism.

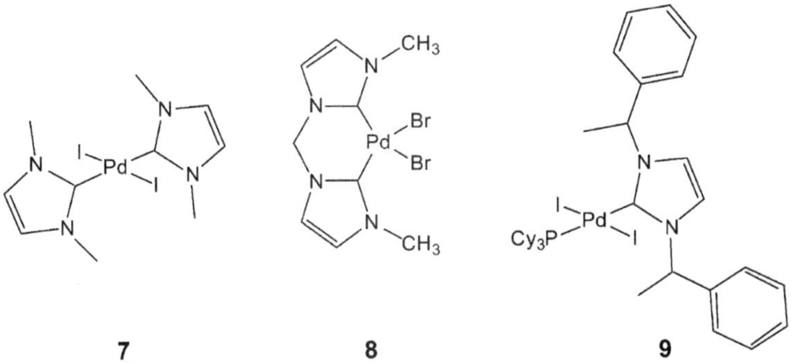

Scheme 11 Examples of active palladium-NHC complexes

Bis-chelating NHC complexes like **8** have also been successfully used for the activation and oxidation of methane to methanol in CF_3COOH in the presence of peroxodisulphate [200, 201]. The methanol is deactivated by esterification and therefore protected from further oxidation reactions. The analogous platinum NHC complexes could be synthesized by a new synthetic route and structurally characterized [202]. They have proven to be geometrically very similar to the palladium complexes [203]; the differences in the observed (and calculated) bond lengths and angles are not significant. Unfortunately the bis-chelated platinum NHC complexes are not stable under the reaction conditions used for the palladium complexes and attempts are under way to better stabilize the platinum complexes. Since we first reported the bis-chelated palladium NHC complexes several other reports appeared in the literature [204–207], showing that it is an area of current interest. Several experimental and theoretical projects in our group are currently directed towards the goal of solving the obvious mechanistic questions and we hope to report them soon.

References

1. Doering WvE, Hoffmann AK (1954) J Am Chem Soc 76:6162
2. Fischer EO, Maasboel A (1964) Angew Chem 76:645
3. Fischer EO (1974) Angew Chem 86:651
4. Igau A, Grutzmacher H, Baceiredo A, Bertrand G (1988) J Am Chem Soc 110:6463
5. Bourissou D, Bertrand G (1999) Adv Organomet Chem 44:175
6. Arduengo AJ III, Harlow RL, Kline M (1991) J Am Chem Soc 113:2801
7. Arduengo AJ III, Harlow RL, Kline M (1991) J Am Chem Soc 113:361
8. Arduengo AJ III (1999) Acc Chem Res 32:913
9. Wanzlick HW, Schoenherr HJ (1968) Angew Chem Int Ed Engl 7:141
10. Oefele K (1968) J Organomet Chem 12: P42
11. Schrock RR (1974) J Am Chem Soc 96:6796
12. Wanzlick HW, Kleiner HJ (1963) Chem Ber 96:3024
13. Wanzlick HW, Esser F, Kleiner HJ (1963) Chem Ber 96:1208
14. Wanzlick HW (1962) Angew Chem 74:129
15. Wanzlick HW, Kleiner HJ (1961) Angew Chem 73:493
16. Wanzlick HW, Schikora E (1961) Chem Ber 94:2389
17. Koecher C, Herrmann WA (1997) J Organomet Chem 532:261
18. Herrmann WA, Goossen LJ, Artus GRJ, Koecher C (1997) Organometallics 16:2472
19. Herrmann WA, Goossen LJ, Koecher C, Artus GRJ (1997) Angew Chem Int Ed Engl 35:2805
20. Herrmann WA, Koecher C, Goossen LJ, Artus GRJ (1996) Chem Eur J 2:1627
21. Herrmann WA, Elison M, Fischer J, Koecher C, Artus GRJ (1996) Chem Eur J 2:772
22. Herrmann WA, Elison M, Fischer J, Koecher C, Artus GRJ (1995) Angew Chem Int Ed Engl 34:2371
23. Denk MK, Thadani A, Hatano K, Lough AJ (1997) Angew Chem Int Ed Engl 36:2607
24. Alder RW, Blake ME, Bortolotti C, Bufali S, Butts CP, Linehan E, Oliva JM, Orpen AG, Quayle MJ (1999) Chem Commun (Camb) 1049
25. Alder RW, Blake ME, Bortolotti C, Bufali S, Butts CP, Linehan E, Oliva JM, Orpen AG, Quayle MJ (1999) Chem Commun (Camb) 241
26. Alder RW, Blake ME (1997) Chem Commun (Camb) 1513
27. Alder RW, Allen PR, Murray M, Orpen AG (1996) Angew Chem Int Ed Engl 35:1121
28. Alder RW, Butts CP, Orpen AG (1998) J Am Chem Soc 120:11526
29. Herrmann WA, Weskamp T, Bohm VPW (2001) Adv Organomet Chem 48:1
30. Weskamp T, Bohm VPW, Herrmann WA (2000) J Organomet Chem 600:12
31. Herrmann WA (2002) Angew Chem Int Ed Engl 41:1290
32. Weiss K, Fischer EO (1973) Chem Ber 106:1277
33. Casey CP, Burkhardt TJ (1973) J Am Chem Soc 95:5833
34. Casey CP, Boggs RA, Anderson RL (1972) J Am Chem Soc 94:8947
35. Wulff WD, Tang PC, Chan KS, McCallum JS, Yang DC, Gilbertson SR (1985) Tetrahedron 41:5813
36. Dötz KH (1975) Angew Chem 87:672
37. Dötz KH (1984) Angew Chem 96:573
38. Schrock RR (1979) Acc Chem Res 12:98
39. Takai K, Hotta Y, Oshima K, Nozaki H (1980) Bull Chem Soc Jpn 53:1698
40. Takai K, Hotta Y, Oshima K, Nozaki H (1978) Tetrahedron Lett 2417
41. Parshall GW, Herskovitz T, Tebbe FN, English AD, Zeile JV (1979) Fundam Res Homogeneous Catal 3:95
42. Tebbe FN, Parshall GW, Ovenall DW (1979) J Am Chem Soc 101:5074

43. Tebbe FN, Parshall GW, Reddy GS (1978) J Am Chem Soc 100:3611
44. Peet WG, Tebbe FN, Parshall GW (1978) Res Disclosure 168:21
45. Bunz UHF, Adams RD (1999) J Organomet Chem 578:1
46. Binger P, Mueller P, Wenz R, Mynott R (1990) Angew Chem 102:1070
47. Jiao H, Costuas K, Gladysz JA, Halet J-F, Guillemot M, Toupet L, Paul F, Lapinte C (2003) J Am Chem Soc 125:9511
48. Jiao H, Gladysz JA (2001) New J Chem 25:551
49. Le Bras J, Jiao H, Meyer WE, Hampel F, Gladysz JA (2000) J Organomet Chem 616:54
50. Adams RD (1989) Chem Rev 89:1703
51. Barluenga J (1996) Pure Appl Chem 68:543
52. Barluenga J (1999) Pure Appl Chem 71:1385
53. Sierra MA (2000) Chem Rev 100:3591
54. Davies MW, Johnson CN, Harrity JPA (2001) J Org Chem 66:3525
55. Harvey DF, Sigano DM (1996) Chem Rev 96:271
56. de Meijere A (1996) Pure Appl Chem 68:61
57. Aumann R, Nienaber H (1997) Adv Organomet Chem 41:163
58. Dötz KH, Tomuschat P (1999) Chem Soc Rev 28:187
59. Barluenga J, Martinez S, Suarez-Sobrino AL, Tomas M (2001) J Am Chem Soc 123:11113
60. Barluenga J, Lopez S, Trabanco AA, Fernandez-Acebes A, Florez J (2000) J Am Chem Soc 122:8145
61. Barluenga J, Tomas M, Ballesteros A, Santamaria J, Brillet C, Garcia-Granda S, Pinera-Nicolas A, Vazquez JT (1999) J Am Chem Soc 121:4516
62. Barluenga J, Tomas M, Rubio E, Lopez-Pelegrin JA, Garcia-Granda S, Perez Priede M (1999) J Am Chem Soc 121:3065
63. Schmalz HG (1994) Angew Chem 106:311
64. Fruehauf H-W (1997) Chem Rev 97:523
65. Frohlich N, Frenking G (1999) Phys Organomet Chem 2:173
66. Frenking G, Froehlich N (2000) Chem Rev 100:717
67. Frenking G, Pidun U (1997) J Chem Soc Dalton Trans 1653
68. Jiang W, Fuertes MJ, Wulff WD (2000) Tetrahedron 56:2183
69. Block TF, Fenske RF (1977) J Organomet Chem 139:235
70. Block TF, Fenske RF (1977) J Am Chem Soc 99:4321
71. Block TF, Fenske RF, Casey CP (1976) J Am Chem Soc 98:441
72. Goddard RJ, Hoffmann R, Jemmis ED (1980) J Am Chem Soc 102:7667
73. Volatron F, Eisenstein O (1986) J Am Chem Soc 108:2173
74. Nakatsuji H, Ushio J, Han S, Yonezawa T (1983) J Am Chem Soc 105:426
75. Ushio J, Nakatsuji H, Yonezawa T (1984) J Am Chem Soc 106:5892
76. Marynick DS, Kirkpatrick CM (1985) J Am Chem Soc 107:1993
77. Cundari TR, Gordon MS (1992) Organometallics 11:55
78. Cundari TR, Gordon MS (1992) J Am Chem Soc 114:539
79. Cundari TR, Gordon MS (1991) J Am Chem Soc 113:5231
80. Taylor TE, Hall MB (1984) J Am Chem Soc 106:1576
81. Carter EA, Goddard WA III (1986) J Am Chem Soc 108:4746
82. Marquez A, Fernandez Sanz J (1992) J Am Chem Soc 114:2903
83. Wang C-C, Wang Y, Liu H-J, Lin K-J, Chou L-K, Chan K-S (1997) J Phys Chem A 101:8887
84. Cases M, Frenking G, Duran M, Sola M (2002) Organometallics 21:4182
85. Jacobsen H, Ziegler T (1996) Inorg Chem 35:775
86. Jacobsen H, Ziegler T (1995) Organometallics 14:224
87. Jacobsen H, Schreckenbach G, Ziegler T (1994) J Phys Chem 98:11406

88. Ehlers AW, Dapprich S, Vyboishchikov SF, Frenking G (1996) Organometallics 15:105
89. Beste A, Kramer O, Gerhard A, Frenking G (1999) Eur J Inorg Chem 2037
90. Frenking G (2001) J Organomet Chem 635:9
91. Froehlich N, Pidun U, Stahl M, Frenking G (1997) Organometallics 16:442
92. Vyboishchikov SF, Frenking G (1998) Chem Eur J 4:1439
93. Vyboishchikov SF, Frenking G (1998) Chem Eur J 4:1428
94. Torrent M, Duran M, Sola M (1998) Organometallics 17:1492
95. Bernasconi CF, Ali M (2000) J Am Chem Soc 122:7152
96. Bernasconi CF, Ali M, Lu F (2000) J Am Chem Soc 122:1352
97. Bernasconi CF, Ali M (1999) J Am Chem Soc 121:11384
98. Bernasconi CF, Ali M (1999) J Am Chem Soc 121:3039
99. Bernasconi CF, Leyes AE, Ragains ML, Shi Y, Wang H, Wulff WD (1998) J Am Chem Soc 120:8632
100. Bernasconi CF, Leyes AE (1997) J Am Chem Soc 119:5169
101. Bernasconi CF, Sun W (2002) J Am Chem Soc 124:2299
102. Bernasconi CF, Perez GS (2000) J Am Chem Soc 122:12441
103. Gleichmann MM, Doetz KH, Hess BA (1996) J Am Chem Soc 118:10551
104. Ziegler T, Rauk A (1979) Inorg Chem 18:1755
105. Ziegler T, Rauk A (1979) Inorg Chem 18:1558
106. Ziegler T, Rauk A (1977) Theor Chim Acta 46:1
107. Bickelhaupt FM, Nibbering NMM, Van Wezenbeek EM, Baerends EJ (1992) J Phys Chem 96:4864
108. Kitaura K, Morokuma K (1976) Int J Quant Chem 10:325
109. Dapprich S, Frenking G (1995) J Phys Chem 99:9352
110. CDA 2.1 by S. Dapprich and G. Frenking, Marburg 1994. The program is available via anonymous ftp server: ftp.chemie.uni-marburg.de (/pub/cda)
111. Li J, Schreckenbach G, Ziegler T (1994) J Phys Chem 98:4838
112. Dewar MJS (1951) Bull Soc Chim France C71
113. Chatt J, Duncanson LA (1953) J Chem Soc 2939
114. Parr RG, Szentpaly Lv, Liu S (1999) J Am Chem Soc 121:1922
115. Spangler D, Wendoloski JJ, Dupuis M, Chen MML, Schaefer HF III (1981) J Am Chem Soc 103:3985
116. Fox HH, Schofield MH, Schrock RR (1994) Organometallics 13:2804
117. Schoeller WW, Rozhenko AJB, Alijah A (2001) J Organomet Chem 617–618:435
118. Bernardi F, Bottoni A, Miscione GP (2000) Organometallics 19:5529
119. Marquez A, Fernandez Sanz J (1992) J Am Chem Soc 114:10019
120. Bader RFW (1994) Atoms in molecules: a quantum theory. Oxford University Press, New York
121. Bader RFW, Laidig KE (1991) Theochem 80:75
122. Bader RFW (1985) Acc Chem Res 18:9
123. Reed AE, Curtiss LA, Weinhold F (1988) Chem Rev 88:899
124. Cremer D, Kraka E (1984) Angew Chem 96:612
125. Arduengo AJ III, Davidson F, Dias HVR, Goerlich JR, Khasnis D, Marshall WJ, Prakasha TK (1997) J Am Chem Soc 119:12742
126. Fischer J (1996) Dissertation, Technische Universität München
127. Denk M, Lennon R, Hayashi R, West R, Belyakov AV, Verne HP, Haaland A, Wagner M, Metzler N (1994) J Am Chem Soc 116:2691
128. Denk M, Green JC, Metzler N, Wagner M (1994) J Chem Soc Dalton Trans 2405
129. Herrmann WA, Denk M, Behm J, Scherer W, Klingan FR, Bock H, Solouki B, Wagner M (1992) Angew Chem 104:1489

130. Arduengo AJ III, Dias HVR, Dixon DA, Harlow RL, Klooster WT, Koetzle TF (1994) J Am Chem Soc 116:6812
131. Arduengo AJ III, Goerlich JR, Marshall WJ (1995) J Am Chem Soc 117:11027
132. Lehmann JF, Urquhart SG, Ennis LE, Hitchcock AP, Hatano K, Gupta S, Denk MK (1999) Organometallics 18:1862
133. Boehme C, Frenking G (1996) J Am Chem Soc 118:2039
134. Boehme C, Frenking G (1998) Organometallics 17:5801
135. Heinemann C, Mueller T, Apeloig Y, Schwarz H (1996) J Am Chem Soc 118:2023
136. McGuinness DS, Yates BF, Cavell KJ (2002) Organometallics 21:5408
137. Gleiter R, Hoffmann R (1968) J Am Chem Soc 90:5457
138. Dixon DA, Arduengo AJ III (1991) J Phys Chem 95:4180
139. Cioslowski J (1993) Int J Quantum Chem Quantum Chem Symp 27:309
140. Koopmans T (1933) Physica 1:104
141. Kutzelnigg W (1984) Angew Chem 96:262
142. Gobbi A, Frenking G (1994) J Am Chem Soc 116:9287
143. Kutzelnigg W (1980) Isr J Chem 19:193
144. Schindler M, Kutzelnigg W (1982) J Chem Phys 76:1919
145. Hansen AE, Bouman TD (1985) J Chem Phys 82:5035
146. Heinemann C, Herrmann WA, Thiel W (1994) J Organomet Chem 475:73
147. Heinemann C, Thiel W (1994) Chem Phys Lett 217:11
148. Arduengo AJ III, Bock H, Chen H, Denk M, Dixon DA, Green JC, Herrmann WA, Jones NL, Wagner M, West R (1994) J Am Chem Soc 116:6641
149. Schmidt B (2003) Angew Chem Int Ed Engl 42:4996
150. Schwab P, Grubbs RH, Ziller JW (1996) J Am Chem Soc 118:100
151. Schwab P, France MB, Ziller JW, Grubbs RH (1995) Angew Chem Int Ed Engl 34:2039
152. Nguyen ST, Grubbs RH, Ziller JW (1993) J Am Chem Soc 115:9858
153. Furstner A, Ackermann L, Gabor B, Goddard R, Lehmann CW, Mynott R, Stelzer F, Thiel OR (2001) Chem Eur J 7:3236
154. Herrmann WA, Schattenmann W, Weskamp T (1999) Ger Offen (Aventis Research & Technologies GmbH & Co KG, Germany). De p 12 pp
155. Weskamp T, Kohl FJ, Hieringer W, Gleich D, Herrmann WA (1999) Angew Chem Int Ed Engl 38:2416
156. Huang J, Schanz H-J, Stevens ED, Nolan SP (1999) Organometallics 18:5375
157. Scholl M, Trnka TM, Morgan JP, Grubbs RH (1999) Tetrahedron Lett 40:2247
158. Sanford MS, Love JA, Grubbs RH (2001) J Am Chem Soc 123:6543
159. Sanford MS, Ulman M, Grubbs RH (2001) J Am Chem Soc 123:749
160. Adlhart C, Hinderling C, Baumann H, Chen P (2000) J Am Chem Soc 122:8204
161. Adlhart C, Volland MAO, Hofmann P, Chen P (2000) Helv Chim Acta 83:3306
162. Adlhart C, Chen P (2002) Angew Chem Int Ed Engl 41:4484
163. Adlhart C, Chen P (2003) Helv Chim Acta 86:941
164. Bernardi F, Bottoni A, Miscione GP (2003) Organometallics 22:940
165. Hansen SM, Volland MAO, Rominger F, Eisentrager F, Hofmann P (1999) Angew Chem Int Ed Engl 38:1273
166. Licandro E, Maiorana S, Vandoni B, Perdicchia D, Paravidino P, Baldoli C (2001) Synlett 757
167. Furstner A (1998) Top Organomet Chem 1:37
168. Cavallo L (2002) J Am Chem Soc 124:8965
169. Albert K, Gisdakis P, Roesch N (1998) Organometallics 17:1608
170. Suresh CH, Koga N (2004) Organometallics 23:76
171. Mizoroki T, Mori K, Ozaki A (1971) Bull Chem Soc Jpn 44:581

172. Heck RF, Nolley JP Jr (1972) J Org Chem 37:2320
173. Shibasaki M, Vogl EM (1999) In: Jacobsen EN, Pfaltz A, Yamamoto H (eds) Comprehensive asymmetric catalysis, vol 1. Springer, Berlin Heidelberg New York, p 457
174. Shibasaki M, Vogl EM (1999) J Organomet Chem 576:1
175. de Meijere A, Meyer FE (1994) Angew Chem 106:2473
176. Miyaura N, Yanagi T, Suzuki A (1981) Synthetic Commun 11:513
177. Miyaura N, Suzuki A (1979) J Chem Soc Chem Commun 866
178. Stanforth SP (1998) Tetrahedron 54:263
179. Suzuki A (1999) J Organomet Chem 576:147
180. Tamao K, Sumitani K, Kumada M (1972) J Am Chem Soc 94:4374
181. Corriu RJP, Masse JP (1972) J Chem Soc Chem Commun 144
182. Kumada M (1980) Pure Appl Chem 52:669
183. Sonogashira K, Tohda Y, Hagihara N (1975) Tetrahedron Lett 4467
184. Cassar L (1975) J Organomet Chem 93:253
185. Dieck HA, Heck FR (1975) J Organomet Chem 93:259
186. Hartwig JF (1998) Acc Chem Res 31:852
187. Louie J, Hartwig JF (1995) Tetrahedron Lett 36:3609
188. Yang BH, Buchwald SL (1999) J Organomet Chem 576:125
189. Guram AS, Rennels RA, Buchwald SL (1995) Angew Chem Int Ed Engl 34:1348
190. Calo V, Del Sole R, Nacci A, Shingaro E, Scordari F (2000) Eur J Org Chem 869
191. McGuinness DS, Green MJ, Cavell KJ, Skelton BW, White AH (1998) J Organomet Chem 565:165
192. Schwarz J, Bohm VPW, Gardiner MG, Grosche M, Herrmann WA, Hieringer W, Raudaschl-Sieber G (2000) Chem Eur J 6:1773
193. Herrmann WA, Reisinger C-P, Spiegler M (1998) J Organomet Chem 557:93
194. Tulloch AAD, Danopoulos AA, Cafferkey SM, Kleinhenz S, Hursthouse MB, Tooze RP (2000) Chem Commun (Camb) 1247
195. McGuinness DS, Cavell KJ (2000) Organometallics 19:741
196. McGuinness DS, Cavell KJ, Yates BF, Skelton BW, White AH (2001) J Am Chem Soc 123:8317
197. McGuinness DS, Saendig N, Yates BF, Cavell KJ (2001) J Am Chem Soc 123:4029
198. McGuinness DS, Yates BF, Cavell KJ (2001) Chem Commun (Camb) 355
199. McGuinness DS, Cavell KJ, Skelton BW, White AH (1999) Organometallics 18:1596
200. Muehlhofer M, Strassner T, Herrmann WA (2002) Angew Chem Int Ed Engl 41:1745
201. Maletz G, Schmidt F, Reimer A, Strassner T, Muehlhofer M, Mihalios D, Herrmann W (2003) Ger Offen (Sued-Chemie AG, Germany). De p 10 pp
202. Muehlhofer M, Strassner T, Herdtweck E, Herrmann WA (2002) J Organomet Chem 660:121
203. Herdtweck E, Muehlhofer M, Strassner T (2003) Acta Cryst E59:m970
204. Garrison JC, Simons RS, Tessier CA, Youngs WJ (2003) J Organomet Chem 673:1
205. Quezada CA, Garrison JC, Tessier CA, Youngs WJ (2003) J Organomet Chem 671:183
206. Simons RS, Custer P, Tessier CA, Youngs WJ (2003) Organometallics 22:1979
207. Garrison JC, Simons RS, Kofron WG, Tessier CA, Youngs WJ (2001) Chem Commun (Camb) 1780

Topics Organomet Chem (2004) 13: 21–57
DOI 10.1007/978-3-540-40910-6

The Multifaceted Chemistry of Variously Substituted α,β-Unsaturated Fischer Metalcarbenes

Yao-Ting Wu · Armin de Meijere (✉)

Institut für Organische und Biomolekulare Chemie, Georg-August-Universität Göttingen, Tammannstrasse 2, 37077 Göttingen, Germany
ameijer1@gwdg.de

Abstract The insertion of an alkyne into an α,β-unsaturated Fischer metalcarbene complex leads to a 1-metalla-1,3,5-hexatriene. This usually undergoes subsequent insertion of a carbon monoxide molecule, and the resulting dienylketene complex, in a 6π-electrocyclization, yields an alkoxycyclohexadienone or its tautomeric hydroquinone monoether. The overall process is a [3+2+1] cocyclization and constitutes the so-called Dötz reaction. With a dialkylamino instead of the alkoxy group on the carbene center, or an additional dialkylamino group on C3 of an alkoxycarbene complex, the 1-metalla-1,3,5-hexatrienes resulting from alkyne insertion more generally do not undergo CO insertion, but direct 6π-electrocyclization and subsequent reductive elimination to yield five- rather than six-membered rings. 1-Dialkylamino-1-arylcarbenemetals thus yield indenes, and 1-alkoxy-3-dialkylamino-propenylidenemetal complexes with alkynes furnish 3-alkoxy-5-dialkylaminocyclopentadienes, which essentially are protected cyclopentenones and even doubly protected cyclopentadienones. The multifunctionality of these cyclopentadienes makes them highly versatile building blocks for organic synthesis. Synthetically useful cyclopentenones are also obtained from 1-cyclopropyl-1-alkoxycarbenemetals with alkynes. Yet, with certain combinations of substituents and conditions, the amino-substituted metallatrienes can also undergo CO insertion with subsequent cyclization to five-membered rings, twofold alkyne and CO insertion with subsequent intramolecular [4+2] cycloaddition to yield cyclopenta[*b*]pyranes, or even threefold alkyne insertion with subsequent twofold cyclization to yield spiro[4.4]nonatrienes. Variously amino-substituted α,β-unsaturated Fischer carbenes can also give rise to pyrrolidines, pyridines, and pyrroles. Normal, i.e., 1-alkoxy-substituted, α,β-unsaturated Fischer carbene complexes react with acceptor-substituted alkenes and alkadienes to yield donor–acceptor-substituted vinylcyclopropanes or cyclopentenes and cycloheptadienes, respectively. Enantiocontrolled formal [3+2] cycloadditions of chirally modified alkoxycarbenemetals with imines can be achieved to yield, after hydrolysis of the alkoxypyrrolines, 1,2,5-trisubstituted pyrrolidin-3-ones with high enantiomeric excesses.

Keywords Fischer carbenes · Template synthesis · Cocyclization · Cycloaddition · Cyclopentadienes · Cyclopentenones · Domino reactions

1
Introduction

When E. O. Fischer et al. discovered the straightforward access to alkoxycarbene complexes of chromium and other transition metals about four decades ago [1], it was not obvious that they would soon start to become an important item in the toolbox for organometallics and organic synthesis [2, 3]. One of the most important features of Fischer carbene complexes is the distinctly electron-deficient nature of the carbene carbon due to the strong electron-withdrawing effect of the pentacarbonylmetal fragment. It makes such a carbon atom more electrophilic than the carbon atom of any carbonyl group and, as a consequence, an alkenyl or an alkynyl moiety in an α,β-unsaturated Fischer carbene complex is more active toward any sort of nucleophile than the carbonyl carbon atom in a corresponding ester, amide, and/or thioester [4]. As electrophilic species, such α,β-unsaturated Fischer carbene complexes, unlike carbonyl compounds, readily undergo insertion with alkynes, and in certain cases even alkenes, to furnish reactive intermediates from which a large variety of different products can be formed [5, 6]. In particular, the formal [3+2+1] cycloaddition of an α,β-unsaturated (or an α-aryl-substituted) Fischer carbene complex, an alkyne, and a carbon monoxide molecule to form a six-membered ring – the so-called Dötz reaction – has convincingly been applied toward the preparation of a large variety of natural products and other interesting molecules (see Chap. 4 in this book) [7–9]. Yet, a number of α,β-unsaturated Fischer carbene complexes, especially *β*-amino-substituted ones, follow different reaction pathways to yield five-membered carbo- and heterocycles without or with carbon monoxide insertion, as well as more complex bicyclic, spirocyclic, and tricyclic structures. In view of all the different reaction modes accessible to them, *α*,*β*-unsaturated Fischer carbene complexes can be regarded as true chemical multitalents [10, 11].

2
Synthesis of *α*,*β*-Unsaturated Fischer Carbene Complexes

2.1
From (Pentacarbonyl)metallaacylates

α,*β*-Unsaturated Fischer carbene complexes **3** are prepared from lithiated alkynes (or alkenes, arenes) **1** according to a variant of the classical route of Fischer et al. (Scheme 1) [12–14]. Treatment of **1** with hexacarbonylmetals affords a (pentacarbonyl)metallaacylate **2**, which can be trapped with hard alkylating agents (especially Meerwein salts) to form stable Fischer carbene complexes **3**. The key intermediates **2** are also accessible from acid chlorides **4** and pentacarbonylmetallates **5** [15].

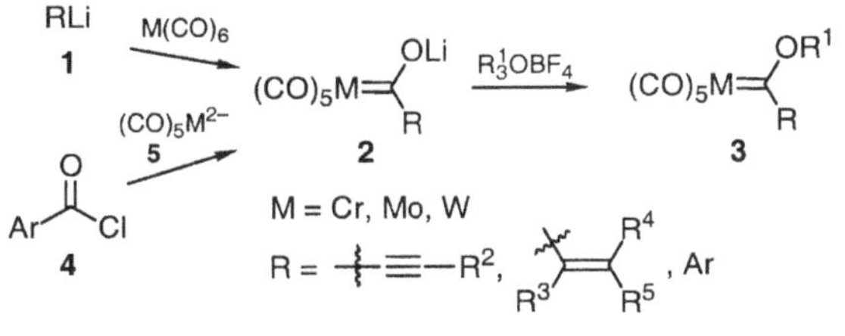

Scheme 1 Synthesis of *α*,*β*-unsaturated Fischer carbene complexes **3** from (pentacarbonyl)-metallaacylates **2** [12–15]

2.2
From Alkyl-Substituted Fischer Carbene Complexes

Due to the high *α*-C,H acidity in the alkoxyethylidene complexes **6** (e.g., pK_a=8 (R=Me)) [16], transformations via an enolate analog are possible and have been used to introduce additional functionality into the carbene side chain to access various Fischer carbene complexes [3]. The *α*,*β*-unsaturated complex **8** could be obtained from **6** (R=Et) by an aldol-type condensation with benzaldehyde **7** in the presence of triethylamine and trimethylsilyl chloride (Scheme 2) [17]. This reaction proceeds completely diastereoselectively to yield only the *trans*-isomer. Analogously, binuclear complexes have been prepared from **6** and 1,3- and 1,4-phthaldialdehyde in good yields [17]. This type of condensation has

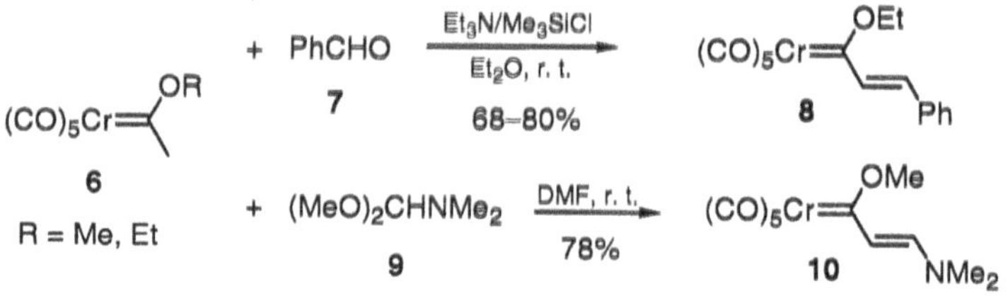

Scheme 2 Preparation of ethenylcarbene complexes **8** and **10** by aldol condensations [17–18]

also been used to access β-amino-substituted α,β-unsaturated Fischer carbene complexes like **10** [18].

The possibility of being involved in olefin metathesis is one of the most important properties of Fischer carbene complexes. [2+2] Cycloaddition between the electron-rich alkene **11** and the carbene complex **12** leads to the intermediate metallacyclobutane **13**, which undergoes [2+2] cycloreversion to give a new carbene complex **15** and a new alkene **14** [19]. The (methoxy)phenylcarbenetungsten complex is less reactive in this mode than the corresponding chromium and molybdenum analogs (Scheme 3).

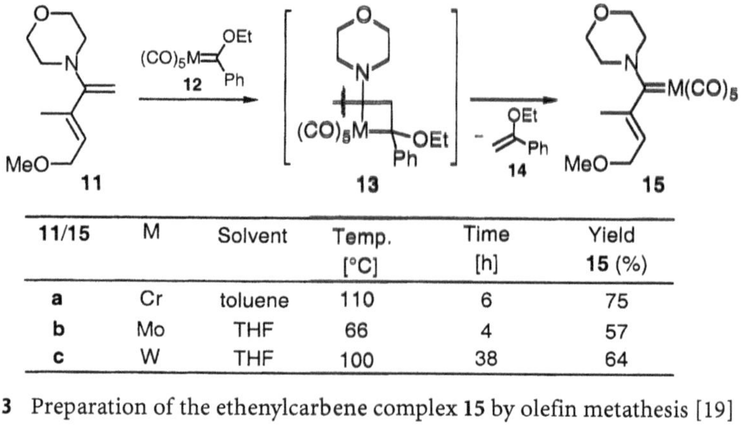

11/15	M	Solvent	Temp. [°C]	Time [h]	Yield **15** (%)
a	Cr	toluene	110	6	75
b	Mo	THF	66	4	57
c	W	THF	100	38	64

Scheme 3 Preparation of the ethenylcarbene complex **15** by olefin metathesis [19]

2.3 From Alkynylcarbene Complexes

In view of the strong electron-withdrawing influence of the pentacarbonylmetal moiety on the carbene ligand, it is obvious that in alkynyl-substituted complexes of type **23**, the triple bond is highly activated toward nucleophilic attack by a variety of reagents. Thus, 1,3-dipolar cycloadditions of nitrones such as **18** readily occur to yield the 2,3-dihydroisoxazolidinyl carbene complexes **16** highly chemo- and regioselectively (reaction mode A in Scheme 4) [20, 21]. Compared to a corresponding propargylic acid ester, the complexes **23** undergo this type of reaction faster. The triple bond reactivity of **23** is also drastically

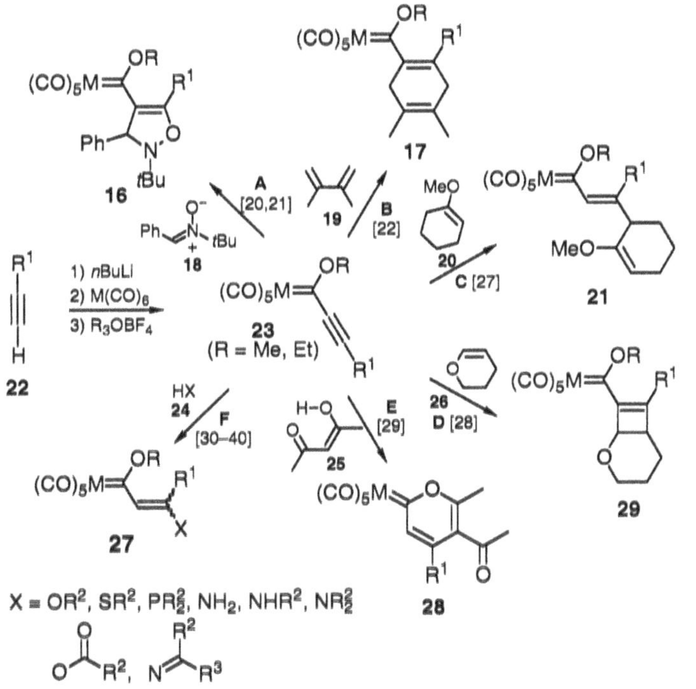

Scheme 4 Access to various α,β-unsaturated carbene complexes from alkynylcarbene complexes **23**. **A:** 1,3-Dipolar cycloaddition. **B:** Diels–Alder reaction. **C:** Ene reaction. **D:** [2+2] Cycloaddition. **E:** Michael-type addition followed by cyclization. **F:** Michael-type additions

enhanced for Diels–Alder reactions. Treatment of alkynyl Fischer carbene complexes **23** with a diene like **19** affords [4+2] cycloaddition products **17** in good to excellent yields (mode B) [22]. The investigations concerning the dienophilicity of 1-alkynylcarbene complexes of type **23** and regioselectivities in their Diels–Alder reactions with dienes extend well into the 1990s [23, 24]. Since 1-alkynylcarbene complexes **23** are significantly better dienophiles than the corresponding esters, they react at lower temperature, require shorter reaction times, and give better chemical yields [25, 26]. With enol ethers like **20** they undergo ene reactions to α,β-unsaturated complexes like **21** (mode C) [27] or [2+2] cycloadditions to cyclobutenyl complexes like **29** (mode D) [28]. These two modes can be competing with each other, depending on the substitution pattern on the enol ether and the substituents (R^1) on the complexes **23** [28].

In the presence of a catalytic amount of triethylamine, a readily enolizable carbonyl compound like acetylacetone (**25**) can undergo a Michael-type addition onto the triple bond of **23** with C–C bond formation, and subsequent 1,2-addition of the hydroxy group with elimination of an alcohol (MeOH or EtOH) to eventually yield a pyranylidene complex **28** (mode E) [29]. The most versatile access to β-donor-substituted ethenylcarbene complexes **27** is by Michael-type additions of nucleophiles, including alcohols [30–32], primary and secondary amines [30, 33–35], ammonia [30, 36], imines [37], phosphines [38, 39], thiols [30], and carboxylic acids [40] to alkynylcarbene complexes **23** (mode F). In some cases, like the addition of weaker nucleophiles such as alcohols and thiols, reaction rates and chemical yields can be improved by the presence of a catalytic amount of the corresponding sodium alkoxide or thiolate, respectively [30, 41].

This reaction mode of alkynylcarbene complexes of type **23** undoubtedly provides the most convenient access to β-amino-substituted α,β-unsaturated Fischer carbene complexes **27** ($X=NH_2, NHR^2, NR^2_2$). Fischer et al. reported the very first such addition of an amine to an alkynylcarbene complex of type **23** and observed a temperature-dependent competition between 1,4- and 1,2-addition [12]. In a later systematic study, de Meijere et al. found that in addition to the 1,4-addition products **30**, 1,2-addition–elimination (formal substitution) **31** and 1,4-addition–elimination products **32** can be formed (Scheme 5) [33]. The ratio of the three complexes **30**, **31**, and **32** largely depends on the polarity of the solvent, the reaction temperature, and the substituents on the alkyne (R^1) as well as the amine (R^2). If complexes **30** are desired, they can be obtained as single products or at least as the major products by careful choice of reaction conditions. Formation of the {[2-(dialkylamino)ethenyl]carbene}chromium complexes **30** is favored at low temperatures (–115 to 20°C) [41]. Room temperature is sufficient to give high yields of **30** from most complexes **23** and secondary amines. The complexes **30** are usually obtained as (*E*)-isomers with the exception of those with bulky substituents R^1 (e.g., $R^1=t$Bu [30] or $R^1=SiMe_3$ [42]). It is particularly favorable that these carbene complexes **30**, especially the ones with chromium, are easily accessible in a one-pot procedure from terminal alkynes **15**, hexacarbonylchromium, triethyloxonium tetrafluoroborate, and a secondary amine, generally in good to excellent yields [43, 44]. Formation of certain 1,2-addition–elimination products of type **31** is favored at low temperature [12, 45, 46]. (3-Dialkylaminoallenylidene)chromium complexes **32** were found as by-products, or even main products [30, 33], when bulky or highly basic secondary amines were added to the alkynylcarbene complexes **23** in polar solvents and at high temperature. With lithium amides, these metallacumulenes **32** could be produced as the sole products [33].

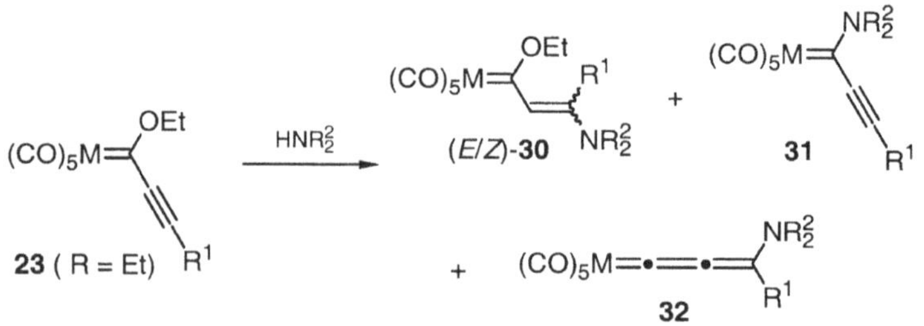

(CO)$_5$M=C(OEt)–C≡C–R^1 **23** (R = Et) —HNR^2_2→ (CO)$_5$M=C(OEt)–CH=C(R^1)NR$^2_2$ (*E*/*Z*)-**30** + (CO)$_5$M=C(NR$^2_2$)–C≡C–R^1 **31** + (CO)$_5$M=C=C=C(NR$^2_2$)R^1 **32**

Scheme 5 Access to β-amino-substituted α,β-unsaturated Fischer carbene complexes **30** by Michael-type addition of amines to alkynylcarbene complexes **23** (R=Et) [30, 33]

1,3-Diamino-substituted complexes of type **37** were first obtained by Fischer et al. [12] in two steps via the 1,2-addition–elimination product **34** from dimethylamine and **35** (Scheme 6). The (3-aminoallenylidene)chromium complexes **36**, which can be prepared either from **33** [47,48] or directly from **35** [33], can also be transformed to 1,3-bis(dialkylamino)-substituted complexes of type **37** (e.g., R^2=*i*Pr) by treatment with dimethylamine in excellent yields [33]. Although the complex **37** is accessible by further reaction of the complex **34** with dimethylamine, and **34** itself stems from the reaction of **35** with dimethylamine, the direct transformation of **33** to **37** could not be achieved [12]. In spite of this, heterocyclic carbene complexes with two nitrogens were obtained by reactions of alkynylcarbene complexes **35** with hydrazine [49] and 1,3-diamines [50].

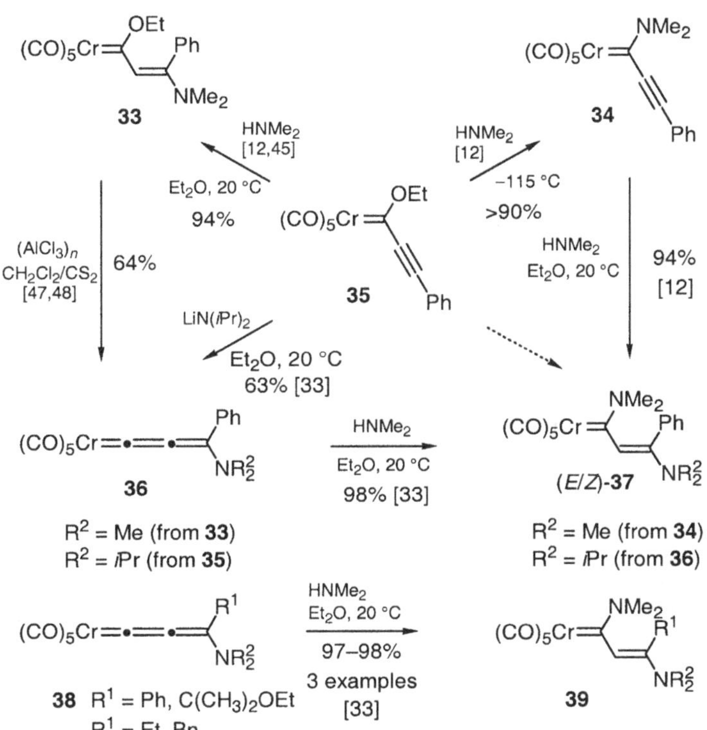

Scheme 6 Chemical relationships among the complexes **33**, **34**, **35**, **36**, and **37** [12, 33, 45, 47, 48]

In contrast to other terminal alkynes, the lithiated dimethylaminoethyne **40** does not give the corresponding alkynylcarbene but the cyclopropenylidene complex **41** (Scheme 7) [51]. Further addition of dimethylamine to **41** affords the substitution product **42** in excellent yield. This 2,3-bis(dimethylamino-cyclopropenylidene)pentacarbonylchromium (**42**) is extremely stable, and it cannot be transformed to the corresponding carbonyl compound, 2,3-bis (dimethylamino)cyclopropenone, by oxidation with ceric ammonium nitrate (CAN) [52] or dimethyl sulfoxide (DMSO) [53].

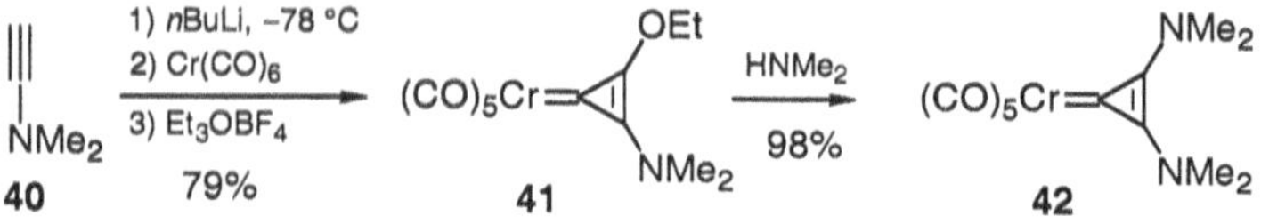

1) *n*BuLi, –78 °C
2) $Cr(CO)_6$
3) Et_3OBF_4
79%
NMe_2 40
$(CO)_5Cr$ OEt NMe_2 41
$HNMe_2$ 98%
$(CO)_5Cr$ NMe_2 NMe_2 42

Scheme 7 Synthesis of 2,3-bis(dimethylamino)cyclopropenylidene complex **42** [51]

3
Cocyclizations of *α*,*β*-Unsaturated Fischer Carbene Complexes with Alkynes

Most of the formal cycloaddition reactions of *α*,*β*-unsaturated Fischer carbene complexes **43** with alkynes **44** arise from a primary insertion product of type **45** (Scheme 8). The subsequent reactions of **45** depend mainly on the nature – electronic as well as steric – and pattern of substituents in **45**, brought in by the starting materials **43** and **44**. The first discovered reaction mode of **45** with an alkoxy group at the carbene center is that with carbonyl insertion and subsequent cyclization leading to alkoxycyclohexadienones or their enol tautomers, hydroquinone monoalkyl ethers, commonly known as the Dötz reaction or Dötz benzannelation reaction (see corresponding chapter in this book). Direct cyclization of **45** with subsequent reductive elimination, leading to five-membered rings may also occur, and five-membered ring products may also be formed with carbonyl insertion. In certain cases, **45** inserts another alkyne, and the resulting intermediate continues with carbonyl insertion or alkyne insertion before undergoing cyclization or oligomerization. All of these reaction modes may be classified as formal $[k+m+n]$ cycloadditions, in which k, m, and n represent the respective number of atoms from the carbene ligand (k), the alkyne (m), and the carbonyl ligand (n). In the following subsections those cases with $k>1$, i.e., more than one atom from the carbene complexes participating in the cocyclizations, which do not lead to six-membered rings, will be described.

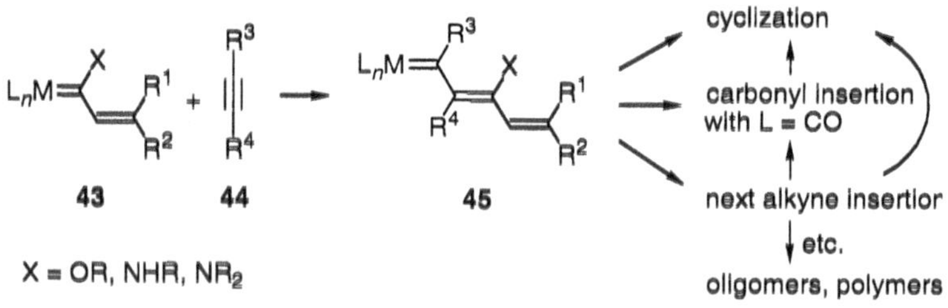

L_nM X R^1 R^2 43 + R^3 R^4 44 → L_nM R^3 X R^4 R^1 R^2 45
cyclization
carbonyl insertion with L = CO
next alkyne insertion
etc.
oligomers, polymers
X = OR, NHR, NR_2

Scheme 8 Various modes of reaction of ethenylcarbene complexes **43** with alkynes **44** [11]

3.1
Formal [3+2] Cycloadditions

In 1986 Yamashida et al. found that the reaction of the (morpholino)phenylcarbene complex **46** with symmetric alkynes **47** gave the morpholinylindene derivatives **48** and **49**, as well as the indanones **50** derived from the latter by hydrolysis, in excellent yields (Scheme 9) [54]. This contrasts with the behavior of the corresponding (methoxy)phenylcarbene complex, which solely undergoes the Dötz reaction [55]. This transformation of the amino-substituted complex **46** apparently does not involve a CO insertion, which is an important feature of the Dötz benzannelation.

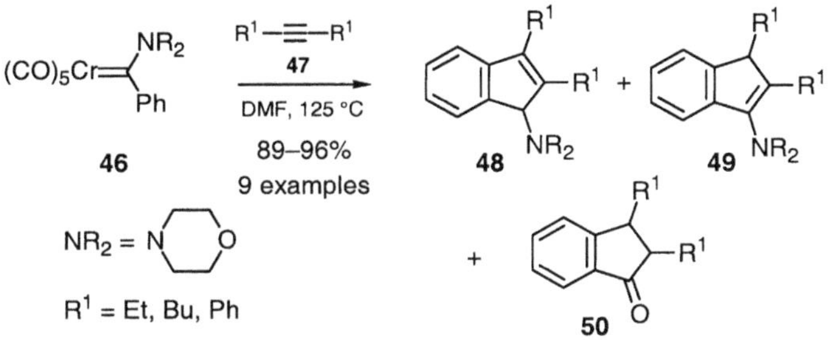

Scheme 9 Formation of indene derivatives from the complex **46** and alkynes **47** [54, 55]

The non-CO-inserted products, the indenes **48/49**, almost certainly are formed by reductive elimination from chromadihydronaphthalenes **52**, which arise by 6π-electrocyclization from the alkyne-insertion intermediates **51** (Scheme 10). According to the study of Wulff et al. [56], an electron-donating dialkylamino group stabilizes a 1-chroma-1,3,5-triene **51**, and increases the electron density at the chromium atom. This in turn strengthens the Cr–CO bond and reduces the tendency of a *cis*-CO ligand to undergo insertion. The same selectivity for the formation of five-membered rings without CO insertion had also been observed by Dötz et al. [57].

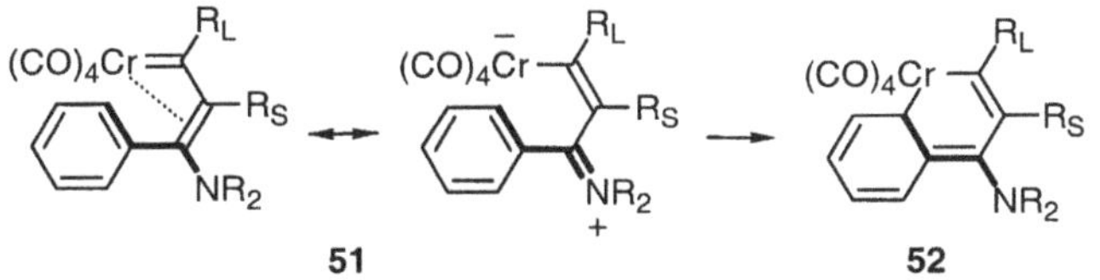

Scheme 10 Suppression of the CO insertion by the electron-donating ability of a dialkyamino moiety [54–56]

The formation of a formal [3+2] cycloaddition product **56** upon reaction of the ethoxystyryltungsten complex **53** with 1-diethylaminopropyne, as observed by Aumann et al., shed some light on the mechanism (Scheme 11). The intermediate 2,4-bisdonor-substitued 1-tungsta-1,3,5-hexatriene **54**, formed by initial insertion of the alkyne into the carbene complex **53**, could be isolated in 40% yield [58]. It readily underwent 6π-electrocyclization at ambient temperature with a half-life of 14 h to give the zwitterionic cyclopentene derivative **55** which, upon treatment with hydrochloric acid, afforded the corresponding cyclopentenone **56** with loss of the pentacarbonyltungsten fragment [59].

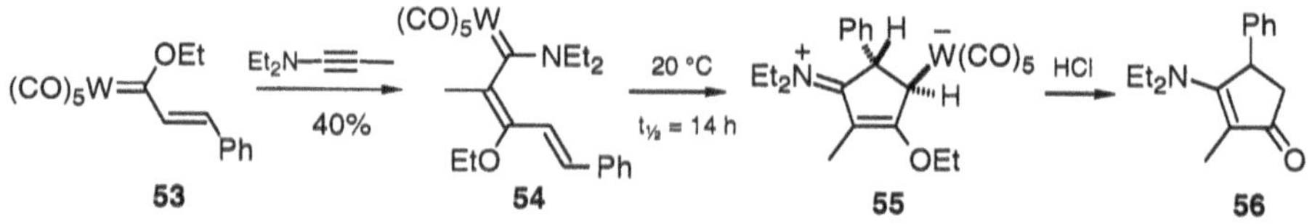

Scheme 11 Formation of the cyclopentenyl zwitterion derivative **55** from a 1-tungsta-2-diethylamino-1,3,5-hexatriene **54** [58, 59]

What later became a widely applicable, high yielding synthesis of 5-dialkylamino-3-ethoxy-1,3-cyclopentadienes of type **60** originally was observed only for the reaction of 3-cyclopropyl-substituted 3-dialkylamino-1-ethoxypropenylidenechromium complexes of type **57** (R^1=*c*Pr) with alkynes (Scheme 12) in THF [60] or in *n*-hexane [61]. This unique behavior was attributed to the well-known electron-donating property of the cyclopropyl group, which apparently disfavors the insertion of carbon monoxide at the stage of the alkyne insertion product **58**, and favors the 6π-electrocyclization to yield an intermediate chromacyclohexadiene **59**. The latter, just like the intermediate **52**, undergoes reductive elimination to yield the five-membered ring **60a(b)**. As de Meijere et al. subsequently found out, this reaction mode becomes quite general with almost any kind of substituent – except for very bulky ones, which

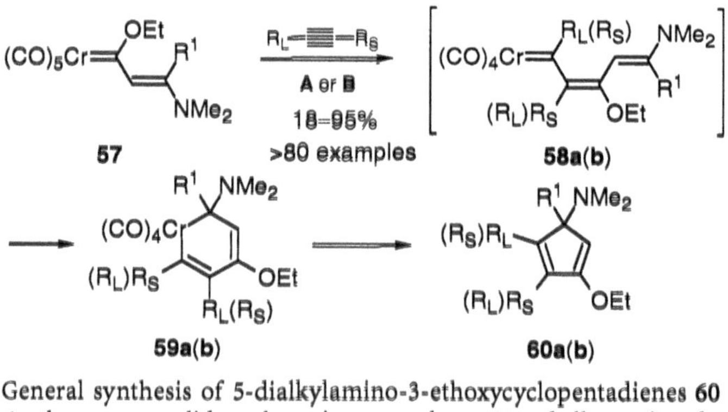

Scheme 12 General synthesis of 5-dialkylamino-3-ethoxycyclopentadienes **60** from 3-dialkylamino-1-ethoxypropenylidenechromium complexes **57** and alkynes in a donor solvent. Conditions **A**: pyridine, 55–80 °C, 1.5–4 equiv. of the alkyne; **B**: MeCN, 80 °C, slow addition of 2–4 equiv. of the alkyne. For further details see Table 1 [43, 44, 60, 61]

Table 1 Selected examples of 5-dialkylamino-3-ethoxycyclopentadienes **60a(b)** obtained from 3-dialkylamino-1-ethoxypropenylidenechromium complexes **57** and alkynes in a donor solvent. For details see Scheme 12 [43, 44, 60, 61]

Entry	R^1	Condition	R_L	R_S	Yield (%) **60a/60b**
1	Me	A	Me	Me	82/0
2	Me	A	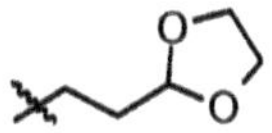 O O	H	86/0
3	*n*Pr	A	Me	Me	95/0
4	*n*Pr	A	Ph	Ph	80/0
5	*n*Pr	A	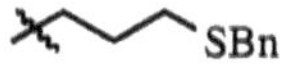 SBn	H	75/11
6	*n*Pr	B	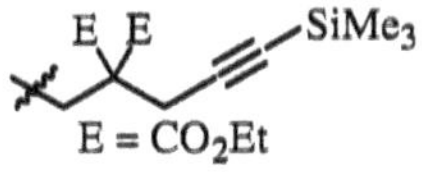 E E $SiMe_3$ E = CO_2Et	H	81/0
7	*n*Pr	A	Me	Me	77/0
8	*c*Pr	A	Me	Me	84/0
9	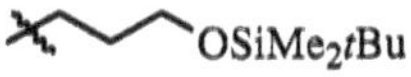 $OSiMe_2tBu$	A	Me	Me	91/0
10	 SBn	A	Me	H	78/0
11	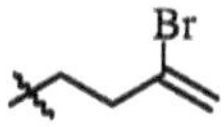 Br	A	Me	Me	79/0
12	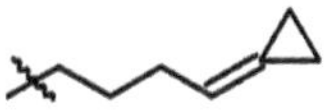	A	Me	Me	69/0
13	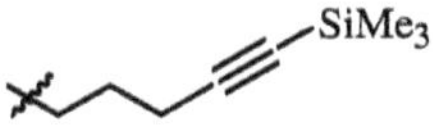 $SiMe_3$	A	$SiMe_3$	H	48/6
14	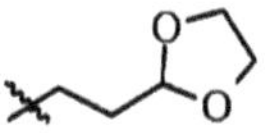 O O	A	Me	H	88/0

lead to different types of products (see below) – when the reaction of 57 with an alkyne is carried out in a donor-type solvent such as pyridine or acetonitrile (Scheme 12 and Table 1) [43, 44]. The regioselectivity largely depends on the relative steric bulk of the substituents R^1 in the complexes **57** and R_L, R_S in the alkynes, and in the former they have more influence than in the latter [44]. Other factors, including concentrations of the complexes **57** and applied alkynes, and the electronic properties of substituents on the alkynes, do not play important roles [62].

Cocyclizations of internal alkynes and carbene complexes **57** with larger substituents R^1 (e.g., R^1=*i*Pr) not only lead to formation of an increased proportion of the regioisomers **60b**, but also to that of the isomeric cyclopentadienes **61**, which would result from **60a** by 1,2-migration of the dimethylamino group via the bridged zwitterionic intermediate **62** (Scheme 13) [44]. The fact that isomeric cyclopentadienes **61** are formed only when the less sterically demanding substituent R_S in the incoming alkyne has an electron-withdrawing effect is in line with this assumption, and not with a 1,5-migration of the dimethylamino functionality.

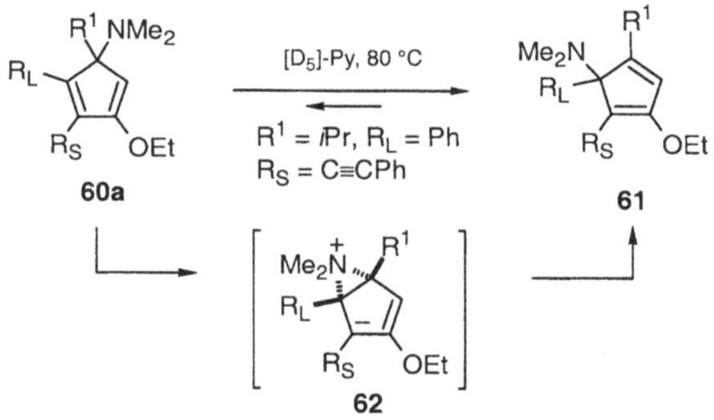

Scheme 13 Possible mode of formation of the cyclopentadiene **61** isomeric with **60a** by 1,2-migration of the dimethylamino group via a bridged zwitterionic intermediate **62** [44]

Variously substituted 5-amino-3-ethoxycyclopentadienes **66** have been applied toward the preparation of more complex structures to demonstrate their versatility in organic synthesis. When dienyl-substituted cyclopentadienes of type **66** (R_L, R_S=cycloalkene) are generated from the reaction of correspondingly substituted complexes of type **57** with conjugated 1,5-dien-3-ynes, the trisannelated benzene derivatives **63** were obtained by a sequence of 6π-electrocyclization, twofold 1,n-hydrogen shifts, elimination of a dimethylamine, 1,5-hydrogen shift, and finally hydrolysis (Scheme 14) [63, 64]. Compared to the traditional approaches to trisannelated benzene derivatives of type **63** by aldol condensation [65–69], this method has the advantages of milder conditions and the provision of additional functionality. It remains to be tested whether skeletons with two annelated small rings would be accessible by this new method. Because of their enol ether moieties, the cyclopentadienes **66** can be easily hydrolyzed to the corresponding cyclopentenones **67** in excellent yields by treatment with a catalytic amount of hydrochloric acid [44]. With this in mind, de Meijere et al. developed very short and direct accesses to bicyclo[3.3.0]oct-2-en-4-ones **64** and 8-azabicyclo[3.3.0]octenones **65** by intramolecular aldol reactions of dicarbonyl compounds derived from cyclopentenones **67** with an acetal-protected aldehyde or ketone carbonyl group in the substituent R^1 or R, respectively [70]. This type of transformation has been applied toward short syntheses of angular triquinanes like **68** in an enantiomerically pure form [71], as well as other complex oligocycles [63, 64].

The dialkylamino, especially the dimethylamino, group in a cyclopentenone of type **67** can be alkylated with methyl iodide to yield a quaternary ammonium salt. Upon treatment with a base, these quaternary ammonium salts undergo

Hofmann elimination to correspondingly substituted cyclopentadienones which, depending on the nature and the nucleophilicity of the base as well as the nature of the substituents R_L and R_S, undergo [2+2] or [4+2] cyclodimerization or in situ Michael addition to yield compounds **69**, **70**, and **71**, respectively (Scheme 14) [44,70].

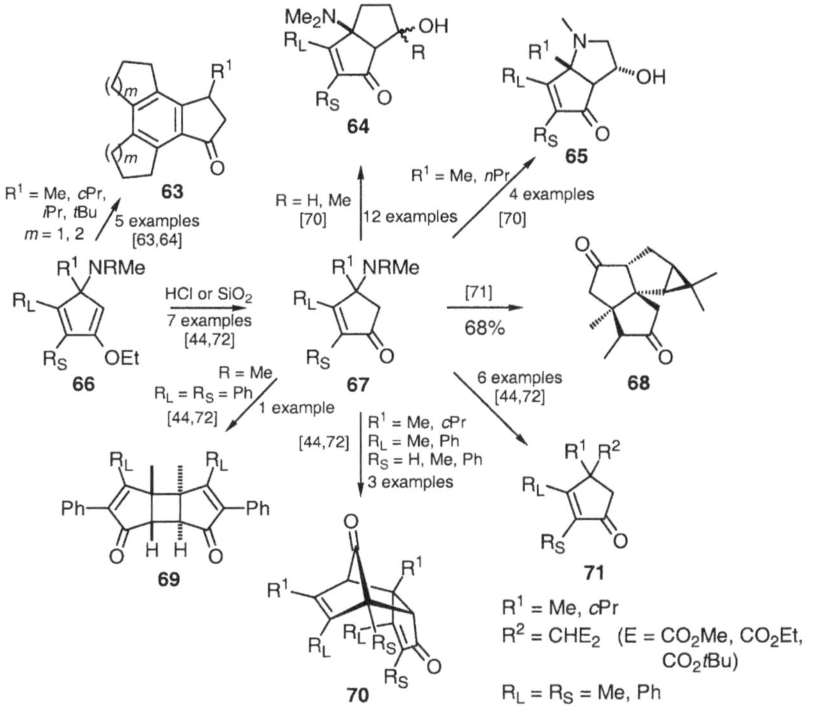

Scheme 14 Some applications of 5-amino-3-ethoxycyclopentadienes **66** to the syntheses of cyclopentanoid skeletons [44,63,64,70–72]

Recently, Aumann et al. reported that rhodium catalysts enhance the reactivity of 3-dialkylamino-substituted Fischer carbene complexes **72** to undergo insertion with enynes **73** and subsequent formation of 4-alkenyl-substituted 5-dialkylamino-2-ethoxycyclopentadienes **75** via the transmetallated carbene intermediate **74** (Scheme 15, Table 2) [73]. It is not obvious whether this transformation is also applicable to complexes of type **72** with substituents other than phenyl in the 3-position. One alkyne **73**, with a methoxymethyl group instead of the alkenyl or phenyl, i.e., propargyl methyl ether, was also successfully applied [73].

Alkylideneaminocarbene complexes **76**, which are aza analogs of alkenylcarbene complexes, upon reaction with alkynes primarily give formal [3+2] cycloadducts analogous to the 1-aminocarbene complexes (Scheme 16) [74,75]. Aumann et al. proposed that this should be considered as a formal 1,3-dipo-

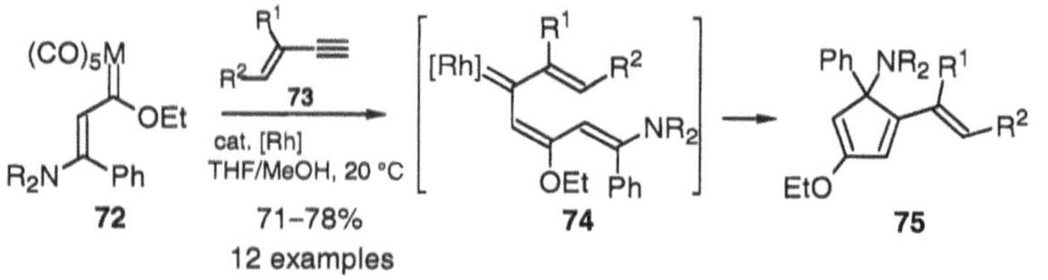

Scheme 15 Formation of 4-alkenyl(phenyl)-substituted 5-dialkylamino-2-ethoxycyclopentadienes **75** via transmetallated alkyne-inserted rhodium-carbene complexes **74** [73]. For further details see Table 2

Table 2 Formation of cyclopentadienyl derivatives **75** via transmetallated alkyne-inserted rhodium-carbene complexes (see Scheme 15)

Entry	M	NR_2	R^1	R^2	$[(COD)RhCl]_2$ Yield (%)	$[(CO)_2RhCl]_2$ Yield (%)	$RhCl_3 \cdot 3H_2O$ Yield (%)
1	Cr	NMe_2	Me	H			76
2	W	NMe_2	Me	H	53	74	78
3	W	NEt_2	Me	H			74
4	Cr	Morpholine	Me	H			74
5	W	Morpholine	Me	H	58	75	76
6	Cr	NMe_2	$-(CH_2)_4-$				72
7	W	NMe_2	$-(CH_2)_4-$				73
8	W	NEt_2	$-(CH_2)_4-$		61		71
9	W	Morpholine	$-(CH_2)_4-$		60		71
10	W	(+)-Ephedrine	Me	H			78
11	W	(+)-Prolinole	Me	H			75
12	W	NHMe	Me	H		0	0
13	W	NHEt	Me	H		0	0
14	W	NMe_2	$-(CH)_4-$				76
15	W	NEt_2	$-(CH)_4-$				77

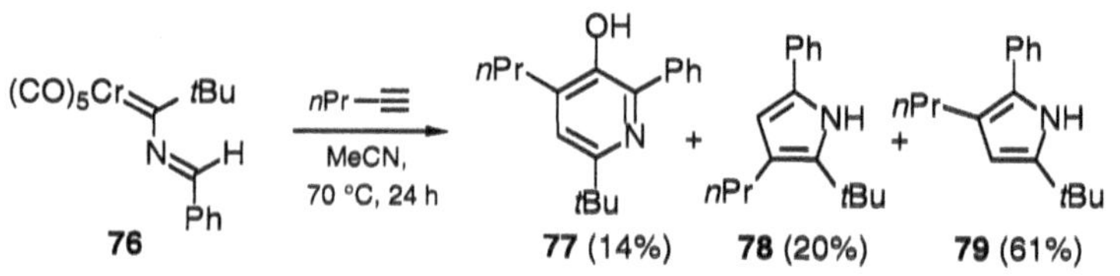

Scheme 16 Formation of pyrroles **78** and **79** from the benzylideneaminocarbene complex **76** and 1-pentyne [74, 75]

lar cycloaddition. The product distribution from the reaction of **76** with 1-pentyne to a certain extent depends on the solvent used [74]. When hexane is applied instead of acetonitrile, the ratio of the formal [3+2+1] **77** to formal [3+2] cycloadducts **78** and **79** does not significantly change, but the ratio of the regioisomers **78** and **79** does.

The formation of the tricarbonylchromium-complexed fulvene **81** from the 3-dimethylamino-3-(2′-trimethylsilyloxy-2′-propyl)propenylidene complex **80** and 1-pentyne also constitutes a formal [3+2] cycloaddition, although the mechanism is still obscure (Scheme 17) [76]. The η^6-complex **81** must arise after an initial alkyne insertion, followed by cyclization, 1,2-shift of the dimethylamino group, and subsequent elimination of the trimethylsilyloxy moiety. Particularly conspicuous here are the alkyne insertion with opposite regioselectivity as compared to that in the Dötz reaction, and the migration of the dimethylamino functionality, which must occur by an intra- or intermolecular process. The mode of formation of the cyclopenta[*b*]pyran by-product **82** will be discussed in the next section.

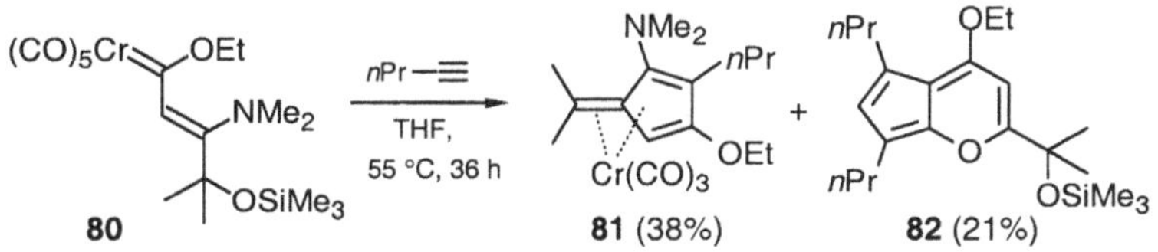

Scheme 17 Formation of the (tricarbonylchromium)-complexed fulvene **81** and the cyclopenta[*b*]pyran **82** from the 3-dimethylamino-3-(2′-trimethylsilyloxy-2′-propyl)propenylidene complex **80** and 1-pentyne [76]

3.2 [3+4+1] and [3+2+2+1] Cocyclizations

Reaction of the dihydropyranyl-substituted complex **83** with a conjugated internal alkynone **84** affords the Dötz-type formal [3+2+1] cycloadduct **86** in only 6% yield. The major product is the tricycle **85** as the result of a formal [3+4+1] cycloaddition with incorporation of the ynone carbonyl group (Scheme 18) [77].

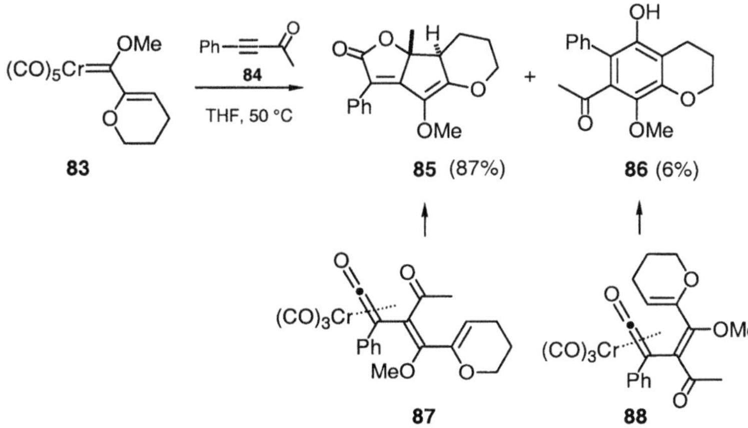

Scheme 18 Formation of tricyclic product **85** via a von Halban–White-type cyclization [77]

This crisscross or von Halban–White-type cyclization product is formed from the (*E*)-configured intermediate **87**, which cannot undergo the 6π-electrocyclization like the (*Z*)-configured isomer **88**, to yield the benzannelation product **86** [78, 79]. While the diastereoselectivity of the alkyne insertion must have been controlled by the electronic and not the steric factors of the substituents on the alkyne, the *anti*-configuration of the tricyclic system **85** was confirmed by an X-ray structure analysis [77].

Steric effects must play a major role in determining the configurations of 2-donor-substituted ethenylcarbenechromium complexes **89** obtained by Michael-type additions onto alkynylcarbene complexes, and of their alkyne-insertion products. With bulky substituents in the 2′-position, complexes **89** are mostly (*Z*)-configured and yield (*Z,E*)-configured 1-chroma-1,3,5-hexatrienes which cannot easily undergo 6π-electrocyclization. They rather insert another molecule of the alkyne **90** and carbon monoxide to give **93** and **94**, respectively, which undergo intramolecular [4+2] cycloaddition and subsequent elimination of HY to the regioisomeric cyclopenta[*b*]pyrans **91** and **92** in yields up to 90% (Scheme 19, Table 3) [80]. In most cases, the isomers **91** are formed as major or even single products. However, the second alkyne insertion into complexes **89** can occur with incomplete regioselectivity, thus the two isomeric products can be formed. High chemical yields in this kind of transformation are obtained from complexes **89** with a tertiary or a bulky secondary substituent (R^1), a weak donor substituent X (e.g., OEt is better than NMe_2), and a good donor group Y (e.g., NR_2>OR≥SR) [41]. This new synthesis of cyclopenta[*b*]pyrans from easily prepared starting materials is superior to previously developed accesses to these so-called pseudoazulenes, which show unusual photophysical properties. Besides strong absorptions in the UV region, they also exhibit a broad absorption band in the visible light region with extinction coefficients ε of about 1,000.

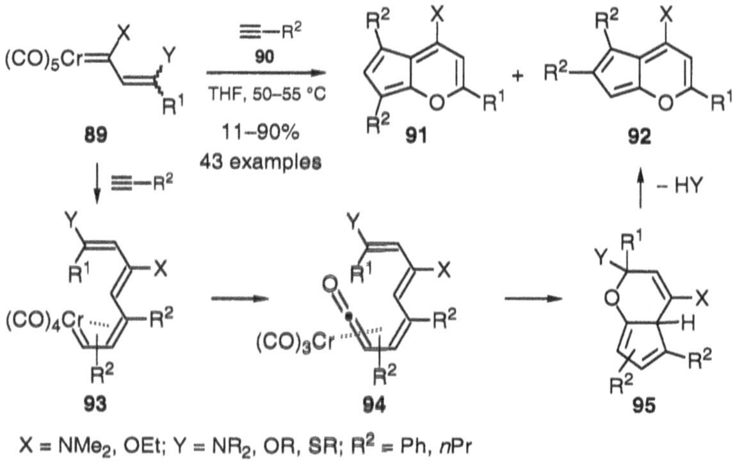

X = NMe_2, OEt; Y = NR_2, OR, SR; R^2 = Ph, *n*Pr

Scheme 19 Formation of cyclopenta[*b*]pyrans **91** and **92** by a [3+2+2+1] cocyclization [41, 80]. For further details see Table 3

Table 3 Selected examples of cyclopenta[*b*]pyrans **91** and **92** formed by [3+2+2+1] cocyclizations (see Scheme 19)

Entry	R^1	X	Y	R^2	Product	Yield (%)
1	Ph	OEt	NMe_2	Ph	**91a**	24
2	Ph	OEt	NBn_2	Ph	**91a**	48
3	Ph	OEt	$N(CH_2CH{=}CH_2)_2$	Ph	**91a**	43
4	Ph	OEt	N(*i*Pr)$_2$	Ph	**91a**	17
5	Ph	OEt	NMe_2	*n*Pr	**91b**	19
6	Ph	OEt	N(*i*Pr)$_2$	*n*Pr	**91b**	11
7	$C(CH_3)_2OEt$	NMe_2	NBn_2	Ph	**91c**	39
8	$C(CH_3)_2OEt$	OEt	NMe_2	Ph	**91d**	51
9	$C(CH_3)_2OEt$	OEt	NBn_2	Ph	**91d**	68
10	$C(CH_3)_2OEt$	OEt	OEt	Ph	**91d**	27
11	$C(CH_3)_2OEt$	OEt	OPh	Ph	**91d**	28
12	$C(CH_3)_2OEt$	OEt	OBn	Ph	**91d**	18
13	$C(CH_3)_2OSiMe_3$	NMe_2	NBn_2	Ph	**91e**	28
14	$C(CH_3)_2OSiMe_3$	OEt	NMe_2	Ph	**91f**	90
15	$C(CH_3)_2OSiMe_3$	OEt	NBn_2	Ph	**91f**	78
16	$CHCH_3OSi$*t*$BuPh_2$	OEt	NMe_2	Ph	**91g/92g**	39/2
17	$CHCH_3OSi$*t*$BuPh_2$	OEt	NBn_2	Ph	**91g/92g**	74/22
18	OEt	OEt	NMe_2	*n*Pr	**91h**	33
19	OEt	OEt	NMe_2	Ph	**91i**	84
20	O O	OEt	NBn_2	Ph	**91j/92j**	44/37
21	adamantyl	OEt	NMe_2	Ph	**91k**	56
22	adamantyl	OEt	NBn_2	Ph	**91k**	47

3.3
[3+2+2+2] Cocyclizations

The novel highly substituted spiro[4.4]nonatrienes **98** and **99** are produced by a [3+2+2+2] cocyclization with participation of three alkyne molecules and the (2′-dimethylamino-2′-trimethylsilyl)ethenylcarbene complex **96** (Scheme 20). This transformation is the first one ever observed involving threefold insertion of an alkyne and was first reported in 1999 by de Meijere et al. [81]. The structure of the product was eventually determined by X-ray crystal structure analysis of the quaternary ammonium iodide prepared from the regioisomer **98** (Ar=Ph) with methyl iodide. Interestingly, these formal [3+2+2+2] cycloaddition products are formed only from terminal arylacetylenes. In a control experiment with the complex **96** ^{13}C-labeled at the carbene carbon, the ^{13}C label was found only at the spiro carbon atom of the products **98** and **99** [42].

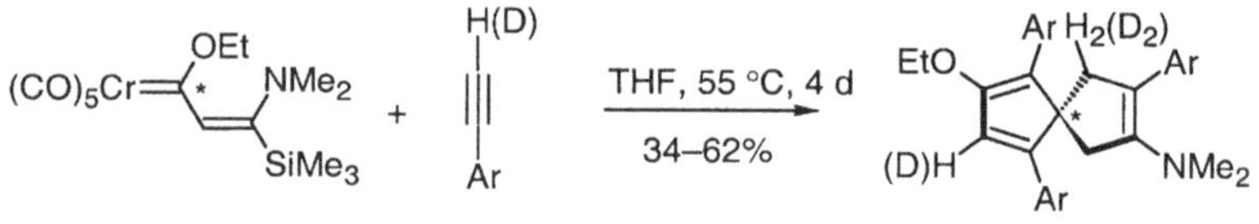

Ar	Yield **98+99** (%)	Ratio **98:99**
C_6H_5	62	1.8:1
4-PhC_6H_4	34	4.5:1
4-$MeOC_6H_4$	37	2.5:1
3,5-$(Me)_2C_6H_3$	48	1.6:1

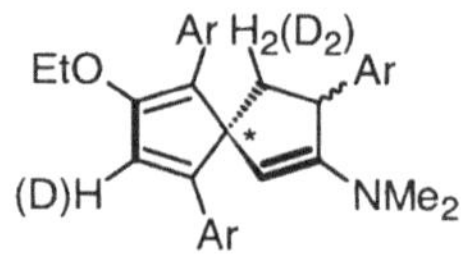

Scheme 20 Formation of highly substituted spiro[4.4]nonatrienes **98** and **99** from the (2′-dimethylamino-2′-trimethylsilyl)ethenylcarbene complex **96** and arylacetylenes **97** [42, 81]

Terminally deuterium-labeled phenylacetylene was also used to elucidate the possible mechanism of this reaction. In view of all these results, a rationalization for the loss of the trimethylsilyl and the migration of the ethoxy group from its original position in the complex **96** has been put forward. Due to the contribution of the conjugated diarylcyclopentadiene moiety in **98** and **99**, these molecules showed intense fluorescence with a relatively high quantum yield of 46%.

3.4 [2+2+1] Cocyclizations

Strikingly, (2′-dialkylamino)ethenylcarbene complexes **100** (type **57**, but with a morpholinyl or dibenzylamino group) can also undergo a [2+2+1] cocyclization with insertion of carbon monoxide like in the classical Dötz reaction, yet with only two carbons of complexes **100** participating in the formation of the ring, thus yielding a methylenecyclopentenone **101** or **102**. After insertion of an alkyne and a CO molecule, the resulting dienylketene complex **103**, due to its 1,5-dipolar properties, undergoes a 1,5-cyclization rather than a 6π-electrocyclization to form the five-membered ring **104** (Scheme 21) [82, 83]. Depending on the reaction conditions and the nature of the amino substituent, either the aminomethylenecyclopentenone **101** or the 2-(aminoalkenyl)cyclopentenone **102** is formed by a hydrogen shift and loss of the tricarbonylchromium fragment. The products **101** are obtained as mixtures of (*E*)- and (*Z*)-isomers, with their ratios depending on the nature of the substituents. In all cases, the (*Z*)-isomers of **101** were obtained as the major products. Starting from the enantiomerically pure (*S*)-5-(*tert*-butyldimethylsilyl)-1-octyne, the cyclopentenone **105**, which is of type **102**, was prepared along such a route and applied to a short synthesis of the natural product (–)-oudenone **106** with 92% *ee* [83] (Table 4).

Under the same conditions, but in moist solvents, complexes of type **100** with terminal alkynes **90** gave 2-acyl-3-amino- **107** and 2-acyl-3-ethoxycy-

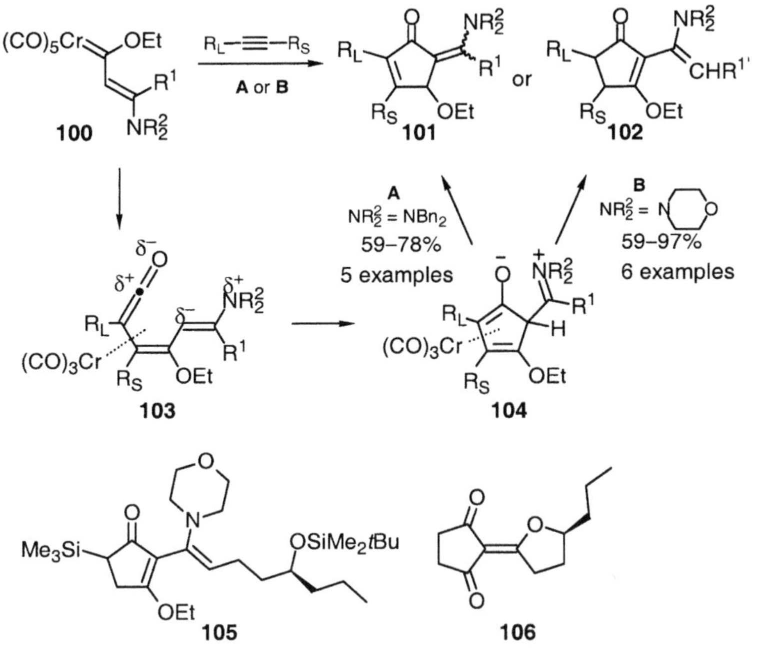

Scheme 21 Formation of 5-(aminomethylene)cyclopentenones **101** and 2-(1′-aminoalkenyl)-cyclopentenones **102** by formal [2+2+1] cycloadditions. Conditions **A**: THF, 50–55 °C. **B**: THF/MeCN (9/1), 65 °C [82, 83]. For further details see Table 4

Table 4 Selected examples of cyclopentenones **101** and **102** formed from complexes **100** (see Scheme 21)

Entry	R^1	Condition	R_L	R_S	Product	Yield (%)
1	*n*Pr	**A**	Ph	Ph	**101a**	68
2	*n*Pr	**A**	Me	Me	**101b**	78
3	*n*Pr	**A**	$-(CH_2)_6-$		**101c**	75
4	*c*Pr	**A**	Ph	Ph	**101d**	59
5	*c*Pr	**A**	$-(CH_2)_6-$		**101e**	62
6	*n*Pr	**B**	*n*Pr	H	**102a**	82
7	*n*Pr	**B**	Ph	H	**102b**	97
8	Me	**B**	$SiMe_2tBu$	H	**102c**	72
9	$OSiMe_2tBu$	**B**	$SiMe_2tBu$	H	**102d**	69

clopentenones **108** (Scheme 22). The latter are also formed via the intermediates **104** (NR^2_2=NMe_2; R_S=H) and subsequent hydrolysis [84]. Formation of **107** (NR^2_2=morpholine), however, not only requires hydrolysis, but also a formal shift of the morpholine group which probably occurs by 1,4-addition of morpholine to **108** with subsequent 1,4-elimination of ethanol [85].

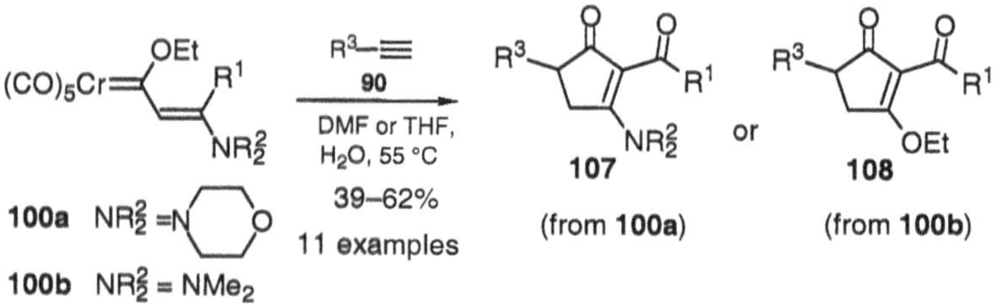

Scheme 22 Formation of 2-acyl-3-amino- **107** and 2-acyl-3-ethoxycyclopentenones **108** in moist solvents [84, 85]. For further details see Table 5

Table 5 Selected examples of cyclopentenone derivatives **107** and **108** formed from complexes **100a,b** in moist solvents (see Scheme 22)

Entry	R^1	NR^2_2	R^3	Product	Yield (%)
1	*n*Pr	Morpholine	*n*Pr	**107a**	68
2	*n*Pr	Morpholine	*t*Bu	**107b**	78
3	*n*Pr	Morpholine	$SiMe_2$*t*Bu	**107c**	75
4	Ph	Morpholine	*n*Pr	**107d**	59
5	Ph	Morpholine	*t*Bu	**107e**	62
6	Ph	Morpholine	$SiMe_2$*t*Bu	**107f**	82
7	*i*Pr	NMe_2	*t*Bu	**107g/108g**	15/47

3.5 [5+2] Cocyclizations

The reactions of the isopropyl-substituted 3-dimethylaminopropenylidenechromium complex **109** with terminal alkynes **90** bearing a bulky substituent (e.g., R=*tert*-butyl, mesityl, adamantyl etc.), in the presence of moist pyridine, yield 2-(acylmethylene)pyrrolidines **110** (Scheme 23) [84]. The dihydroazepinetricarbonylchromium complexes **111** were found to be the key

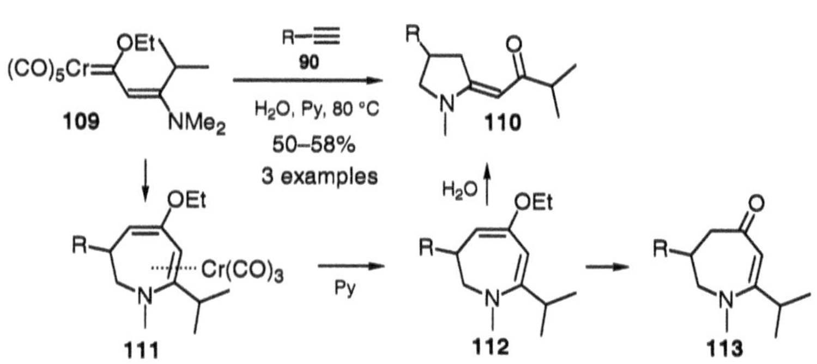

Scheme 23 Formation of tetrahydroazepinones **113** and methylenepyrrolidines **111** by a formal [5+2] cycloaddition with C–H activation [85]

intermediates in this transformation. The complexes **111** could be synthesized from the same starting materials in the presence of 1 equiv. of triphenylphosphine in THF. The formation of these unusual complexes **111** was proposed to occur with initial insertion of the alkyne into the complex **109**, activation by the chromium fragment of a carbon–hydrogen bond in the dimethylamino functionality, and insertion into it, thus leading to ring closure to give **111** [84]. The structure of **111** was rigorously proved by X-ray crystal structure analysis of a derivative with R=mesityl. It shows that the tricarbonylchromium fragment is η^5-coordinated with the aminodienyl unit of the dihydroazepine. Treatment of the complex **111** with anhydrous pyridine afforded decomplexed dihydroazepines **112** which were isolated as the corresponding ketones **113**. However, in the presence of moist pyridine, **111** underwent hydrolysis with ring contraction to yield methylenepyrrolidines **110**.

3.6 [5+2+1] Cocyclizations

Barluenga et al. reported interesting transformations of the 2-oxabicyclo-[3.2.0]heptenyl-substituted complex **116**, which was prepared by a [2+2] cycloaddition of the ethynylcarbene complex **114** to dihydrofuran **115** (Scheme 24). Upon heating the tricyclic complex **116** at 65 °C, CO insertion with subsequent 6π-electrocyclization in the sense of an intramolecular Dötz reaction occurs, to yield the tetracyclic catechol monoether **117** [86]. This result is quite surprising since 1-metalla-1,3,5-hexatrienes usually undergo 6π-electrocyclization without CO insertion (cf. Scheme 32). On the other hand, the complex **116** upon intermolecular reaction with a terminal alkyne, CO insertion, and subsequent cyclization of an intermediate trienylketenyl derivative gave bisannelated methoxycyclooctatrienones **118** [87]. This overall transformation constitutes a formal [5+2+1] cycloaddition or – in other terms – a vinylogous Dötz reaction.

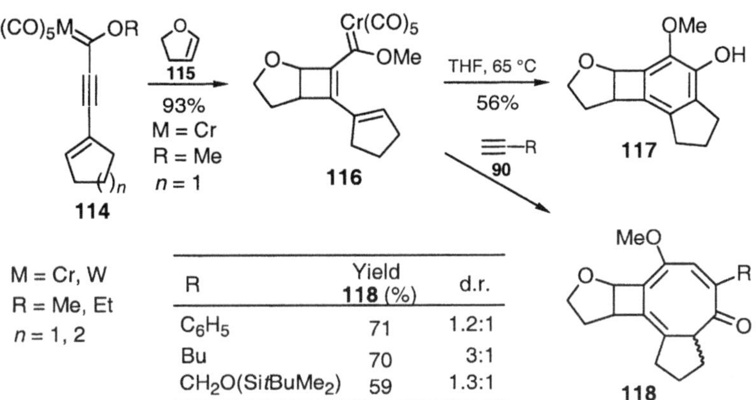

R	Yield **118** (%)	d.r.
C_6H_5	71	1.2:1
Bu	70	3:1
$CH_2O(SitBuMe_2)$	59	1.3:1

Scheme 24 Formation of a bisannelated methoxycyclooctatrienone **118** by a formal [5+2+1] cycloaddition [86, 87]

3.7
[4+2] Cocyclizations

Upon heating an alkenylidenechromium complex **119** substituted with a secondary 3-amino group, in the presence of a terminal alkyne **90** in THF, 4-(1*H*)-pyridinylidene complexes of type **120** were formed with a high degree of regioselectivity (Scheme 25) [76, 88]. This reaction mode is completely different from that of an alkenylidenechromium complex with a tertiary amino substituent in the 3-position. The formation of **120** can be rationalized by way of a 4π-electrocyclization to yield a 3-aminoalkenyl-3-ethoxycyclopropenylpentacarbonylchromium complex **121**. The alkenylcyclopropene derivative **121** would be expected to undergo a regioselective intramolecular addition of the amino group onto the cyclopropene double bond, with attack at the least substituted carbon atom to give a bicyclic zwitterionic intermediate **122**. Ring expansion with opening of the three-membered ring and migration of the carbonylchromium residue would lead to **123**, from which 1,4-elimination of ethanol would provide the pyridinylidene complex **120**. The pentacarbonylchromium fragment can be removed from these by treatment with HBF_4 to afford the corresponding pyridinium salts [88].

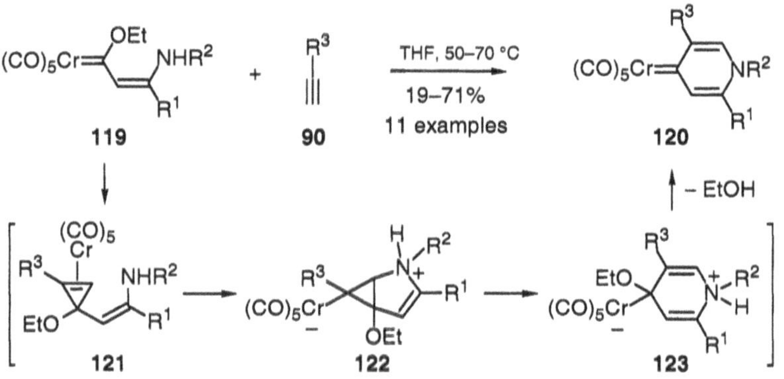

R^1 = *n*Bu, Ph, $CMe_2(OSiMe_3)$; R^2 = Me, *i*Pr, Ph, Bn; R^3 = *n*Bu, Ph

Scheme 25 Formation of 4-(1*H*)-pyridinylidene complexes **120** by a formal [4+2] cycloaddition [76, 88]

3.8
Cocyclizations with Aza- and Phosphaalkynes

Aumann et al. showed that 1,2,4-tridonor-substituted naphthalenes, such as **126**, are accessible from 3-donor-substituted propenylidenecarbene complexes **124** containing a (*Z*)-positioned 3-phenyl substituent and isocyanide (Scheme 26). These transformations constitute formal [5+1] cycloadditions [39, 89, 90]. Since isocyanides are strongly coordinating ligands on chromium, at least

2 equiv. has to be applied for this reaction, which in most cases proceeds under mild conditions, even at 20 °C, and affords good to excellent yields (72–96%). The proposed key intermediates, the ketenimine complexes **125** (with coordination of $(CO)_4Cr(RNC)$ at the imine moiety), cannot be isolated, but rapidly undergo 6π-electrocyclization and subsequent tautomerization to form naphthalenes **126**. (*Z*)-Configured ketenimines **125** with an acylamino substituent in the 4-position and the complex **127**, however, can be isolated in excellent yields from the reaction of the corresponding complexes **124** with *tert*-butyl isocyanide. Upon heating, these phenylethenylketimines of type **125** and complex **127**, still (*Z*)-configured, also gave naphthalenes **126** in excellent yields.

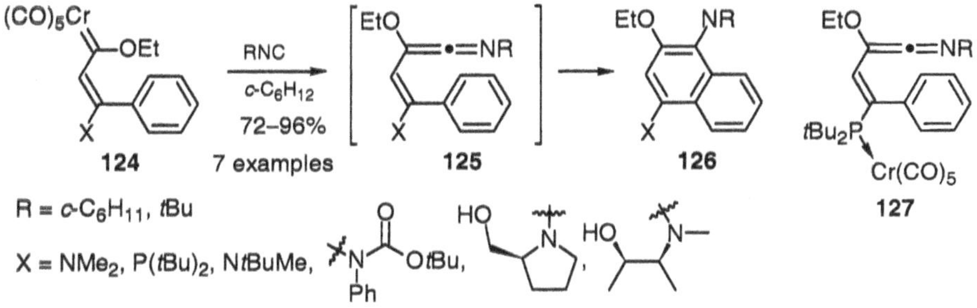

Scheme 26 Cocyclizations of 3-phenyl-substituted propenylidenechromium complexes **124** with isocyanides [39, 89, 90]

Kinetically stabilized phosphaalkynes have also been applied as reaction partners for α,β-unsaturated Fischer carbene complexes. Thus, reaction of the (1-naphthyl)carbenechromium complex **128** with 3,3-dimethyl-1-phosphabutyne (**129**) afforded the substituted 3-phosphaphenanthrenetricarbonylchromium complex **130** in very good yield (Scheme 27) [91]. A kinetic investigation disclosed that **129** reacts six times faster than its carbon analog, 3,3-dimethylbutyne, in this same transformation. According to an X-ray crystal structure analysis, one carbonyl group of the $Cr(CO)_3$ unit in **130** is nearly eclipsed with the phosphorus atom, apparently in order to minimize steric interactions between the ring substituents and the carbonyl ligands. The coordination of the phosphinine ring to the tricarbonylchromium moiety is very strong. The

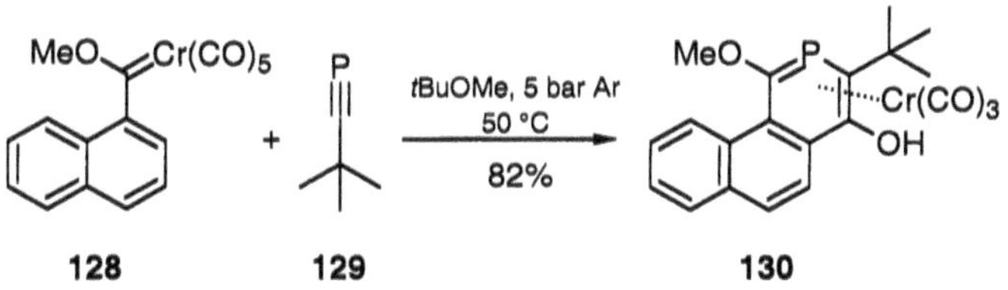

Scheme 27 Cocyclization of the 1-naphthylcarbene complex **128** with *tert*-butylphosphaalkyne **129** [91]

complex has to be heated in refluxing toluene to give the decomplexed 3-phosphaphenanthrene, which can also be obtained in significantly better yield (95%) by treatment of the complex **130** with CO under pressure (30 bar) at 70 °C.

3.9 Cocyclizations of In Situ Generated Alkenylcarbene Complexes

The insertion of alkynes into a chromium–carbon double bond is not restricted to Fischer alkenylcarbene complexes. Numerous transformations of this kind have been performed with simple alkylcarbene complexes, from which unstable α,β-unsaturated carbene complexes were formed in situ, and in turn underwent further reactions in several different ways. For example, reaction of the 1-methoxyethylidene complex **6a** with the conjugated enyne-ketimines and -ketones **131** afforded pyrrole [92] and furan **134** derivatives [93], respectively. The alkyne-inserted intermediate **132** apparently undergoes 6π-electrocyclization and reductive elimination to afford enol ether **133**, which yields the cycloaddition product **134** via a subsequent hydrolysis (Scheme 28). This transformation also demonstrates that Fischer carbene complexes are highly selective in their reactivity toward alkynes in the presence of other multiple bonds (Table 6).

131 132 133 134

A: THF, 65 °C
B: dioxane, 100 °C

X = NR (9–74%)
X = O (79–84%)

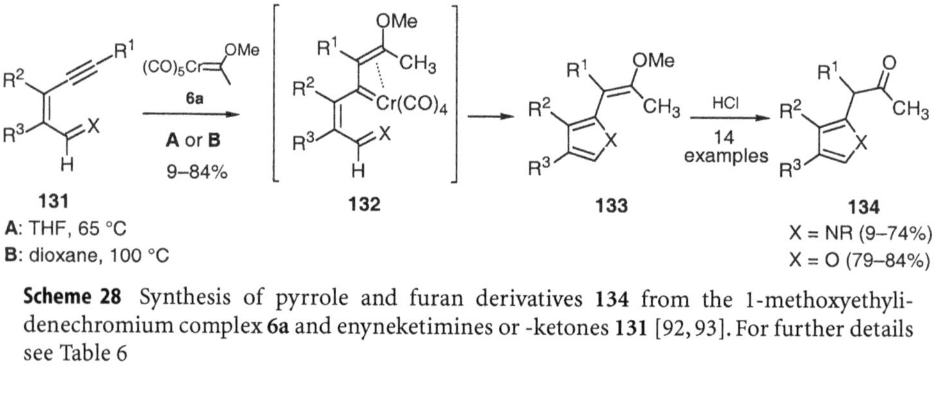

Scheme 28 Synthesis of pyrrole and furan derivatives **134** from the 1-methoxyethylidenechromium complex **6a** and enyneketimines or -ketones **131** [92, 93]. For further details see Table 6

Table 6 Synthesis of pyrrole and furan derivatives **134** (see Scheme 28)

Entry	Conditions	R^1	R^2	R^3	X	Product	Yield (%)
1	**B**	*n*Bu	$-(CH_2)_3-$		O	**134a**	84
2	**B**	H	$-(CH_2)_3-$		O	**134b**	79
3	**A**	*n*Bu	Ph	H	$NNMe_2$	**134c**	62
4	**A**	*n*Bu	Ph	H	NTs	**134d**	37
5	**A**	*n*Bu	Ph	H	NMs	**134e**	35
6	**A**	*n*Bu	Ph	H	NBn	**134f**	9
7	**A**	*n*Bu	H	Et	$NNMe_2$	**134g**	74

Combinations of alkyne insertion and subsequent intramolecular [2+1] cycloaddition to produce 1-(2-oxopropyl)-3-oxabicyclo[3.1.0]hexanes and their azaanalogs from 1-methoxyethylidenecarbenechromium complex **6a** and nonconjugated enynes have been reported [94, 95]. In view of this reaction mode, Harvey et al. used 1,3-dien-8-ynes **136** instead of nonconjugated enynes to generate 1,6-dialkenylbicyclo[3.1.0]hexanes **137**, which immediately underwent Cope rearrangement to furnish hexahydroazulenes **138** (Scheme 29) [96]. The 1-methoxyalkylidenemolybdenum complexes **135b** gave better yields (up to 87%) than their chromium analogs **135a**. The diastereomers of **138** with the methoxy and the R^3 substituents on the same side of the seven-membered ring were obtained as major products.

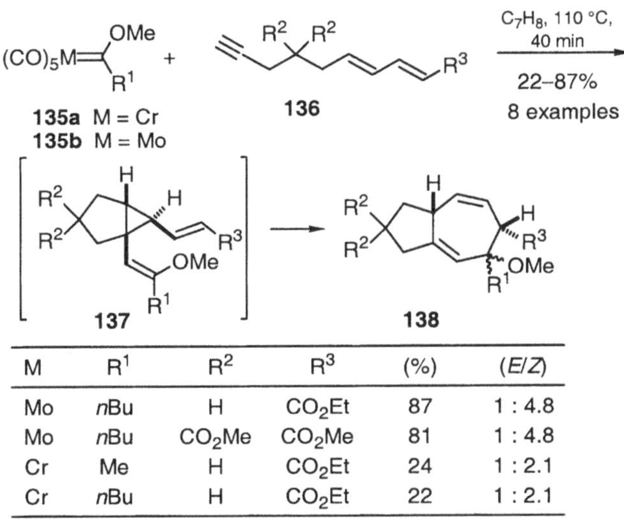

M	R^1	R^2	R^3	(%)	(*E*/*Z*)
Mo	*n*Bu	H	CO_2Et	87	1 : 4.8
Mo	*n*Bu	CO_2Me	CO_2Me	81	1 : 4.8
Cr	Me	H	CO_2Et	24	1 : 2.1
Cr	*n*Bu	H	CO_2Et	22	1 : 2.1

Scheme 29 Synthesis of substituted hexahydroazulenes **138** from simple 1-methoxyalkylidene complexes **135** and 1,3-dien-8-ynes **136** [96]

The η^2-(allylamino)methylcarbenetetracarbonylchromium complex **139** is formed upon heating of the corresponding pentacarbonyl complex in THF (Scheme 30). The 1-allylaminocarbene complex **139** also undergoes insertion of diphenylacetylene and subsequent intramolecular cyclopropanation to form the 2-(tricarbonylchromiumphenyl)-substituted azabicyclo[4.1.0]heptenes **140** as well as their ring-enlargement products **141**. The $Cr(CO)_3$ unit sits on the more electron-rich phenyl moiety in **140**. The amount of ring-enlargement product **141** varies with the nature of the substituent R on the nitrogen, with benzyl apparently facilitating this ring enlargement [97–99].

The simple cyclopropylmethoxycarbenechromium complex **142** reacts with alkynes to afford cyclopentenones **143** and **144** via the cyclopentadiene intermediate **145**, which is hydrogenated with the aid of the chromium(0) residue and water (Scheme 31) [100–103]. Formation of **145** can be regarded as

R	140 (%)	141 (%)
Bn	24	12
Me	51	0
allyl	18	9

Scheme 30 Formation of 2-(tricarbonylchromiumphenyl)-substituted 1-phenyl-4-azabicyclo[4.1.0]hexanes **140** and their ring-enlargement products **141** from the 1-(*N*-allylamino)-ethylidenetetracarbonylchromium complex **139** [97–99]

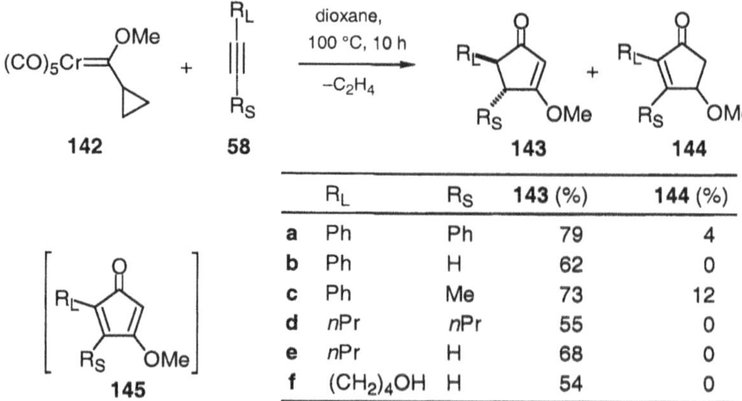

	R_L	R_S	143 (%)	144 (%)
a	Ph	Ph	79	4
b	Ph	H	62	0
c	Ph	Me	73	12
d	*n*Pr	*n*Pr	55	0
e	*n*Pr	H	68	0
f	$(CH_2)_4OH$	H	54	0

Scheme 31 Formation of cyclopentenones **143** and **144** by a formal [4+2+1–2] cocyclization from the cyclopropylmethoxycarbenechromium complex **142** and alkynes [100–103]

a formal [4+2+1–2] cocyclization of the complex **142**, an alkyne, and a carbon monoxide molecule with ring opening of the cyclopropyl moiety and loss of an ethene molecule. The cyclopentenone isomers **143** are always obtained as the major products in this reaction, and in most cases the isomers **144** were not observed. In contrast to the chromium complex **142**, its tungsten analog does not give cyclopentenones **143** or **144**, but cycloheptadienone derivatives under the same reaction conditions or even at a higher temperature [104]. It is obvious that ethene is not split off in this case.

4
Cyclizations and Other Intramolecular Rearrangements of Carbene Complexes

1-Metalla-1,3,5-hexatrienes **146** at ambient temperature undergo a formal 5-*endo-trig*-cyclization to give, after hydrolysis of the enol ether moiety in the corresponding cyclopentadiene, the bicyclo[3.3.0]oct-3-en-2-one **148** (m=1) or 7-ethoxy-9-dimethylaminobicyclo[4.3.0]nona-7,9-diene (**150**) (m=2) (Scheme 32). Depending on the size of the cycloalkenyl substituent in the 3-position, however, these transformations proceed according to completely different mechanisms. The 3-cyclopentenyl-substituted propenylidenemetal complexes **146a** undergo a rapid intramolecular insertion of the carbon–carbon into the metal–carbene double bond leading to ring-annelated zwitterionic η^1-cyclopenteniminium complexes **147** [105], which are transformed to **148** upon treatment with pyridine and subsequent hydrolysis. On the other hand, complexes **146b** in pyridine prefer to undergo 6π-electrocyclization, probably after loss of one CO ligand, to furnish the pyridine-stabilized chromacyclohexadiene **149** which, by reductive elimination, gives the cyclization product **150** [64, 106, 107]. Even noncoordinating solvents (e.g., toluene) can be used for this cyclization of the cyclohexenyl-substituted complexes **146b**, and none of a zwitterionic intermediate of type **147** could ever be detected.

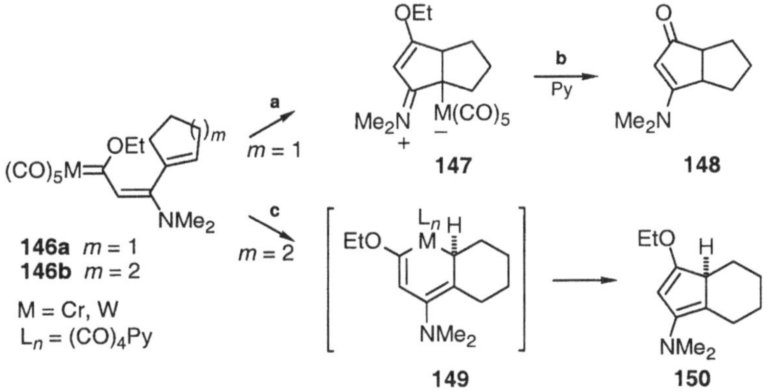

Scheme 32 Two types of cyclization of 3-cycloalkenyl-substituted 1-metalla-1,3,5-hexatrienes **146**. *a*: Et_2O, 20 °C, 12 h, 89%. *b*: Py, C_6D_6, 70 °C, 5 h, 50% conversion. *c*: M=Cr, [D_5]-Py, 20 °C, 18 h, (or 80 °C, 1 h), 100% conversion; M=W, [D_5]-Py, 20 °C, >48 h, 100% conversion [64, 105–107]

Due to the two electron-donating groups in the bicyclic product **150** and the unhydrolyzed precursor of **148**, they should be quite reactive dienes in Diels-Alder reactions. However, such [4+2] cycloadditions were observed only for the cyclohexane-annelated cyclopentadienes **151b**, which equilibrate with the more reactive isomers **154** by 1,5-hydrogen shifts (Scheme 33). The [4+2] cycload-

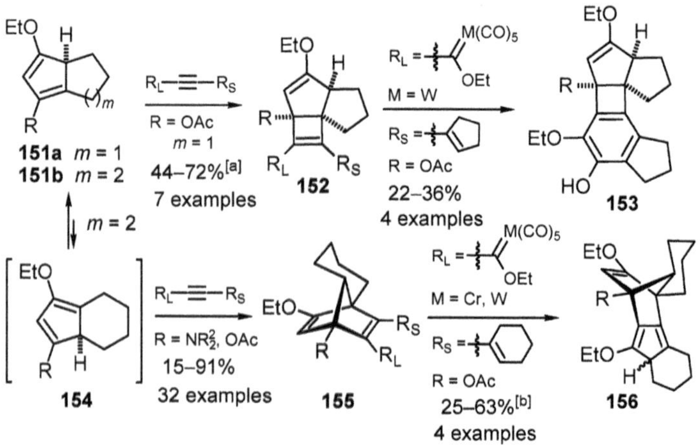

Scheme 33 [2+2] and [4+2] cycloadditions of cyclopentadienes **151** with alkynes. [*a*] One-pot reaction from the corresponding complex **114** (M=Cr, R=Et, *n*=1) and acetic acid. [*b*] One-pot reaction from the corresponding complex **114** (M=Cr, W, R=Et, *n*=2) and acetic acid [106–110]. For further details see Table 7

Table 7 [2+2] and [4+2] cycloadditions of cyclopentadienes **151** with alkynes (see Scheme 33)

Entry	R	R_L	R_S	Product	Yield (%)
1	$MeCO_2$	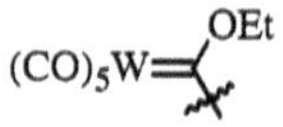 $(CO)_5W{=}C(OEt)$–	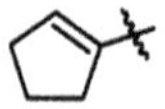 1-cyclopentenyl	**152a** **153a**	61[a] 36[a]+9[b]
2	$BnCO_2$	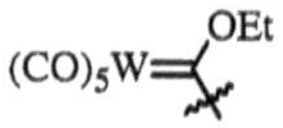 $(CO)_5W{=}C(OEt)$–	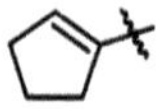 1-cyclopentenyl	**152b** **153b**	72[a] 18[a]+10[b]
3	PhCO	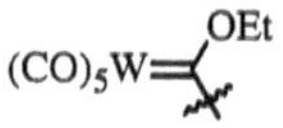 $(CO)_5W{=}C(OEt)$–	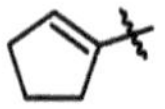 1-cyclopentenyl	**152c** **153c**	54[a] 22[a]+20[b]
4	NMe_2	4-CF_3-C_6H_4	H	**155a**	85
5	NMe_2	Ph	H	**155b**	88
6	NMe_2	–C≡C–Ph	Ph	**155c**	60
7	NMe_2	4-EtO_2C-C_6H_4	H	**155d**	91
8	NMe_2	1-cyclopentenyl	H	**155e**	43
9	NMe_2	2-isopropenyl	H	**155f**	73
10	NMe_2	2-thienyl	H	**155g**	34
11	NMe_2	–CH=CH–C≡C–*c*Pr	*c*Pr	**155h**	46
12	Pyrrolidine	CO_2Et	Ph	**155i**	78
13	OAc	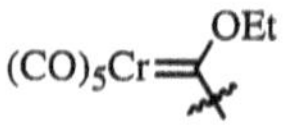 $(CO)_5Cr{=}C(OEt)$–	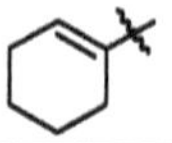 1-cyclohexenyl	**156**	63[a,c]

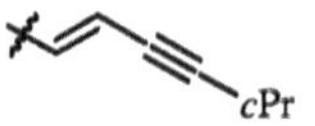

[a] One-pot reaction from the corresponding complexes of type **114** and acetic acid.
[b] The yield of the corresponding hydrolysis product.

ditions of **154** with dienophiles, even with simple alkynes, yield norbornadiene derivatives **155** which, due to through-space interaction between the enol ether moiety and the other carbon–carbon double bond, underwent rapid hydrolysis upon workup and chromatographic purification to yield the corresponding ketones, except for the cases when R_L was a strongly electron-withdrawing group [107, 108]. When cyclopentadiene **154** (R=OAc) was treated with an enynylcarbene complex, the primary adduct **155** underwent a further intramolecular cyclization to yield **156** [109]. In contrast to **151b**, the cyclopentane-annelated cyclopentadienes **151a** prefer to undergo a [2+2] cycloaddition to form tricycles of type **152**. When an enynylcarbene complex is applied as the alkyne in this case, a benzannelation product **153** derived from **152** is eventually formed [110] (Table 7).

An analogous cyclization to eventually form five-membered rings has also been observed for 1-metalla-1,3,5-hexatrienes with an additional heteroatom within the chain, such as in the complexes **157**. These are obtained by Michael additions of imines to alkynylcarbene complexes in good to excellent yields (reaction type F in Scheme 4), and their configurations were determined to be *Z* (≥91%) in all cases. Upon warming in THF solution, complexes **157** underwent cyclization with reductive elimination to furnish 2*H*-pyrroles **158** in up to 97% yield (Scheme 34). With two cyclopropyl substituents at the terminus in

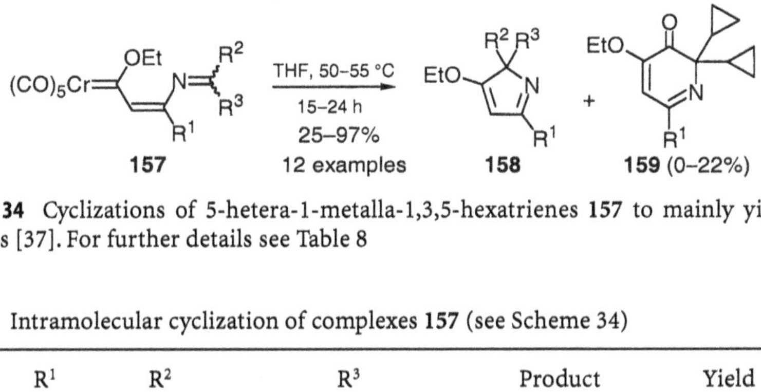

$(CO)_5Cr$=C(OEt)–CH=C(R^1)–N=C(R^2)(R^3) **157** —[THF, 50–55 °C, 15–24 h, 25–97%, 12 examples]→ **158** + **159** (0–22%)

Scheme 34 Cyclizations of 5-hetera-1-metalla-1,3,5-hexatrienes **157** to mainly yield 2*H*-pyrroles [37]. For further details see Table 8

Table 8 Intramolecular cyclization of complexes **157** (see Scheme 34)

Entry	R^1	R^2	R^3	Product	Yield (%)
1	Ph	*p*-$MeOC_6H_4$	*p*-$MeOC_6H_4$	**158a**	25
2	*n*Pr	*p*-$MeOC_6H_4$	*p*-$MeOC_6H_4$	**158b**	57
3	*c*Pr	*p*-$MeOC_6H_4$	*p*-$MeOC_6H_4$	**158c**	65
4	*t*Bu	*p*-$MeOC_6H_4$	*p*-$MeOC_6H_4$	**158d**	78
5	Ph	Ph	*c*Pr	**158e**	62
6	*n*Pr	Ph	*c*Pr	**158f**	88
7	*c*Pr	Ph	*c*Pr	**158g**	97
8	*t*Bu	Ph	*c*Pr	**158h**	85
9	Ph	*c*Pr	*c*Pr	**158i/159i**	45/21
10	*n*Pr	*c*Pr	*c*Pr	**158j/159j**	63/22
11	*c*Pr	*c*Pr	*c*Pr	**158k/159k**	81/0
12	*t*Bu	*c*Pr	*c*Pr	**158l/159l**	92/0

the complex **157** (R^2=R^3=*c*Pr), pyridones **159** were by-products. The formation of the latter must arise after initial carbonyl insertion into the chromium–carbon bond in **157** and a subsequent 6π-electrocyclization. With a cyclopropyl or a *tert*-butyl substituent at the 3-position (R^1=*c*Pr, *t*Bu), the CO-insertion products **159** were not observed. The tungsten complexes of type **157** also yield the products **158**, but require much longer reaction times (7 days) (Table 8).

(2-Aminoalkenyl)carbenechromium complexes **160** with a primary amino group behave quite differently compared to the ones with a secondary or a tertiary amino group (Scheme 35). Upon heating complexes of type **160** in THF, they rearrange to (η^1-1-aza-1,3-butadiene)pentacarbonylchromium complexes **161** which can be isolated in yields of 52–69% [36]. The mechanism of this rearrangement can only be speculated about. It may start with a 1,5-hydride shift, followed by a reductive elimination with a concomitant shift of the pentacarbonylchromium fragment from carbon to nitrogen. In the presence of alkynes **90**, the pentacarbonylchromium-coordinated 1-azabutadienes **161** can undergo a [4+2] cycloaddition and a subsequent 1,4-elimination of ethanol to produce disubstituted pyridines **162**. This rationalization also holds for the formation of pyridines directly from (β-aminoethenyl)carbenechromium complexes **160** and alkynes [88].

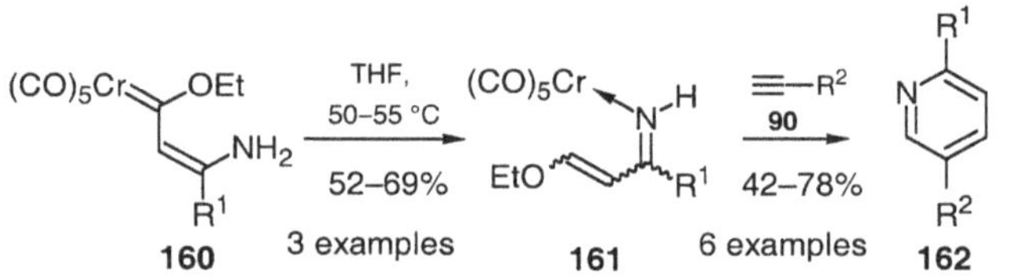

Scheme 35 Formation of 2,5-disubstituted pyridines **162** from α,β-unsaturated complexes with a primary 3-amino group **160** and alkynes **90** [36, 88]

5 Reaction of α,β-Unsaturated Fischer Carbene Complexes with Alkenes, Butadienes, Enamines, and Imines

It is well known that the reaction of Fischer carbene complexes and alkenes with electron-withdrawing substituents affords donor–acceptor-substituted cyclopropanes by a [2+2] cycloaddition with subsequent reductive elimination, rather than the products of an alkene metathesis (cf. Scheme 3) [111–114]. According to Reissig et al., heating of an α,β-unsaturated complex **163** with an electron-deficient alkene **164** not only leads to the expected cyclopropanes **165**, but also to cyclopentenes **166** (Scheme 36) [115, 116] predominantly as the *trans*-isomers with respect to the groups OMe and EWG on the cyclopropane ring in **165** as well as Ar and EWG in **166**. Most probably, the latter products are formed from **165** by a vinylcyclopropane to cyclopentene rearrangement. A systematic study indicated that the yields of cyclopentenes **166** were higher upon longer reaction times, with donor aryl groups (Ar=pyrrolyl) in the 3-position of **163** and in noncoordinating solvents. Thus, **166** cannot be formed from **165** by a purely thermal vinylcyclopropane rearrangement, but by participation of the chromium fragment via an intermediate of type **167**.

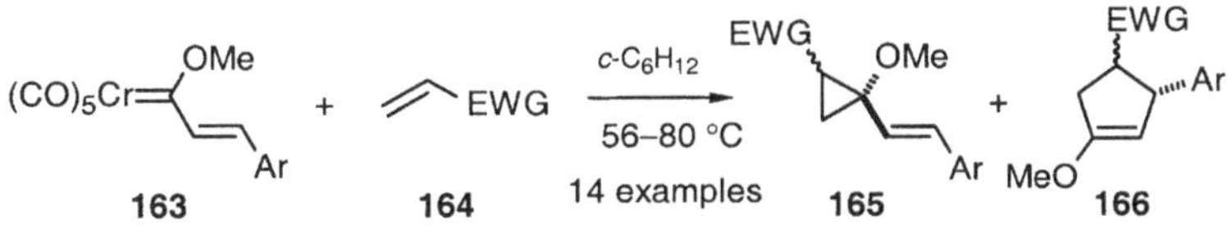

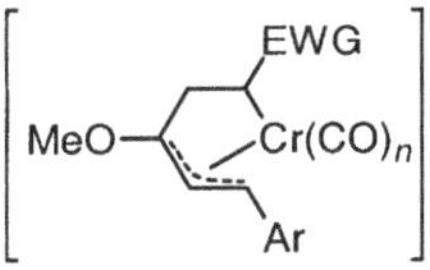

$(CO)_5Cr$ **163** + **164** (c-C_6H_{12}, 56–80 °C, 14 examples) → **165** + **166**

167

	Ar	EWG	**165** (%)	**166** (%)
a	Ph	CO_2Me	81	–
b	Ph	$CONMe_2$	45	–
c	Ph	CN	59	–
d	Ph	$PO(OMe)_2$	73	–
e	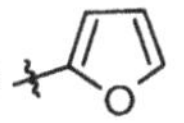	CO_2Me	40	19
f	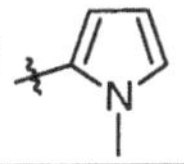	CO_2Me	–	67

Scheme 36 Synthesis of donor–acceptor-substituted cyclopropanes **165** and cyclopentenes **166** from complexes **163** and acceptor-substituted alkenes **164** [115, 116]

In accordance with this, the reaction of the electron-donor-substituted butadienes **170** (R=Ph, OMe) with the arylcarbene complexes **163** yields divinylcyclopropane intermediates **168** with high chemoselectivity for the electron-rich double bond in **170**, which readily undergo a [3,3]-sigmatropic rearrangement to give the *cis*-6,7-disubstituted 1,4-cycloheptadiene derivatives **169** (Scheme 37) [117, 118]. When the methoxycarbonyl-substituted butadiene **170** (R=CO_2Me) was treated with **163** in the same way, the cyclopentene derivatives **172**, the substitution pattern of which is completely different from that of the cyclopentenes **166**, were obtained. In accordance with the high diastereoselectivity in this reaction, the formation of **172** is attributed to a Diels–Alder reaction of the electron-deficient 1-chroma-1,3-dienes **163** acting as a 4π-component, with the silyloxy-substituted double bond of **170** acting as the 2π-component, yielding the chromacyclohexene intermediate **171**, which then undergoes reductive elimination to furnish **172**.

Another interesting example is provided by the phenylethynylcarbene complex **173** and its reactions with five-, six-, and seven-membered cyclic enamines **174** to form bridgehead-substituted five-, six-, and seven-membered cycloalkane-annelated ethoxycyclopentadienes with high regioselectivity under mild reaction conditions (Scheme 38) [119, 120]. In these transformations the phenylethynylcarbene complex **173** acts as a C_3 building block in a formal [3+2] cycloaddition. Like in the Michael additions (reaction route F in Scheme 4), the cyclic electron-rich enamines **174** as nucleophiles attack the

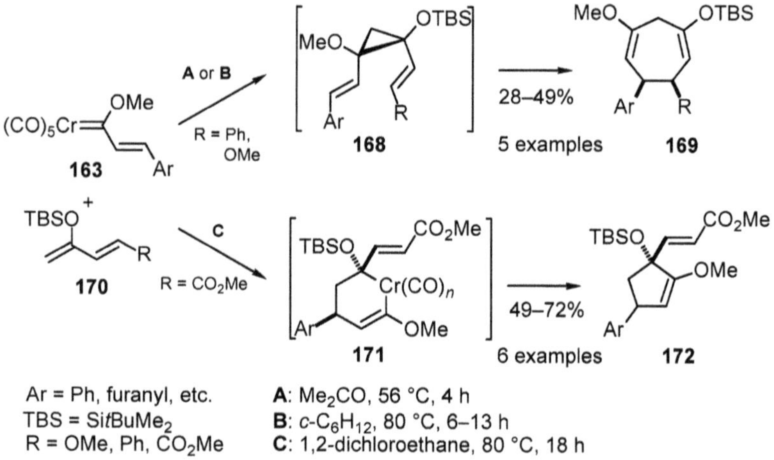

Ar = Ph, furanyl, etc.
TBS = Si*t*BuMe$_2$
R = OMe, Ph, CO$_2$Me

A: Me$_2$CO, 56 °C, 4 h
B: *c*-C$_6$H$_{12}$, 80 °C, 6–13 h
C: 1,2-dichloroethane, 80 °C, 18 h

Scheme 37 Electronic effects of substituents on butadienes **170** determine the formation of cycloheptadienes **169** or cyclopentenes **172** [117, 118]

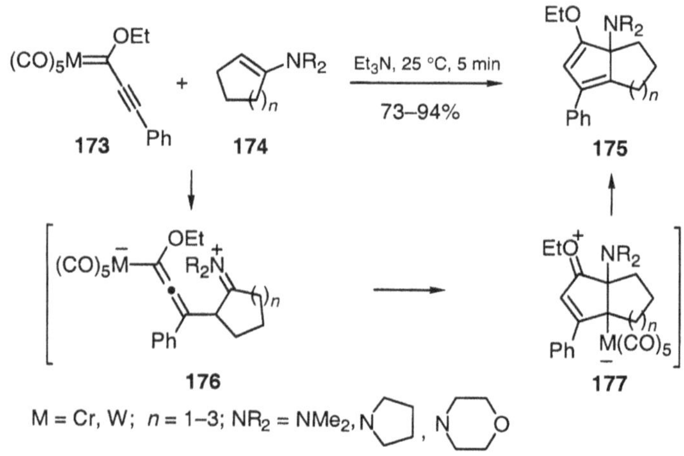

Scheme 38 Formation of five-, six-, and seven-membered cycloalkane-annelated ethoxycyclopentadienes **175** from the phenylethynylcarbene complex **173** and cyclic enamines **174** [119, 120]

electron-deficient triple bond in **173** to give the zwitterionic intermediates **176**, which undergo 1,5-cyclization. Elimination of the pentacarbonylmetal fragment in **177** then furnishes the cyclopentadiene derivatives **175**. This type of ring annelation has been applied to assemble the tetracyclic skeleton of steroids.

The dihydronaphthalene-annelated pyranylidene complex **178**, prepared according to reaction route E in Scheme 4 from β-tetralone and complex **35**, upon treatment with the pyrrolidinocyclopentene **174** (n=1) or -cyclohexene **174** (n=2) at room temperature gave the tetracyclic compounds **179** in excellent yields. In this case, the carbenechromium complex **178**, just like α-pyrone, undergoes a [4+2] cycloaddition with the enamine **174** and the cycloadduct fragments to form a cyclohexadiene moiety as well as hexacarbonylchromium (Scheme 39) [29]. *cis*-Elimination of pyrrolidine in the cycloadduct **179a** with the 6-6-6-5 tetracycle is obviously slow, but the 6-6-6-6 tetracyclic analog **179b** eliminates pyrrolidine extremely fast, especially in the presence of silica gel, to yield 11-phenyl-1,2,3,4,5,6-hexahydrochrysene.

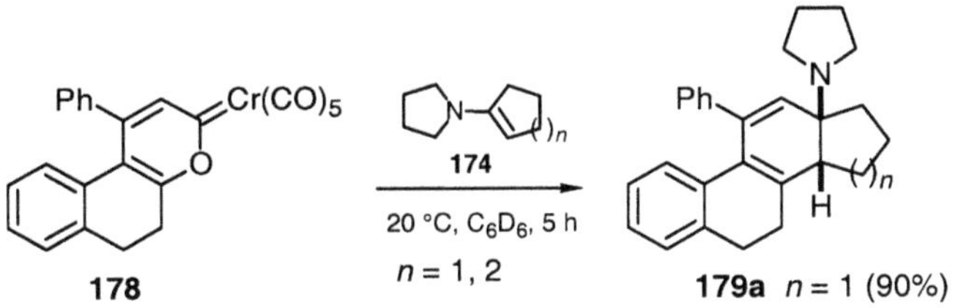

Scheme 39 Synthesis of tetracyclic skeletons **179** from the dihydronaphthalene-annelated pyranylidene complex **178** and cyclic enamines **174** [29]

Recently, Akiyama et al. reported an enantiocontrolled [3+2] cycloaddition of chirally modified Fischer alkenylcarbene complexes **180** with aldimines **181** under Lewis-acid catalysis ($Sn(OTf)_2$) to afford enantiomerically pure 1,2,5-trisubstituted 3-alkoxypyrrolines **182** (Scheme 40) [121]. The mode of formation of these products **182** was proposed to be a [4+2] cycloaddition, with the complexes **180** acting as a 1-metalla-1,3-diene with subsequent reductive elimination. Upon hydrolysis under acidic conditions, the enol ethers give the enantiomerically pure 3-pyrrolidinones **183** (Table 9).

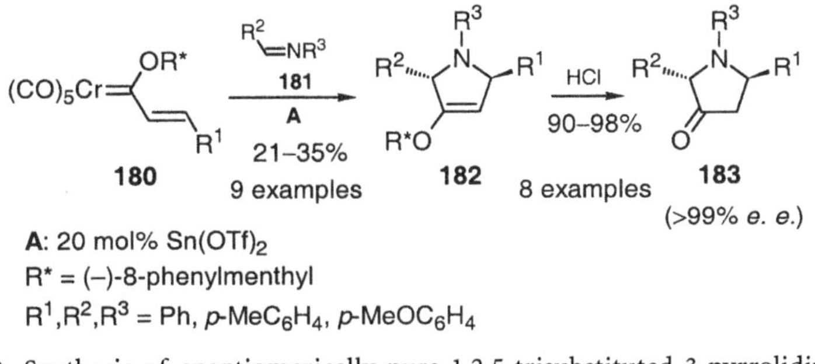

A: 20 mol% $Sn(OTf)_2$
R* = (–)-8-phenylmenthyl
R^1,R^2,R^3 = Ph, *p*-MeC_6H_4, *p*-$MeOC_6H_4$

Scheme 40 Synthesis of enantiomerically pure 1,2,5-trisubstituted 3-pyrrolidinones **183** from chirally modified 1-alkoxypropenylidene complexes **180** and aldimines [121]. For further details see Table 9

Table 9 Synthesis of enantiomerically pure 1,2,5-trisubstituted 3-pyrrolidinones **183** from chirally modified 1-alkoxypropenylidene complexes **180** and aldimines (see Scheme 40)

Entry	R^1	R^2	R^3	Product	Yield (%)	Product	Yield (%)
1	Ph	Ph	Ph	**182a**	34	**183a**	96
2	Ph	Ph	p-MeC_6H_4	**182b**	31	**183b**	91
3	Ph	Ph	p-$MeOC_6H_4$	**182c**	30	**183c**	93
4	Ph	p-MeC_6H_4	Ph	**182d**	35	**183d**	92
5	Ph	p-$MeOC_6H_4$	Ph	**182e**	30	**183e**	90
6	Ph	p-ClC_6H_4	Ph	**182f**	27	**183f**	95
7	p-ClC_6H_4	Ph	Ph	**182g**	27	**183g**	98

6 Conclusion

In 20 years of usage, α,β-unsaturated Fischer carbene complexes demonstrated their multitalented versatility in organic synthesis, yet new reaction types are still being discovered every year. In view of their facile preparation and multifold reactivity, their versatile chemistry will undoubtedly be further developed and applied in years to come. The application of chirally modified Fischer carbene complexes in asymmetric synthesis has only begun, and it will probably be an important area of research in the near future.

Acknowledgements The work of our own group described herein has been supported by the "Volkswagen-Stiftung", the State of Niedersachsen, the "Gesellschaft für technische Zusammenarbeit", the "Studienstiftung des deutschen Volkes", and the "Fonds der Chemischen Industrie" as well as Bayer, BASF AG, Chemetall GmbH, Degussa, Höchst, and Hüls AG through generous gifts of chemicals. A. d. M. is indebted to the group of dedicated and enthusiastic young chemists who, over the years, have made this research flourish. The authors are grateful to Dr. B. Knieriem, Göttingen, for his careful proofreading of the final manuscript.

References

1. Fischer EO, Maasböl A (1964) Angew Chem 76:645; Angew Chem Int Ed Engl 3:580
2. Dötz KH (1986) In: Braterman PS (ed) Reactions of coordinated ligands. Plenum, New York, p 285
3. Wulff WD (1989) In: Liebeskind LS (ed) Advances in metal-organic chemistry, vol 1. JAI, London, p 209
4. Carbene complexes are isolobal with the corresponding organic carbonyl compounds: Hoffmann R (1982) Angew Chem 94:725; Angew Chem Int Ed Engl 21:711
5. Dörwald FZ (1999) Metal carbenes in organic synthesis. Wiley-VCH, Weinheim
6. Hegedus LS (1995) Organische synthese mit Übergangsmetallen. VCH, Weinheim
7. Dötz KH, Pruskil I, Mühlemeier J (1982) Chem Ber 115:1278
8. Dötz KH, Kuhn W (1983) Angew Chem 95:750; Angew Chem Int Ed Engl 22:732

9. Fischer H, Mühlemeier J, Märkl R, Dötz KH (1982) Chem Ber 115:1355
10. de Meijere A (1996) Pure Appl Chem 68:61
11. de Meijere A, Schirmer H, Duetsch M (2000) Angew Chem 112:4124; Angew Chem Int Ed 39:3964
12. Alkynyl complexes: Fischer EO, Kalder HJ (1977) J Organomet Chem 131:57
13. Alkenyl complexes: Hallett MR, Painter JE, Quayle P, Ricketts D (1998) Tetrahedron Lett 39:2851
14. Aryl complexes: Raubenheimer HG, Kruger GJ, Van Lombard A, Linford L, Viljonen JC (1985) Organometallics 4:275
15. Semmelhack MF, Lee GR (1987) Organometallics 6:1839
16. Casey CP, Anderson RL (1974) J Am Chem Soc 96:1230
17. Aumann R, Heinen H (1987) Chem Ber 120:537
18. Lattuada L, Licandro E, Papagni A, Maiorana S, Villa AC, Guastini C (1988) J Chem Soc Chem Commun 1092
19. Barluenga J, Aznar F, Martín A (1995) Organometallics 14:1429
20. Chan KS (1991) J Chem Soc Perkin Trans I 2602
21. Chan KS, Yeung ML, Chan WK, Wang RJ, Mak TCW (1995) J Org Chem 60:1741
22. Wulff WD, Yang DC (1984) J Am Chem Soc 106:7565
23. Barluenga J, Aznar F, Barluenga S (1995) J Chem Soc Chem Commun 1973
24. Barluenga J, Aznar F, Barluenga S, Fernández M, Martín A, García-Granda S, Piñera-Nicolás A (1998) Chem Eur J 4:2280
25. Wulff WD, Tang PC, Chan KS, McCallum JS, Yang DC, Gilbertson SR (1985) Tetrahedron 41:5813
26. Dötz KH, Kuhn W (1985) J Organomet Chem 286:C23
27. Wulff WD, Faron KL, Su J, Springer JP, Rheingold AL (1999) J Chem Soc Perkin Trans I 197
28. Faron KL, Wulff WD (1988) J Am Chem Soc 110:8727
29. Aumann R, Meyer AG, Fröhlich R (1996) J Am Chem Soc 118:10853
30. Duetsch M, Stein F, Lackmann R, Pohl E, Herbst-Irmer R, de Meijere A (1992) Chem Ber 125:2051
31. Aumann R (1992) Chem Ber 125:2773
32. Camps F, Llebaria A, Moreto JM, Ricart S, Vinas JM, Ros J, Yanez R (1991) J Organomet Chem 401:C17
33. Stein F, Duetsch M, Pohl E, Herbst-Irmer R, de Meijere A (1993) Organometallics 12:2556
34. Aumann R, Hinterding P (1993) Chem Ber 126:421
35. Llebaria A, Moretó JM, Ricart S, Ros J, Viñas JM, Yáñez R (1992) J Organomet Chem 440:79
36. Duetsch M, Stein F, Funke F, Pohl E, Herbst-Irmer R, de Meijere A (1993) Chem Ber 126:2535
37. Funke F, Duetsch M, Stein F, Noltemeyer M, de Meijere A (1994) Chem Ber 127:911
38. Aumann R, Jasper B, Läge M, Krebs B (1994) Chem Ber 127:2475
39. Aumann R, Jasper B, Fröhlich R (1995) Organometallics 14:231
40. Aumann R, Jasper B, Läge M, Krebs B (1994) Organometallics 13:3502
41. Schirmer H, Stein F, Funke F, Duetsch M, Wu YT, Noltemeyer M, Belgardt T, Knieriem B, de Meijere A (2004) Chem Eur J (submitted for publication)
42. Schirmer H, Flynn B, de Meijere A (2000) Tetrahedron 56:4977
43. Preliminary communication see: Flynn BL, Funke FJ, Silveira CC, de Meijere A (1995) Synlett 1007
44. Wu YT, Flynn B, Schirmer H, Funke F, Müller S, Labahn T, Nötzel M, de Meijere A (2004) Eur J Org Chem 724

45. Fischer EO, Kreissl FR (1972) J Organomet Chem 35:C47
46. Chan KS, Wulff WD (1986) J Am Chem Soc 108:5229
47. Fischer EO, Kalder HJ, Frank A, Köhler FH, Huttner G (1976) Angew Chem 88:683; Angew Chem Int Ed Engl 15:623
48. Aumann R, Jasper B, Fröhlich R (1995) Organometallics 14:3173
49. Aumann R, Jasper B, Fröhlich R (1995) Organometallics 14:2447
50. Polo R, Moretó JM, Schick U, Ricart S (1998) Organometallics 17:2135
51. de Meijere A, Müller S, Labahn T (2001) J Organomet Chem 617–618:318
52. Chan KS, Wulff WD (1986) J Am Chem Soc 108:5229
53. Camps F, Moreto JM, Ricart S, Vinas JM, Molins E, Miravitlles C (1989) Chem Commun 1560
54. Yamashita A (1986) Tetrahedron Lett 27:5915
55. Yamashita A, Toy A (1986) Tetrahedron Lett 27:3471
56. Chan KS, Peterson GA, Brandvold TA, Faron KL, Challener CA, Hyldahl C, Wulff WD (1987) J Organomet Chem 334:9
57. Grotjahn DB, Dötz KH (1991) Synlett 381
58. Aumann R, Heinen H, Hinterding P, Sträter N, Krebs B (1991) Chem Ber 124:1229
59. Aumann R, Heinen H, Dartmann M, Krebs B (1991) Chem Ber 124:2343
60. Duetsch M, Lackmann R, Stein F, de Meijere A (1991) Synlett 324
61. Flynn B, de Meijere A (1999) J Org Chem 64:400
62. Flynn B, Schirmer H, Duetsch M, de Meijere A (2001) J Org Chem 66:1747
63. Wu YT, de Meijere A (unpublished results)
64. Wu YT (2003) Dissertation, Universität Göttingen
65. Elmorsey SS, Pelter A, Smith K (1991) Tetrahedron Lett 32:4175
66. Ranganathan S, Muraleedharan KM, Bharadwaj P, Madhusudanan KP (1998) Chem Commun 2239
67. Mahmoodi NO, Hajati N (2002) J Chin Chem Soc 49:91
68. Shirai H, Amano N, Hashimoto Y, Fukui E, Ishii Y, Ogawa M (1991) J Org Chem 56:2253
69. Boorum MM, Scott LT (2002) In: Astruc D (ed) Modern arene chemistry. Wiley-VCH, Weinheim, p 20
70. Schirmer H, Funke FJ, Müller S, Noltemeyer M, Flynn B, de Meijere A (1999) Eur J Org Chem 2025
71. Milic J, Schirmer H, Flynn B, Noltemeyer M, de Meijere A (2002) Synlett 875
72. Schirmer H (1999) Dissertation, Universität Göttingen
73. Göttker-Schnetmann I, Aumann R (2000) Organometallics 20:346
74. Dragisich V, Murray CK, Warner BP, Wulff WD, Yang DC (1990) J Am Chem Soc 112:1251
75. Aumann R, Heinen H, Krüger C, Betz P (1990) Chem Ber 123:599
76. Stein F, Deutsch M, Noltemeyer M, de Meijere A (1993) Synlett 486
77. Brandvold TA, Wulff WD, Rheingold AL (1990) J Am Chem Soc 112:1645
78. Schmid H, Hochweber M, von Halban H (1948) Helv Chim Acta 31:1899
79. Cannon JR, Patrick VA, Raston CL, White AH (1978) Aust J Chem 31:1265
80. Stein F, Duetsch M, Lackmann R, Noltemeyer M, de Meijere A (1991) Angew Chem 103:1669; Angew Chem Int Ed Engl 30:1658
81. Schirmer H, Duetsch M, Stein F, Labahn T, Knieriem B, de Meijere A (1999) Angew Chem 111:1369; Angew Chem Int Ed Engl 38:1285
82. Duetsch M, Vidoni S, Stein F, Funke F, Noltemeyer M, de Meijere A (1994) J Chem Soc Chem Commun 1679
83. Flynn BL, Silveira CC, de Meijere A (1995) Synlett 812
84. Schirmer H, Labahn T, Flynn B, Wu YT, de Meijere A (1999) Synlett 2004

85. Flynn BL, Funke FJ, Noltemeyer M, de Meijere A (1995) Tetrahedron 51:11141
86. Barluenga J, Aznar F, Palomero MA, Barluenga S (1999) Org Lett 1:541
87. Barluenga J, Aznar F, Palomero MA (2000) Angew Chem 112:4514; Angew Chem Int Ed Engl 39:4346
88. Aumann R, Hinterding P (1992) Chem Ber 125, 2765
89. Aumann R, Jasper B, Goddard R, Krüger C (1994) Chem Ber 127:717
90. Aumann R (1993) Chem Ber 126:1867
91. Dötz KH, Tiriliomis A, Harms K (1993) Tetrahedron 49:5577
92. Zhang Y, Herndon JW (2003) Org Lett 5:2043
93. Herndon JW, Wang H (1998) J Org Chem 63:4564
94. Harvey DF, Lund KP, Neil DA (1992) J Am Chem Soc 114:8424
95. Katz TJ, Yang GXQ (1991) Tetrahedron Lett 32:5895
96. Harvey DF, Lund KP (1991) J Am Chem Soc 113:5066
97. Parlier A, Yefsah R, Rudler M, Rudler H, Daran JC, Vaissermann J (1990) J Organomet Chem 381:191
98. Parlier A, Rudler H, Yefsah R, Daran JC, Knobler C (1988) J Chem Soc Chem Commun 635
99. Alvarez C, Parlier A, Rudler H, Yefsah R, Daran JC, Knobler C (1989) Organometallics 8:2253
100. Herndon JW, Tumer SU, McMullen LA, Matasi JJ, Schnatter WFK (1994) In: Liebeskind LS (ed) Advances in metal-organic chemistry, vol 3. JAI, London, p 51
101. Herndon JW, Tumer SU, Schnatter WFK (1988) J Am Chem Soc 110:3334
102. Herndon JW, Tumer SU (1989) Tetrahedron Lett 30:295
103. Several 4-alkoxy-2,3-diphenyl- and one 4-methoxy-2,3-bis(trimethylsilyl)-substituted cyclopentadienones have been isolated as reasonably stable compounds, see Herndon JW, Patel PP (1997) Tetrahedron Lett 38:59
104. Herndon JW, Chatterjee G, Patel PP, Matasi JJ, Tumer SU, Harp JJ, Reid MD (1991) J Am Chem Soc 113:7808
105. Aumann R, Fröhlich R, Prigge J, Meyer O (1999) Organometallics 18:1369
106. Wu YT, Schirmer H, Noltemyer M, de Meijere A (2001) Eur J Org Chem 2501
107. Wu HP, Aumann R, Fröhlich R, Wibbeling B, Kataeva O (2001) Chem Eur J 7:5084
108. Concerning the space interaction in norbornadiene, see: Hoffmann R (1971) Acc Chem Res 4:1
109. Wu HP, Aumann R, Fröhlich R, Saarenketo P (2001) Chem Eur J 7:700
110. Wu HP, Aumann R, Fröhlich R, Wibbeling B (2000) Eur J Org Chem 1183
111. Fischer EO, Dötz KH (1970) Chem Ber 103:1273
112. Dötz KH, Fischer EO (1972) Chem Ber 105:1356
113. Fischer EO, Dötz KH (1972) Chem Ber 105:3966
114. Wienand A, Reißig HU (1990) Organometallics 9:3133
115. Wienand A, Reißig HU (1991) Chem Ber 124:957
116. Hoffmann M, Reißig HU (1995) Synlett 625
117. Hoffmann M, Buchert M, Reißig HU (1997) Angew Chem 109:281; Angew Chem Int Ed Engl 36:283
118. Hoffmann M, Buchert M, Reißig HU (1999) Chem Eur J 5:876
119. Meyer AG, Aumann R (1995) Synlett 1011
120. Aumann R, Meyer AG, Fröhlich R (1996) Organometallics 15:5018
121. Kagoshima H, Okamura T, Akiyama T (2001) J Am Chem Soc 123:7182

Topics Organomet Chem (2004) 13: 59–121
DOI 10.1007/978--540-40910-6

Cycloaddition Reactions of Group 6 Fischer Carbene Complexes

José Barluenga (✉) · Félix Rodríguez · Francisco J. Fañanás · Josefa Flórez

Instituto Universitario de Química Organometálica "Enrique Moles", Unidad Asociada al CSIC, Universidad de Oviedo, Julián Clavería 8, 33006 Oviedo, Spain
barluenga@uniovi.es, frodriguez@uniovi.es, fjfv@uniovi.es, jflorezg@uniovi.es

Abstract Group 6 heteroatom-stabilised carbene complexes (Fischer carbene complexes) offer many interesting possibilities to build rings (carbocycles and heterocycles) not readily accessible through conventional methods. In this chapter, a summary of cycloaddition reactions involving group 6 Fischer carbene complexes is presented. Firstly, two-component coupling reactions where a substrate reacts with the carbene complex to afford three- to nine-membered carbo- or heterocycles are considered. Next, cyclisation processes where more than two components are involved in the formation of the final ring are summarised. Finally, a few examples of tandem cycloaddition reactions are presented in order to highlight the amazing possibilities that Fischer carbene complexes offer for the efficient synthesis of complex molecules.

Keywords Fischer carbene complexes · Cycloaddition reactions · Carbocycles · Heterocycles

Abbreviations

Ac	Acetyl
BHT	2,6-di-*tert*-butyl-4-methylphenol
Bn	Benzyl
cod	1,5-Cyclooctadiene
Cp	Cyclopentadienyl
de	Diastereoisomeric excess
DMF	*N,N*-Dimethylformamide
ee	Enantiomeric excess
Fc	Ferrocenyl
2-Fu	2-Furyl
LDA	Lithium diisopropylamide
MCPBA	Metachloroperbenzoic acid
PMDTA	*N,N,N',N',N''P*-Pentamethyldiethylenetriamine
RT	Room temperature
TBDMS, TBS	*Tert*-butyldimethylsilyl
Tf	Trifluoromethanesulphonyl
TFA	Trifluoromethanesulphonic acid
THF	Tetrahydrofuran
TIPS	Triisopropylsilyl
TMS	Trimethylsilyl

1 Introduction

Fischer carbene complexes have proved to be very efficient and extraordinarily versatile starting materials for carrying out a wide range of cycloaddition reactions, which provide a great array of carbocyclic and heterocyclic ring systems with a high degree of selectivity in most cases. The need to employ stoichiometric amounts of a group 6 transition metal is, perhaps, the major drawback of these synthetically useful molecules and this, most likely, has been hampering their general use in organic synthesis. Nevertheless, efforts to perform the chemistry of Fischer carbene complexes using catalytic amounts of the metal are under way and some limited success has been achieved.

In this chapter, an important part of the chemistry of group 6 Fischer carbene complexes will be discussed. Particularly, those processes in which cyclic compounds are formed will be described in detail [1]. The chapter is organised by looking firstly at the number of reacting components taking part in the cycloaddition process and then at the size of the ring being formed [2]. The characteristic reactions of either heteroatom-stabilised (X=OR, NR_2) or non-heteroatom-stabilised (X=alkyl, aryl) alkyl- (**1**), alkenyl- (**2**), aryl- (**3**) and alkynylcarbene (**4**) complexes of a group 6 metal (Cr, Mo, W) are presented in this work (Fig. 1).

The type of cycloaddition reaction is identified by the topological notation which will be used in a formal sense to describe the number of atoms provided

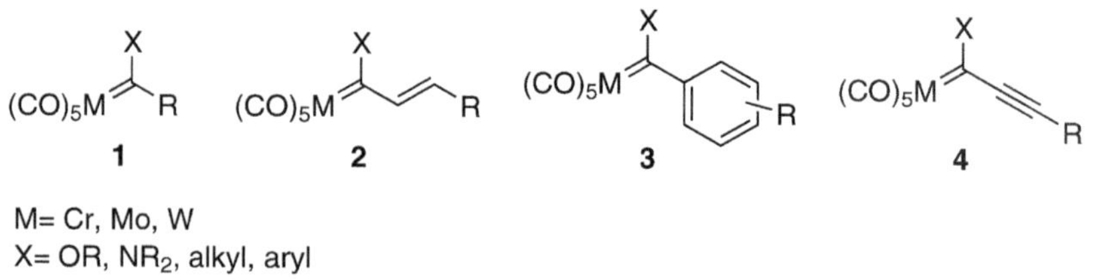

M= Cr, Mo, W
X= OR, NR_2, alkyl, aryl

Fig. 1 Heteroatom-stabilised (X=OR, NR_2) and non-heteroatom-stabilised (X=alkyl, aryl) alkyl- (**1**), alkenyl- (**2**), aryl- (**3**) and alkynylcarbene (**4**) complexes of group 6 metals

by each fragment to the final cycloadduct, regardless of the mechanism and the number of steps involved [3]. The subscripts C=carbene complex and S=substrate refer to the corresponding reagent. The Dötz benzannulation reaction ($[3_C+2_S+1_{CO}]$, CO=carbonyl ligand), the photochemical reactions of carbene complexes with organic substrates such as imines, alkenes or azo compounds ($[2_S+1_C+1_{CO}]$) besides the photochemical benzannulation reactions ($[5_C+1_{CO}]$), and the cycloaddition reactions involving *β*-donor-substituted alkenylcarbene complexes will not be included in this chapter as they are covered elsewhere in this book.

2 Two-Component Cycloaddition Reactions

2.1 $[2_S+1_C]$ Cycloaddition Reactions: Cyclopropanation of Alkenes and Dienes with Fischer Carbene Complexes

The cyclopropanation reaction of an unsaturated substrate is one of the most important strategies to access three-membered ring derivatives. The use of Fischer carbene complexes to perform this kind of cyclisation has become an important tool in organic synthesis [4]. In the next few sections the most significant features of this chemistry are briefly described.

2.1.1 Cyclopropanation of Alkenes

The ability of Fischer carbene complexes to transfer their carbene ligand to an electron-deficient olefin was discovered by Fischer and Dötz in 1970 [5]. Further studies have demonstrated the generality of this thermal process, which occurs between (alkyl)-, (aryl)-, and (alkenyl)(alkoxy)carbene complexes and different electron-withdrawing substituted alkenes [6] (Scheme 1). For certain substrates, a common side reaction in these processes is the insertion of the carbene ligand into an olefinic C–H bond [6, 7]. In addition, it has been observed that steric hindrance caused by either the number or the size of the substituents of the alkene is a limitation of the cyclopropanation reaction [6c]. The diastereoselectivities of these carbene transfer reactions are generally low, leading to the corresponding cyclopropanes as nearly equimolecular mixtures of *cis* and *trans* isomers/epimers at the carbon arising from the carbene carbon atom. Nevertheless, better diastereoselectivities were attained when the cyclopropanation reactions involve a conjugated system either in the carbene ligand or in the alkene [8] (Scheme 1). The mechanism to explain the cyclopropanation of electron-deficient olefins with Fischer carbene complexes was initially proposed by Casey and Cesa [9], and involves dissociation of a CO ligand, coordination of the alkene, generation of a 16-electron metalacyclobutane intermediate, and finally reductive elimination of the metal fragment (Scheme 1).

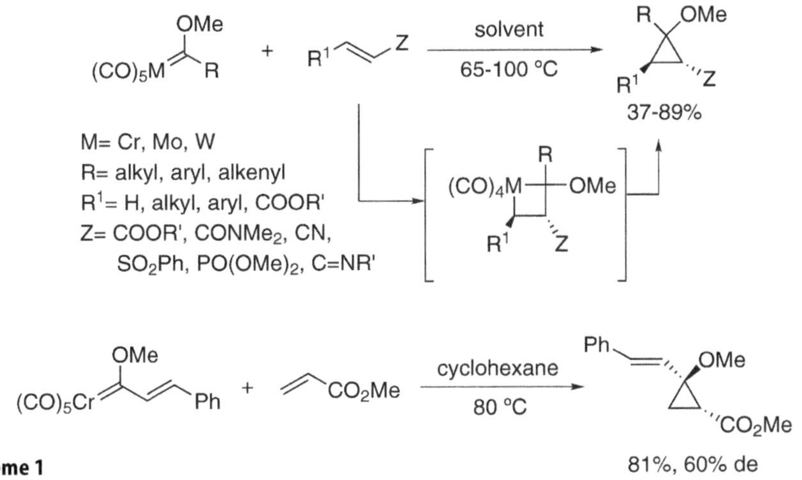

M= Cr, Mo, W
R= alkyl, aryl, alkenyl
R^1= H, alkyl, aryl, COOR'
Z= COOR', $CONMe_2$, CN, SO_2Ph, $PO(OMe)_2$, C=NR'

solvent, 65-100 °C: 37-89%

cyclohexane, 80 °C: 81%, 60% de

Scheme 1

Alkenes substituted with electron-donating groups can also be cyclopropanated under thermal conditions in an intermolecular fashion with alkoxycarbene complexes [10] (Scheme 2). In most cases this reaction must be carried out under high pressure of carbon monoxide in order to avoid the formation of the corresponding olefin metathesis products [11]. These $[2_S+1_C]$ cycloaddition reactions are assumed to involve nucleophilic addition of the electron-rich alkene to the electrophilic carbene carbon atom to produce a zwitterionic intermediate which further undergoes ring closing [12]. Moreover, the diastereoselectivity of this reaction, generally low, is clearly improved by the use of alkenylcarbene complexes [10a] (Scheme 2).

Although the intramolecular cyclopropanation of simple alkenes easily occurs in those cases where a five- or six-membered ring is formed in addition to the three-membered ring [13], the intermolecular version of this process was described by Barluenga et al. in 1997 [14c]. Thus, this reaction has shown a high

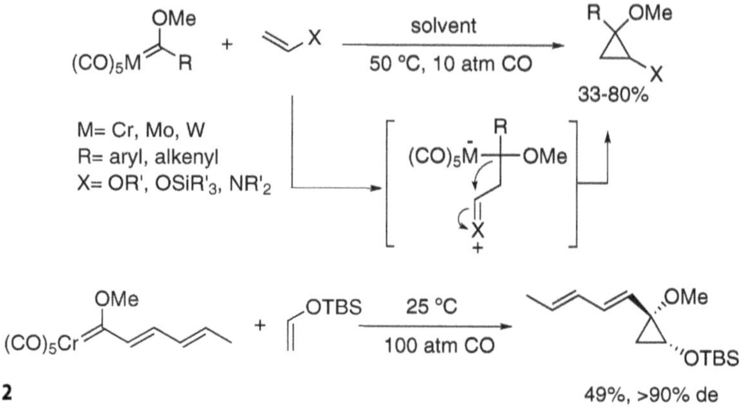

Scheme 2

degree of diastereoselectivity with different substituted alkoxy(alkenyl)- and alkoxy(2-heteroaryl)carbene complexes of chromium and terminal, acyclic and cyclic 1,2-disubstituted simple olefins. In addition, a good functional group tolerance at the allylic position of the olefin is observed [14] (Scheme 3). A mechanism similar to that described for the electron-poor olefins and which involves the initial formation of a chelated tetracarbonyl complex intermediate is proposed to account for the experimental results. The cyclopropane stereochemistry can be explained on the basis of steric interactions between the alkenyl substituent of the carbene ligand and the olefin alkyl chain which will favour a relative *trans* disposition of these groups [14b,c] (Scheme 3). The use of 2-iodoethoxy-substituted alkenylcarbene complexes allows the easy preparation of cyclopropanol derivatives by removing the 2-iodoethyl moiety of the corresponding cyclopropane derivative by treatment with *t*BuLi at low temperature [14a].

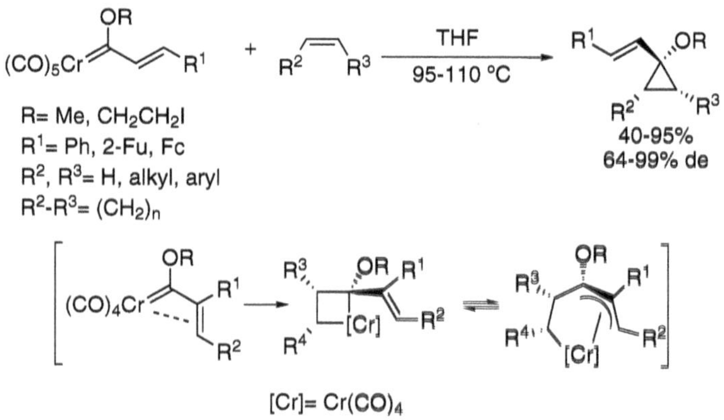

Scheme 3

The first examples of alkene cyclopropanation reactions with alkynylcarbene complexes were reported by Barluenga et al. in 2002 [15]. These intermolecular processes involve the treatment of singular simple olefins, such as fulvenes [15a] and strained olefins [15b], with methoxy(alkynyl)carbene complexes.

The cyclopropanation reaction with aminocarbene complexes has been much less studied than the corresponding reaction with alkoxy-derived carbene complexes. Indeed, these reagents have shown scarce ability to effectively transfer their carbene ligands to an alkene and, in general, electron-deficient olefins react with aminocarbene complexes to form open-chain products resulting from a formal C_{sp2}-H insertion [6c, 16]. Only one example involving the reaction of pyrrolo-derived carbene complexes and electron-deficient olefins leading to cyclopropane derivatives has been reported [17] (Scheme 4). In this context, very recently an example involving the intermolecular cyclopropanation of a simple alkene with an aminocarbene complex has been described [18] (Scheme 4). Moreover, two examples of intramolecular cyclopropanation of simple alkenes with chromium- [13b] and tungstencarbene complexes [13d] are known.

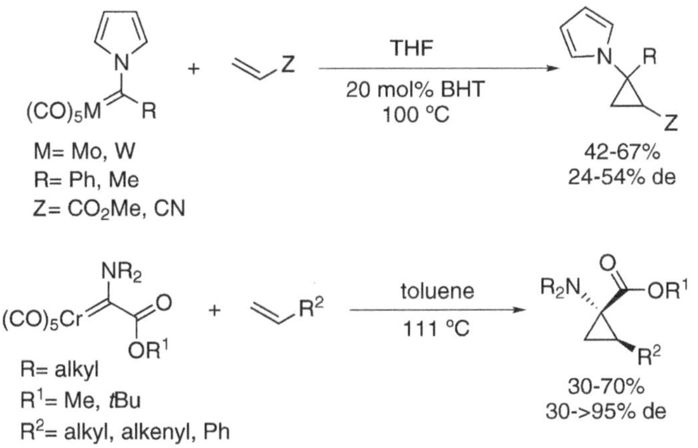

$(CO)_5M$ ... R + Z; THF, 20 mol% BHT, 100 °C → 42-67%, 24-54% de

M= Mo, W
R= Ph, Me
Z= CO_2Me, CN

NR_2, $(CO)_5Cr$, OR^1 + R^2; toluene, 111 °C → R_2N, OR^1, R^2; 30-70%, 30->95% de

R= alkyl
R^1= Me, *t*Bu
R^2= alkyl, alkenyl, Ph

Scheme 4

Non-heteroatom-stabilised Fischer carbene complexes also react with alkenes to give mixtures of olefin metathesis products and cyclopropane derivatives which are frequently the minor reaction products [19]. Furthermore, non-heteroatom-stabilised vinylcarbene complexes, generated in situ by reaction of an alkoxy- or aminocarbene complex with an alkyne, are able to react with different types of alkenes in an intramolecular or intermolecular process to produce bicyclic compounds containing a cyclopropane ring [20].

Asymmetric versions of the cyclopropanation reaction of electron-deficient olefins using chirally modified Fischer carbene complexes, prepared by exchange of CO ligands with chiral bisphosphites [21a] or phosphines [21b], have been tested. However, the asymmetric inductions are rather modest [21a] or not quantified (only the observation that the cyclopropane is optically active is reported) [21b]. Much better facial selectivities are reached in the cyclopropanation of enantiopure alkenyl oxazolines with aryl- or alkyl-substituted alkoxycarbene complexes of chromium [22] (Scheme 5).

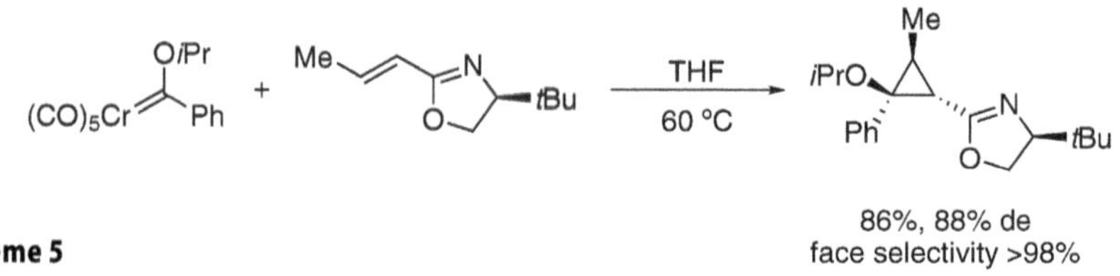

Scheme 5

Catalytic cyclopropanation of alkenes has been reported by the use of diazoalkanes and electron-rich olefins in the presence of catalytic amounts of pentacarbonyl(η^2-*cis*-cyclooctene)chromium [23a,b] (Scheme 6) and by treatment of conjugated ene-yne ketone derivatives with different alkyl- and donor-substituted alkenes in the presence of a catalytic amount of pentacarbonylchromium tetrahydrofuran complex [23c]. These $[2_S+1_C]$ cycloaddition reactions catalysed by a Cr(0) complex proceed at room temperature and involve the formation of a non-heteroatom-stabilised carbene complex as intermediate.

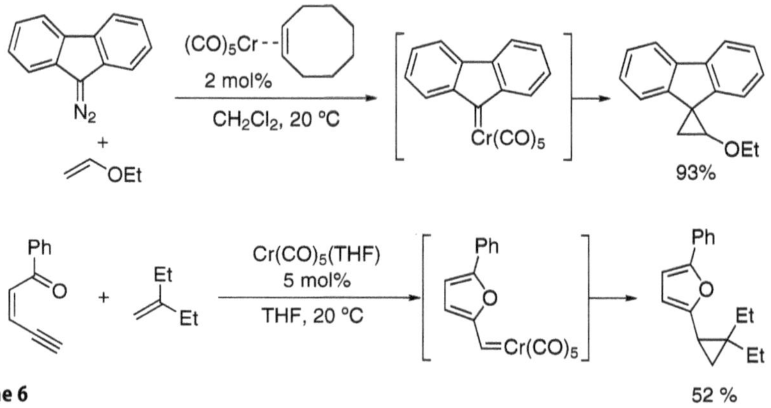

Scheme 6

2.1.2
Cyclopropanation of 1,3-Dienes

The reactions of Fischer carbene complexes with 1,3-dienes (carbodienes or heterodienes) lead to the formation of cyclic products with different ring sizes depending upon both the nature of the reaction partners and the reaction conditions. Between these synthetically useful transformations are found $[2_C+2_S]$, $[3_C+2_S]$, $[4_S+1_C]$, $[3_S+3_C]$, $[4_S+2_C]$, $[4_S+3_C]$ and $[2_S+1_C+1_{CO}]$ cycloaddition reactions which will be summarised further on, in addition to the $[2_S+1_C]$ cycloaddition processes here described.

Electron-deficient 1,3-dienes are known to react when heated with methoxy(aryl)- or methoxy(alkyl)carbene complexes to afford vinylcyclopropane derivatives with high regioselectivity and diastereoselectivity [8a, 24]. Cyclopropanation of the double bond not bearing the acceptor functional group and formation of the diastereoisomer with the methoxy group *cis*-positioned with respect to the olefinic moiety are both largely favoured. One example is shown in Scheme 7. Even trisubstituted 1,3-dienes undergo this $[2_S+1_C]$ cycloaddition reaction [24a].

Electron-rich 1,3-dienes react smoothly with Fischer carbene complexes, but these reactions have been reported to produce cyclopropanes only in very isolated examples [25]. Methoxy(phenyl)carbene complex reacts with Danishefsky's dienes to produce, with low diastereoselectivity, the vinylcyclopropane resulting from the regioselective transfer of the carbene ligand to the more electron-rich double bond of the diene [25b] (Scheme 7). The reaction must be carried out under pressure of carbon monoxide to minimise the formation of side products. Methoxy(alkenyl)carbene complexes of chromium also react with this type of 1,3-diene, affording initially divinylcyclopropanes that in most cases undergo Cope rearrangement in the reaction medium to give seven-membered rings, as will be described in a following section. The cyclopropanation reaction of 4-substituted 2-(*tert*-butyldimethylsiloxy)-1,3-butadiene with acetoxy(methyl)- and acetoxy(alkenyl)carbene complexes of chromium has also been reported [26].

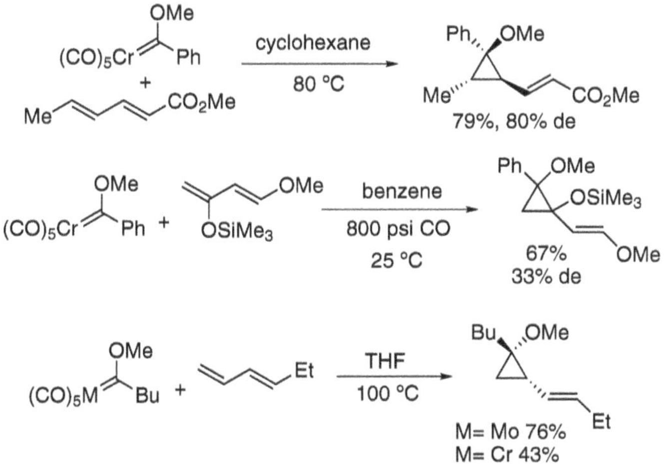

OMe
$(CO)_5Cr$ Ph
\+
Me CO$_2$Me
cyclohexane
80 °C
Ph OMe
Me CO$_2$Me
79%, 80% de

OMe
$(CO)_5Cr$ Ph + OSiMe$_3$ OMe
benzene
800 psi CO
25 °C
Ph OMe
OSiMe$_3$
OMe
67%
33% de

OMe
$(CO)_5M$ Bu + Et
THF
100 °C
Bu OMe
Et
M= Mo 76%
M= Cr 43%

Scheme 7

Simple 1,3-dienes also undergo a thermal monocyclopropanation reaction with methoxy(alkyl)- and methoxy(aryl)carbene complexes of molybdenum and chromium [27]. The most complete study was carried out by Harvey and Lund and they showed that this process occurs with high levels of both regio- and diastereoselectivity. The chemical yield is significantly higher with molybdenum complexes [27a] (Scheme 7). Tri- and tetrasubstituted 1,3-dienes and 3-methylenecyclohexene (diene locked in an *s-trans* conformation) fail to react [28]. The monocyclopropanation of electronically neutral 1,3-dienes with non-heteroatom-stabilised carbene complexes has also been described [29].

2.2
[2_C+1_S] Cycloaddition Reactions: Synthesis of Cyclopropylcarbene Complexes

Stabilised sulphur ylides react with alkenylcarbene complexes to form a mixture of different products depending on the reaction conditions. However, at −40 °C the reaction results in the formation of almost equimolecular amounts of vinyl ethers and diastereomeric cyclopropane derivatives. These cyclopropane products are derived from a formal [2_C+1_S] cycloaddition reaction and the mechanism that explains its formation implies an initial 1,4-addition to form a zwitterionic intermediate followed by cyclisation. Oxidation of the formed complex renders the final products [30] (Scheme 8).

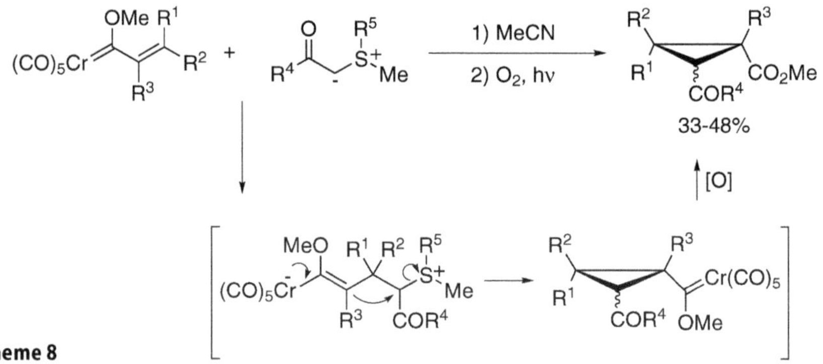

Scheme 8

Alkenylcarbene complexes react with in situ-generated iodomethyllithium or dibromomethyllithium, at low temperature, to produce cyclopropylcarbene complexes in a formal [2_C+1_S] cycloaddition reaction. This reaction is highly diastereoselective and the use of chiral alkenylcarbene complexes derived from (−)-8-phenylmenthol has allowed the enantioselective synthesis of highly interesting 1,2-disubstituted and 1,2,3-trisubstituted cyclopropane derivatives [31] (Scheme 9). As in the precedent example, this reaction is supposed to proceed through an initial 1,4-addition of the corresponding halomethyllithium derivative to the alkenylcarbene complex, followed by a spontaneous γ-elimination of lithium halide to produce the final cyclopropylcarbene complexes.

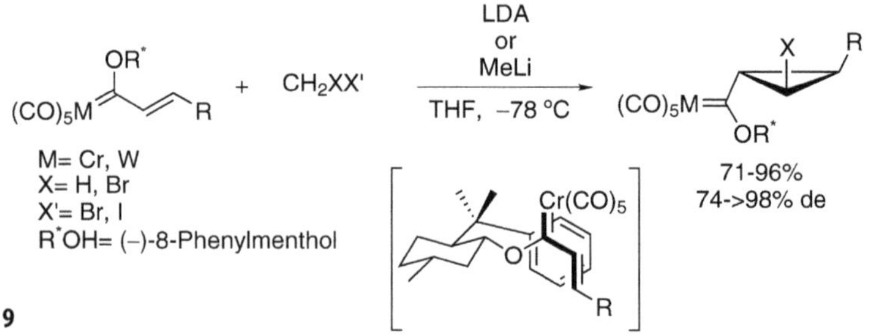

Scheme 9

The asymmetric induction that has been observed in this reaction can be explained in terms of the model shown in Scheme 9. In the most stable conformation the appropriately positioned phenyl group shields selectively the *Re,Re* face of the chromadiene by π,π-orbital overlap forcing the nucleophile to attack preferentially on the opposite side.

2.3
$[2_C+2_S]$ Cycloaddition Reactions: Synthesis of Cyclobutenylcarbene Complexes

The [2+2] cycloaddition reaction is the most versatile method to access four-membered rings [32]. This process may proceed under thermal, photochemical or metal-catalysed conditions. However, the thermally induced reaction can only be applied to a limited extent, especially in the reaction of ester-functionalised acetylene derivatives with enol ethers [33]. Nonetheless, if Fischer alkynylcarbene complexes are used as ester analogues, such [2+2] cycloaddition reaction gives the corresponding cyclobutene derivatives under milder experimental conditions according to the overall reaction shown in Scheme 10 [34]. The first example of this kind of reaction was observed by Wulff and Faron during their investigations on the Diels–Alder reaction of 2,3-bis(*tert*-butyldimethylsilyloxy)-1,3-butadiene with alkynylcarbene complexes of chromium [35]. After this initial discovery several examples of [2+2] cycloaddition reactions involving enol ethers, silyl enol ethers, vinyl acetates and ketene acetals were published [36].

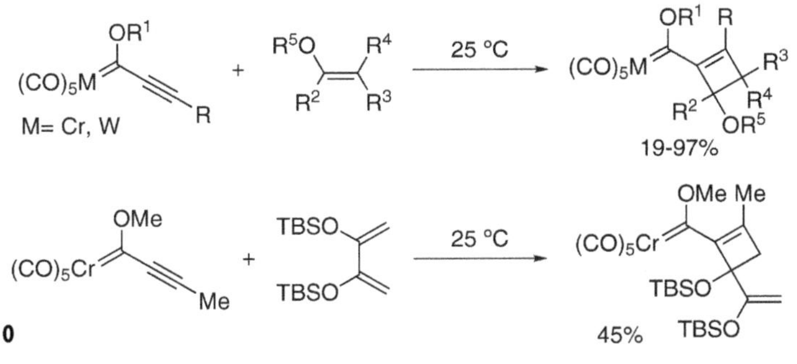

$(CO)_5M$ (M= Cr, W) alkynylcarbene (OR^1, R) + $R^5O(R^2)C=C(R^4)R^3$ → (25 °C) cyclobutenylcarbene complex, 19-97%

$(CO)_5Cr$ carbene (OMe, Me) + 2,3-bis(TBSO)-1,3-butadiene → (25 °C) cyclobutenylcarbene complex, 45%

Scheme 10

Apart from these oxygen-substituted electron-rich olefins, it has been reported that nitrogen-substituted olefins such as lactims and alkenyl imidates react with alkynylcarbene complexes through domino reactions, in which a [2+2] process is involved, to give cyclobutene-containing biscarbene complex derivatives [37]. While the [2+2] cycloaddition reaction of alkynylcarbene complexes with electron-rich olefins has been widely studied, the analogous reaction using alkenylcarbene complexes remains almost unexplored and only two examples have been reported so far. Thus, α-*exo*-methylene-2-oxacyclopentylidene complexes of chromium and tungsten undergo [2+2] cycloaddition processes with enol ethers under mild thermal conditions to give spirocyclobutanes in good yields and as single diastereoisomers [38] (Scheme 11). The other example of an alkenylcarbene complex involved in a [2+2] cycloaddition implies the reaction of an ynamine and a tungsten alkenylcarbene complex leading to a new cyclobutenylcarbene complex as a side product and in very low yield [39] (Scheme 11).

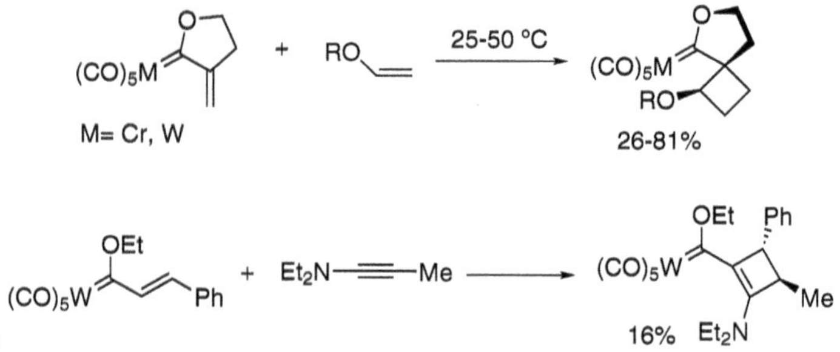

Scheme 11

An unusual example of a formal [2+2] cycloaddition process is that described by Aumann et al. who referred to the reaction of alkyl-substituted carbene complexes with α,β-unsaturated *N,N*-disubstituted acid amides in the presence of $POCl_3/Et_3N$ [40]. This reaction is initiated by the transformation of the acid amides into the more reactive iminium chlorides. A 1,4-addition of the conjugated base of the carbene complex to the iminium chloride generates an open-chain carbene complex derivative, which undergoes a cyclisation process to afford a cyclobutene complex derivative. This intermediate evolves by HCl elimination followed by a [1,3]-migration of the metal fragment to furnish the final aminocarbene derivatives (Scheme 12). Interestingly, in those cases where R^1=H, the reaction follows a different pathway affording a mixture of open-chain, formal $[3_S+2_C]$ and formal $[3_S+2_S+1_C]$ products in low yield [40].

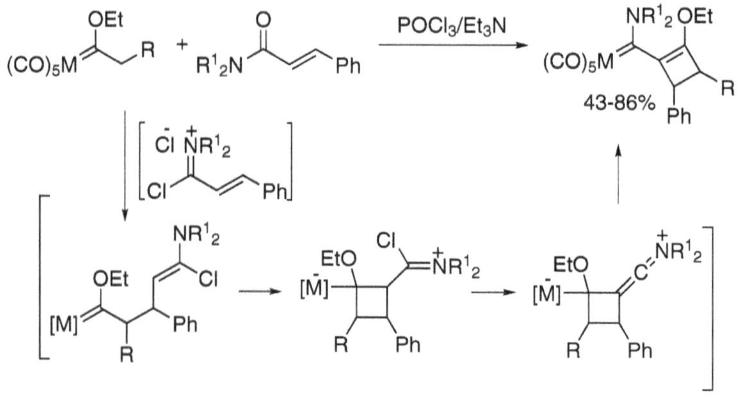

Scheme 12 $[M]= M(CO)_5$, M= Cr, W

2.4
$[3_S+1_C]$ Cycloaddition Reactions

The $[3_S+1_C]$ cycloaddition reaction with Fischer carbene complexes is a very unusual reaction pathway. In fact, only one example has been reported. This process involves the insertion of alkyl-derived chromium carbene complexes into the carbon–carbon σ-bond of diphenylcyclopropenone to generate cyclobutenone derivatives [41] (Scheme 13). The mechanism of this transformation involves a CO dissociation followed by oxidative addition into the cyclopropenone carbon–carbon σ-bond, affording a metalacyclopentenone derivative which undergoes reductive elimination to produce the final cyclobutenone derivatives.

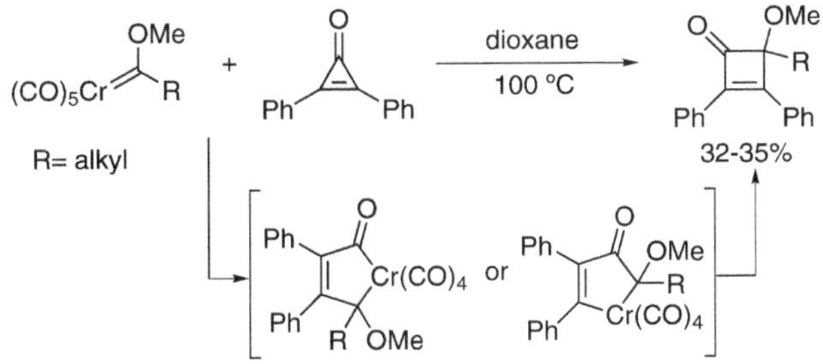

Scheme 13

2.5
$[3_S+2_C]$ Cycloaddition Reactions

The 1,3-dipolar cycloadditions are a powerful kind of reaction for the preparation of functionalised five-membered heterocycles [42]. In the field of Fischer carbene complexes, the α,β-unsaturated derivatives have been scarcely used in cycloadditions with 1,3-dipoles in contrast with other types of cycloadditions [43]. These complexes have low energy LUMOs, due to the electron-acceptor character of the pentacarbonyl metal fragment, and hence, they react with electron-rich dipoles with high energy HOMOs.

Although most of the examples of $[3_S+2_C]$ cycloaddition reactions with carbene complexes are referred to as 1,3-dipolar processes, we should include in this section another kind of "non-dipolar" transformation dealing with the reaction of pentacarbonyl(methoxymethylcarbene)chromium with a base followed by treatment with an epoxide in the presence of boron trifluoride. This reaction gives cyclic carbene complexes in a process that can be considered a $[3_S+2_C]$ cycloaddition [44] (Scheme 14).

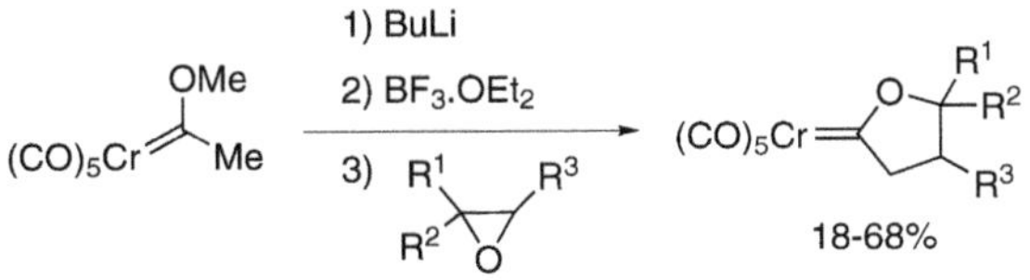

Scheme 14

2.5.1
Alkynylcarbene Complexes in 1,3-Dipolar Cycloadditions

The first $[3_S+2_C]$ cycloaddition reaction using a Fischer carbene complex was accomplished by Fischer et al. in 1973 when they reported the reaction of the pentacarbonyl(ethoxy)(phenylethynyl)carbene complex of tungsten and diazomethane to give a pyrazole derivative [45]. But it was 13 years later when Chan and Wulff demonstrated that in fact this was the first example of a 1,3-dipolar cycloaddition reaction [46,47a]. The introduction of a bulky trimethylsilyl group on the diazomethane in order to prevent carbene-carbon olefination leads to the corresponding pyrazole carbene complexes in better yields (Scheme 15).

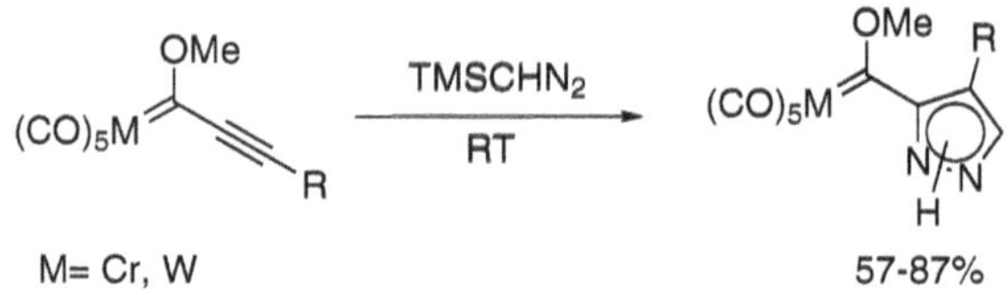

OMe (CO)$_5$M R TMSCHN$_2$ RT OMe R (CO)$_5$M N N H

Scheme 15 M= Cr, W 57-87%

(Alkoxy)alkynylcarbene complexes have been shown to react with nitrones to give dihydroisoxazole derivatives [47]. Masked 1,3-dipoles such as 1,3-thiazolium-4-olates also react with alkynylcarbene complexes to yield thiophene derivatives. The initial cycloadducts formed in this reaction are not isolated and they evolve by elimination of isocyanate to give the final products [48]. The analogous reaction with munchnones or sydnones as synthetic equivalents of

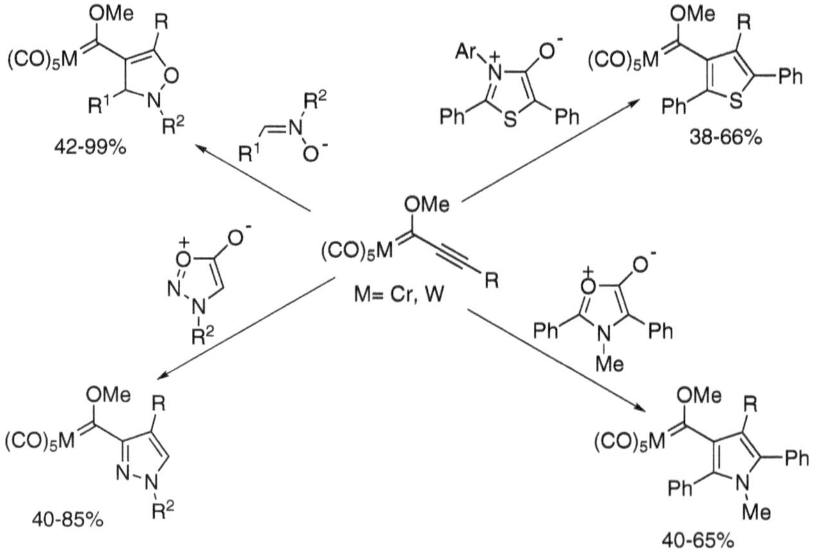

OMe R (CO)$_5$M O R^1 N R^2 42-99% R^2 R^1 N O$^-$ Ar N$^+$ O$^-$ Ph S Ph OMe R (CO)$_5$M Ph Ph S 38-66%

OMe (CO)$_5$M R M= Cr, W O$^+$ O$^-$ N N R^2 O$^+$ O$^-$ Ph N Ph Me

OMe R (CO)$_5$M N N R^2 40-85% OMe R (CO)$_5$M Ph Ph N Me 40-65%

Scheme 16

azomethine ylides and imines, respectively, leads to pyrrole or pyrazole carbene complexes. In these cases, the final products are those derived from carbon dioxide extrusion and are obtained as single regioisomers [49] (Scheme 16).

2.5.2
Alkenylcarbene Complexes in 1,3-Dipolar Cycloadditions

Diazo compounds react with alkenylcarbene complexes to yield the corresponding [3+2] cycloadduct as a single regioisomer but as a mixture of diastereoisomers [50]. However, chiral α,β-unsaturated carbene complexes derived from (–)-8-phenylmenthol react with different diazo compounds to give the corresponding pyrazoline derivatives as single diastereoisomers [51]. In the same way, the cycloaddition reaction of these chiral carbene complexes has been successfully performed with other 1,3-dipoles. Thus, the reaction with nitrilimines leads, after oxidation of the pentacarbonylchromium fragment, to Δ^2-pyrazoline derivatives as single diastereoisomers [52]. Moreover, the reaction with azomethine ylides also produces the $[3_S+2_C]$ adducts as single regioisomers in a highly diastereoselective fashion. Interestingly, this latter reaction has been used as the key step in the total synthesis of the pharmaceutically useful compound (+)-rolipran [53]. Another proof of the potential of chiral α,β-unsaturated carbene complexes derived from (–)-8-phenylmenthol can be found in the formal $[3_S+2_C]$ cycloaddition reaction of these complexes and *N*-alkylidene glycine ester anions. This reaction is thought to proceed through an initial 1,4-addition of the enolate to the α,β-unsaturated carbene followed by a 5-*endo-trig* ring closure. The cycloadducts obtained in this reaction are precursors of interesting enantiomerically highly enriched proline derivatives [54] (Scheme 17).

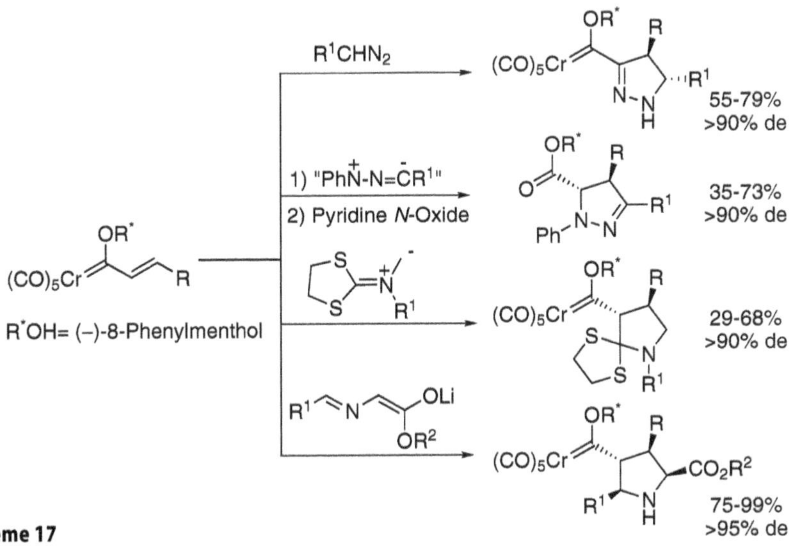

Scheme 17

2.6 $[3_C+2_S]$ Cycloaddition Reactions

Fischer carbene complexes are valuable C3 building blocks for the formal $[3_C+2_S]$ carbo- and heterocyclisation reactions [55]. Thus, not only the traditional α,β-unsaturated but also aryl and iminocarbene complexes have been used to get a great variety of compounds derived from the $[3_C+2_S]$ reaction with different C2 counterparts.

2.6.1 Iminocarbene Complexes as C3 Building Blocks

Iminocarbene complexes of chromium and tungsten are useful isolable synthetic equivalents to nitrile ylides having the advantage that the range of 1,3-dipolarophiles is not limited to electron-acceptor substrates and can be extended to electronically neutral as well as to electron-rich systems [56] (Scheme 18).

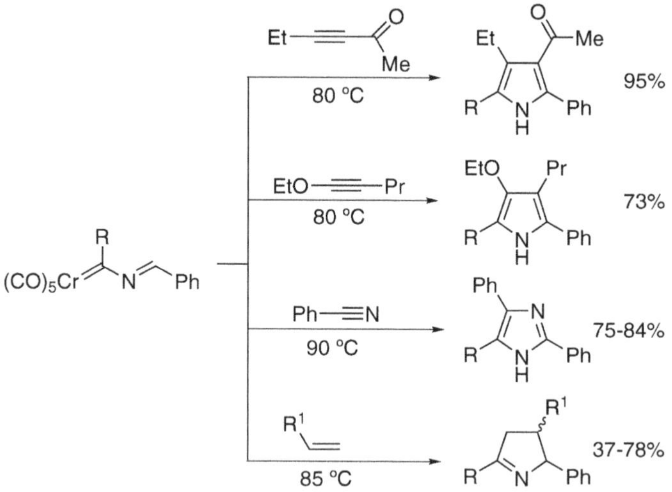

Scheme 18

The regioselectivity observed in these reactions can be correlated with the resonance structure shown in Fig. 2. The reaction with electron-rich or electron-poor alkynes leads to intermediates which are the expected on the basis of polarity matching. In Fig. 2 is represented the reaction with an ynone leading to a metalacycle intermediate (formal $[4_C+2_S]$ cycloadduct) which produces the final products after a reductive elimination and subsequent isomerisation. Also, these reactions can proceed under photochemical conditions. Thus, Campos, Rodríguez et al. reported the cycloaddition reactions of iminocarbene complexes and alkynes [57, 58], alkenes [57] and heteroatom-containing double bonds to give 2*H*-pyrrole, 1-pyrroline and triazoline derivatives, respectively [59].

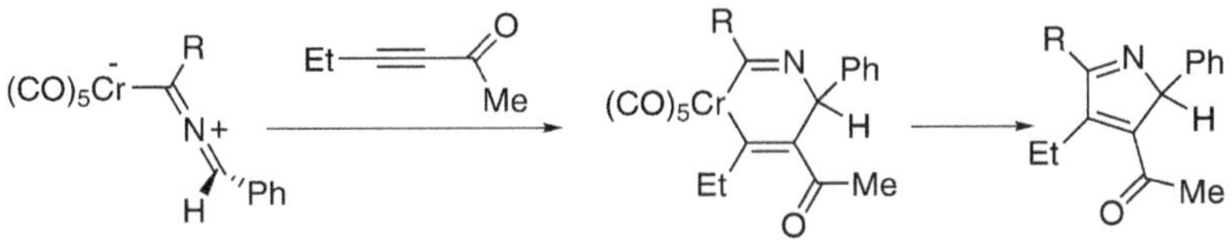

Fig. 2 Reaction of an iminocarbene complex of chromium with an ynone

2.6.2
Arylcarbene Complexes as C3 Building Blocks

The reaction of alkoxyarylcarbene complexes with alkynes mainly affords Dötz benzannulated $[3_C+2_S+1_{CO}]$ cycloadducts. However, uncommon reaction pathways of some alkoxyarylcarbene complexes in their reaction with alkynes leading to indene derivatives in a formal $[3_C+2_S]$ cycloaddition process have been reported. For example, the reaction of methoxy(2,6-dimethylphenyl)chromium carbene complex with 1,2-diphenylacetylene at 100 °C gives rise to an unusual indene derivative where a sigmatropic 1,5-methyl shift is observed [60]. Moreover, a related (4-hydroxy-2,6-dimethylphenyl)carbene complex reacts in benzene at 100 °C with 3-hexyne to produce an indene derivative. However, the expected Dötz cycloadduct is obtained when the solvent is changed to acetonitrile [61] (Scheme 19). Also, Dötz et al. have shown that the introduction of an isocyanide ligand into the coordination sphere of the metal induces the preferential formation of indene derivatives [62].

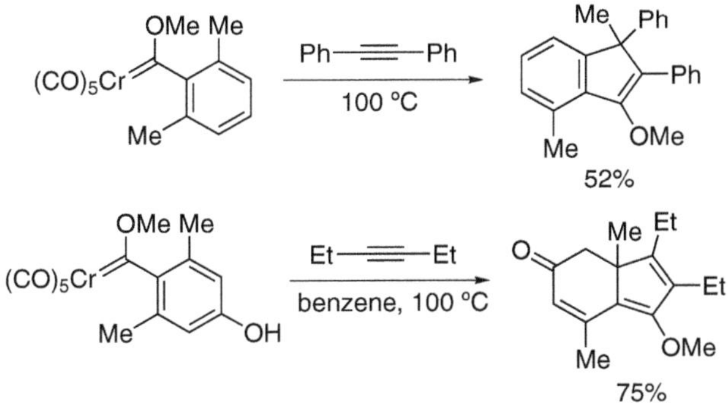

Scheme 19

Interestingly, amino(aryl)carbene complexes react with alkynes to give exclusively $[3_C+2_S]$ cycloaddition derivatives in high yields. This behaviour is totally different from the analogous alkoxy(aryl)carbene complexes as these preferentially lead to Dötz cycloadducts. Thus, Yamashita et al. found that morpholinophenylcarbene complex reacts with symmetrical alkynes to produce, after hydrolysis, the corresponding indanone derivatives [63] (Scheme 20). The dialkylaminofuranylcarbene complexes [64] and amidoarylcarbene complex derivatives [65] react in a similar way.

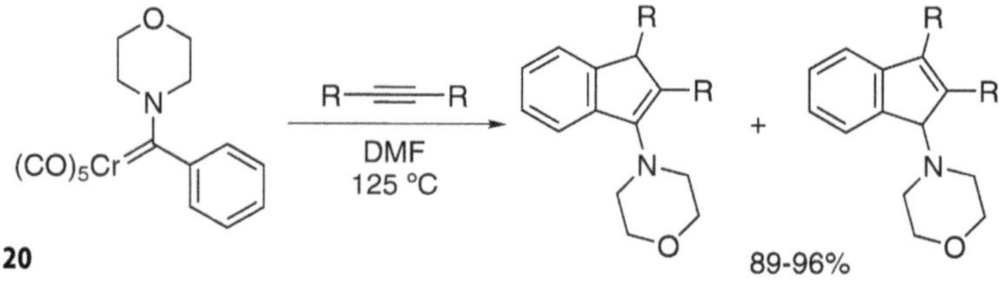

(CO)$_5$Cr, R—≡—R, DMF 125 °C, +, 89-96%

Scheme 20

2.6.3
Alkynylcarbene Complexes as C3 Building Blocks

α,β-Unsaturated carbene complexes have two electrophilic positions, so they may react with nucleophiles by the carbene carbon in a 1,2-addition fashion or by the β-carbon in a Michael-type or 1,4-addition way. Thus, compounds such as hydrazines, which possess two nucleophilic centres, react with alkynyl carbene complexes to formally produce the cycloaddition products coming from a double 1,2- and 1,4-addition process [66]. When the reaction is performed using the electron-deficient acetylhydrazine or phenylhydrazine, the intermediate cyclic carbene complex is not isolated and the reaction produces the corresponding pyrazole derivatives in high yields (Scheme 21).

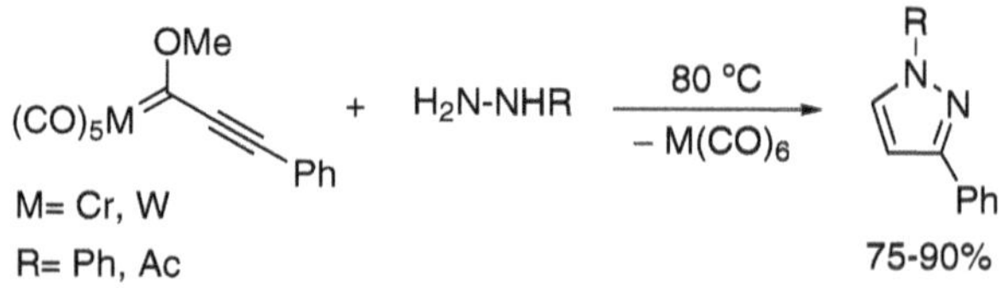

OMe, (CO)$_5$M, Ph, + H$_2$N-NHR, 80 °C, – M(CO)$_6$, M= Cr, W, R= Ph, Ac, 75-90%

Scheme 21

Fused cyclopentadiene derivatives are easily obtained by the reaction of alkynylcarbene complexes and cyclic enamines of five-, six- or seven-membered rings derived from secondary amines [67]. The overall [3_C+2_S] cycloaddition process is highly regioselective and proceeds under very mild reaction conditions. The reaction pathway is initiated by Michael-type addition of the nucleophilic tertiary cycloalkenylamine to the electrophilic alkynylcarbene complex resulting in the formation of a zwitterionic allene-type intermediate. This undergoes intramolecular hydrogen transfer to give a 1-metalatriene which cyclises to a cyclopentadiene complex yielding the final products after decomplexation and isomerisation (Scheme 22).

Non-enolizable imines such as 9-fluorene imines react with alkynylcarbene complexes to afford mixtures of mesoionic pyrrolium carbonyltungstates and dihydropyrrole derivatives [68] (Scheme 23). Although both compounds can be considered as [3_C+2_S] cycloadducts, formation of each of them follows a very different pathway. However, the first intermediate of the reaction is common for both compounds and supposes the conjugated addition of the imine to the alkynylcarbene complex to form a zwitterionic intermediate. A cyclisation

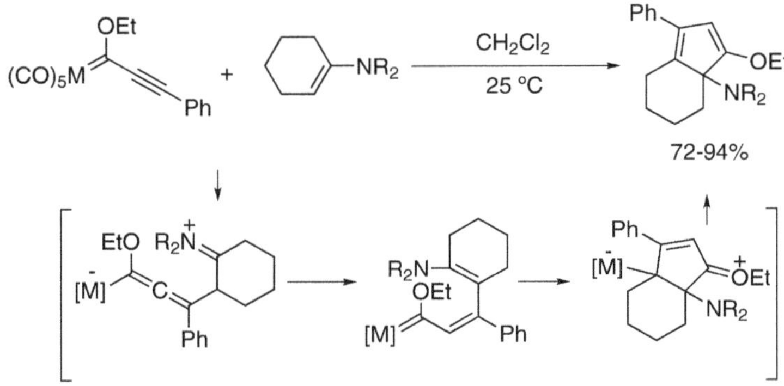

Scheme 22 [M]= $M(CO)_5$, M= Cr, W

favoured by a [1,2]-migration of the metallic fragment leads to the mesoionic pyrrolium carbonyltungstates. On the other hand, formation of the dihydropyrrole derivatives follows a more complicated reaction sequence involving the formation of a four-membered ring intermediate followed by a metathesis step and rearrangement (Scheme 23).

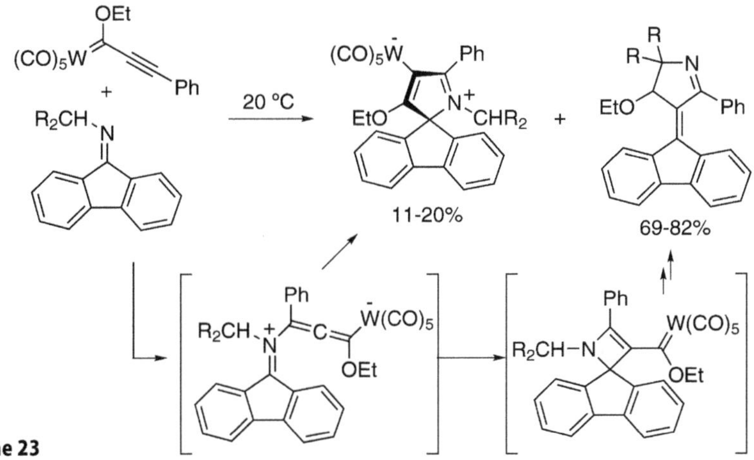

Scheme 23

A particular case of a $[3_C+2_S]$ cycloaddition is that described by Sierra et al. related to the tail-to-tail dimerisation of alkynylcarbenes by reaction of these complexes with C_8K (potassium graphite) at low temperature and further acid hydrolysis [69] (Scheme 24). In fact, this process should be considered as a $[3_C+2_C]$ cycloaddition as two molecules of the carbene complex are involved in the reaction. Remarkable features of this reaction are: (i) the formation of radical anion complexes by one-electron transfer from the potassium to the carbene complex, (ii) the tail-to-tail dimerisation to form a biscarbene anion intermediate and finally (iii) the protonation with a strong acid to produce the final product. Also, alkynylcarbene complexes react with alkenyl *N*-H imidates to give 2*H*-pyrrole complexes in a process which formally represents a [3_C+2_S] cycloaddition reaction. However, these compounds are obtained as minor products of the reaction and in very low yield (6–8%) [70].

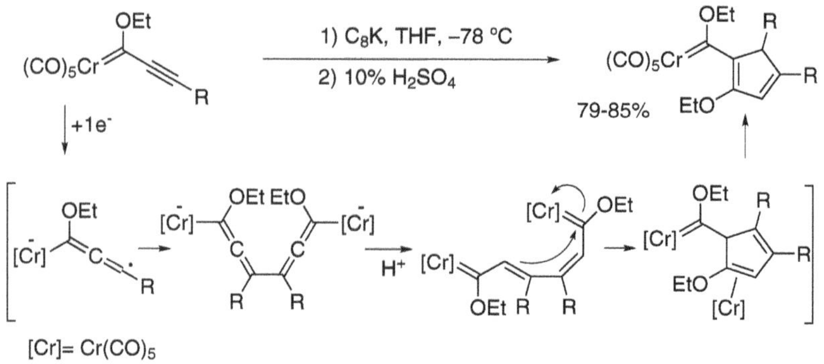

Scheme 24

2.6.4
Alkenylcarbene Complexes as C3 Building Blocks

The utility of alkenylcarbene complexes as C3 building blocks in the [3_C+2_S] cycloaddition reaction has been demonstrated by the wide variety of five-membered hetero- and carbocycles obtained when these complexes are treated with several C2 building block reagents. This impressive chemistry will be briefly discussed in the next few sections.

2.6.4.1
Reaction with Alkynes

In the same way as arylcarbene complexes, alkenylcarbene complexes typically react with alkynes to provide [3_C+2_S+1_{CO}] Dötz cycloadducts (see Chap. "Chromium-Templated Benzannulation Reactions", p. 123 in this book). However, some isolated examples involving the formation of five-membered rings through [3_C+2_S] cycloaddition processes have been reported [71]. In this context, de Meijere et al. found that β-donor-substituted alkenylcarbene complexes react with alkynes to give cyclopentene derivatives [71a]. This topic is also discussed in detail in Chap. "The Multifaceted Chemistry of Variously Substituted α,β-Unsaturated Fischer Metalcarbenes", p. 21 of this book.

2.6.4.2
Reaction with Electron-Poor Alkenes

The reaction of alkenylcarbene complexes and electron-poor alkenes normally leads to mixtures of the expected [2_S+1_C] vinylcyclopropane derivatives (see

Sect. 2.1.1) and [3_C+2_S] cyclopentene derivatives. The product distribution can be controlled by choosing the appropriate reaction conditions [72]. Moreover, the cyclopentene derivatives are the exclusive products from the coupling of β-pyrrolyl-substituted carbene complexes [72b,c] (Scheme 25). The crucial intermediate chromacyclobutane is formed in an initial step by a [2+2] cycloaddition. This chromacyclobutane rearranges to give the η^3-complex when non-coordinating solvents are used. Finally, a reductive elimination leads to the formal [3_C+2_S] cyclopentene derivatives.

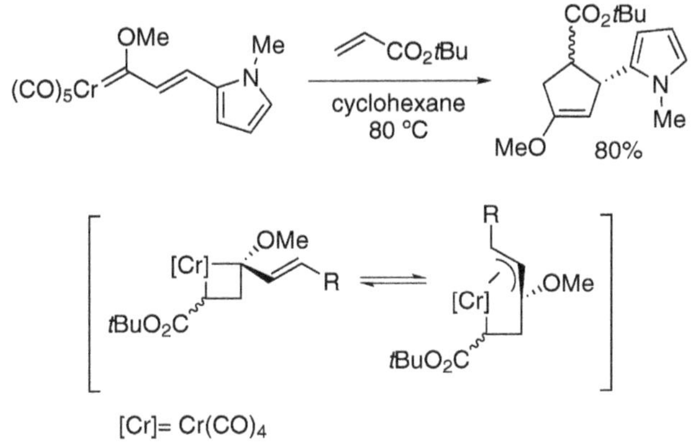

Scheme 25 [Cr]= $Cr(CO)_4$

2.6.4.3
Reaction with Electron-Rich Siloxy-Substituted 1,3-Dienes

Coupling of alkenylcarbene complexes and siloxy-substituted 1,3-dienes affords vinylcyclopentene derivatives through a formal [3_C+2_S] cycloaddition process. This unusual reaction is explained by an initial [4_C+2_S] cycloaddition of the electron-poor chromadiene system as the 4π component and the terminal double bond of the siloxydiene as the dienophile. The chromacyclohexene intermediate evolves by a reductive elimination of the metal fragment to generate the [3_C+2_S] cyclopentene derivatives [73] (Scheme 26).

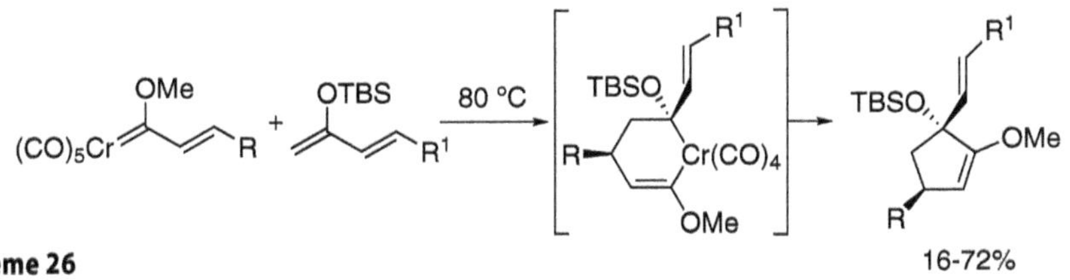

Scheme 26

2.6.4.4
Reaction with Electronically Neutral 1,3-Dienes

While studying the intermolecular cyclopropanation of simple alkenes with alkenylcarbene complexes, Barluenga et al. observed that the reaction between these complexes and electronically neutral 1,3-dienes results in the formation of mixtures of $[3_C+2_S]$ and $[4_S+1_C]$ cycloadducts [74a]. The reaction seems to be highly dependent on the solvent [74b] and temperature of the reaction, and selective formation of the $[3_C+2_S]$ cyclopentene derivative can be achieved by performing the reaction in toluene at 80 °C. Moreover, high asymmetric induction is observed when chiral alkenylcarbene complexes derived from (–)-8-phenylmenthol are used (Scheme 27). The mechanism proposed for this reaction follows a pathway analogous to that described before for the reaction of siloxy-substituted 1,3-dienes. Thus, the alkenylcarbene complex acts as a 1-chroma-1,3-diene in a Diels–Alder-type cycloaddition reaction to give a chromacyclohexene, which generates the final products after reductive elimination of the metal fragment. Interestingly, the observed diastereofacial selection cannot be explained by the model previously proposed (see Scheme 9), in which the phenyl group of the chiral auxiliary shields the *Re,Re* face of the alkenyl moiety allowing the substrates to approach from the *Si,Si* face. In this case, the necessary *s-cis* conformation of the chromadiene makes the dienophile react from the *Si-Re* face of the chromadiene (Scheme 27).

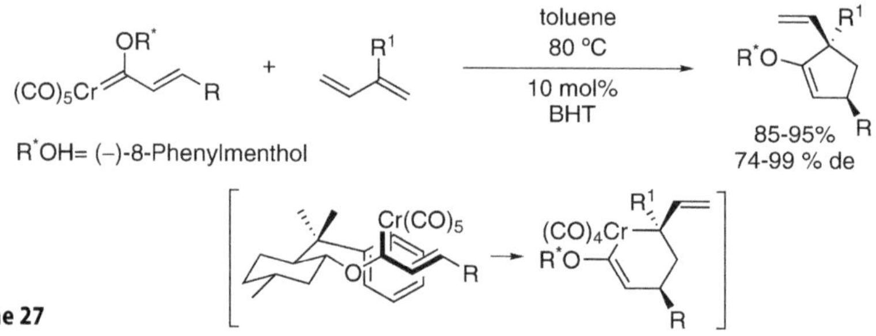

OR* (CO)5Cr R + R1 — toluene, 80 °C, 10 mol% BHT → R*O, R1, R; 85-95%, 74-99 % de

R*OH= (–)-8-Phenylmenthol

[Cr(CO)5, O, R → (CO)4Cr, R*O, R1, R]

Scheme 27

2.6.4.5
Reaction with 1-Amino-1-Aza-1,3-Dienes

Fischer alkenylcarbene complexes undergo cyclopentannulation to alkenyl *N,N*-dimethylhydrazones (1-amino-1-azadienes) to furnish $[3_C+2_S]$ substituted cyclopentenes in a regio- and diastereoselective way along with minor amounts of $[4_S+1_C]$ pyrrole derivatives. Enantiopure carbene complexes derived from (–)-8-(2-naphthyl)menthol afford mixtures of *trans,trans*-cyclopentenes and *cis,cis*-cyclopentenes with excellent face selectivity [75]. The mechanism proposed for the formation of these cyclopentene derivatives is outlined in Scheme 28. The process is initiated by nucleophilic 1,2-attack of the C_β carbon of the hydrazone on the less hindered face of the Cr=C double bond to generate a zwitterionic intermediate which may undergo a [1,2]-$Cr(CO)_5$ shift-promoted ring closure. Formation of one or the other diastereoisomer of the final product depends on the orientation of the azadiene moiety during this cyclisation step. Finally, hydrogen transfer to chromium followed by reductive elimination leads to the final *trans,trans*- or *cis,cis*-cyclopentenes.

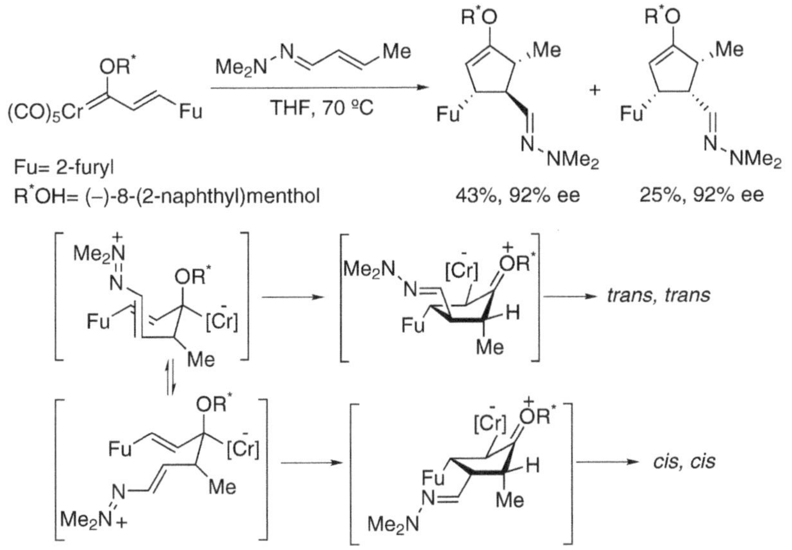

Fu= 2-furyl
R*OH= (–)-8-(2-naphthyl)menthol

THF, 70 °C

43%, 92% ee

25%, 92% ee

trans, trans

cis, cis

Scheme 28 [Cr]= $(CO)_5Cr$

2.6.4.6
Reaction with Imines

The reaction of alkenylcarbene complexes and imines in the presence of a Lewis acid generates pyrroline derivatives as a result of a $[3_C+2_S]$ cyclisation process [76]. This reaction has been extended to an asymmetric version by the use of chiral alkenylcarbene complexes derived from several chiral alcohols. However, the best results are found when (–)-8-phenylmenthol-derived complexes are used and catalytic amounts of $Sn(OTf)_2$ are added to the reaction. In these conditions high levels of *trans*/*cis* selectivity are achieved and the hydrolysis of the major *trans* diastereoisomers allows the preparation of optically pure 2,5-disubstituted-3-pyrrolidinone derivatives (Scheme 29).

The diastereofacial selectivity of this asymmetric $[3_C+2_S]$ process is explained following a model similar to that described in Sect. 2.6.4.4 for the reaction of chiral alkenylcarbene complexes and 1,3-dienes. Thus, the proposed mechanism that explains the stereochemistry observed assumes a [4+2] cycloaddition reaction between the chromadiene system and the C=N double bond of the imine. The necessary *s-cis* conformation of the complex makes the imine

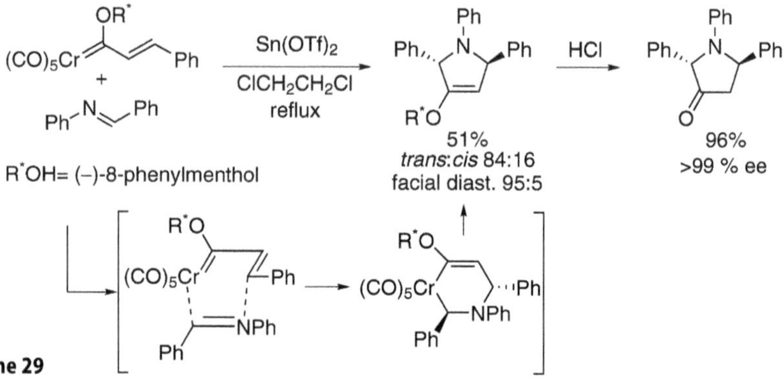

Scheme 29

approach from the less hindered *Si,Re* face of the complex to give a chromacyclohexene which, after reductive elimination, leads to the observed major 2*S*,5*R* isomer of the final product (Scheme 29).

2.6.4.7 Reaction with Enamines

Diastereoselective and enantioselective $[3_C+2_S]$ carbocyclisations have been recently developed by Barluenga et al. by the reaction of tungsten alkenylcarbene complexes and enamines derived from chiral amines. Interestingly, the regiochemistry of the final products is different for enamines derived from aldehydes and those derived from ketones. The use of chiral non-racemic enamines allows the asymmetric synthesis of substituted cyclopentenone derivatives [77] (Scheme 30).

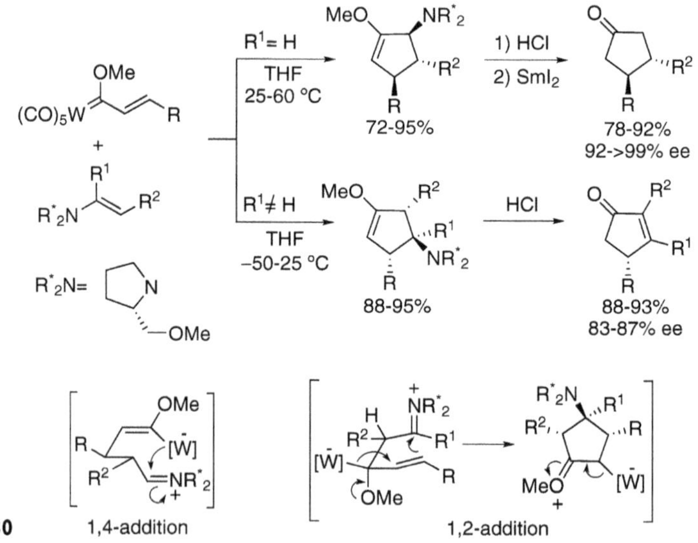

Scheme 30

The mechanism for aldehyde-derived enamines involves a Michael-type 1,4-addition of the enamine to the alkenylcarbene complex to generate a zwitterionic intermediate which evolves to the final product by cyclisation. On the other hand, ketone-derived enamines react through an initial 1,2-addition to the carbene carbon to generate a different zwitterionic intermediate. Then, a [1,2]-$W(CO)_5$ shift-promoted ring closure produces a new intermediate which, after elimination of the metal moiety, furnishes the corresponding cyclopentene derivatives (Scheme 30).

2.6.4.8 Reaction with Ynamines

The insertion reaction between alkenylcarbene complexes and electron-rich alkynes such as 1-alkynylamines (ynamines) leads to mixtures of two regioisomeric cyclopentyl derivatives [78]. Thus, if the insertion occurs on the carbon–metal bond a new aminocarbene complex is produced which evolves to a cyclopentenylmetal derivative. On the other hand, if the insertion reaction occurs on the carbon=carbon double bond of the alkenyl complex, the reaction gives a 1-metala-4-amino-1,3,5-triene complex which finally generates a different regioisomer of the cyclopentenylmetal derivative (Scheme 31).

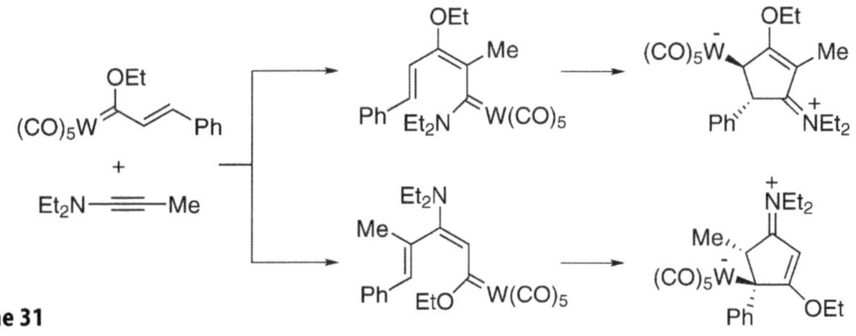

Scheme 31

2.6.4.9 Reaction with Methyl Ketone Lithium Enolates

An interesting strategy for the diastereoselective synthesis of five-membered carbocycles was achieved by the reaction of alkenylcarbene complexes and lithium enolates derived from simple methyl ketones [79]. The use of more or less coordinating solvents (THF or Et_2O) or the presence of cosolvents such as PMDTA allows the selective synthesis of one or the other diastereoisomer of the final cyclopentene derivative (Scheme 32).

The α-substitution in the alkenylcarbene complex seems to be crucial to direct the reaction to the five-membered rings. The mechanism proposed for this transformation supposes an initial 1,2-addition of the enolate to the carbene carbon atom to generate a zwitterionic intermediate. Cyclisation promoted by

[1,2]-$(CO)_5M$ migration followed by loss of the metal fragment and decoordination leads to the final cyclopentene derivatives. Formation of one or the other diastereoisomer depending on the solvent used for this reaction seems to be closely related to coordinative effects of the lithium ions to the oxygen atoms of these intermediates, favouring the orientation of the carbonyl group in a particular conformation (Scheme 32).

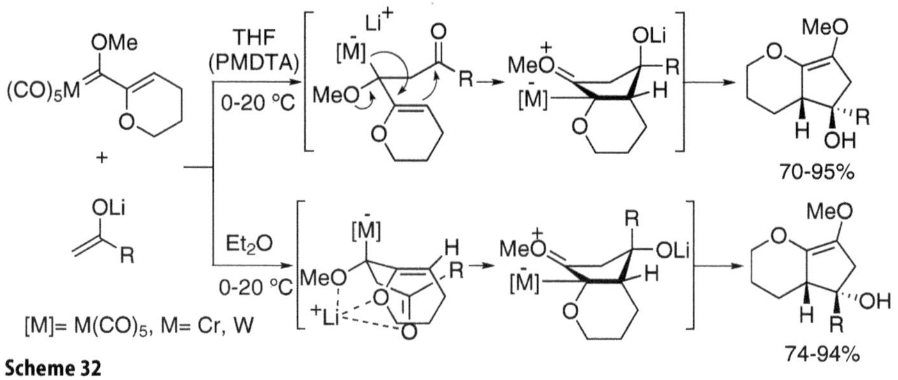

[M]= M(CO)5, M= Cr, W

Scheme 32

2.6.4.10 Reaction with Isonitriles

Isonitriles react with alkenylcarbene complexes to form initially at 0 °C a 3-ethoxy-3-styrylketeneimine complex, which on warming to room temperature leads to the formation of a cyclic 3-ethoxy-2,5-dihydro-2-pyrrolylidene complex. Finally, on heating to 100 °C a pyrrole derivative is produced [80] (Scheme 33).

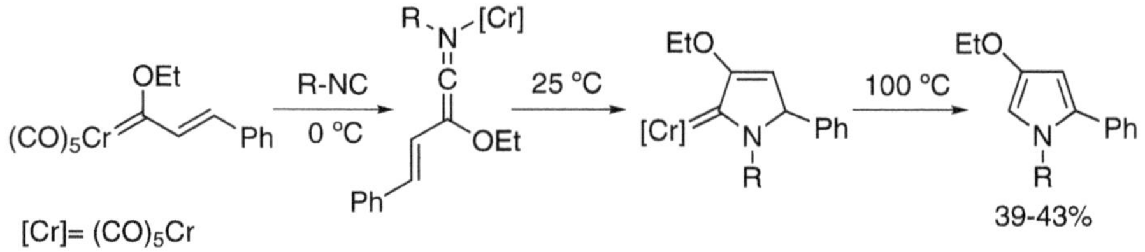

[Cr]= (CO)5Cr

Scheme 33

2.7 [4_S+1_C] Cycloaddition Reactions

The participation of carbene/carbenoid metal complexes in [4_S+1_C] cycloaddition reactions is very infrequent [81]. In fact, only a few examples involving Fischer carbene complexes have been reported in recent years [82]. A remarkable [4_S+1_C] cycloaddition process was reported by Herndon et al. when they reacted alkyl-derived chromium carbene complexes and cyclobutenediones to obtain furanone derivatives [83] (Scheme 34). The mechanism of this reaction involves the oxidative addition of the carbene to the acyl–acyl carbon–carbon σ-bond to finally produce a chromacyclohexenedione derivative which, after reductive elimination, generates the final products.

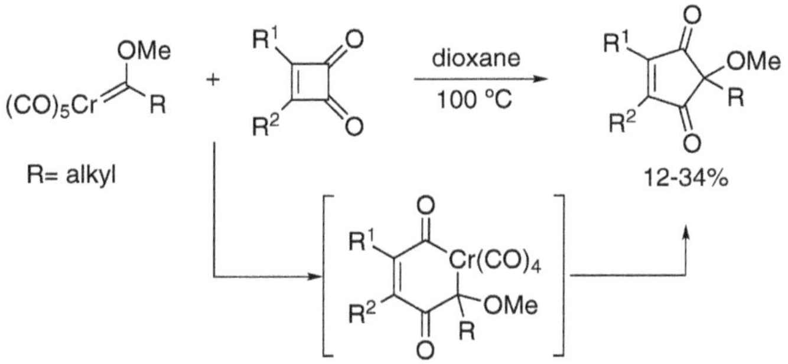

Scheme 34

The reaction of 1,3-diamino-1,3-dienes with aryl or α,β-disubstituted alkenylcarbene complexes leads to the formation of formal [4_S+1_C] cyclopentenones [25a] (Scheme 35). In the case of alkenylcarbene complexes, the substitution of the double bond of the complex in both α- and β-carbons seems to play a fundamental role as reactions performed in the same conditions but using alkenylcarbene complexes with other substitution patterns leads to compounds of a different nature ([4+3], [4+2] and [2+1] cycloadducts).

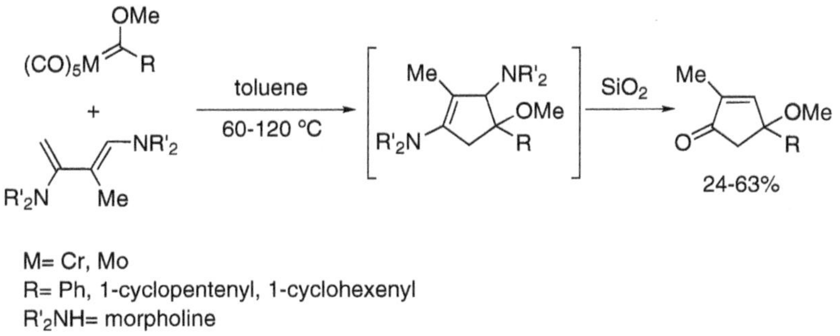

Scheme 35

The reaction of *N,N*-dimethylhydrazones (1-amino-1-azadienes) and alkenylcarbene complexes mainly produces [3_C+2_S] cyclopentene derivatives (see Sect. 2.6.4.5). However, a minor product in this reaction is a pyrrole derivative which can be considered as derived from a [4_S+1_C] cycloaddition process [75]. In this case, the reaction is initiated by the nucleophilic 1,2-addition of the nitrogen lone pair to the metal–carbon double bond followed by cyclisation and elimination of the corresponding alcohol (Scheme 36). If we compare this mechanism to that proposed for the formation of the major [3_C+2_S] product we may realise that formation of one or the other compound depends on the initial nucleophilic 1,2-addition (nitrogen or C_β attack) (compare to mechanism in Scheme 28, Sect. 2.6.4.5).

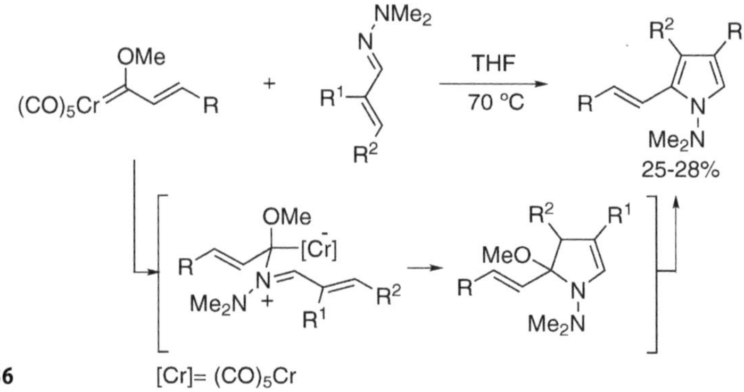

Scheme 36

Also, 2-aza-1,3-dienes react with arylcarbene complexes to undergo a formal [4_S+1_C] cycloaddition reaction to furnish pyrrolidinone derivatives in good yield [84a,b]. The formation of these cycloadducts is explained by initial [2+2] cycloaddition of the metal carbene to the electron-rich C=C double bond of the azadiene to form an intermediate metalacyclobutane. Transformation of this species into the final adducts can follow two pathways: (i) [1,3]-metal migration to form a 1-metala-3-azacyclohexene followed by reductive elimination and (ii) reductive metal elimination followed by three- to five-membered ring expansion of the resulting *N*-cyclopropylimine intermediate [84a] (Scheme 37). In a similar way, 1-aza-1,3-dienes react with arylcarbene complexes to furnish pyrrole derivatives through a formal [4_S+1_C] cycloaddition process, probably by a tandem cyclopropanation and ring enlargement [8a, 84c].

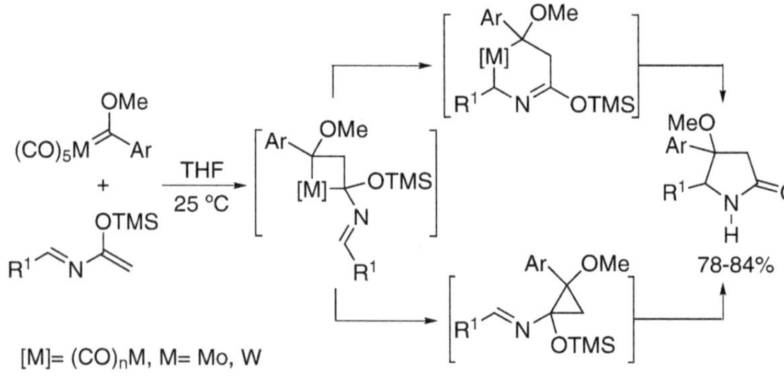

Scheme 37

Another example of a $[4_S+1_C]$ cycloaddition process is found in the reaction of alkenylcarbene complexes and lithium enolates derived from alkynyl methyl ketones. In Sect. 2.6.4.9 it was described how, in general, lithium enolates react with alkenylcarbene complexes to produce $[3_C+2_S]$ cycloadducts. However, when the reaction is performed using lithium enolates derived from alkynyl methyl ketones and the temperature is raised to 65 °C, a new formal $[4_S+1_C]$ cyclopentenone derivative is formed [79] (Scheme 38). The mechanism proposed for this transformation supposes the formation of the $[3_C+2_S]$ cycloadducts as depicted in Scheme 32 (see Sect. 2.6.4.9). This intermediate evolves through a retro-aldol-type reaction followed by an intramolecular Michael addition of the allyllithium to the ynone moiety to give the final cyclopentenone derivatives after hydrolysis. The role of the pentacarbonyltungsten fragment seems to be crucial for the outcome of this reaction, as experiments carried out with isolated intermediates in the absence of tungsten complexes do not afford the $[4_S+1_C]$ cycloadducts (Scheme 38).

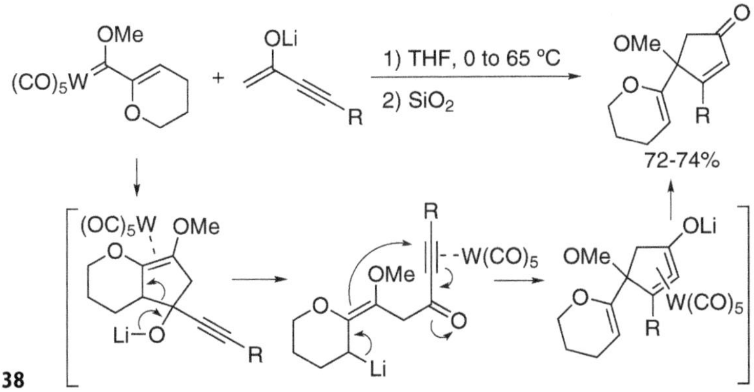

Scheme 38

$[4_S+1_C]$ Cycloadducts have also been obtained in the reaction of alkenylcarbene complexes with electronically neutral 1,3-dienes by appropriate choice of the reaction conditions (see for comparison Sect. 2.6.4.4). Thus, performing the reaction in THF at 120 °C in a sealed flask the formal $[4_S+1_C]$ cyclopentene derivative is generated in moderate yield [74a, 85] (Scheme 39). The key step

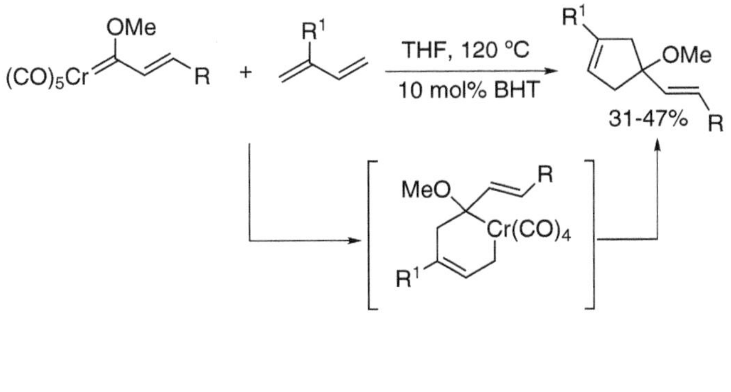

Scheme 39

in the proposed mechanism for this transformation involves a metala-Diels–Alder reaction in which the Cr=C acts as dienophile to produce a chromacyclohexene derivative intermediate which, after reductive elimination, leads to the final $[4_S+1_C]$ cyclopentene derivatives.

At this point the catalytic process developed by Dötz et al. using diazoalkanes and electron-rich dienes in the presence of catalytic amounts of pentacarbonyl(η^2-*cis*-cyclooctene)chromium should be mentioned. This reaction leads to cyclopentene derivatives in a process which can be considered as a formal $[4_S+1_C]$ cycloaddition reaction. A Fischer-type non-heteroatom-stabilised chromium carbene complex has been observed as an intermediate in this reaction [23a].

2.8 $[3_C+3_S]$ Cycloaddition Reactions

Despite the fact that transition metal complexes have found wide application in the synthesis of carbo- and heterocycles, [3+3] cyclisation reactions mediated or assisted by transition metals remain almost unexplored [3, 86]. However, a few examples involving Fischer carbene complexes have been reported. In all cases, this complex is α,β-unsaturated in order to act as a C3-synthon and it reacts with different types of substrates acting as C3-synthons as well.

All around this chapter, we have seen that α,β-unsaturated Fischer carbene complexes may act as efficient C3-synthons. As has been previously mentioned, these complexes contain two electrophilic positions, the carbene carbon and the β-carbon (Fig. 3), so they can react via these two positions with molecules which include two nucleophilic positions in their structure. On the other hand, alkenyl- and alkynylcarbene complexes are capable of undergoing [1,2]-migration of the metalpentacarbonyl allowing an electrophilic-to-nucleophilic polarity change of the carbene ligand β-carbon (Fig. 3). These two modes of reaction along with other processes initiated by [2+2] cycloaddition reactions have been applied to [3+3] cyclisation processes and will be briefly discussed in the next few sections.

2.8.1 Reaction of α,β-Unsaturated Fischer Carbene Complexes with 1,3-Dinucleophiles

Alkynylcarbene complexes react with β-dicarbonyl compounds and catalytic amounts of a base to generate formal [3+3] pyranylidene derivatives [87]. The

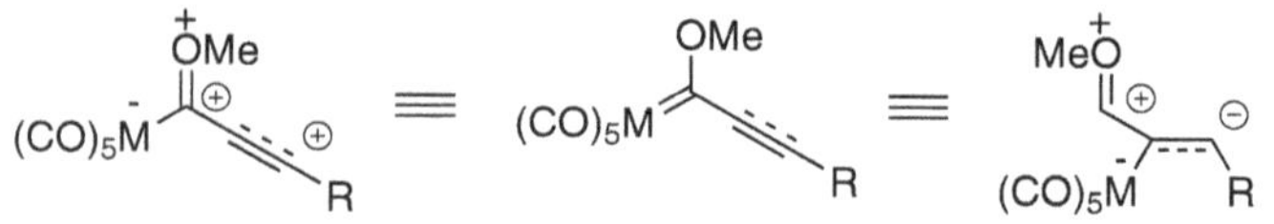

Fig. 3 The 1,2-migration of the metalpentacarbonyl of alkenyl- and alkynylcarbene complexes

reaction is initiated by the addition of the enolate to the β-position leading to an intermediate which evolves through an intramolecular exchange of the alkoxy group (Scheme 40).

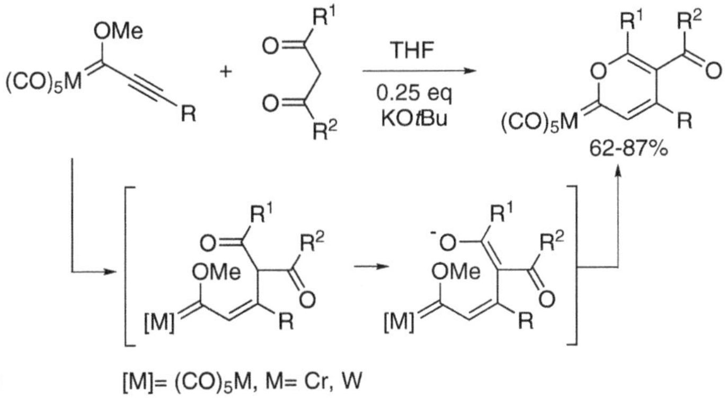

Scheme 40 [M]= (CO)$_5$M, M= Cr, W

In a similar way, 1,3-dinitrogen systems such as diamines, amidines, guanidines, aminothiazoles, aminopyridines, ureas and thioureas react with alkynylcarbene complexes generating the corresponding heterocycles. Of particular interest is the reaction with ureas, as the process can be applied to the easy synthesis of pyrimidine derivatives [88] (Scheme 41).

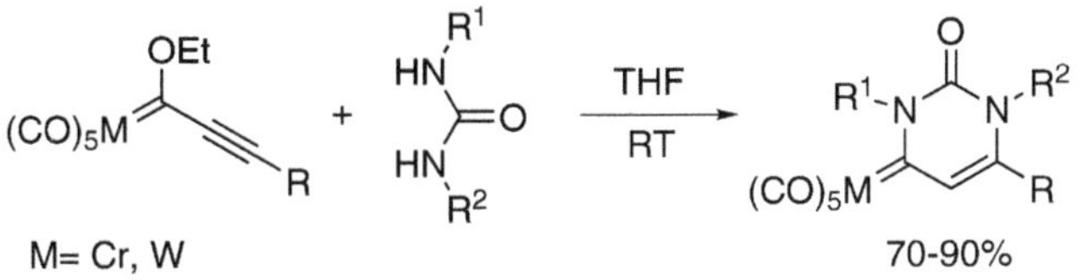

Scheme 41 M= Cr, W

β-Oxygen-functionalised sp^3 organolithium compounds react with alkenylcarbene complexes to generate the corresponding cyclic carbene complexes [89] (Scheme 42). This sequence involves initial Michael addition of the β-alkoxide organolithium reagent to give an anionic adduct which subsequently undergoes a spontaneous intramolecular alkoxide exchange.

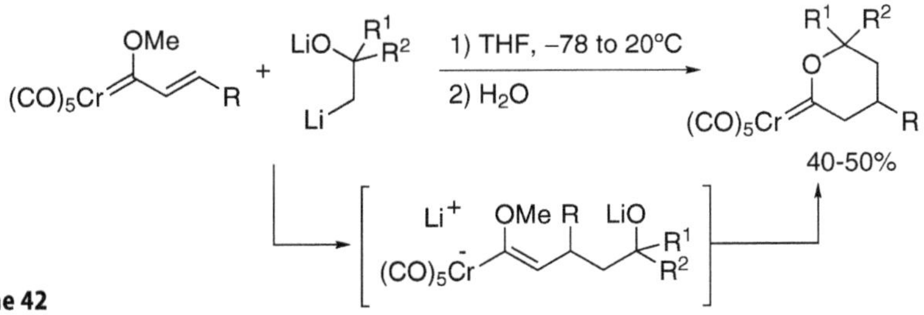

Scheme 42

In a reaction closely related to the latter, pyranylidene derivatives are obtained by the intermolecular radical coupling of alkynyl- or alkenylcarbene complexes and epoxides. Good diastereoselectivities are observed when cyclic epoxides are used. Moreover, the best results are reached by the generation of the alkyl radical using titanocene monochloride dimer [90] (Scheme 43).

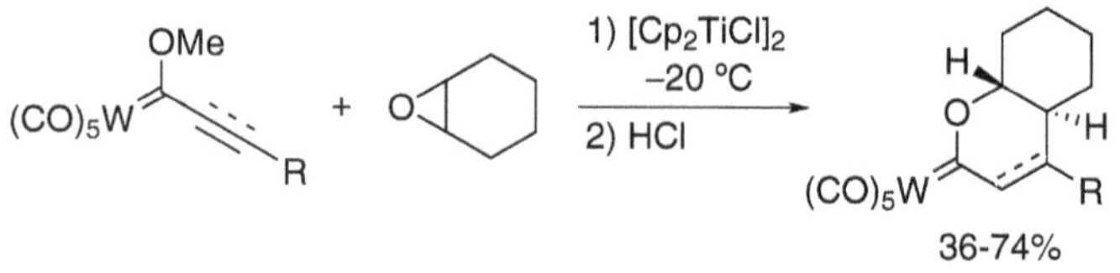

Scheme 43

The potential of Fischer carbene complexes in the construction of complex structures from simple starting materials is nicely reflected in the next example. Thus, the reaction of alkenylcarbene complexes of chromium and tungsten with cyclopentanone and cyclohexanone enamines allows the diastereo- and enantioselective synthesis of functionalised bicyclo[3.2.1]octane and bicyclo[3.3.1]nonane derivatives [12] (Scheme 44). The mechanism of this transformation is initiated by a 1,4-addition of the C_β-enamine to the alkenylcarbene complex. Further 1,2-addition of the $C_{\beta'}$ of the newly formed enamine to the carbene carbon leads to a metalate intermediate which can

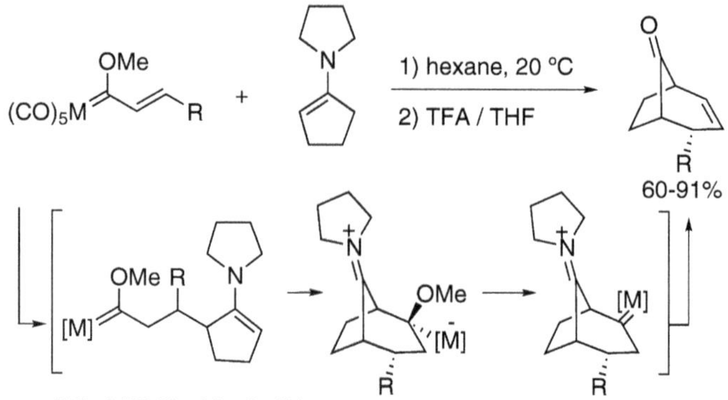

Scheme 44

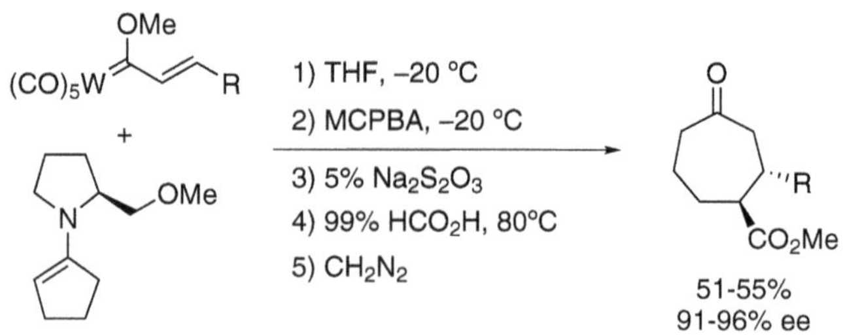

Scheme 45

be isolated. This metalate may suffer an acid-induced elimination of methanol to form a non-heteroatom-stabilised carbene species, which then undergoes β-hydrogen elimination and reductive elimination to yield the final products after hydrolysis of the imonium function. Applying this α,β,β'-annulation reaction it is possible to access enantioenriched 3,4-disubstituted cycloheptanones in a one-pot process from chiral cyclopentanone enamines (Scheme 45).

2.8.2 [1,2]-Metalpentacarbonyl-Promoted [3+3] Cycloaddition Reactions

Alkynylcarbene complexes react with imines derived from furan-, benzofuran-, *N*-substituted pyrrole- and *N*-substituted indole-2-carboxaldehydes to give the corresponding formal [3+3] cyclic derivative [91] (Scheme 46). This carbocyclisation process can be explained by assuming a [1,2]-migration of the pentacarbonylmetal fragment as the key step. Thus, an initial 1,2-addition of the C3 carbon of the ring generates a zwitterionic intermediate. Further [1,2]-$M(CO)_5$ shift promotes cyclisation and finally, hydrogen transfer and reductive elimination of the metal furnishes the final products.

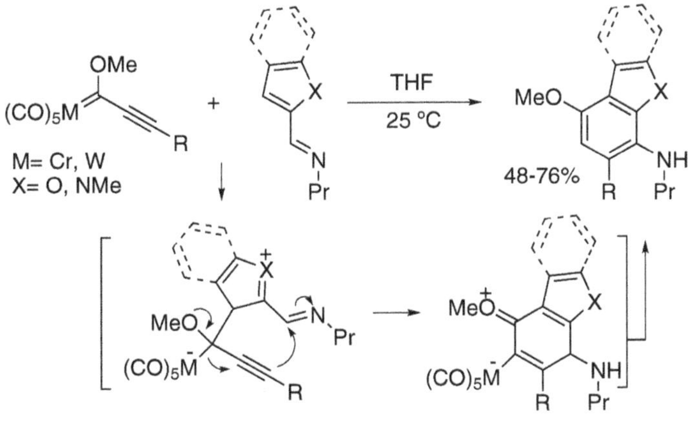

Scheme 46

Interestingly, the analogous reaction performed with alkenylcarbene complexes and pyrrole-2-carboxaldehyde imine leads to other kinds of formal [3+3] cycloadducts. These compounds are obtained as single regio- and diastereoisomers [91] (Scheme 47). This heterocyclisation resembles the precedent [3+3] carbocyclisation of alkynylcarbene complexes, except that the unsubstituted ring nitrogen is now involved rather than the ring C3 atom. In this case, the sequence is initiated by a 1,2-addition of the N–H of the pyrrole to the carbene carbon affording a zwitterionic intermediate. Further cyclisation induced by [1,2]-$M(CO)_5$ shift followed by hydrogen transfer and reductive elimination of the metal leads to the final cycloadducts (Scheme 47).

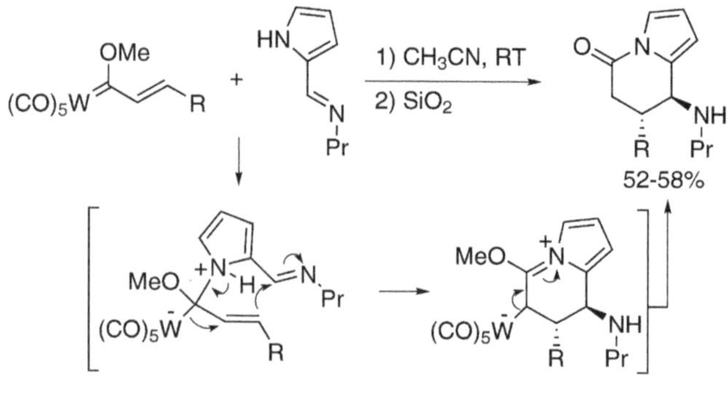

Scheme 47

2.8.3
[3+3] Cycloaddition Reactions Initiated by a [2+2] Process

The reaction of ethyl 2,2-diethoxyacrylate with alkynylalkoxycarbene complexes affords 6-ethoxy-2*H*-2-pyranylidene metal complexes [92] (Scheme 48). The mechanism that explains this process is initiated by a [2+2] cycloaddition reaction (see Sect. 2.3), followed by a cyclobutene ring opening to generate a tetracarbonylcarbene complex. This complex can be isolated and on standing for one day at room temperature renders the final 6-ethoxy-2*H*-pyranylidene pentacarbonyl complex. This last transformation requires the formal transfer of one carbonyl group and one proton from the diethoxy methylene moiety to the metal and to the C3 2*H*-pyranylidene ring, respectively, with concomitant cyclisation. Further studies on this unusual transformation have been extensively performed by Moretó et al. [93].

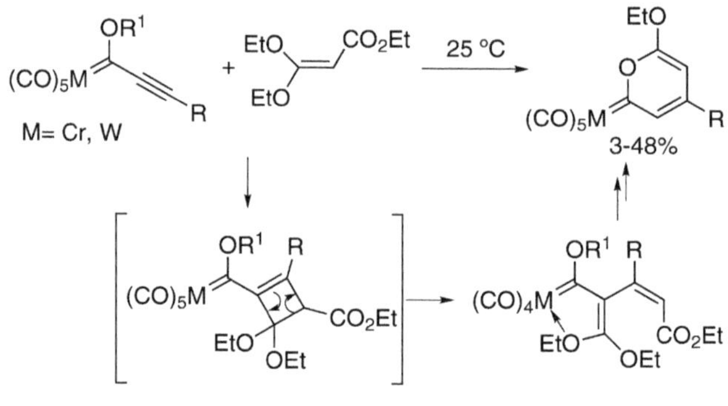

Scheme 48

In a similar process, tertiary enaminones react with alkynylcarbene complexes to give the corresponding pyranylidene complexes following a reaction pathway analogous to that described above. First, a [2+2] cycloaddition reaction between the alkynyl moiety of the carbene complex and the C=C double bond of the enamine generates a cyclobutene intermediate, which evolves by a conrotatory cyclobutene ring opening followed by a cyclisation process [94] (Scheme 49).

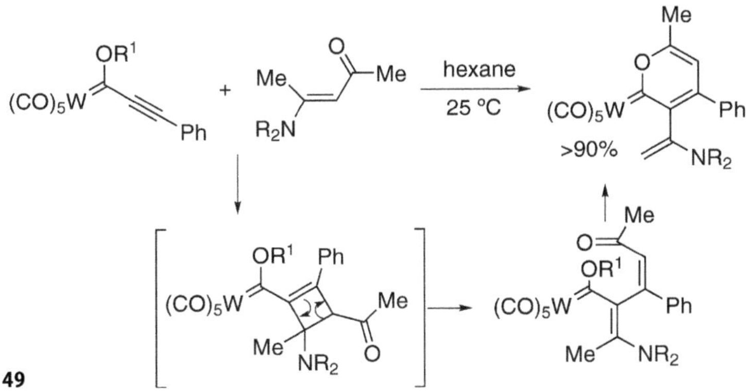

Scheme 49

2.8.4
[3+3] Benzannulation Processes

Highly strained cyclic compounds such as cyclopropenone derivatives react with alkyl-derived chromium complexes to afford $[3_S+1_C]$ cycloadducts (see Sect. 2.4). However, the use of alkenyl- or arylcarbene complexes leads to a mixture of two regioisomers of a benzannulation product which can be considered as derived from a [3+3] cycloaddition reaction [41] (Scheme 50). The reaction is initiated by the insertion of the metalcarbene into the cyclopropenone carbon–carbon σ-bond to generate two possible metalacyclopentenone derivatives. The first one evolves through a 1,3-shift of the metallic moiety to give a metalacycloheptadienone derivative which, after reductive elimination, leads to one of the regioisomers observed in the reaction. The other regioisomer can

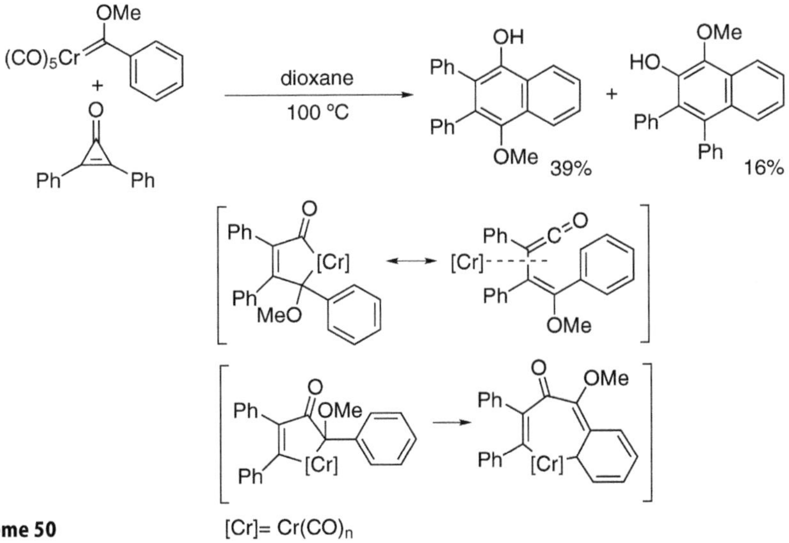

Scheme 50 [Cr]= $Cr(CO)_n$

arise from the other metalacyclopentenone formed in the first step of the reaction. A resonance form of this compound is the vinyl ketene complex which, after electrocyclisation and isomerisation, produces the major regioisomer observed in the reaction.

2.9 $[4_S+2_C]$ Cycloaddition Reactions

The Diels–Alder reaction of activated olefins is considered as one of the most useful and predictable reactions in organic synthesis. The electron-acceptor character of the pentacarbonylmetal fragment makes α,β-unsaturated carbene complexes ideal substrates for the $[4_S+2_C]$ cycloaddition reaction with dienes.

2.9.1 Alkenylcarbene Complexes as C2 Building Blocks

2.9.1.1 Alkoxy Alkenylcarbene Complexes

It has been established that alkoxy alkenylcarbene complexes participate as dienophiles in Diels–Alder reactions not only with higher rates but also with better regio- and stereoselectivities than the corresponding esters [95]. This is clearly illustrated in Scheme 51 for the reactions of an unsubstituted vinyl complex with isoprene. This complex reacts to completion at 25 °C in 3 h whereas the cycloaddition reaction of methyl acrylate with isoprene requires 7 months at the same temperature. The rate enhancement observed for this complex is comparable to that for the corresponding aluminium chloride-catalysed reactions of methyl acrylate and isoprene (Scheme 51).

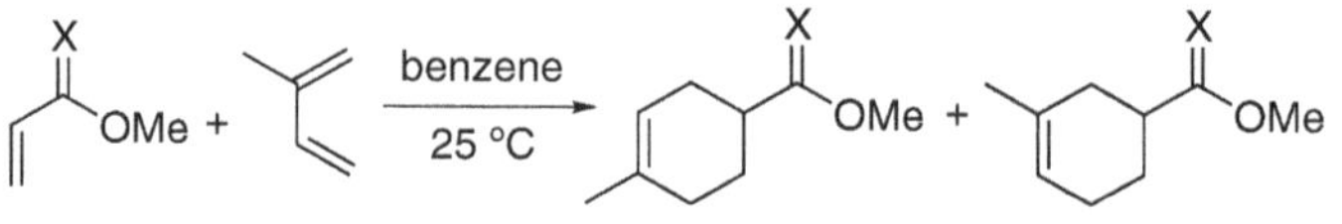

X, OMe + ; benzene, 25 °C ; X, OMe + X, OMe

X		
X= $Cr(CO)_5$	3h	92:8, 70%
X=O	7 months	70:30, 54%
X= O ($AlCl_3$)	3 h	95:5, 50%

Scheme 51

The Diels–Alder reaction of simple alkoxy alkenylcarbene complexes leads to mixtures of *endo* and *exo* cycloadducts, with the *endo* isomer generally being the major one [96, 97]. Asymmetric examples of *endo* Diels–Alder reactions have also been reported by the use of chiral auxiliaries both on the carbene complex and the diene. Thus, the reaction of cyclopentadiene with chiral alkenylcarbene complexes derived from (–)-menthol proceeds to afford a 4:1

endo:*exo* mixture. The diastereomeric excess found for the *endo* isomer is 75% [97] (Scheme 52). On the other hand, chiral 2-amino-1,3-dienes derived from (*S*)-methoxymethylpyrrolidine react with alkoxy alkenylcarbene complexes of tungsten providing the corresponding *endo* cycloadducts as the major products and with high enantioselectivities in most cases [98] (Scheme 52).

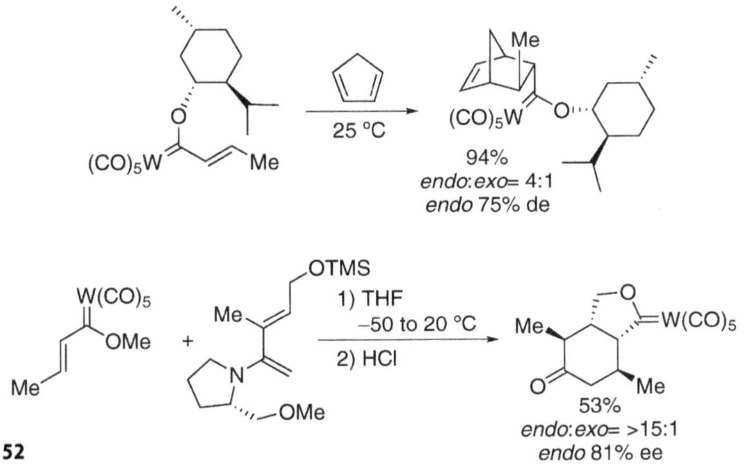

Scheme 52

However, *exo*-selective Diels–Alder reactions are found when α,β-unsaturated exocyclic carbene complexes are used as dienophiles. The fixed *s-cis* conformation of the vinylcarbene moiety of the complex seems to be responsible for the *exo* selectivity observed in this reaction. Moreover, the reaction of optically active carbene complexes with 2-morpholino-1,3-butadienes allows the asymmetric synthesis of spiro compounds [99] (Scheme 53).

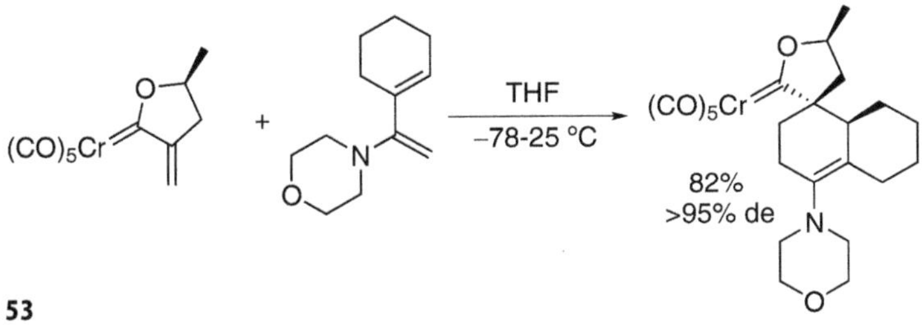

Scheme 53

2.9.1.2 Metaloxy Alkenylcarbene Complexes

Titanoxy alkenylcarbene complexes have been used as dienophiles in their reaction with cyclopentadiene to give predominantly the *exo* cycloadduct in high yield. The unexpected formation of the *exo* isomer is attributed to the steric environment of the dienophile in opposition to the stereoelectronic factors usually identified with *endo* selectivity [100] (Scheme 54).

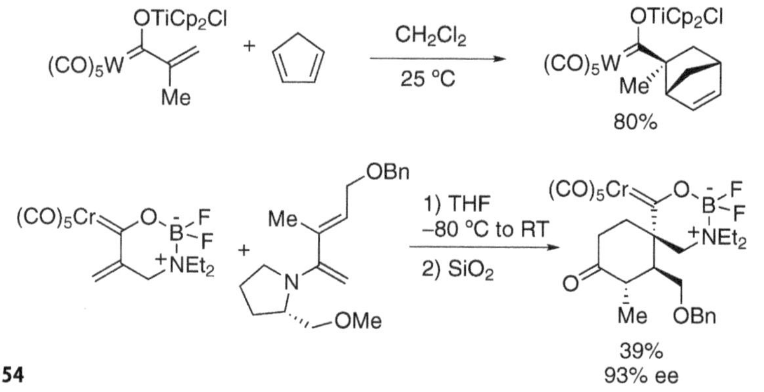

Scheme 54

Barluenga et al. have described novel vinylcarbene complexes containing a cyclic BF_2 chelated structure which temporarily fixes the *s-cis* conformation of the exocyclic C=C and Cr=C double bonds. These boroxycarbene complexes behave as dienophiles with 2-amino-1,3-butadienes in a remarkably regio- and *exo*-selective way. Moreover, high degrees of enantioselectivity are reached by the use of chiral 2-aminodienes derived from (*S*)-methoxymethylpyrrolidine [101] (Scheme 54).

2.9.1.3
Amino Alkenylcarbene Complexes

The reactivity of α,β-unsaturated aminocarbene complexes in Diels–Alder processes is much lower than that of the corresponding alkoxycarbene complexes. Despite this low reactivity it has been possible to determine the high *exo* selectivity of processes involving the reaction of aminocarbene complexes and acyclic dienes. An important improvement on the reactivity of aminocarbene complexes was achieved by derivatisation of the nitrogen with an electron-withdrawing *N*-benzoyl group. The best results were found for tetracarbonyl complexes in which the benzoyl carbonyl oxygen is chelated to the metal. The high degree of *exo* selectivity also observed in these cases was explained as a consequence of the severe close contacts between the apical CO ligands and the diene in the *endo* but not the *exo* transition state [97, 102] (Scheme 55).

An asymmetric version of this reaction was achieved by the use of complexes derived from chiral imidazolidinones. For example, the reaction of Danishefsky's diene with these chiral complexes occurs with both high *exo:endo* selectivity and high facial selectivity at the dienophile [103] (Scheme 56).

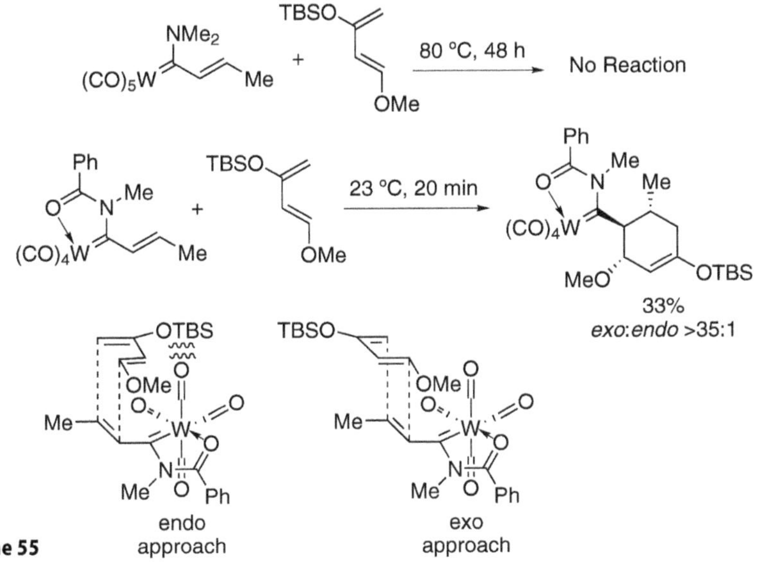

Scheme 55

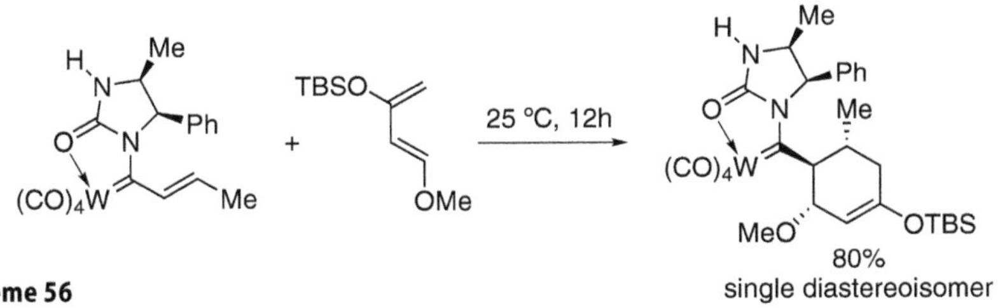

Scheme 56

2.9.2
Alkynylcarbene Complexes as C2 Building Blocks

2.9.2.1
Alkoxy Alkynylcarbene Complexes

Alkoxy alkynylcarbene complexes undergo Diels–Alder reactions with neutral and electron-rich dienes [36f, 104] and also with 1-aza- and 2-aza-1,3-butadiene derivatives [84a, 105] (Scheme 57).

2.9.2.2
Amino Alkynylcarbene Complexes

Following the same tendency as alkenylcarbene complexes, the substitution of the alkoxy group for an amino group in alkynylcarbene derivatives greatly decreases the rate of Diels–Alder reactions [102, 104b]. In fact, substituted

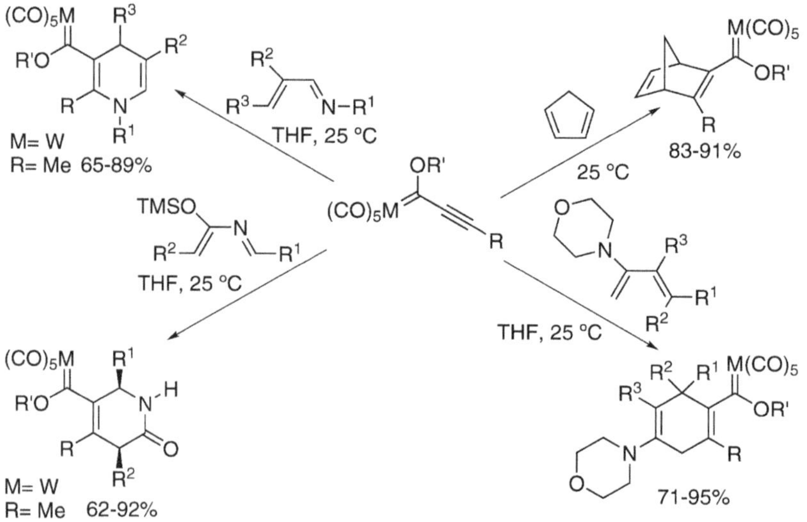

Scheme 57

acetylenic aminocarbene complexes failed to react in intermolecular processes. Only unsubstituted amino alkynylcarbene complexes react with cyclopentadiene to produce the corresponding $[4_S+2_C]$ cycloadduct [106]. Significant asymmetric induction can be achieved by the use of alkynylcarbene complexes derived from chiral pyrrolidines. However, this reaction seems to be highly dependent on the substituents of the diene, and the highest diastereoselectivities are found in the reaction with 2-triisopropylsiloxy-1,3-pentadiene whilst modest selectivities are reached with cyclopentadiene and α-triisopropylsiloxyvinyl cyclohexene [107] (Scheme 58).

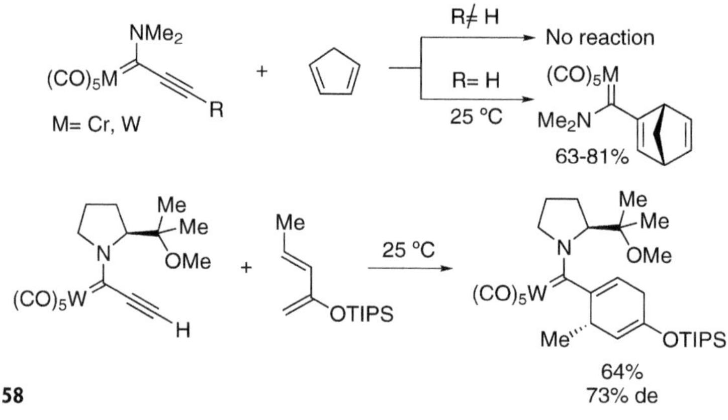

Scheme 58

2.10 $[4_C+2_S]$ Cycloaddition Reactions

Intermolecular $[4_C+2_S]$ cycloaddition reactions where the diene moiety is contained in the carbene complex are less frequent than the $[4_S+2_C]$ cycloadditions summarised in the previous section. However, 2-butadienylcarbene complexes, generated by a [2+2]/cyclobutene ring opening sequence, undergo Diels–Alder reactions with typical dienophiles [34, 35] (Scheme 59). Also, Wulff et al. have described the application of pyranylidene complexes, obtained by a [3+3] cycloaddition reaction (see Sect. 2.8.1), in the inverse-electron-demand Diels–Alder reaction with enol ethers and enamines [87a]. Later, this strategy was applied to the synthesis of steroid-like ring skeletons [87b] (Scheme 59).

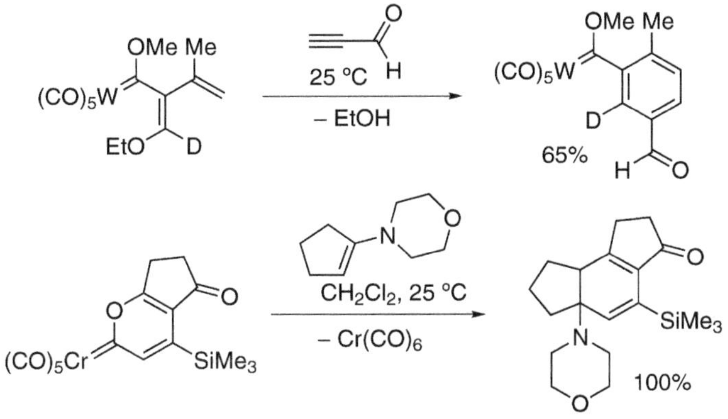

Scheme 59

2.11 Intramolecular [4+2] Cycloaddition Reactions

For clarity, the reactions contained in this section can be divided into three categories according to the structure of the carbene complexes (Fig. 4): (i) those in which the dienophile and the diene are tethered through the heteroatom and the carbene carbon of the complex (type 1), (ii) those in which the dienophile and the diene are part of the same carbon chain (type 2), and finally (iii) those where the diene and the dienophile belong to different ligands within the complex (type 3).

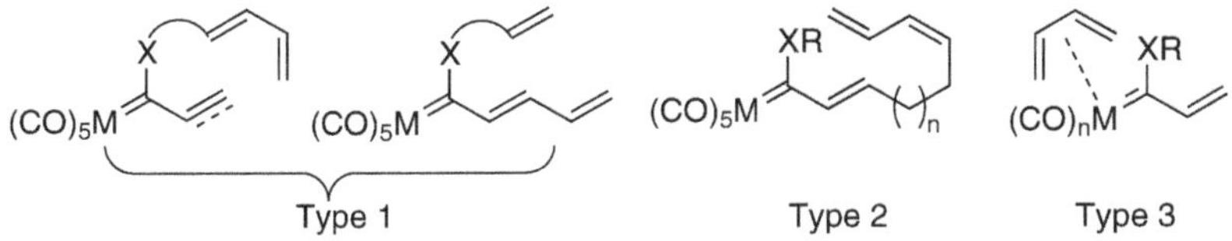

Fig. 4 Categories of intramolecular [4+2] cycloaddition reactions (for details see text)

2.11.1
Type 1 Intramolecular [4+2] Cycloadditions

Carbene complexes containing either the dienophile or the diene functionality bonded directly to the carbene carbon undergo intramolecular [4+2] cycloadditions under mild conditions [108] (Scheme 60).

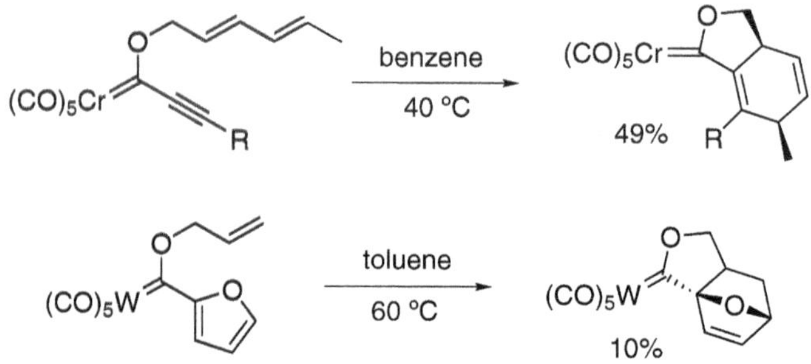

Scheme 60

2.11.2
Type 2 Intramolecular [4+2] Cycloadditions

Carbene complexes which have an all-carbon tether between the diene and the dienophile react via intramolecular Diels–Alder reaction to give the corresponding bicyclic compound. The stereoselectivities of these reactions are comparable to those observed for the Lewis acid-catalysed reactions of the corresponding methyl esters and much higher than those of the thermal reactions of the methyl esters which are completely unselective. Moreover, the *cis*-substituted complexes undergo *endo*-selective reactions where the corresponding reaction of the ester fails [109] (Scheme 61).

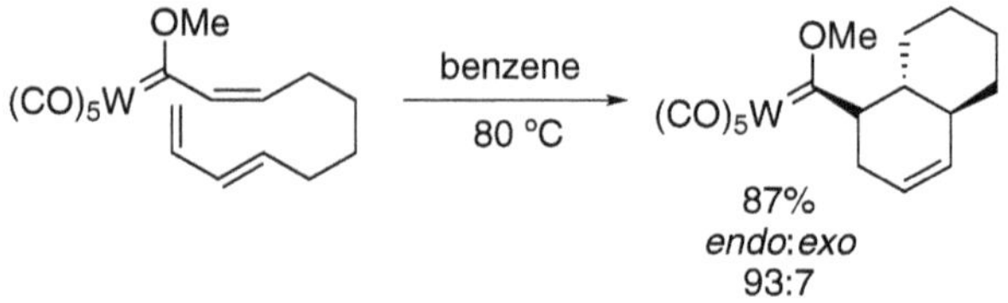

Scheme 61

2.11.3
Type 3 Intramolecular [4+2] Cycloadditions

Mathey et al. have described a quite unusual intramolecular [4+2] cycloaddition process. In this reaction the diene and the dienophile are part of two different ligands within the same complex. Thus, *cis*-(vinyl ethoxycarbene)(1-phenyl-3,4-dimethylphosphole)tetracarbonylchromium complex reacts at room temperature to afford the corresponding intramolecular Diels–Alder cycloadduct [110] (Scheme 62).

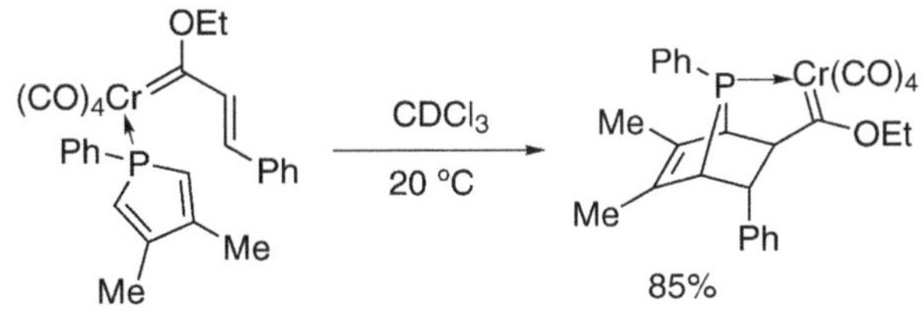

Scheme 62

2.12 [5_C+1_S] Cycloaddition Reactions

Several examples of [5_C+1_S] cycloaddition reactions have been described involving in all cases a 1,3,5-metalahexatriene carbene complex as the C5-synthon and a CO or an isocyanide as the C1-synthon. Thus, Merlic et al. described the photochemically driven benzannulation of dienylcarbene complexes to produce *ortho* alkoxyphenol derivatives when the reaction is performed under an atmosphere of CO, or *ortho* alkoxyanilines when the reaction is thermally performed in the presence of an isonitrile [111] (Scheme 63). In related works, Barluenga et al. carried out analogous reactions under thermal conditions [36a, c, 47a]. Interestingly, the dienylcarbene complexes are obtained in a first step by a [2+2] or a [3_S+2_C] process (see Sects. 2.3 and 2.5.1). Further reaction of these complexes with CO or an isonitrile leads to highly functionalised aromatic compounds (Scheme 63).

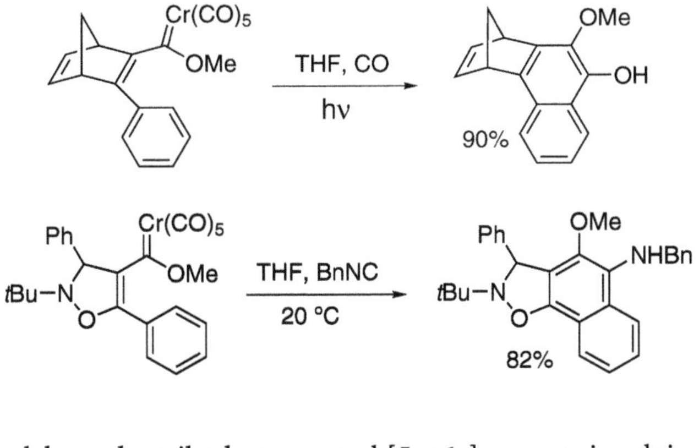

Scheme 63

Mathey et al. have described an unusual [5_C+1_S] process involving the reaction of a transient terminal phosphinidene complex [$PhP=W(CO)_5$] with a butadienyl carbene complex yielding a 1-phenyl-1,2-dihydrophosphine P-$W(CO)_5$ complex [112].

2.13
$[5_S+1_{CO}]$ Cycloaddition Reactions

The coupling of carbene complexes with conjugated enediynes provides benzannulated compounds which incorporate five atoms of the endiyne and a CO ligand from the carbene complex [113] (Scheme 64). The formation of these products has been explained as follows: firstly, selective coupling of the less hindered alkyne moiety of the endiyne to the carbene complex gives rise, after further insertion of a CO ligand, to an enyne-ketene intermediate; then a Moore cyclisation affords a chromium-complexed diradical species which produces the final product by hydrogen abstraction (from the solvent or by intramolecular hydrogen atom transfer) followed by formation of the furan ring upon acid treatment.

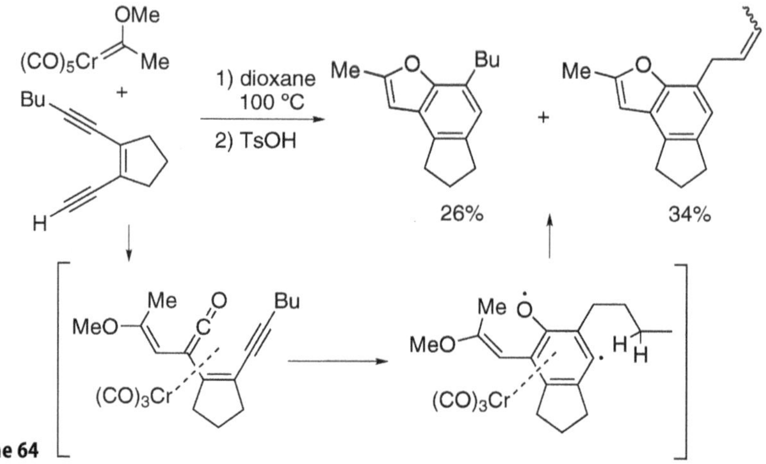

Scheme 64

2.14
$[4_S+3_C]$ Cycloaddition Reactions

2.14.1
Alkenylcarbene Complexes as C3 Building Blocks

Electronically rich 1,3-butadienes such as Danishefsky's diene react with chromium alkenylcarbene complexes affording seven-membered rings in a formal $[4_S+3_C]$ cycloaddition process [73a, 95a]. It is important to remark on the role played by the metal in this reaction as the analogous tungsten carbene complexes lead to $[4_S+2_C]$ cycloadducts (see Sect. 2.9.1.1). Formation of the seven-membered ring is explained by an initial cyclopropanation of the most electron-rich double bond of the diene followed by a Cope rearrangement of the formed divinylcyclopropane (Scheme 65). Amino-substituted 1,3-butadienes also react with chromium alkenylcarbene complexes to produce the corresponding seven-membered rings [25a, 114]. Applying this strategy, Barluenga et al. developed an asymmetric synthesis of substituted cyclohepta-1,3-diones using chiral 2-amino-1,3-butadienes derived from (*S*)-2-methoxymethylpyrrolidine [114] (Scheme 65).

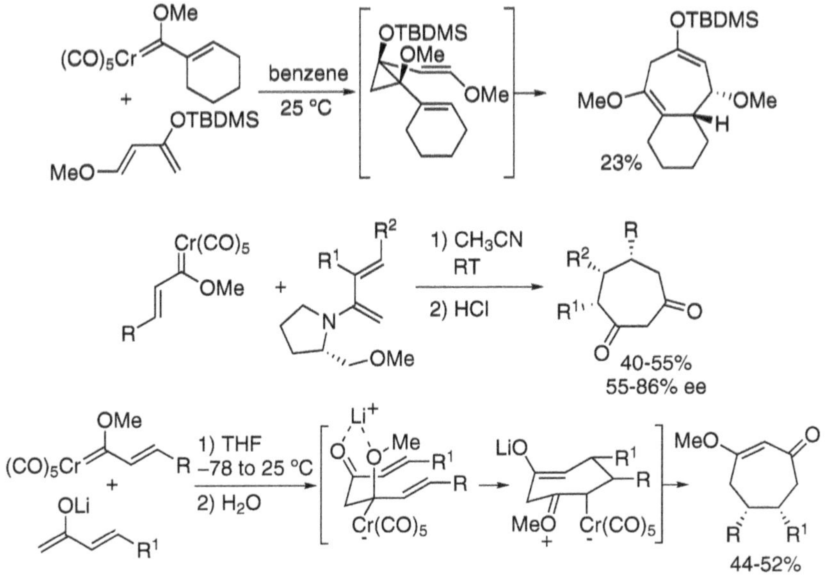

Scheme 65

Seven-membered carbocycles are also available from the reaction of alkenylcarbene complexes of chromium and lithium enolates derived from methyl vinyl ketones [79b] (Scheme 65). In this case, the reaction is initiated by the 1,2-addition of the enolate to the carbene complex. Cyclisation induced by a [1,2]-migration of the pentacarbonylchromium group and subsequent elimination of the metal fragment followed by hydrolysis leads to the final cycloheptenone derivatives (Scheme 65).

[4_S+3_C] Heterocyclisations have been successfully effected starting from 4-amino-1-azadiene derivatives. The cycloaddition of reactive 4-amino-1-aza-1,3-butadienes towards alkenylcarbene complexes goes to completion in THF at a temperature as low as −40 °C to produce substituted 4,5-dihydro-3*H*-azepines in 52–91% yield [115] (Scheme 66). Monitoring the reaction by NMR allowed various intermediates to be determined and the reaction course outlined in Scheme 66 to be established. This mechanism features the following points in the chemistry of Fischer carbene complexes: (i) the reaction is initiated at −78 °C by nucleophilic 1,2-addition and (ii) the key step cyclisation is triggered by a [1,2]-$W(CO)_5$ shift.

A chiral version of this [4+3] heterocyclisation was achieved using chiral, non-racemic carbene complexes derived from menthol and oximes as depicted in Scheme 67 [115]. This reaction requires the use of one equivalent of another simple carbene complex in order to remove the oxygen of the oxime functionality at some point during the reaction process. Significantly, the major diastereoisomer crystallises readily from methanol, allowing the isolation of the azepine in enantiomerically pure form.

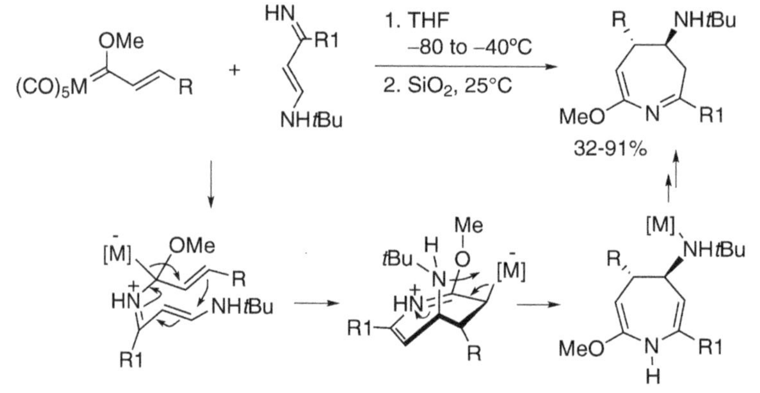

Scheme 66 [M] = $(CO)_5M$, M= Cr, W

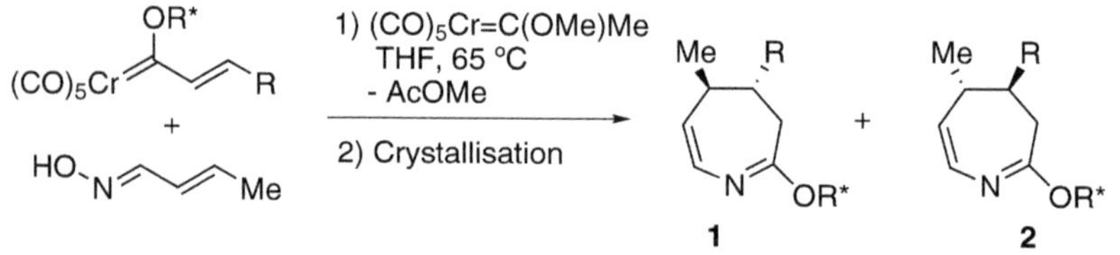

	HOR*	R	Yield	1 + 2 Ratio	Isolated Product	Isolated Yield (d.e.)
a	(-)-Menthol	Ph	87	70:30	**1a**	50% (>97%)
b	(-)-Menthol	2-Furyl	90	72:28	**1b**	48% (>97%)
c	(+)-Menthol	Ph	80	30:70	**2c**	46% (>97%)

Scheme 67

2.14.2 Alkynylcarbene Complexes as C3 Building Blocks

Tungsten alkynyl Fischer carbene complexes are excellent dienophile partners in the classical Diels–Alder reaction with 1-azadienes (see Sect. 2.9.2.1). On the contrary, the chromium-derived complexes exhibit a different behaviour and they react through a $[4_S+3_C]$ heterocyclisation reaction to furnish azepine derivatives [116] (Scheme 68). The reaction is initiated by a 1,2-addition of the nitrogen lone pair to the carbene carbon followed by a [1,2]-$Cr(CO)_5$ shift-promoted cyclisation which generates a metalated zwitterionic intermediate. Interestingly, this intermediate crystallises and its structure could be determined unambiguously by X-ray analysis.

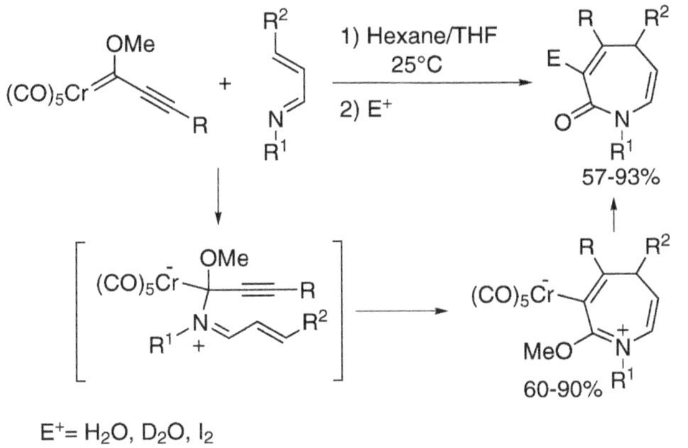

Scheme 68 $E^+ = H_2O, D_2O, I_2$

In a related process, alkynylcarbene complexes react with imines derived from *N*-unsubstituted pyrrole-2-carboxaldehyde to furnish zwitterionic pyrrolodiazepine derivatives through a formal $[4_S+3_C]$ heterocyclisation reaction [91]. Although the imines involved in these reactions resemble the 1-azadienes described in the last paragraph, the mechanism of the process is different. Also, it has been shown how the corresponding *N*-substituted pyrrole derivatives led to [3+3] cycloadducts (see Sect. 2.8.2). In this case the reaction is initiated by an NH Michael-type addition to the carbene complex followed by an intramolecular 1,2-addition of the imine nitrogen to generate a zwitterionic intermediate. Finally, a [1,3]-migration of the metal fragment leads to the final products (Scheme 69).

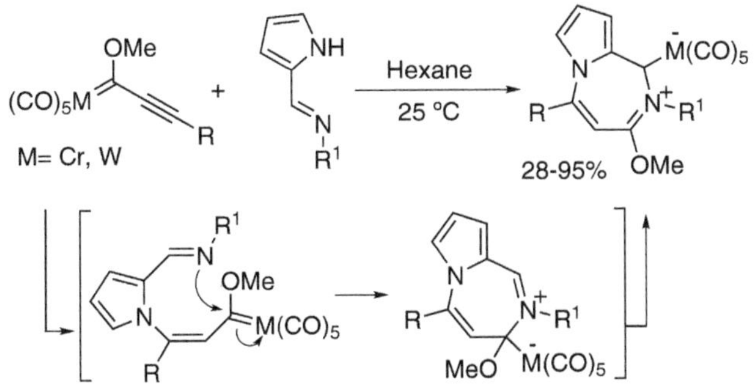

Scheme 69

The cyclopropanation of fulvenes has been effected with alkynylcarbene complexes (see Sect. 2.1.1). However, this reaction is inhibited in the presence of CO and under these conditions a formal $[4_S+3_C]$ cycloadduct is formed [15a]

(Scheme 70). The formation of these products likely involves two key steps: (i) 1,2-addition of fulvene to the carbene carbon and (ii) regioselective cyclisation promoted by [1,2]-$W(CO)_5$ shift.

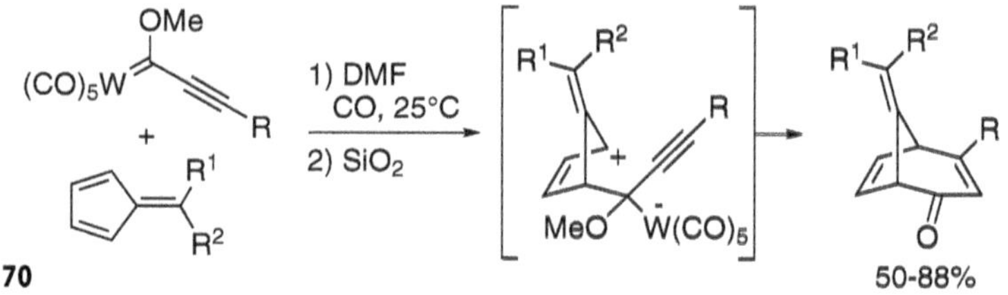

Scheme 70

2.15
[6_S+2_C] Cycloaddition Reactions

Aumann et al. have observed an unusual formal [6_S+2_C] cycloaddition reaction when they performed the reaction between an alkynylcarbene complex and 1-aminobenzocyclohexenes. The solvent used in this reaction exerts a crucial influence on the reaction course and products of different nature are obtained depending on the solvent chosen. However, in pentane this process leads to cyclooctadienylcarbene complexes in a reaction which can be formally seen as a [6_S+2_C] cycloaddition [117] (Scheme 71). The formation of these compounds is explained by an initial [2+2] cycloaddition reaction which leads to a cyclobutenylcarbene derivative which, under the reaction conditions, undergoes a cyclobutene ring opening to furnish the final products.

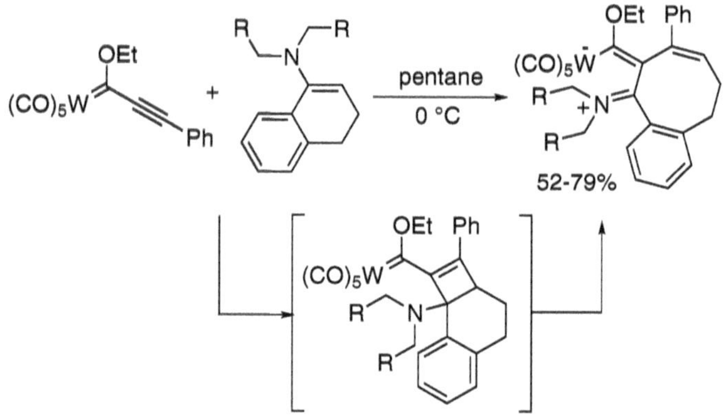

Scheme 71

2.16
$[6_S+3_C]$ Cycloaddition Reactions

The unconventional structure of fulvenes with a unique C=C bond conjugation leads to unusual cycloaddition reactions with other unsaturated systems. For example, alkenylcarbene complexes react with fulvenes leading to indanone or indene derivatives which can be considered as derived from a $[6_S+3_C]$ cycloaddition process [118] (Scheme 72). The reaction pathway is well explained by an initial 1,2-addition of the fulvene to the carbene carbon followed by [1,2]-$Cr(CO)_5$-promoted cyclisation.

Scheme 72 X= OMe, OCH_2CH_2I, $N(CH_2)_4$, NMe_2

3
Three-Component Cycloaddition Reactions

3.1
$[2_S+2_{S'}+1_C]$ Cycloaddition Reactions

The reaction of methyl acrylate and acrylonitrile with pentacarbonyl[(*N*,*N*-dimethylamino)methylene]chromium generates trisubstituted cyclopentanes through a formal $[2_S+2_S+1_C]$ cycloaddition reaction, where two molecules of the olefin and one molecule of the carbene complex have been incorporated into the structure of the cyclopentane [17b] (Scheme 73). The mechanism of this reaction implies a double insertion of two molecules of the olefin into the carbene complex followed by a reductive elimination.

Iwasawa et al. also developed a new reaction involving a three-component coupling process which affords five-membered heterocycles. This $[2_S+2_{S'}+1_C]$ cycloaddition reaction supposes the consecutive addition of an alkynyllithium derivative to a Fischer carbene complex followed by the addition of a third component which can be an aldehyde, an imine, an isocyanate, or CO_2 [119] (Scheme 74).

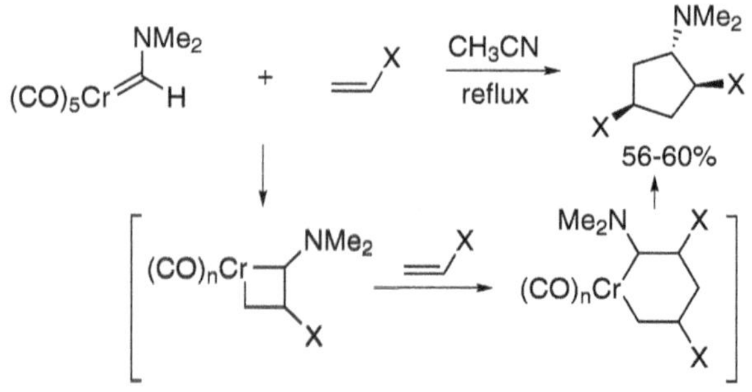

Scheme 73 X= CO_2Me, CN

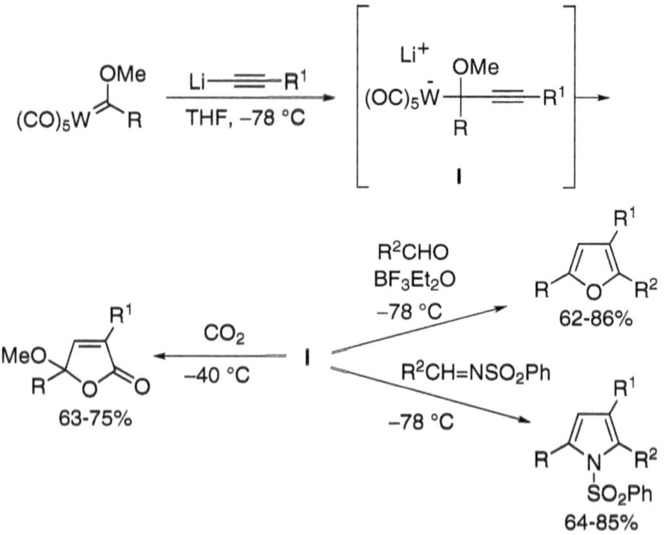

Scheme 74

Highly substituted cyclopentanols are diastereoselectively obtained by the successive reaction of chromium carbene complexes with β-substituted lithium enolates and then with allylmagnesium bromide [120]. The ring skeleton of the cyclopentanols combines the carbene ligand, the enolate framework and two carbons of the allyl unit. The mechanism that accounts for the formation of this $[2_S+2_{S'}+1_C]$ cycloadduct involves initial 1,2-addition of the lithium enolate to the carbene complex which generates a lithium 1-methoxy-3-oxoalkyl pentacarbonylchromate intermediate. Subsequent addition of the organomagnesium reagent to the corresponding ketone functional group produces a 5-hexenylchromate intermediate which undergoes an intramolecular carbometalation reaction to give, after hydrolysis, the final cyclopentanol derivatives (Scheme 75).

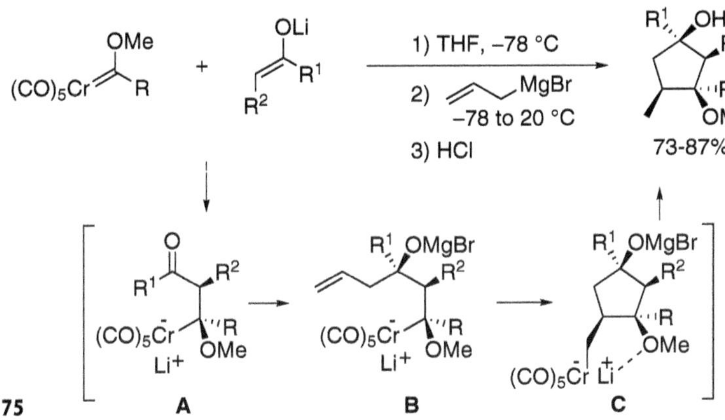

Scheme 75

3.2
[2_C+2_S+1_{CO}] Cycloaddition Reactions

The reactions of aminocarbene complexes with alkynes were widely investigated by Rudler et al. Thus, the reaction of these complexes and diphenylacetylene in refluxing benzene leads to formal [2_C+2_S+1_{CO}] cycloaddition products. The reaction implies the consecutive insertion of the alkyne into the carbene complex followed by insertion of a carbonyl ligand and finally production of ylide derivatives [121] (Scheme 76). These ylide complexes undergo, upon moderate heating, rearrangement as a result of a nitrogen-to-carbon migration of an alkyl group. Oxidation of the ylide complexes with dimethyldioxirane leads to new lactame complexes.

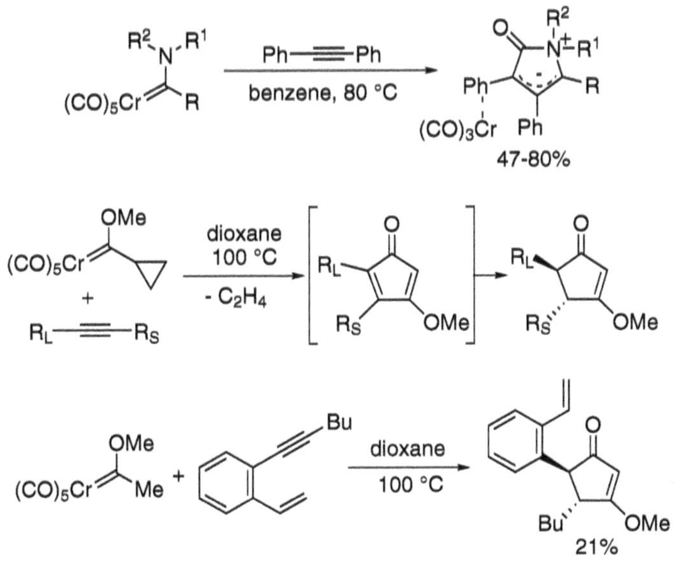

Scheme 76

Other examples of $[2_C+2_S+1_{CO}]$ cycloaddition reactions have been described by Herndon et al. by the use of chromium cyclopropyl(methoxy)carbenes. These complexes react with alkynes releasing ethene and forming cyclopentadienone derivatives, which evolve to cyclopentenone derivatives in the presence of chromium(0) and water [122] (Scheme 76). This reaction has been extended to intramolecular processes and also to the synthesis of some natural products [123]. These authors have also described another process involving a formal $[2_C+2_S+1_{CO}]$ cycloaddition reaction. Thus, the reaction of methyl and cyclopropylcarbene complexes with phenylacetylene derivatives does not afford the expected benzannulated products, and several regioisomers of cyclopentenone derivatives are the only products isolated [124] (Scheme 76).

3.3 $[3_C+2_S+2_S]$ Cycloaddition Reactions

The reaction of alkenylcarbene complexes and alkynes in the presence of Ni(0) leads to cycloheptatriene derivatives in a process which can be considered as a $[3_C+2_S+2_S]$ cycloaddition reaction [125]. As shown in Scheme 77, two molecules of the alkyne and one molecule of the carbene complex are involved in the formation of the cycloheptatriene. This reaction is supposed to proceed through the initial formation of a nickel alkenylcarbene complex. A subsequent double regioselective alkyne insertion produces a new nickel carbene complex, which evolves by an intramolecular cyclopropanation reaction to form a norcaradiene intermediate. These species easily isomerise to the observed cycloheptatriene derivatives (Scheme 77).

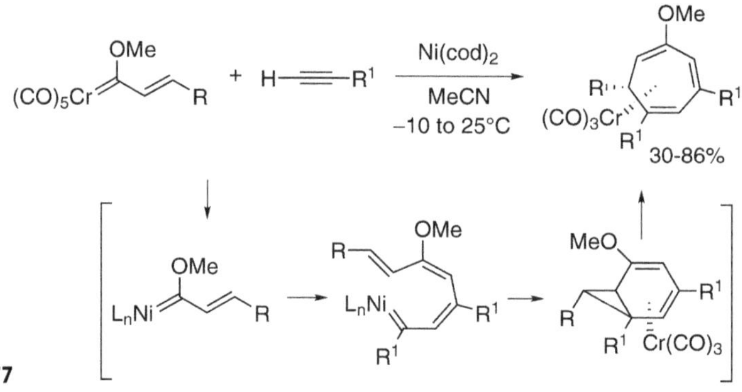

OMe
$(CO)_5Cr$ R + H—≡—R^1 $Ni(cod)_2$ MeCN −10 to 25°C
OMe R R^1 $(CO)_3Cr$ R^1 30-86%
OMe L_nNi R → R OMe L_nNi R^1 R^1 → MeO R R^1 R^1 $Cr(CO)_3$

Scheme 77

3.4 $[4_C+2_S+1_{CO}]$ Cycloaddition Reactions

Chromium cyclopropylcarbene complexes react with alkynes to provide cyclopentenone derivatives in a formal $[2_C+2_S+1_{CO}]$ cycloaddition process (see Sect. 3.2). However, tungsten and molybdenum cyclopropylcarbene complexes react with alkynes to afford cycloheptadienone derivatives in a sequence which can be considered as a $[4_C+2_S+1_{CO}]$ cycloaddition reaction [126] (Scheme 78). Interestingly, this reaction can be directed to one or another diastereoisomer simply by changing the metal (W or Mo) of the starting carbene complex. The mechanism of this reaction starts with the insertion of the alkyne into the carbene complex to generate a new non-heteroatom-stabilised carbene. From here, two possible pathways can be envisaged, which differ in their timing of CO insertion vs. cyclopropane ring opening steps. The first option resembles the mechanism of the Dötz reaction, and thus the insertion of CO leads to a vinylketene derivative which then evolves by oxidative addition into a cyclopropane C–C bond followed by reductive elimination. The second option implies an initial ring opening of the cyclopropyl group to generate a new complex, which then inserts CO to generate the same intermediate as before and finally produces the cycloheptenone derivatives by reductive elimination (Scheme 78).

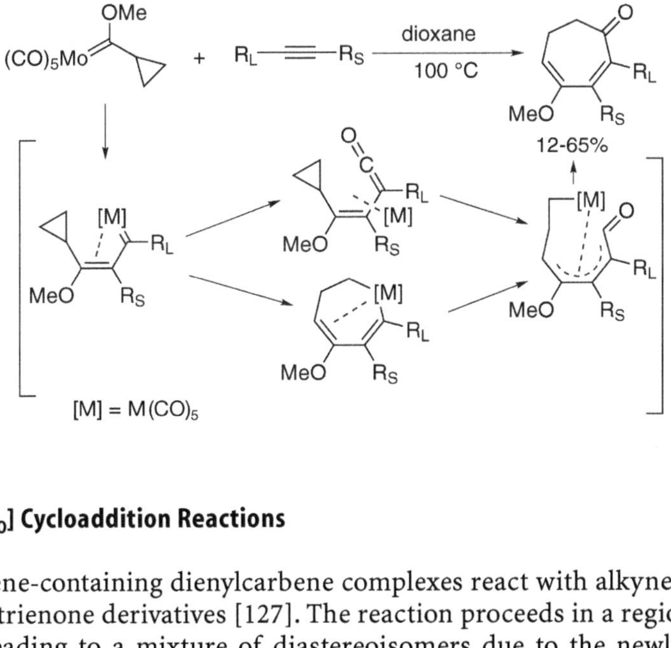

OMe
$(CO)_5Mo$ + R_L—≡—R_S dioxane 100 °C
MeO R_S R_L O
12-65%
[M] R_L MeO R_S
O C R_L [M] MeO R_S
[M] R_L MeO R_S
[M] O R_L MeO R_S
$[M] = M(CO)_5$

Scheme 78

3.5
$[5_C+2_S+1_{CO}]$ Cycloaddition Reactions

Cyclobutene-containing dienylcarbene complexes react with alkynes to form cyclooctatrienone derivatives [127]. The reaction proceeds in a regioselective fashion leading to a mixture of diastereoisomers due to the newly created stereogenic centre (Scheme 79). This process can be viewed as a variation of the

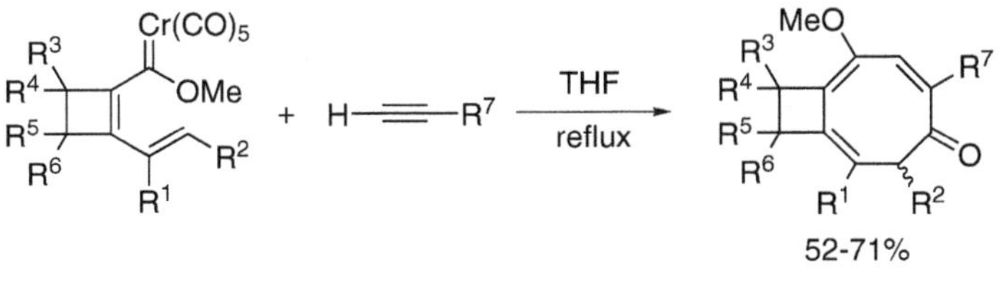

$Cr(CO)_5$ OMe R^3 R^4 R^5 R^6 R^1 R^2 + H—≡—R^7 THF reflux
MeO R^3 R^4 R^5 R^6 R^7 O R^1 R^2
52-71%

Scheme 79

Dötz reaction, since both an alkyne and CO are inserted. However, the additional double bond present in the starting complex participates in the subsequent electrocyclic ring closure, giving rise to eight-membered carbocycles.

4 Four-Component Cycloaddition Reactions

4.1 $[2_S+2_{S'}+1_C+1_{CO}]$ Cycloaddition Reactions

Aryl- and alkenylcarbene complexes are known to react with alkynes through a $[3_C+2_S+1_{CO}]$ cycloaddition reaction to produce benzannulated compounds. This reaction, known as the "Dötz reaction", is widely reviewed in Chap. "Chromium-Templated Benzannulation Reactions", p. 123 of this book. However, simple alkyl-substituted carbene complexes react with excess of an alkyne (or with diynes) to produce a different benzannulated product which incorporates in its structure two molecules of the alkyne, a carbon monoxide ligand and the carbene carbon [128]. As referred to before, this $[2_S+2_{S'}+1_C+1_{CO}]$ cycloaddition reaction can be carried out with diyne derivatives, showing these reactions give better yields than the corresponding intermolecular version (Scheme 80).

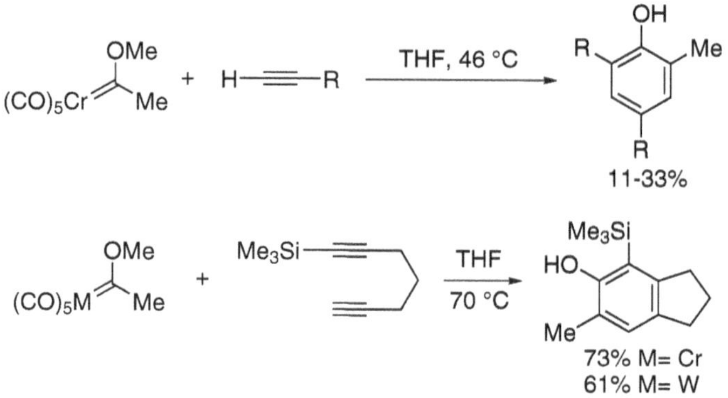

OMe (CO)5Cr Me + H—≡—R THF, 46 °C → OH R Me R 11-33%

OMe (CO)5M Me + Me3Si THF 70 °C → Me3Si HO Me 73% M= Cr 61% M= W

Scheme 80

Another example of a $[2_S+2_{S'}+1_C+1_{CO}]$ cycloaddition reaction was observed by Barluenga et al. in the sequential coupling reaction of a Fischer carbene complex, a ketone enolate and allylmagnesium bromide [120]. This reaction produces cyclopentanol derivatives in a $[2_S+2_{S'}+1_C]$ cycloaddition process when β-substituted lithium enolates are used (see Sect. 3.1). However, the analogous reaction with β-unsubstituted lithium enolates leads to the diastereoselective synthesis of 1,3,3,5-tetrasubstituted cyclohexane-1,4-diols. The ring skeleton of these compounds combines the carbene ligand, the enolate framework, two carbons of the allyl unit and a carbonyl ligand. Overall, the process can be considered as a formal $[2_S+2_{S'}+1_C+1_{CO}]$ cycloaddition reaction (Scheme 81). A plausible explanation for the formation of these cyclohexanediol derivatives involves initial 1,2-addition of the lithium enolate to the carbene complex to generate a lithium 1-methoxy-3-oxoalkyl pentacarbonylchromate intermediate. Subsequent addition of the organomagnesium reagent to the corresponding ketone functional group produces a 5-hexenylchromate derivative, which undergoes migratory insertion of carbon monoxide to provide a lithium acyl tetracarbonylchromate intermediate. These species lead to the final 5-methylenecyclohexane-1,4-diols after intramolecular insertion of the carbene carbon atom into the secondary vinylic C–H bond and subsequent protonation (Scheme 81).

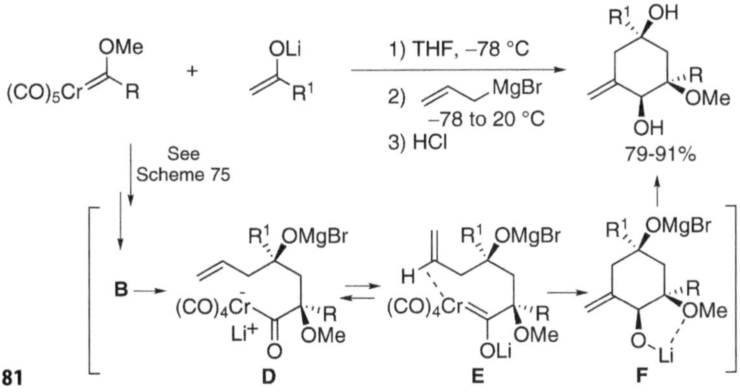

Scheme 81

4.2 $[2_S+2_S+2_S+1_C]$ Cycloaddition Reactions

It has been shown how alkenylcarbene complexes participate in nickel(0)-mediated $[3_C+2_S+2_S]$ cycloaddition reactions to give cycloheptatriene derivatives (see Sect. 3.3). However, the analogous reaction performed with alkyl- or arylcarbene complexes leads to similar cycloheptatriene derivatives, but in this case the process can be considered a $[2_S+2_S+2_S+1_C]$ cycloaddition reaction as three molecules of the alkyne and one molecule of the carbene complex are incorporated into the structure of the final product [125] (Scheme 82). The mechanism of this transformation is similar to that described in Scheme 77 for the $[3_C+2_S+2_S]$ cycloaddition reactions.

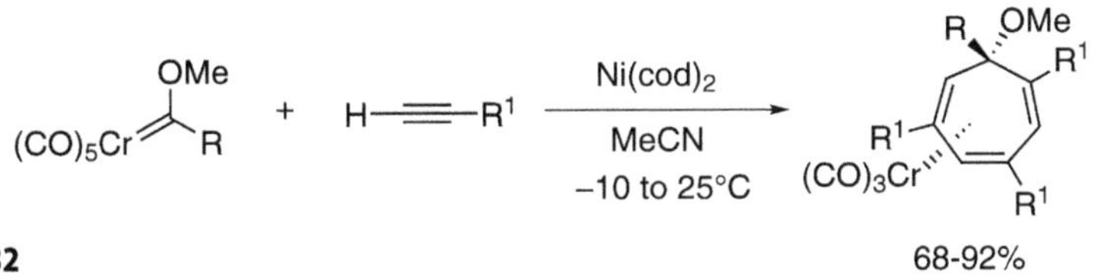

Scheme 82

5
Tandem Cycloaddition Reactions

In recent years the strategic use of tandem reactions has been well recognised as a powerful method for increasing molecular complexity and thereby synthetic efficiency [129]. In the field of Fischer carbene complexes this strategy has also been widely applied to the synthesis of complex structures. Some of these sequences have been mentioned in previous sections within this chapter (for example [2+2]/[5_C+1_S] [36a,c], [3_C+2_S]/[5_C+1_S] [47a] and [4_S+2_C]/[5_C+1_S] [111]). Other interesting tandem sequences are those involving an initial [4_S+2_C] cycloaddition followed by several intramolecular cyclisations [104a, 130]. In the present section we would like to summarise only a few recent examples of tandem cycloaddition processes involving Fischer carbene complexes, which are intended to highlight the incredible potential of these complexes to give access to complex structures from simple starting materials.

5.1
[3_C+3_S]/[2_S+1_C] Sequences

β-Oxygen-functionalised sp^3 organolithium compounds react with alkenylcarbene complexes to generate the corresponding cyclic carbene complexes in a formal [3+3] process (see Sect. 2.8.1). In those cases where the organolithium derivative contains a double bond in an appropriate position, tricyclic ether derivatives are the only products isolated. These compounds derive from an intramolecular cyclopropanation of the corresponding cyclic carbene complex intermediate [89] (Scheme 83).

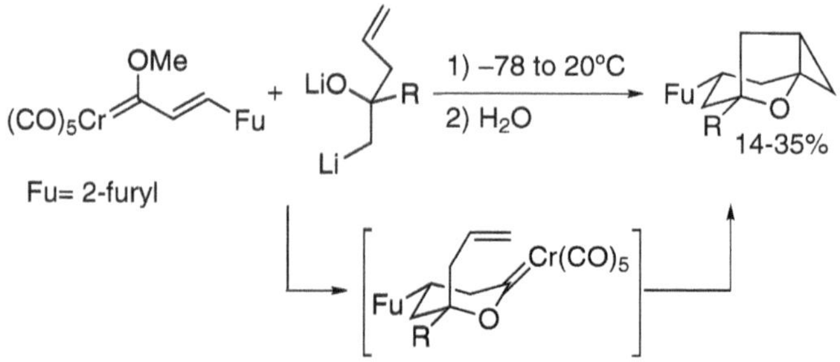

OMe
$(CO)_5Cr$ Fu
+ LiO R Li
1) –78 to 20°C
2) H_2O
Fu R O
14-35%
Fu= 2-furyl
Fu R O $Cr(CO)_5$

Scheme 83

5.2
[4_S+2_C]/[2_S+1_C] Sequences

Aumann et al. have described the synthesis of biscarbene complexes by the reaction of 1-alkylimidates with two equivalents of a tungsten alkynylcarbene complex [131]. An initial [4_S+2_C] cycloaddition generates an intermediate which further reacts with a second molecule of the alkynylcarbene complex through a [2+2] cycloaddition to produce the final azabicyclo[4.2.0]octa-3,7-diene biscarbene derivatives (Scheme 84).

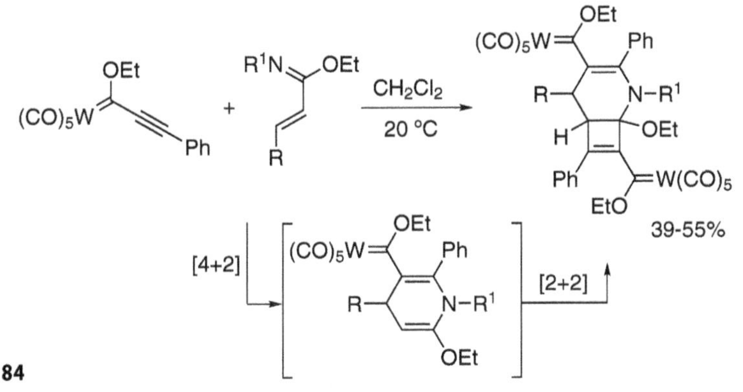

Scheme 84

5.3 $[2_C+2_S+1_{CO}]/[2_S+1_C]$ Sequences

Alkynylcarbene complexes react with strained and hindered olefins yielding products that incorporate up to four different components by the formation of five new carbon–carbon bonds [15b]. This remarkable transformation is explained by an initial [2+2] cycloaddition followed by CO insertion. The resulting intermediate suffers a well precedented [1,3]-migration of the metal fragment to generate a non-heteroatom-stabilised carbene complex intermediate which reacts with a new molecule of the olefin through a cyclopropanation reaction (Scheme 85).

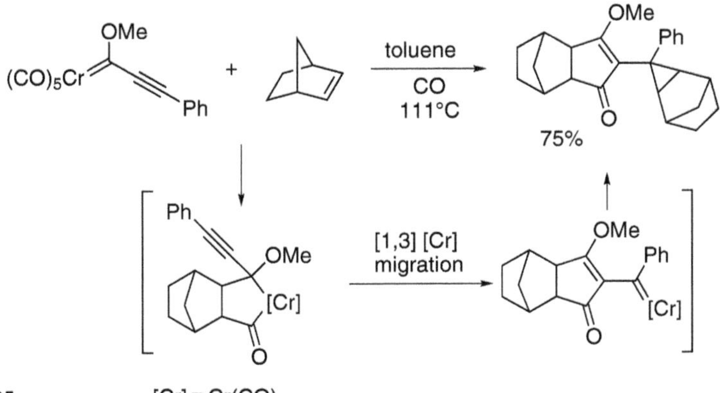

Scheme 85

References

1. (a) Wulff WD (1995) Transition metal carbene complexes: alkyne and vinyl ketene chemistry. In: Abel EW, Stone FGA, Wilkinson G (eds) Comprehensive organometallic chemistry II, vol 12. Pergamon, Oxford, p 469; (b) Wulff WD (1991) Metal-carbene cycloadditions. In: Trost BM, Fleming I (eds) Comprehensive organic synthesis, vol 5. Pergamon, New York, p 1065; (c) Dötz KH, Fischer H, Hofmann P, Kreissl FR, Schubert U, Weiss K (1983) Transition metal carbene complexes. Verlag Chemie, Weinheim
2. (a) Dömling A, Ugi I (2000) Angew Chem Int Ed 39:3168; (b) Bienaymé H, Hulme C, Oddon G, Schmidt P (2000) Chem Eur J 6:3321; (c) Weber L, Illgen K, Almstetter M (1999) Synlett 366; (d) Posner GH (1986) Chem Rev 86:831
3. Frühauf HW (1997) Chem Rev 97:523
4. Reviews: (a) Harvey DF, Sigano DM (1996) Chem Rev 96:271; (b) Wulff WD, Yang DC, Murray CK (1988) Pure Appl Chem 60:137; (c) Brookhart M, Studabaker WB (1987) Chem Rev 87:411
5. Fischer EO, Dötz KH (1970) Chem Ber 103:1273
6. (a) Herndon JW, Tumer SU (1991) J Org Chem 56:286; (b) Harvey DF, Brown MF (1990) Tetrahedron Lett 31:2529; (c) Wienand A, Reissig HU (1990) Organometallics 9:3133; (d) Herndon JW, Tumer SU (1989) Tetrahedron Lett 30:4771; (e) Wienand A, Reissig HU (1988) Tetrahedron Lett 29:2315; (f) Dötz KH, Fischer EO (1972) Chem Ber 105:1356
7. (a) Wienand A, Reissig HU (1990) Angew Chem Int Ed Engl 29:1129; (b) Cooke MD, Fischer EO (1973) J Organomet Chem 56:279
8. (a) Barluenga J, Tomás M, López-Pelegrín JA, Rubio E (1995) J Chem Soc Chem Commun 665; (b) Wienand A, Reissig HU (1991) Chem Ber 124:957
9. Casey CP, Cesa MC (1982) Organometallics 1:87
10. (a) Murray CK, Yang DC, Wulff WD (1990) J Am Chem Soc 112:5660; (b) Dorrer B, Fischer EO, Kalbfus W (1974) J Organomet Chem 81:C20; (c) Fischer EO, Dötz KH (1972) Chem Ber 105:3966. For an intramolecular version of this reaction, see: (d) Casey CP, Hornung NL, Kosar WP (1987) J Am Chem Soc 109:4908
11. For cyclopropanation of enol ethers with in situ-generated acyloxycarbene complexes of chromium and in the absence of CO, see reference [10a]
12. A zwitterionic compound intermediate has been isolated: (a) Barluenga J, Ballesteros A, Bernardo de la Rúa R, Santamaría J, Rubio E, Tomás M (2003) J Am Chem Soc 125:1834; (b) Barluenga J, Ballesteros A, Santamaría J, Bernardo de la Rúa R, Rubio E, Tomás M (2000) J Am Chem Soc 122:12874
13. (a) Barluenga J, Aznar F, Gutiérrez I, Martín JA (2002) Org Lett 4:2719; (b) Söderberg BC, Hegedus LS (1990) Organometallics 9:3113; (c) Casey CP, Shusterman AJ (1985) Organometallics 4:736; (d) Casey CP, Vollendorf NW, Haller KJ (1984) J Am Chem Soc 106:3754; (e) Toledano CA, Rudler H, Daran JC, Jeannin Y (1984) J Chem Soc Chem Commun 574
14. (a) Barluenga J, López S, Trabanco AA, Flórez J (2001) Chem Eur J 7:4723; (b) Barluenga J, López S, Trabanco AA, Fernández-Acebes A, Flórez J (2000) J Am Chem Soc 122:8145; (c) Barluenga J, Fernández-Acebes A, Trabanco AA, Flórez J (1997) J Am Chem Soc 119:7591
15. (a) Barluenga J, Martínez S, Suárez-Sobrino AL, Tomás M (2002) J Am Chem Soc 124:5948; (b) Barluenga J, Fernández-Rodríguez MA, Andina F, Aguilar E (2002) J Am Chem Soc 124:10978
16. (a) Barluenga J, Aznar F, Martín A (1995) Organometallics 14:1429; (b) Sierra MA, Söderberg BC, Lander PA, Hegedus LS (1993) Organometallics 12:3769
17. Merino I, Hegedus LS (1995) Organometallics 14:2522

18. (a) Barluenga J, Aznar F, Gutiérrez I, García-Granda S, Llorca-Baragaño MA (2002) Org Lett 4:4273. For two isolated examples of cyclopropanation of an electron-defficient olefin with iminocarbene complexes, see: (b) Campos PJ, Soldevilla A, Sampedro D, Rodríguez MA (2001) Org Lett 3:4087; (c) Aumann R, Heinen H, Krüger C, Betz P (1990) Chem Ber 123:605
19. (a) Rudler H, Audouin M, Parlier A, Martín-Vaca B, Goumont R, Durand-Réville T, Vaissermann J (1996) J Am Chem Soc 118:12045; (b) Casey CP, Polichnowski SW, Shustermann AJ, Jones CR (1979) J Am Chem Soc 101:7282; (c) Casey CP, Tuinstra HE, Saeman MC (1976) J Am Chem Soc 98:608; (d) Casey CP, Burkhardt TJ (1974) J Am Chem Soc 96:7808
20. See for instance: (a) Harvey DF, Sigano DM (1996) J Org Chem 61:2268; (b) Hoye TR, Vyvyan JR (1995) J Org Chem 60:4184; (b) Hoye TR, Suriano JA (1992) Organometallics 11:2044; (c) Harvey DF, Lund KP, Neil DA (1992) J Am Chem Soc 114:8424; (d) Harvey DF, Brown MF (1990) J Am Chem Soc 112:7806; (e) Parlier A, Rudler H, Platzer N, Fontanille M, Soum A (1987) J Chem Soc Dalton Trans 1041; (f) Parlier A, Rudler H, Yefsah R, Álvarez C (1987) J Organomet Chem 328:C21
21. (a) Barluenga J, Muñiz K, Ballesteros A, Martínez S, Tomás M (2002) Arkivoc 110; (b) Cooke MD, Fischer EO (1973) J Organomet Chem 56:279
22. Barluenga J, Suárez-Sobrino AL, Tomás M, García-Granda S (2001) J Am Chem Soc 123:10494
23. (a) Pfeiffer J, Nieger M, Dötz KH (1998) Eur J Org Chem 1011; (b) Pfeiffer J, Dötz KH (1997) Angew Chem Int Ed Engl 36:2828; (c) Miki K, Nishino F, Ohe K, Uemura S (2002) J Am Chem Soc 124:5260
24. (a) Buchert M, Hoffmann M, Reissig HU (1995) Chem Ber 128:605; (b) Buchert M, Reissig HU (1992) Chem Ber 125:2723; (c) Buchert M, Reissig HU (1988) Tetrahedron Lett 29:2319
25. (a) Barluenga J, Aznar F, Fernández M (1997) Chem Eur J 3:1629; (b) Wulff WD, Yang DC, Murray CK (1988) J Am Chem Soc 110:2653
26. (a) Takeda K, Okamoto Y, Nakajima A, Yoshii E, Koizumi T (1997) Synlett 1181; (b) Takeda K, Sakamura K, Yoshii E (1997) Tetrahedron Lett 38:3257
27. (a) Harvey DF, Lund KP (1991) J Am Chem Soc 113:8916. See also references [6a,d] and: (b) Merlic CA, Bendorf HD (1994) Tetrahedron Lett 35:9529; (c) Söderberg BC, Hegedus LS, Sierra MA (1990) J Am Chem Soc 112:4364
28. For intramolecular cyclopropanation of simple 1,3-carbodienes with acyloxycarbene complexes see reference [13a]
29. Intermolecular process: (a) Fischer EO, Hofmann J (1991) Chem Ber 124:981. Intramolecular process: (b) Harvey DF, Lund KP (1991) J Am Chem Soc 113:5066
30. (a) Alcaide B, Casarrubios L, Domínguez G, Retamosa A, Sierra MA (1996) Tetrahedron 52:13215; (b) Alcaide B, Casarrubios L, Domínguez G, Sierra MA (1994) Organometallics 13:2934
31. (a) Barluenga J, Bernad PL Jr, Concellón JM, Piñera-Nicolás A, García-Granda S (1997) J Org Chem 62:6870; (b) Barluenga J, Bernad PL Jr, Concellón JM (1995) Tetrahedron Lett 36:3937
32. (a) Baldwin JE (1991) Thermal cyclobutane ring formation. In: Trost BM, Fleming I (eds) Comprehensive organic synthesis, vol 5. Pergamon, New York, p 63; (b) de Meijere A (ed) (1997) Houben-Weyl: Methods of organic chemistry; carbocyclic four-membered ring compounds, 4th edn, vol 17e. Georg Thieme, Stuttgart
33. (a) Nicolau KC, Hwang HC, Duggan ME, Reddy KB (1988) Tetrahedron Lett 29:1501; (b) Gollwick K, Fries S (1980) Angew Chem Int Ed Engl 19:832; (c) Doyle TW (1970) Can J Chem 48:1633

34. Wulff WD, Faron KL, Su J, Springer JP, Rheingold AL (1999) J Chem Soc Perkin Trans I 197
35. Faron KL, Wulff WD (1988) J Am Chem Soc 110:8727
36. (a) Barluenga J, Aznar F, Palomero MA (2002) Chem Eur J 8:4149; (b) Wu HP, Aumann R, Frölich R, Wibbeling B (2000) J Org Chem 65:1183; (c) Barluenga J, Aznar F, Palomero MA, Barluenga S (1999) Org Lett 1:541; (d) Camps F, Jordi L, Moretó JM, Ricart S, Castaño AM, Echavarren AM (1992) J Organomet Chem 436:189; (e) Pipoh R, Eldik R, Wang SLB, Wulff WD (1992) Organometallics 11:490; (f) Merlic CA, Xu D (1991) J Am Chem Soc 113:7418; (g) Faron KL, Wulff WD (1990) J Am Chem Soc 112:6419; (h) de Meijere A, Wessjohann L (1990) Synlett 20; (i) Camps F, Llebaría A, Moretó JM, Ricart S, Viñas JM (1990) Tetrahedron Lett 31:2479
37. (a) Aumann R, Zhengkun Y, Frölich R (1998) Organometallics 17:2897; (b) Aumann R, Hildmann B, Frölich R (1998) Organometallics 17:1197
38. Dötz KH, Koch AW, Weyershausen B, Hupfer H, Nieger M (2000) Tetrahedron 56:4925
39. Aumann R, Heinen H, Hinterding P, Sträter N, Krebs B (1991) Chem Ber 124:1229
40. Aumann R, Vogt D, Fu X, Fröhlich R, Schwab P (2002) Organometallics 21:1637
41. Zora M, Herndon JW (1994) Organometallics 13:3370
42. (a) Padwa A (1991) Intermolecular 1,3-dipolar cycloadditions. In: Trost BM, Fleming I (eds) Comprehensive organic synthesis, vol 4. Pergamon, New York, p 1069; (b) Little RD (1991) Thermal cycloadditions. In: Trost BM, Fleming I (eds) Comprehensive organic synthesis, vol 5. Pergamon, New York, p 239
43. Alcaide B, Casarrubios L, Domínguez G, Sierra MA (1998) Curr Org Chem 2:551
44. Lattuada L, Licandro E, Maiorana S, Molinari H, Papagni A (1991) Organometallics 10:807
45. Kreissl FR, Fischer EO, Kreiter CG (1973) J Organomet Chem 57:C9
46. Chan KS, Wulff WD (1986) J Am Chem Soc 108:5229
47. (a) Barluenga J, Aznar F, Palomero MA (2001) Chem Eur J 7:5318; (b) Chan KS, Yeung ML, Li WK, Liu HK, Wang Y (1998) J Org Chem 63:7670; (c) Chan KS, Yeung ML, Chan WK, Wang RJ, Mak TCW (1995) J Org Chem 60:1741; (d) Chan KS (1991) J Chem Soc Perkin Trans I 2602; (e) Kalinin VN, Shilova OS, Kovredov AI, Petrovskii PV, Batsanov AS, Struchkov YT (1989) Organomet Chem USSR 2:268
48. Jung IY, Yoon YJ, Rhee KS, Shin GC, Shin SC (1994) Chem Lett 859
49. (a) Merlic CA, Baur A, Aldrich CC (2000) J Am Chem Soc 122:7398; (b) Choi YH, Kang BS, Yoon YJ, Kim J, Shin SC (1995) Synth Commun 25:2043
50. (a) Baldoli C, Del Buttero P, Licandro E, Maiorana S, Papagni A, Zanotti-Gerosa A (1994) J Organomet Chem 476:C27; (b) Licandro E, Maiorana S, Papagni A, Zanotti-Gerosa A, Cariati F, Bruni S, Moret M, Chiesi-Villa A (1994) Inorg Chim Acta 220:233
51. (a) Barluenga J, Fernández-Marí F, Aguilar E, Viado AL, Olano B, García-Granda S, Moya-Rubiera C (1999) Chem Eur J 5:883; (b) Barluenga J, Fernández-Marí F, Aguilar E, Viado AL, Olano B (1997) J Chem Soc Perkin Trans I 2267
52. (a) Barluenga J, Fernández-Marí F, González R, Aguilar E, Revelli GA, Viado AL, Fañanás FJ, Olano B (2000) Eur J Org Chem 1773; (b) Barluenga J, Fernández-Marí F, Aguilar E, Viado AL, Olano B (1998) Tetrahedron Lett 39:4887
53. Barluenga J, Fernández-Rodríguez MA, Aguilar E, Fernández-Marí F, Salinas A, Olano B (2001) Chem Eur J 7:3533
54. Merino I, Laxmi YRS, Flórez J, Barluenga J, Ezquerra J, Pedregal C (2002) J Org Chem 67:648
55. For a review, see: Herndon JW (2000) Tetrahedron 56:1257
56. (a) Dragisich V, Wulff WD, Hoogsteen K (1990) Organometallics 9:2867; (b) Aumann R, Heinen H, Krüger C, Betz P (1990) Chem Ber 123:599
57. Campos PJ, Sampedro D, Rodríguez MA (2003) J Org Chem 68:4674

58. Campos PJ, Sampedro D, Rodríguez MA (2000) Organometallics 19:3082
59. Campos PJ, Sampedro D, Rodríguez MA (2002) Tetrahedron Lett 43:73
60. Dötz KH, Dietz R, Kappenstein CK, Neugebauer D, Schubert U (1979) Chem Ber 112:3682
61. Bo ME, Wulff WD, Wilson KJ (1996) Chem Commun 1863
62. Dötz KH, Christoffers C (1995) Chem Ber 128:163
63. (a) Yamashita A (1986) Tetrahedron Lett 27:5915. See also: (b) Longen A, Nieger M, Airola K, Dötz KH (1998) Organometallics 17:1538
64. Yamashita A, Toy A, Watt W, Muchmore CR (1988) Tetrahedron Lett 29:3403
65. Grotjahn DB, Kroll FEK, Schäfer T, Harms K, Dötz KH (1992) Organometallics 11:298
66. Aumann R, Jasper B, Frölich R (1995) Organometallics 14:2447
67. (a) Aumann R, Kössmeier M, Jäntti A (1998) Synlett 1120; (b) Aumann R, Meyer AG, Frölich R (1996) Organometallics 15:5018; (c) Meyer AG, Aumann R (1995) Synlett 1011
68. Aumann R, Yu Z, Frölich R, Zippel F (1998) Eur J Inorg Chem 1623
69. Sierra MA, Ramírez-López P, Gómez-Gallego M, Lejon T, Mancheño MJ (2002) Angew Chem Int Ed 41:3442
70. Aumann R, Frölich R, Zippel F (1997) Organometallics 16:2571
71. (a) de Meijere A, Schirmer H, Duetsch M (2000) Angew Chem Int Ed 39:3964; (b) Barluenga J, López LA, Martínez S, Tomás M (2000) Tetrahedron 56:4967; (c) Wulff WD, Bax BM, Brandvold TA, Chan KS, Gilbert AM, Hsung RP (1994) Organometallics 13:102
72. (a) Barluenga J, Tomás M, Suárez-Sobrino AL (2000) Synthesis 935; (b) Hoffmann M, Reissig HU (1995) Synlett 625; (c) see also reference [8b]
73. (a) Hoffmann M, Buchert M, Reissig HU (1999) Chem Eur J 5:876; (b) Hoffmann M, Buchert M, Reissig HU (1997) Angew Chem Int Ed Engl 36:283
74. (a) Barluenga J, López S, Flórez J (2003) Angew Chem Int Ed 42:231; (b) Zaragoza Dörwald F (2003) Angew Chem Int Ed 42:1332
75. Barluenga J, Ballesteros A, Santamaría J, Tomás M (2002) J Organomet Chem 643–644:363
76. (a) Kagoshima H, Okamura T, Akiyama T (2001) J Am Chem Soc 123:7182; (b) Kagoshima H, Akiyama T (2000) J Am Chem Soc 122:11741
77. Barluenga J, Tomás M, Ballesteros A, Santamaría J, Brillet C, García-Granda S, Piñera-Nicolás A, Vázquez JT (1999) J Am Chem Soc 121:4516
78. Aumann R, Heinen H, Dartmann M, Krebs B (1991) Chem Ber 124:2343
79. (a) Barluenga J, Alonso J, Fañanás FJ (2003) J Am Chem Soc 125:2610; (b) Barluenga J, Alonso J, Rodríguez F, Fañanás FJ (2000) Angew Chem Int Ed 39:2460
80. Aumann R, Heinen H (1986) Chem Ber 119:3801
81. For an example of a rhodium carbenoid mediated [4_S+1_C] cycloaddition, see: Schnaubelt J, Marks E, Reissig HU (1996) Chem Ber 129:73
82. Small amounts of cyclopentene derivatives are detected in cyclopropanation reactions of electron-deficient dienes, but they may result from thermal rearrangement of the corresponding vinyl cyclopropanes and not from a direct [4+1] cycloaddition
83. Zora M, Herndon JW (1993) Organometallics 12:248
84. (a) Barluenga J, Tomás M, Ballesteros A, Santamaría J, Suárez-Sobrino A (1997) J Org Chem 62:9229; (b) Fischer EO, Weiss K, Burger K (1973) Chem Ber 106:1581; (c) Danks TN, Velo-Rego D (1994) Tetrahedron Lett 35:9443
85. For a work where [4+1] cycloaddition products are obtained by the use of an electron-poor diene, see reference [17b]
86. Lautens M, Klute W, Tam W (1996) Chem Rev 96:49
87. (a) Wang SLB, Wulff WD (1990) J Am Chem Soc 112:4550. See also: (b) Aumann R, Meyer AG, Frölich R (1996) J Am Chem Soc 118:10853
88. Polo R, Moretó JM, Schick U, Ricart S (1998) Organometallics 17:2135
89. Barluenga J, Monserrat JM, Flórez J (1993) J Chem Soc Chem Commun 1068

90. (a) Merlic CA, Xu D, Nguyen MC, Truong V (1993) Tetrahedron Lett 34:227; (b) Merlic CA, Xu D (1991) J Am Chem Soc 113:9855
91. Barluenga J, Tomás M, Rubio E, López-Pelegrín JA, García-Granda S, Pérez-Priede M (1999) J Am Chem Soc 121:3065
92. Camps F, Moretó JM, Ricart S, Viñas JM, Molins E, Miravitlles C (1989) J Chem Soc Chem Commun 1560
93. (a) Jordi L, Camps F, Ricart S, Viñas JM, Moretó JM, Mejias M, Molins E (1995) J Organomet Chem 494:53; (b) Jordi L, Moretó JM, Ricart S, Viñas JM, Molins E, Miravitlles C (1993) J Organomet Chem 444:C28; (c) Camps F, Jordi L, Moretó JM, Ricart S, Castaño AM, Echavarren AM (1992) J Organomet Chem 436:189
94. (a) Aumann R, Kössmeier M, Roths K, Frölich R (2000) Tetrahedron 56:4935; (b) Aumann R, Roths K, Frölich R (1997) Organometallics 16:5893; (c) Aumann R, Roths K, Läge M, Krebs B (1993) Synlett 667; (d) Aumann R, Roths K, Grehl M (1993) Synlett 669
95. (a) Wulff WD, Bauta WE, Kaesler RW, Lankford PJ, Miller RA, Murray CK, Yang DC (1990) J Am Chem Soc 112:3642; (b) Wulff WD, Yang DC (1983) J Am Chem Soc 105:6726
96. (a) Dötz KH, Christoffers J (1995) Chem Ber 128:157; (b) Adam H, Albrecht T, Sauer J (1994) Tetrahedron Lett 35:557; (c) Dötz KH, Kuhn W, Müller G, Huber B, Alt HG (1986) Angew Chem Int Ed Engl 25:812
97. Wulff WD (1998) Organometallics 17:3116
98. (a) Barluenga J, Aznar F, Martín A, Barluenga S (1997) Tetrahedron 53:9323; (b) Barluenga J, Aznar F, Martín A, Barluenga S, García-Granda S, Paneque-Quevedo AA (1994) J Chem Soc Chem Commun 843. For a related work see also reference [51]
99. Barluenga J, Aznar F, Barluenga S, García-Granda S, Álvarez-Rúa C (1997) Synlett 1040. For a related work, see: Weyershausen B, Nieger M, Dötz KH (1998) Organometallics 17:1602
100. Sabat M, Reynolds KA, Finn MG (1994) Organometallics 13:2084
101. (a) Barluenga J, Canteli RM, Flórez J, García-Granda S, Gutiérrez-Rodríguez A, Martín E (1998) J Am Chem Soc 120:2514; (b) Barluenga J, Canteli RM, Flórez J, García-Granda S, Gutiérrez-Rodríguez A (1994) J Am Chem Soc 116:6949
102. Anderson BA, Wulff WD, Powers TS, Tribbit S, Rheingold AL (1992) J Am Chem Soc 114:10784
103. Powers TS, Jiang W, Su J, Wulff WD (1997) J Am Chem Soc 119:6438
104. (a) Barluenga J, Aznar F, Barluenga S, Fernández M, Martín A, García-Granda S, Piñera-Nicolás A (1998) Chem Eur J 4:2280; (b) Kuhn W, Dötz KH (1985) J Organomet Chem 286:C23; (c) Wulff WD, Yang DC (1984) J Am Chem Soc 106:7565
105. Barluenga J, Tomás M, López-Pelegrín JA, Rubio E (1997) Tetrahedron Lett 38:3981
106. Rahm A, Wulff WD (1993) Organometallics 12:597
107. Rahm A, Rheingold AL, Wulff WD (2000) Tetrahedron 56:4951
108. (a) Dötz KH, Noack R, Harms K, Müller G (1990) Tetrahedron 46:1235; (b) Wulff WD, Tang PC, Chan KS, McCallun JS, Yang DC, Gilbertson SR (1985) Tetrahedron 41:5813
109. (a) Müller G, Jas G (1992) Tetrahedron Lett 33:4417; (b) Wulff WD, Powers TS (1993) J Org Chem 58:2381
110. Huy NHT, Mathey F (1988) Organometallics 7:2233
111. (a) Merlic CA, McInnes DM, You Y (1997) Tetrahedron Lett 38:6787; (b) Merlic CA, Xu D, Gladstone BG (1993) J Org Chem 58:538; (c) Merlic CA, Roberts WM (1993) Tetrahedron Lett 34:7379; (d) Merlic CA, Burns EE, Xu D, Chen SY (1992) J Am Chem Soc 114:8722; (e) Merlic CA, Xu D (1991) J Am Chem Soc 113:7418
112. Huy NHT, Mathey F, Ricard L (1988) Tetrahedron Lett 29:4289

113. (a) Herndon JW, Zhang Y, Wang H, Wang K (2000) Tetrahedron Lett 41:8687; (b) Herndon JW, Zhang Y, Wang H (1998) J Org Chem 63:4562
114. Barluenga J, Aznar F, Martín A, Vázquez JT (1995) J Am Chem Soc 117:9419
115. Barluenga J, Tomás M, Ballesteros A, Santamaría J, Carbajo RJ, López-Ortiz F, García-Granda S, Pertierra P (1996) Chem Eur J 2:88
116. Barluenga J, Tomás M, Rubio E, López-Pelegrín JA, García-Granda S, Pertierra P (1996) J Am Chem Soc 118:695
117. Aumann R, Kössmeier M, Mück-Lichtenfeld C, Zippel F (2000) Eur J Org Chem 37
118. J Barluenga, Martínez S, Suárez-Sobrino AL, Tomás M (2001) J Am Chem Soc 123:11113
119. (a) Iwasawa N, Ochiai T, Maeyama K (1998) J Org Chem 63:3164; (b) Iwasawa N, Ochiai T, Maeyama K (1997) Organometallics 16:5137; (c) Iwasawa N, Maeyama K (1997) J Org Chem 62:1918; (d) Iwasawa N, Maeyama K, Saitou M (1997) J Am Chem Soc 119:1486. For a related process, see: (e) Barluenga J, Trabanco AA, Flórez J, García-Granda S, Llorca MA (1998) J Am Chem Soc 120:12129
120. Barluenga J, Pérez-Sánchez I, Rubio E, Flórez J (2003) Angew Chem Int Ed 42:5860
121. (a) Rudler H, Parlier A, Rudler M, Vaissermann J (1998) J Organomet Chem 567:101; (b) Bouancheau C, Rudler M, Chelain E, Rudler H, Vaissermann J, Daran J-C (1995) J Organomet Chem 496:127; (c) Bouancheau C, Parlier A, Rudler M, Rudler H, Vaissermann J, Daran J-C (1994) Organometallics 13:4708; (d) Chelain E, Goumont R, Hamon L, Parlier A, Rudler M, Rudler H, Daran J-C, Vaissermann J (1992) J Am Chem Soc 114:8088
122. Tumer SU, Herndon JW, McMullen LA (1992) J Am Chem Soc 114:8394
123. For some examples, see: (a) Herndon JW, Zhu J (1999) Org Lett 1:15; (b) Yan J, Zhu J, Matasi JJ, Herndon JW (1999) J Org Chem 64:1291; (c) Matasi JJ, Yan J, Herndon JW (1999) Inorg Chim Acta 296:273; (d) Yan J, Herndon JW (1998) J Org Chem 63:2325
124. Jackson TJ, Herndon JW (2001) Tetrahedron 57:3859
125. Barluenga J, Barrio P, López LA, Tomás M, García-Granda S, Álvarez-Rúa C (2003) Angew Chem Int Ed 42:3008
126. (a) Herndon JW, Zora M, Patel PP, Chatterjee G, Matasi JJ, Tumer SU (1993) Tetrahedron 49:5507; (b) Herndon JW, Zora M (1993) Synlett 363; (c) Herndon JW, Chatterjee G, Patel PP, Matasi JJ, Tumer SU, Harp JJ, Reid MD (1991) J Am Chem Soc 113:7808
127. Barluenga J, Aznar F, Palomero MA (2000) Angew Chem Int Ed 39:4346
128. Wulff WD, Kaesler RW, Peterson GA, Tang P-C (1985) J Am Chem Soc 107:1060
129. Tietze LF, Haunert F (2000) Domino reactions in organic synthesis. An approach to efficiency, elegance, ecological benefit, economic advantage and preservation of our resources in chemical transformations. In: Vögtle F, Stoddart JF, Shibasaki M (eds) Stimulating concepts in chemistry. Wiley-VCH, Weinheim, p 39
130. Bao J, Dragisich V, Wenglowsky S, Wulff WD (1991) J Am Chem Soc 113:9873
131. Aumann R, Hildmann B, Fröhlich R (1998) Organometallics 17:1197

Topics Organomet Chem (2004) 13: 123–156
DOI 10.1007/978-3-540-40910-6

Chromium-Templated Benzannulation Reactions

Ana Minatti · Karl H. Dötz (✉)

Kekulé-Institut für Organische Chemie und Biochemie, Rheinische Friedrich-Wilhelms Universität Bonn, Gerhard-Domagk-Strasse 1, 53121 Bonn, Germany
ana.minatti@gmx.de; doetz@uni-bonn.de

Abstract Since its discovery the chromium-mediated benzannulation reaction has been developed into a unique and useful tool in organic synthesis. In this review, topical aspects of this reaction concerning its mechanism and the chemo-, regio- and stereoselectivity are summerised and discussed in detail. Special attention is paid to the asymmetric benzannulation reaction and, finally, the importance of this reaction as a key step in the total synthesis of natural products is outlined.

Keywords Fischer carbene complex · [3+2+1]-benzannulation reaction · Asymmetric benzannulation · Linear benzannulation

Abbreviations

Ac	Acetyl
Bn	Benzyl
n-Bu	*n*-Butyl
t-Bu	*t*-Butyl
CAN	Ceric ammonium nitrate
cod	Cyclooctadiene
Cp	Cyclopentadienyl
de	Diastereomer excess
DEAD	Diethyl azodicarboxylate
dr	Diastereomer ratio
ee	Enantiomer excess
Et	Ethyl
h	Hour(s)
kcal	Kilocalories
mol	Mole
NMR	Nuclear magnetic resonance
Me	Methyl
Ph	Phenyl
i-Pr	*iso*-Propyl
n-Pr	*n*-Propyl
rac	Racemic
rt	Room temperature
S	Solvent
TBDMSCl	*tert*-Butyldimethylsilyl
TBME	*tert*-Butyl methyl ether
Tf	Trifluoromethanesulphonyl
THF	Tetrahydrofuran
TIPS	Triisopropylsilyl

1 Introduction

The thermal [3+2+1]-benzannulation reaction of α,β-unsaturated Fischer carbene complexes with alkynes was discovered in 1975 in our laboratory along with the reaction of methoxy(phenyl)carbene chromium complex **1** upon gentle warming with tolane [1] (Scheme 1). It established the potential of an organometallic template in the stereocontrolled assembly of different ligands and their activation for C–C bond formation at a low-valent metal centre. This unique type of metal carbene reaction provides one of the most powerful tools to generate densely substituted benzenoid compounds. Within the [3+2+1]-benzannulation the concept of atom economy is convincingly preserved as this type of reaction represents a highly efficient one-pot procedure.

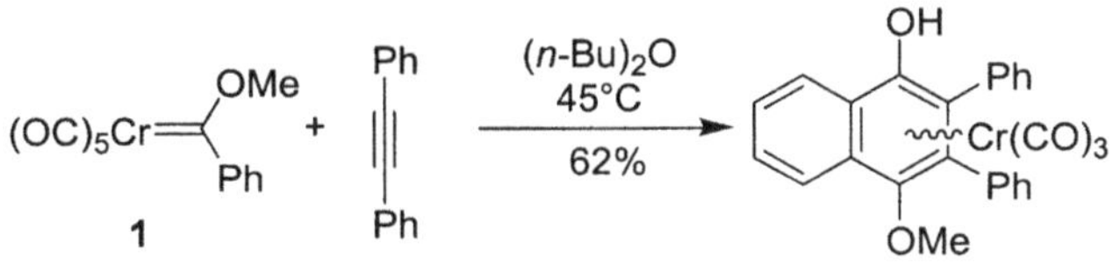

Scheme 1 [3+2+1]-Benzannulation reaction as the first example of a metal-templated coupling of three different ligands

The formal [3+2+1]-cycloaddition involves an α,β-unsaturated carbene ligand (C3-synthon), an alkyne (C2-synthon) and a carbonyl ligand (C1-synthon) and takes place within the coordination sphere of the chromium(0), which acts as a metal template (Scheme 2).

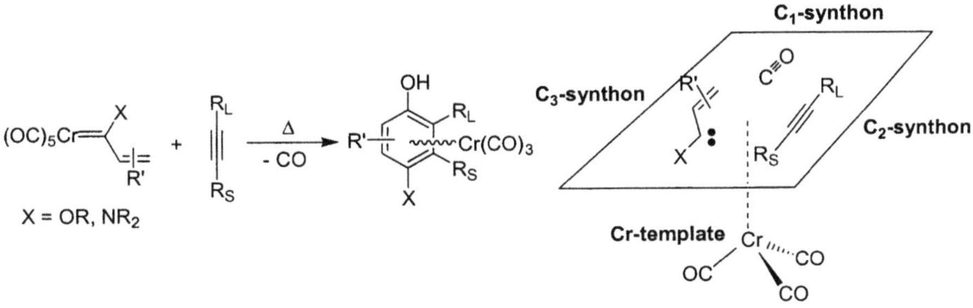

Scheme 2 Atom connectivity in the [3+2+1]-benzannulation reaction

2 Mechanism

2.1 Phenol Formation

Nearly 25 years after its discovery the mechanism of the benzannulation reaction has been theoretically and experimentally elucidated in detail. The most predominant outcome of this reaction is the formation of the 4-methoxyphenol or 4-methoxy-1-naphthol skeleton coordinated to a $Cr(CO)_3$ fragment. Therefore the mechanism leading to this type of product will be discussed first.

The first and rate-determining step involves carbon monoxide dissociation from the initial pentacarbonyl carbene complex **A** to yield the coordinatively unsaturated tetracarbonyl carbene complex **B** (Scheme 3). The decarbonylation and consequently the benzannulation reaction may be induced thermally, photochemically [2], sonochemically [3], or even under microwave-assisted conditions [4]. A detailed kinetic study by Dötz et al. proved that the initial reaction step proceeds via a reversible dissociative mechanism [5]. More recently, density functional studies on the preactivation scenario by Solà et al. tried to propose alkyne addition as the first step [6], but it was shown that this

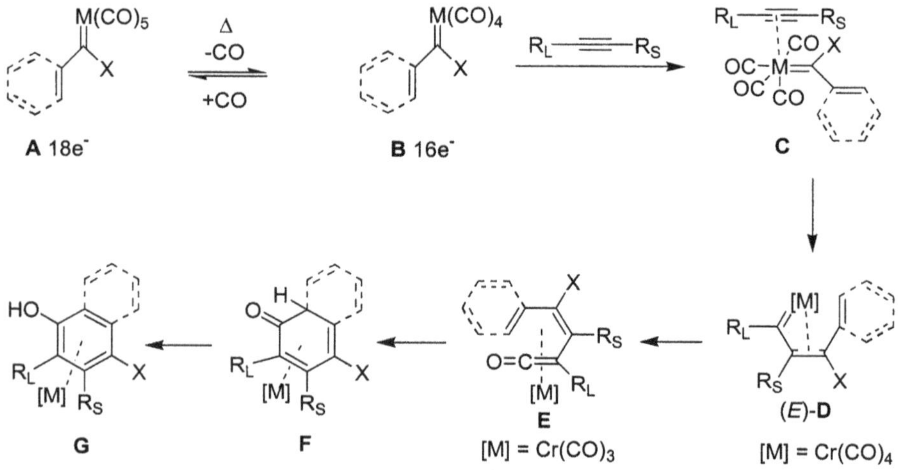

Scheme 3 Mechanism of the benzannulation reaction

associative sequence does not agree with the available experimental data [7]. A (η^1:η^3)-vinylcarbene complex analogue **3**, corresponding to the first reaction intermediate **B** of the benzannulation reaction, has been isolated and characterised [8]. The tetracarbonyl carbene complex **3** was generated upon heating **2** under reflux in tetrahydrofuran in the absence of any alkyne, and the reversibility of this dissociative process was proven by reisolating the starting compound after bubbling CO into the solution at room temperature (Scheme 4). The subsequent step in the benzannulation reaction involves the trapping of the coordinatively unsaturated 16e complex by the alkyne present in the solution to yield **C**. A structural analogue **4** displaying an intramolecular alkyne coordination has been characterised by X-ray analysis [9]. In spite of this promising isolation of an intramolecular alkyne carbene chromium chelate **4**, the expected benzannulation reaction did not take place after heating this formal intermediate. Instead, a formal dimerisation of the carbene complex yielding **5** was observed (Scheme 4).

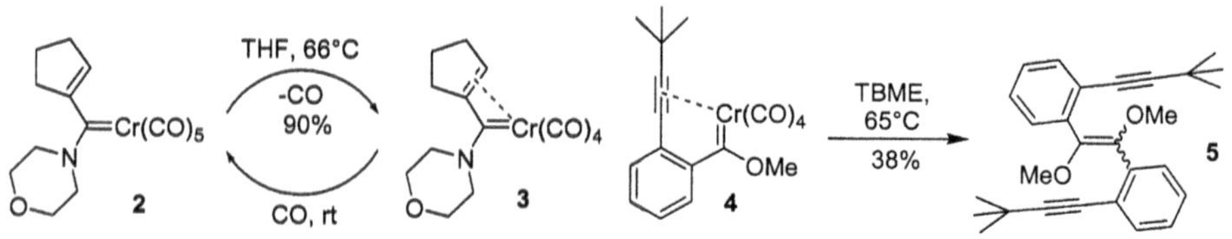

Scheme 4 Isolated intermediates

The subsequent insertion of the alkyne into the metal–carbene bond affords the (η^1:η^3)-vinylcarbene complex **D**, which may exist either as a (*Z*)- or an (*E*)-metallatriene. This intermediate may be considered as a branching point in the benzannulation reaction as three diverging routes starting from this point have been explored.

According to Dötz the η^3-vinylcarbene complex **D** forms a η^4-vinylketene **E** by CO insertion into the chromium–carbene bond, followed by electrocyclic ring closure to yield the η^4-cyclohexadienone **F** [10]. Extended Hückel molecular orbital and recent quantum chemical calculations support the Dötz route [11]. A modification of this theory was proposed by Solà, as recent DFT studies showed that an η^1-coordination of the vinylcarbene **D** is energetically favoured, which subsequently allows formation of a chromahexatriene intermediate through structural rearrangement involving a π-coordination of the terminal C=C bond. This intermediate can be regarded as a five-membered chelate ring if the midpoint of the coordinated double bond is taken as one ring member. A subsequent insertion of a CO ligand was proposed to give the η^4-cyclohexadienone **F** mentioned before [12]. It should be noted that the formation of the η^4-cyclohexadienone **F** via a chromacycloheptadienone was suggested by Casey, but this hypothesis was rejected due to thermodynamic arguments [13]. An exact validation of these mechanistic suggestions requires a benzannulation reaction along which all individual steps can be established by characterisation of the relevant intermediates. Although this goal has not been reached yet, Barluenga succeeded in realising a very similar project referred to as the "first stepwise benzannulation reaction" [14]. The reaction of tetracarbonyl carbene complex **6** with dimethyl acetylene dicarboxylate takes place at –20 °C and after 22 h yields the metallahexatriene **7** (Scheme 5). The coordination of the external double bond to the metal was proven by NMR spectroscopy. Although the decomposition of this compound **7** at room temperature did not lead to the expected phenol product, the metallahexatriene **8** yielded the phenol **9** in a completely selective fashion.

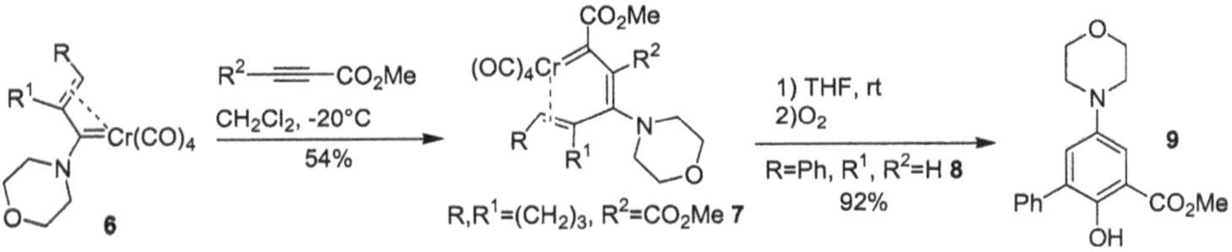

Scheme 5 Isolated intermediates

Structural analogues of the η^4-vinylketene **E** were isolated by Wulff, Rudler and Moser [15]. The enaminoketene complex **11** was obtained from an intramolecular reaction of the chromium pentacarbonyl carbene complex **10**. The silyl vinylketene **13** was isolated from the reaction of the methoxy(phenyl)-carbene chromium complex **1** and a silyl-substituted phenylacetylene **12**, and – in contrast to alkene carbene complex **7** – gave the benzannulation product **14** after heating to 165 °C in acetonitrile (Scheme 6). The last step of the benzannulation reaction is the tautomerisation of the η^4-cyclohexadienone **F** to afford the phenol product **G**. The existence of such an intermediate and its capacity to undergo a subsequent step was validated by Wulff, who synthesised an

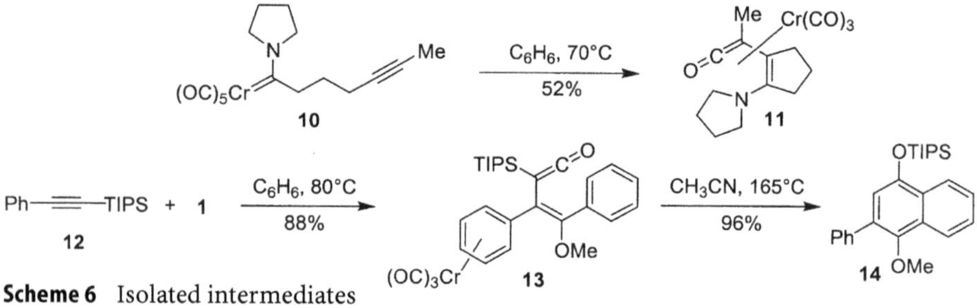

Scheme 6 Isolated intermediates

η^4-cyclohexadienone analogue **16** starting from the molybdenum carbene complex **15** and 3-hexyne. This complex tautomerises in tetrahydrofuran at 70 °C to yield the phenol product **17** [16] (Scheme 7).

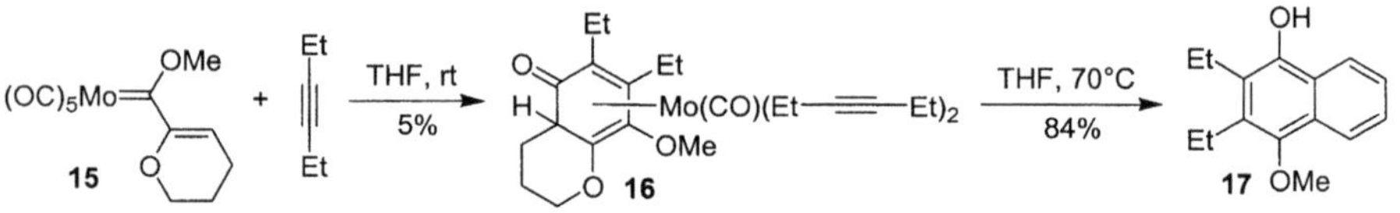

Scheme 7 Isolated intermediates

2.2 Furan and Indene Formation

Chemoselectivity plays an important role in the benzannulation reaction as five-membered rings such as indene or furan derivatives are potential side products. The branching point is again the η^3-vinylcarbene complex **D** intermediate which may be formed either as a (*Z*)- or an (*E*)-metallatriene; the (*E*)-configuration is required for the cyclisation with the terminal double bond. (*Z*)-Metallatriene **D**, however, leads to the formation of furan derivatives **H** (Scheme 8). Studies on the formation of (*E*)- and (*Z*)-isomers discussing stereoelectronic effects have been undertaken by Wulff [17].

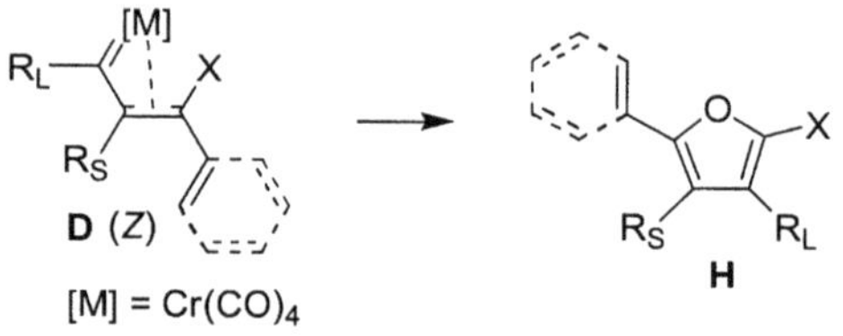

Scheme 8 Formation of furan products

For the indene derivatives **M** two different reaction pathways have been discussed so far, starting from the (*E*)-metallatriene **D**. A strongly coordinating solvent may induce an electrocyclic ring closure yielding the metallacyclohexadiene **K**, and the indene product is obtained after tautomerisation and reductive elimination of the metal centre [14]. The decisive point is the strong coordinating ability of the solvent molecule; non-coordinating solvents are unable to cleave the η^1- or η^3-coordination of the metal centre to the double bond in complex **D**. On the other hand, a direct electrocyclic ring closure may afford the cyclopentadiene **L** which tautomerises to the indene product [16] (Scheme 9).

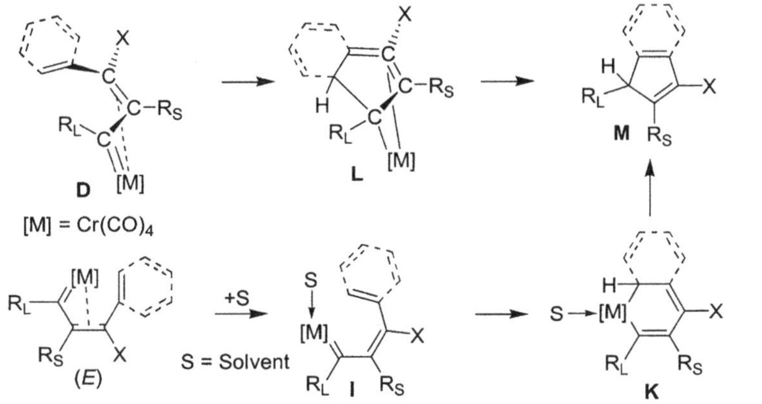

Scheme 9 Formation of indene products

2.3 Allochemical Effect

The distribution of products obtained from the benzannulation reaction may be influenced by the concentration of alkyne substrate [18]. In strongly coordinating solvents the ratio of the phenolic benzannulation product over five-membered cyclisation products increases with the concentration of the alkyne (Scheme 10).

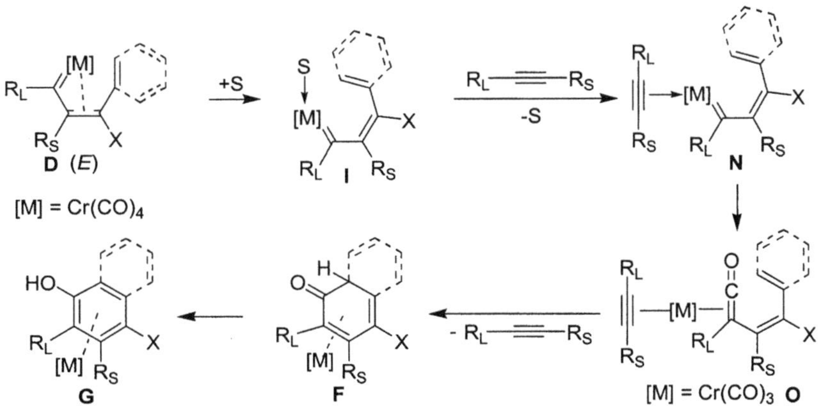

Scheme 10 The allochemical effect

This "allochemical effect" has been explained in terms of an accelerated CO insertion resulting from the coordination of the alkyne [19]. During the insertion of CO, the alkyne can switch from a 2e donor to a 4e donor, resulting in electronic saturation of the metal centre in the intermediate. The effect is distinctly reduced or even non-existent when the reaction is carried out in a non-coordinating solvent.

3
Trends in Chemoselectivity

The chemoselectivity of the [3+2+1]-benzannulation reaction is governed by four general trends:

1. A greatly enhanced chemoselective formation of phenol is observed for alkoxy(alkenyl)carbene complexes compared to alkoxy(aryl)carbene complexes. This behaviour reflects the ease of formation of the η^6-vinylketene complex intermediate **E** starting from alkenylcarbene complexes; for aryl complexes this transformation would require dearomatisation.
2. Phenol over indene formation is favoured in the order chromium>tungsten>molybdenum. The ability for CO insertion during the benzannulation reaction is expected to correlate with the strength of the metal–CO bond [20]. However, this correlation does not hold for molybdenum which is known for the kinetic lability of the Mo–CO bond, indicated by the well-established propensity of molybdenum carbonyl complexes to undergo ligand substitution at higher rates than their homologues [21]. In a coordinating solvent such as acetonitrile, molybdenum is very susceptible to the displacement of the double bond in complex **D** by a solvent molecule, which results in increased amounts of the indene product (Scheme 11).

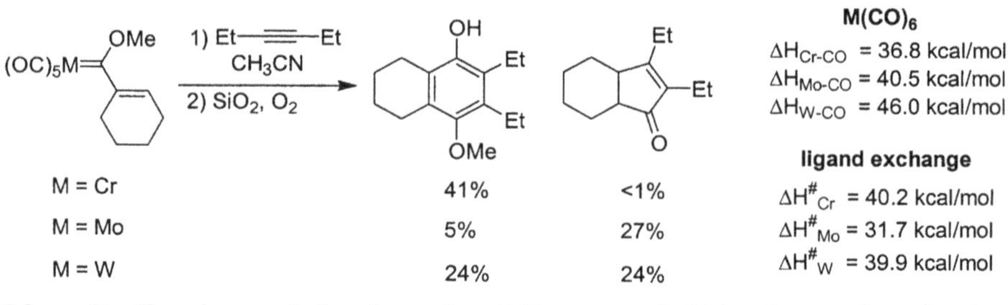

$(OC)_5M$=C(OMe)(cyclohexenyl) — 1) Et—≡—Et, CH_3CN; 2) SiO_2, O_2 →

	Phenol (OH, Et, Et, OMe)	Indenone (Et, Et, O)
M = Cr	41%	<1%
M = Mo	5%	27%
M = W	24%	24%

$M(CO)_6$
$\Delta H_{Cr\text{-}CO}$ = 36.8 kcal/mol
$\Delta H_{Mo\text{-}CO}$ = 40.5 kcal/mol
$\Delta H_{W\text{-}CO}$ = 46.0 kcal/mol

ligand exchange
$\Delta H^{\#}_{Cr}$ = 40.2 kcal/mol
$\Delta H^{\#}_{Mo}$ = 31.7 kcal/mol
$\Delta H^{\#}_{W}$ = 39.9 kcal/mol

Scheme 11 Phenol versus indene formation. ($\Delta H_{M\text{-}CO}$=metal–CO bond strength; $\Delta H^{\#}_{M}$ refers to substitution of CO for PR_3)

3. Phenol formation is favoured in less coordinating and/or less polar solvents; however, for clean reactions affording the $Cr(CO)_3$-coordinated benzannulation products, ethereal solvents are the solvents of choice.
4. Amino(aryl)carbene complexes prefer cyclopentannulation over benzannulation. Amino(alkenyl)carbene complexes may react in a benzannulation reaction.

The superior donor properties of amino groups over alkoxy substituents causes a higher electron density at the metal centre resulting in an increased M–CO bond strength in aminocarbene complexes. Therefore, the primary decarbonylation step requires harsher conditions; moreover, the CO insertion generating the ketene intermediate cannot compete successfully with a direct electrocyclisation of the alkyne insertion product, as shown in Scheme 9 for the formation of indenes. Due to that experience amino(aryl)carbene complexes are prone to undergo cyclopentannulation. If, however, the donor capacity of the aminocarbene ligand is reduced by *N*-acylation, benzannulation becomes feasible [22].

Wulff et al. examined the necessary reaction conditions for α,β-unsaturated aminocarbene complexes to react in a benzannulation reaction [23]. The reaction of dimethylamino(alkenyl)carbene complexes **18** with terminal alkynes in non-coordinating and non-polar solvents afforded phenol products in acceptable yields (Scheme 12).

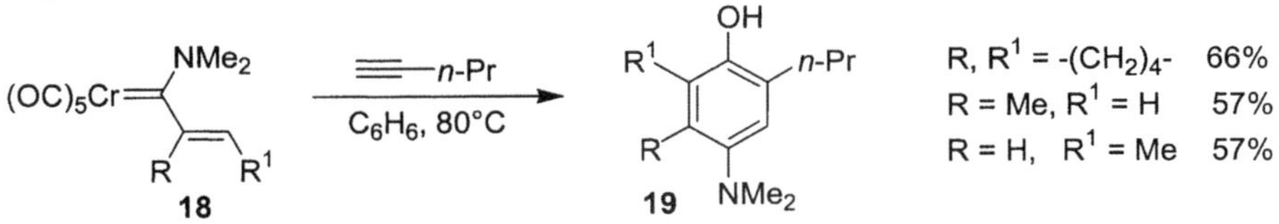

$(OC)_5Cr$=C(NMe$_2$)–CH=C(R)R^1 (**18**) + HC≡C–*n*-Pr, C_6H_6, 80°C → **19** (OH, NMe$_2$, *n*-Pr)

R, R^1 = -$(CH_2)_4$- 66%
R = Me, R^1 = H 57%
R = H, R^1 = Me 57%

Scheme 12 Benzannulation of alkenyl(dimethylamino)carbene complexes

If the dimethylamino group is exchanged for a pyrrolidino **20** or a morpholino moiety the choice of alkyne is not restricted any more, and both electron-rich and electron-poor terminal and internal alkynes may be applied to the benzannulation [24, 42b] (Scheme 13).

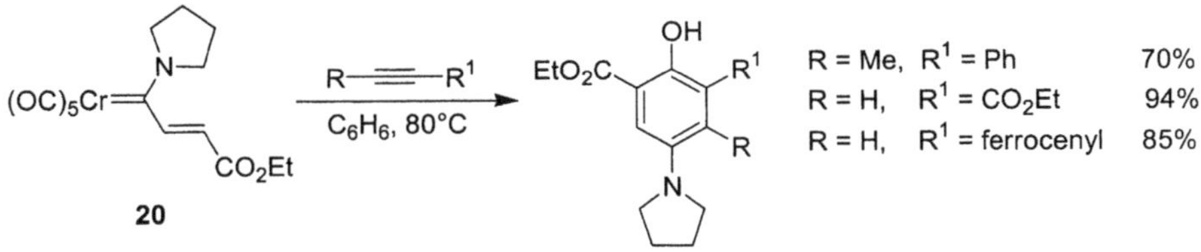

20 (CO_2Et) + R–≡–R^1, C_6H_6, 80°C → EtO_2C, OH, R^1, R, pyrrolidino-substituted arene

R = Me, R^1 = Ph 70%
R = H, R^1 = CO_2Et 94%
R = H, R^1 = ferrocenyl 85%

Scheme 13 Benzannulation of pyrrolidino(alkenyl)carbene complexes

4 Regioselectivity

When the benzannulation is carried out with unsymmetrical alkynes the major regioisomer generally bears the larger alkyne substituent (R_L) next to the phenolic group, suggesting that the regioselectivity is mainly governed by the difference in steric demands of the two alkyne substituents. A reversal of this regiochemistry may be achieved either by an intramolecular version of the benzannulation, where the alkyne is incorporated in the alkoxy chain [25], or by the use of stannyl acetylenes [26] and alkynyl boronates [27] (Scheme 14). There are two possible explanations for the inversion of the regioselectivity for the last two alkynes: (1) the installation of the electron-withdrawing metal centres far away from the electrophilic carbene carbon centre and (2) Lewis acid/base interactions [CO→M] in the η^3-metallatriene intermediate.

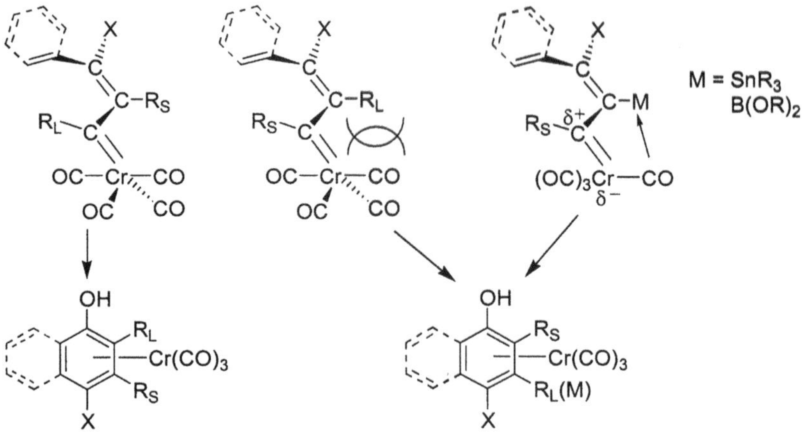

Scheme 14 Normal and inverse regioselectivity

An unexpected varying regiochemistry in intramolecular benzannulation has also been observed in the synthesis of cyclophanes. As mentioned above, there are only two possible regiochemical outcomes in the benzannulation reaction, which differ in the direction of alkyne incorporation. *β*-Tethered vinylcarbene chromium complexes undergo an intramolecular benzannulation reaction with incorporation of the tethered alkyne with normal regioselectivity to give *meta*-cyclophanes [28].

	p	m
n = 10, THF	8%	30%
n = 10, C_6H_6	40%	-
n = 16, THF	56%	1%
n = 16, C_6H_6	48%	-

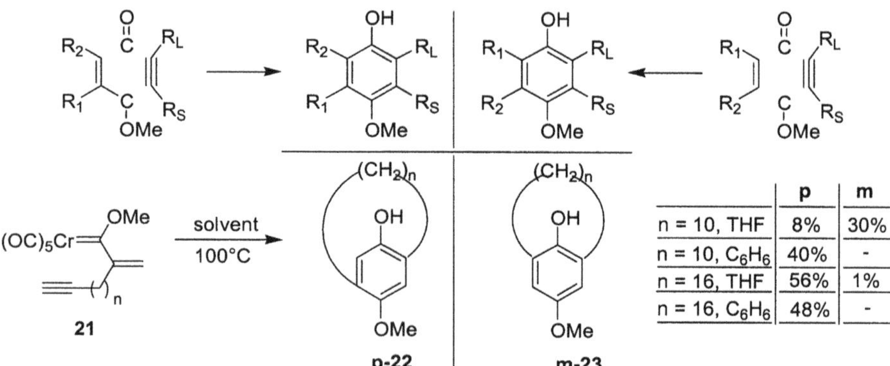

Scheme 15 Inverse regioselectivity via bond cleavage

The formation of a new unanticipated regioisomer was observed in the intramolecular benzannulation reaction of the α-tethered vinylcarbene chromium complex **21** (Scheme 15). The *para*-cyclophane **22** was expected for this reaction on the basis of normal regiochemistry; however, upon warming a carbene complex bearing a bridge of ten methylene units between the alkene and alkyne moieties in tetrahydrofuran, *meta*-cyclophane **23** was obtained in 30% isolated yield [29a]. Its formation requires the cleavage of the carbon–carbon bond between the carbene carbon and the carbon-bearing substituent R_1, which may be consequence of conformational strain within the intermediates. Further studies revealed that the choice of the solvent and the tether length had a strong influence on the outcome of the reaction [29b]. In coordinating solvents macrocycles which are most sensitive to ring strain (n=10) yield *meta*-cyclophanes **23**, whereas non-coordinating solvents facilitate the formation of the expected *para*-cyclophane **22**. Strainless macrocycles (n=16) do not reveal any solvent dependence.

5 Annulation Pattern

Benzannulation of fused arenes raises the question of angular versus linear annulation. The benzannulation of naphthylcarbene ligands generally leads to the phenanthrene skeleton in which both terminal rings obtain an optimum aromaticity [30]; a similar preference was observed even in cases where an *ortho* substitution was applied in order to force the annulation into a linear pathway [31]. However, recent studies indicate that linear benzannulation may become a major competition as observed for carbene complexes derived from dibenzo-substituted five-membered heteroarenes [32] or from helicenes [33]. A surprising linear benzannulation was observed for the dibenzofurylcarbene complex **24** [32a]. The uncoordinated benzo[*b*]naphthol[2,3-*d*]furan **26** was isolated along with the expected angular $Cr(CO)_3$-coordinated benzonaphthofuran **25** (Scheme 16). The formation of a linear $Cr(CO)_3$-coordinated benzannulation

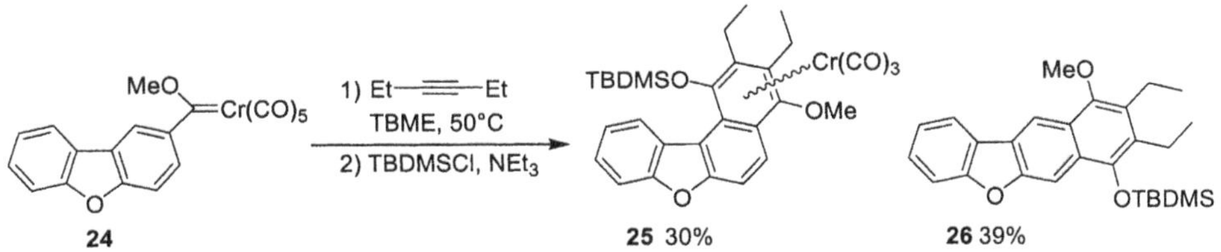

Scheme 16 Angular versus linear benzannulation of (dibenzo)heteroarenes

product was achieved when the central furan ring in the carbene complex was substituted for a thiophene system [32b]. Both types of molecular structures have been widely established by X-ray analysis.

A double linear annulation was observed in the benzannulation reaction of the helical biscarbene complex **27** with 3-hexyne along with a product bearing

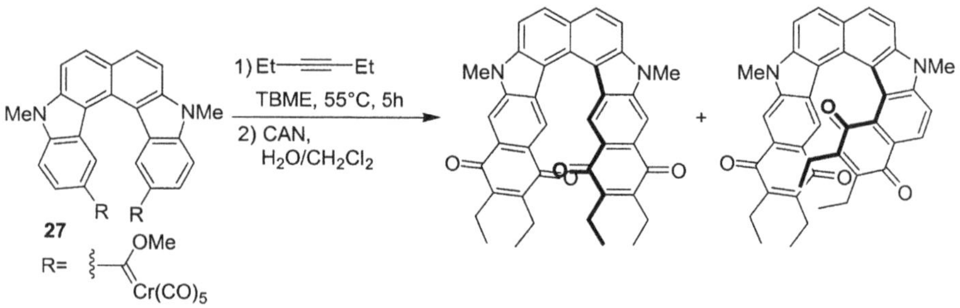

Scheme 17 Angular versus linear benzannulation of helicenes

a mixed annulation pattern (Scheme 17). The two products could be separated by column chromatography [33].

6 Benzannulation with Diarylcarbene Complexes

The electrophilic carbene carbon atom of Fischer carbene complexes is usually stabilised through π-donation of an alkoxy or amino substituent. This type of electronic stabilisation renders carbene complexes thermostable; nevertheless, they have to be stored and handled under inert gas in order to avoid oxidative decomposition. In a typical benzannulation protocol, the carbene complex is reacted with a 10% excess of the alkyne at a temperature between 45 and 60 °C in an ethereal solvent. On the other hand, the non-stabilised and highly electrophilic diphenylcarbene pentacarbonylchromium complex needs to be stored and handled at temperatures below –20 °C, which allows one to carry out benzannulation reactions at room temperature [34]. Recently, the first syntheses of tricyclic carbene complexes derived from diazo precursors have been performed and applied to benzannulation [35a,b]. The reaction of the non-planar dibenzocycloheptenylidene complex **28** with 1-hexyne afforded the $Cr(CO)_3$-coordinated tetracyclic benzannulation product **29** in a completely regio- and diastereoselective way [35c] (Scheme 18).

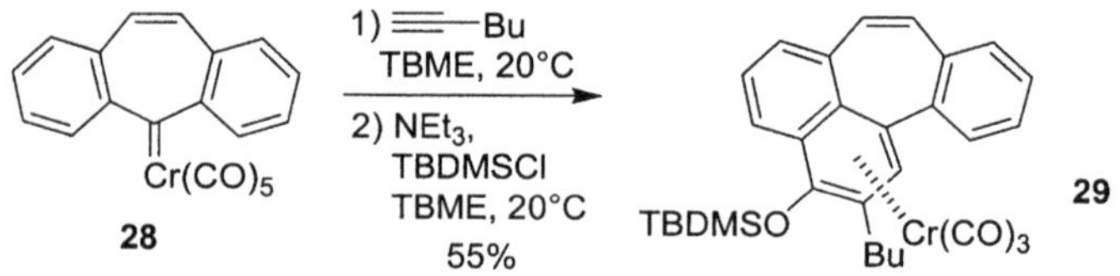

Scheme 18 Benzannulation of diarylcarbene-type complexes

Exo-alkylidene oxacycloalkylidene complexes such as chromium 2-oxacyclopentylidene **30** are reluctant to react thermally with alkynes. Nevertheless, benzannulation can be achieved under photochemical conditions (for a detailed discussion of photochemical reactions of carbene complexes, see Chap. of L. S. Hegedus, page 157). Exploiting this approach, 2,3-dihydro-5-benzofuranol **31** and 2,3-dihydro-6-benzopyranol skeletons, which are encountered as a structural part in a variety of natural products and biologically active compounds, are accessible in moderate to good yields [36] (Scheme 19).

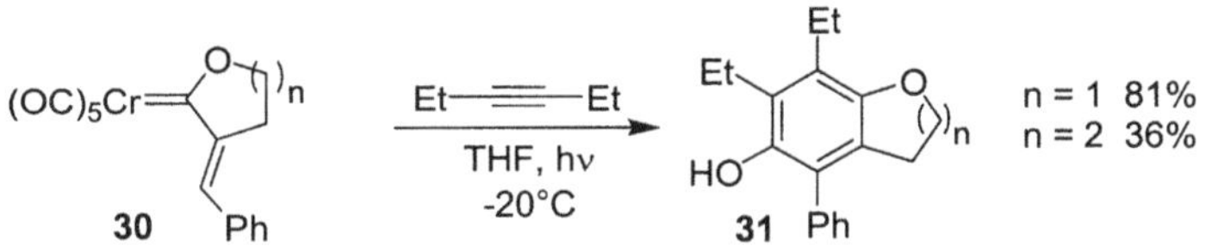

Scheme 19 Photo-induced benzannulation of *exo*-alkylidene oxacycloalkylidene complexes

7
Diastereoselective Benzannulation

Due to the inherent unsymmetric arene substitution pattern the benzannulation reaction creates a plane of chirality in the resulting tricarbonyl chromium complex, and – under achiral conditions – produces a racemic mixture of arene $Cr(CO)_3$ complexes. Since the resolution of planar chiral arene chromium complexes can be rather tedious, diastereoselective benzannulation approaches towards optically pure planar chiral products appear highly attractive. This strategy requires the incorporation of chiral information into the starting materials which may be based on one of three options: a stereogenic element can be introduced in the alkyne side chain, in the carbene carbon side chain or – most general and most attractive – in the heteroatom carbene side chain (Scheme 20).

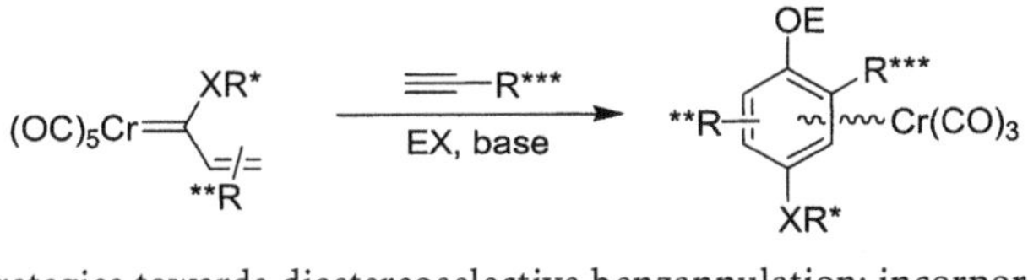

Scheme 20 Strategies towards diastereoselective benzannulation: incorporation of chiral information

7.1
Chiral Alkynes

Chiral alkynes (R***) bearing a chiral propargylic ether functionality show high asymmetric induction, as observed in the benzannulation of the propenyl complex **32** with alkyne **33** [37] (Scheme 21). The degree of optical induction in this reaction depends on the steric bulk of the acetylenic oxygen substituent, and is not the result of chelation of the propargylic oxygen to the metal. Therefore, the propargylic oxygen plays a stereoelectronic role in determining the stereoselectivity, which underlines (and dominates) the steric effects of the propargylic ether protecting group.

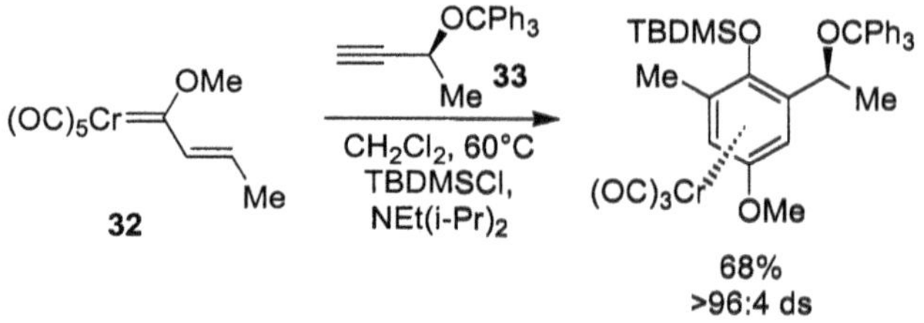

Scheme 21 Diastereoselective benzannulation of chiral propargylic ethers

Benzannulation of the diphenylcarbene ligand by 2-ethynylglucose derivative **34** results in only low diastereoselection, albeit it represents a rather rare example of a low-temperature protocol [38] (Scheme 22).

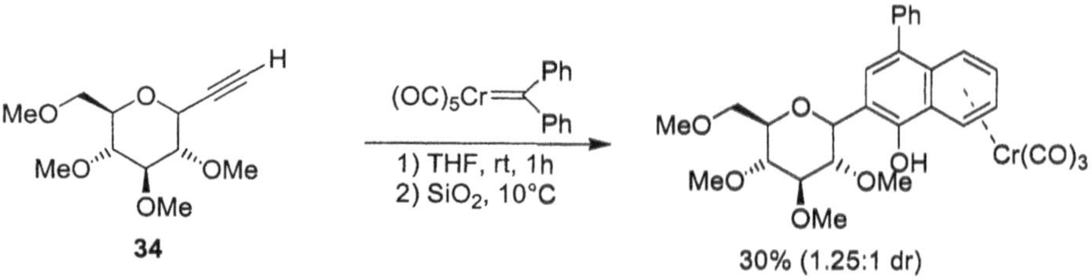

Scheme 22 Room-temperature benzannulation with a 2-ethynylglucose derivative

Attempts to increase the diastereoselectivity by a more rigid cyclopropane backbone were not successful. However, the incorporation of racemic *trans*-cyclopropane carboxylate **35** is completely regioselective, and both diastereomeric products **36** were isolated in a ratio of 4.1:1 [39] (Scheme 23).

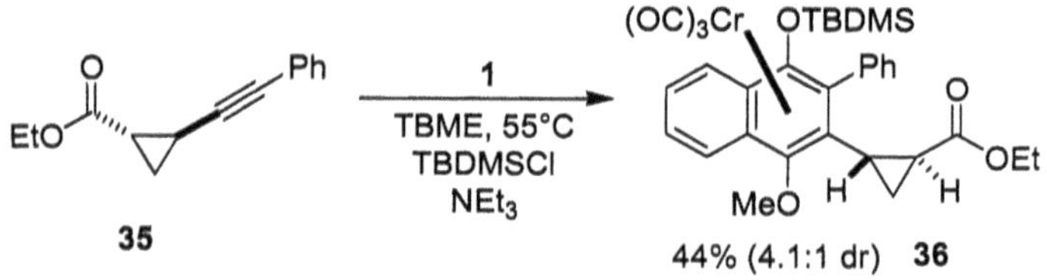

Scheme 23 Benzannulation with a racemic cyclopropane carboxylate

7.2 Chiral Alkoxy or Amino Auxiliaries

The second option involves the incorporation of either chiral amines or chiral alcohols into the heteroatom–carbene side chain (R*), which represents the most versatile approach to diastereoselective benzannulation. The optically pure (2*R*,3*R*)-butane-2,3-diol was used to tether the biscarbene complex **37**. The double intramolecular benzannulation reaction with diphenylbutadiyne allowed introduction of an additional stereogenic element in terms of an axis

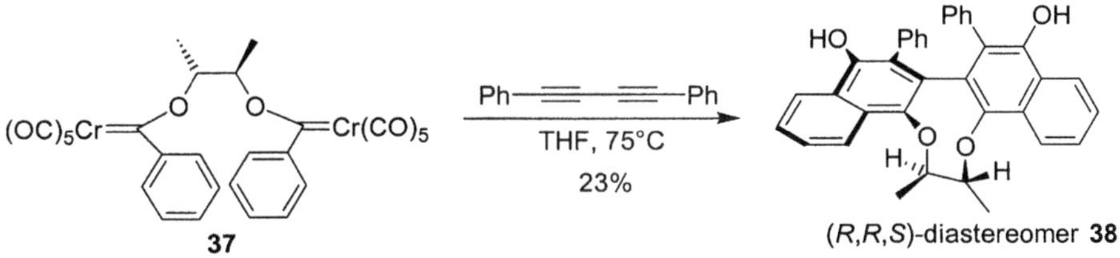

Scheme 24 Diastereoselective biaryl synthesis via double benzannulation

of chirality; a single diastereomer of 2,2′-binaphthol **38** was formed in moderate yield [40] (Scheme 24).

Excellent diastereomeric ratios were achieved with terpene-derived auxiliaries. The pentacarbonyl[(–)-menthyloxycarbene]chromium complex **39** reacted with the sterically hindered 3,3-dimethylbut-1-yne to give tricarbonyl chromium naphthohydroquinone complex **40** in 81% de as the major diastereomer which was also characterised by X-ray analysis [41] (Scheme 25). Surprisingly, the application of other even more sterically demanding terpene auxiliaries or a variation of the alkyne did not improve the diastereomeric ratio [42].

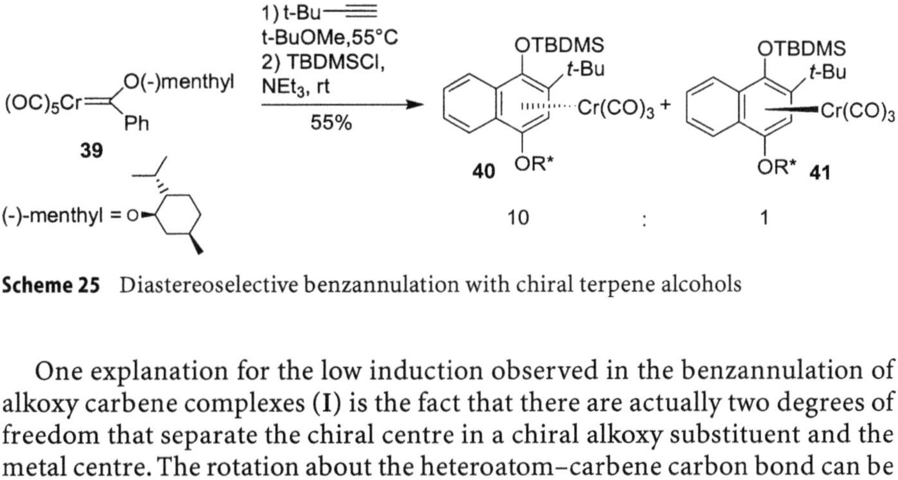

Scheme 25 Diastereoselective benzannulation with chiral terpene alcohols

One explanation for the low induction observed in the benzannulation of alkoxy carbene complexes (I) is the fact that there are actually two degrees of freedom that separate the chiral centre in a chiral alkoxy substituent and the metal centre. The rotation about the heteroatom–carbene carbon bond can be inhibited by switching to aminocarbene complexes (II) as the rotational barrier is increased from 15 to 25–30 kcal/mol due to the resonance delocalisation from nitrogen to the carbene carbon [43]. The other degree of freedom can be removed by using cyclic amino complexes (III) and (IV) (Scheme 26). α,β-Unsaturated carbene complexes derived from (*S*)-prolinol (III) exist as (III)-*syn* and (III)-*anti* isomers. The synthesis and isolation of a single isomer are hampered by α-deprotonation from the aminocarbene complex which is to be expected at different stages of the synthesis, resulting in an equilibration of the rotamers [44]. Due to this uncontrollable isomerisation, which occurs under the benzannulation conditions, these complexes failed to give even modest

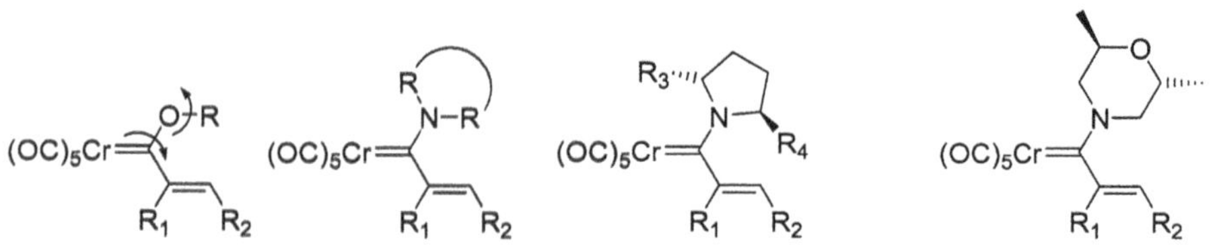

(I) (II) (III)-*anti* $R_3 = H, R_4 = CH_2OR_5$ (IV)
(III)-*syn* $R_3 = CH_2OR_5, R_4 = H$

Scheme 26 Carbene complexes with chiral amino substituents

asymmetric induction. Incorporation of a C_2-symmetric chiral amine such as in complex (**IV**), avoiding the problem of *syn* and *anti* isomers, however, does not improve the diastereoselectivity since the chiral information in the amine is too remote from the metal centre which assists the C–C bond formation to effect any facial selectivity. An exceptionally high asymmetric induction is observed for the heterocyclic carbene complex **42** bearing a (4*R*,5*S*)-5-phenyl-4-methylimidazolidinone auxiliary; in this carbene complex chelate the free rotation around the carbon–carbon bond that connects the imidazole ring to the carbene carbon is blocked as a result of chelation (Scheme 27).

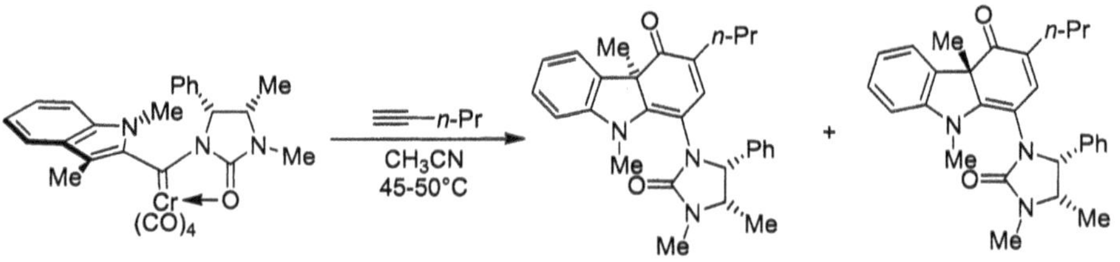

42 61% 96% de 1.2%

Scheme 27 Rigid chiral carbene complex chelates in diastereoselective benzannulation

7.3
Chiral Carbene Carbon Side Chains

Chiral alkoxy and amino cyclohexenylcarbene complexes substituted in the 3- or 6-position (R**), respectively, were examined in the benzannulation reaction [45] (Scheme 28). Tetralin $Cr(CO)_3$ complexes with substituents in the 5- or 8-position, respectively, are formed with different degrees of stereoselectivity and even reversed stereoselection depending on the substitution pattern of the chiral carbene complex. While 5-methyltetralin derivatives (entry 1) are formed with low diastereoselectivity, but with a consistent preference for the *syn*-isomer, the 8-methyltetralin complexes (entries 2–5) show a reversal of the sense of stereoselection and are formed with higher stereoselectivity. Interestingly, the diastereomeric ratio is increased on switching from methoxycarbene complexes (entries 2, 4) to aminocarbene complexes (entries 3, 5). An additional improvement of the stereoselectivity is achieved in the benzannulation reaction of the 3-methoxycyclohexenylcarbene complex with 1-pentyne; the didehydrotetralin $Cr(CO)_3$ complex is also obtained as a result of elimination of methanol.

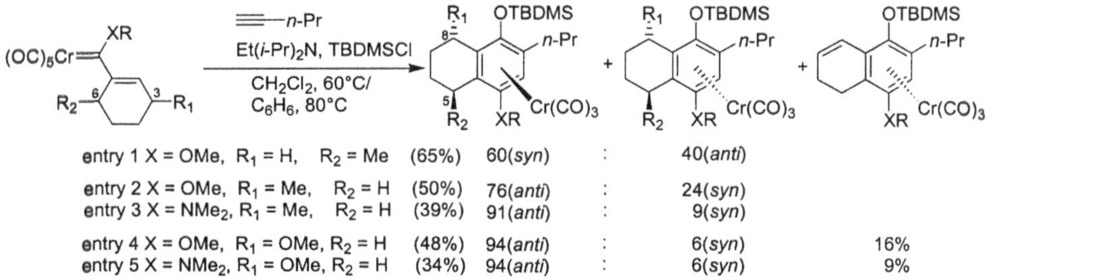

entry	X	R_1	R_2	yield	syn/anti		anti/syn	didehydrotetralin
entry 1	X = OMe	R_1 = H	R_2 = Me	(65%)	60(*syn*)	:	40(*anti*)	
entry 2	X = OMe	R_1 = Me	R_2 = H	(50%)	76(*anti*)	:	24(*syn*)	
entry 3	X = NMe_2	R_1 = Me	R_2 = H	(39%)	91(*anti*)	:	9(*syn*)	
entry 4	X = OMe	R_1 = OMe	R_2 = H	(48%)	94(*anti*)	:	6(*syn*)	16%
entry 5	X = NMe_2	R_1 = OMe	R_2 = H	(34%)	94(*anti*)	:	6(*syn*)	9%

Scheme 28 Benzannulation of chiral cyclohexenylcarbene complexes

The stereoselectivity observed in these reactions is assumed to result from steric interactions in the η^3-vinylcarbene complex intermediate **D**. In the first case (entry 1) an eclipsed interaction between the methyl group in the 6-position of the cyclohexene ring and the methoxy group of the vinylcarbene ligand has to be avoided, favouring the formation of the *syn*-tetralin complex. The minimisation of the steric interaction between the methyl group in the 3-position of the cyclohexene ring and a carbon monoxide ligand favours the formation of the *anti*-isomer in the other cases. The key step in the synthesis of the oxepin derivative **44** is a tandem Dötz–Mitsunobu reaction starting from the enantiomerically pure decalin-derived carbene complex **43** [46] (Scheme 29). The benzannulation reaction with hex-5-yn-1-ol proceeds with a high level of induction, as the complex **43** is isolated as a single diastereomer after ring closure via the Mitsunobu reaction.

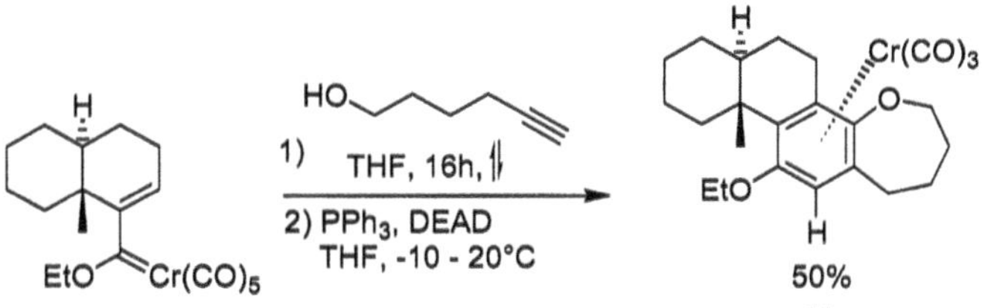

Scheme 29 Tandem benzannulation–Mitsunobu reaction of a chiral decalin-derived carbene complex

A similar tandem Dötz–Mitsunobu reaction has been reported starting from a 1,6-methano[10]annulene carbene complex, but no conclusion could be reached on the influence of the chiral information regarding the stereoselective course of the reaction since the chromium fragment could not be kept coordinated to the benzannulation product [47].

The inherent plane of chirality in the metal carbene-modified cyclophane **45** was also tested in the benzannulation reaction as a source for stereoselectivity [48]. The racemic pentacarbonyl(4-[2.2]metacyclophanyl(methoxy)carbene)-chromium **45** reacts with 3,3-dimethyl-1-butyne to give a single diastereomer of naphthalenophane complex **46** in 50% yield; the sterically less demanding 3-hexyne affords a 2:1 mixture of two diastereomers (Scheme 30). These moderate diastereomeric ratios indicate that [2.2]metacyclophanes do not serve as efficient chiral tools in the benzannulation reaction.

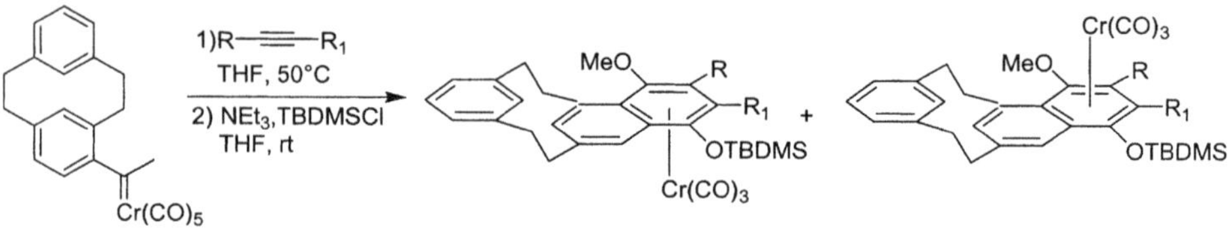

Scheme 30 A chiral [2.2]metacyclophane carbene complex in a benzannulation reaction

A bidirectional benzannulation of the axial-chiral biscarbene complex **47** affords a bis-$Cr(CO)_3$-coordinated biphenanthrene derivative **48**, which combines elements of axial and planar chirality [49] (Scheme 31). Four diastereomers are formed in moderate diastereoselectivity, two of which have been isolated as the major isomers.

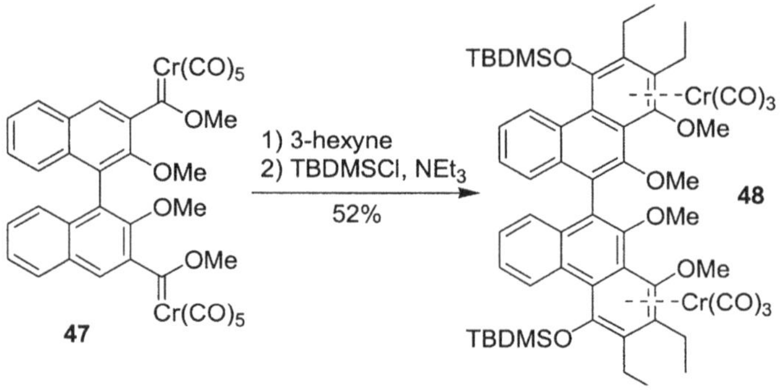

Scheme 31 Bidirectional benzannulation with an axial-chiral biscarbene complex

Upon reaction with 3-hexyne glucal-derived chromium, carbenes undergo benzannulation to afford highly oxygenated chromans coordinated to the chromium tricarbonyl fragment [50]. The diastereoselectivity depends on the nature of the protective groups. Best results were obtained with the TIPS-protected complex **49**, which produced benzochroman **50** as a single isomer along with demetalated hydroquinone **51** (Scheme 32).

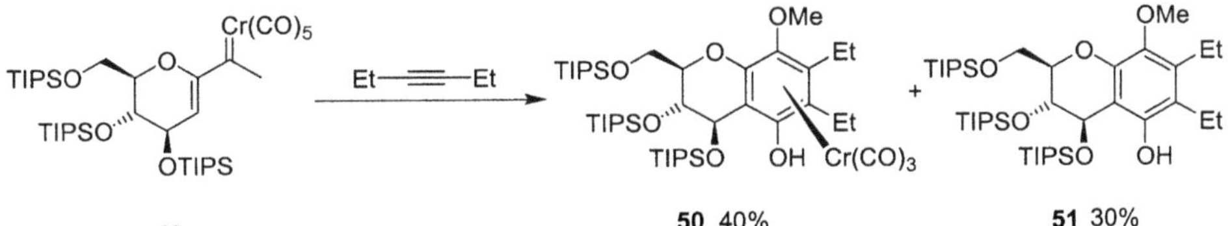

Scheme 32 Benzannulation with glucal-derived chromium carbenes

The use of a stereogenic carbon centre allowed an efficient asymmetric induction in the benzannulation reaction towards axial-chiral intermediates in the synthesis of configurationally stable ring-C-functionalised derivatives of allocolchicinoids [51]. The benzannulation of carbene complex **52** with 1-pentyne followed by oxidative demetalation afforded a single diastereomer **53** (Scheme 33).

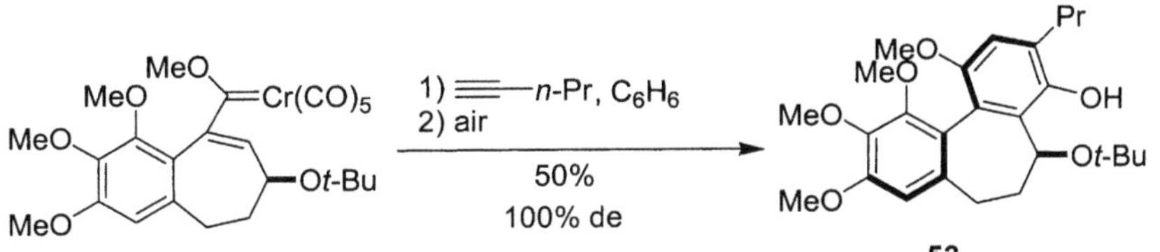

Scheme 33 Diastereoselective benzannulation towards allocolchinoids

8 Carbene Complexes with Different Metal Centres

For a long time the benzannulation reaction has been restricted to metals of group 6, primarily complexes of the pentacarbonylchromium moiety. Carbene complexes of manganese do not undergo benzannulation reaction with alkynes unless the carbene system is activated by introducing a second, electron-deficient metal bound through oxygen to the carbene carbon atom. The carbene complex **54** with an electron-withdrawing titanium(IV)oxy substituent reacts with 1-hexyne under photochemical conditions, or in refluxing toluene, affording the naphthoquinone derivative **55** after oxidative workup [52]. Crystallographic data show that upon Ti(IV) substitution the manganese increases π-donation to the carbonyl ligands and therefore decreases π-donation to the carbene centre. The silyloxycarbene manganese complex **56** bears a tethered alkyne and therefore reacts in an intramolecular benzannulation reaction, yielding the functionalised naphthoquinone **57** [53] (Scheme 34).

Alkoxy(carbene)iron(0) and amino(carbene)iron(0) complexes usually react with alkynes to give η^4-pyrone iron complexes and furans, respectively [54]. Nevertheless the chemoselective formation of naphthols was reported for alkoxy(carbene)iron(0) complexes with the electron-poor alkyne dimethyl

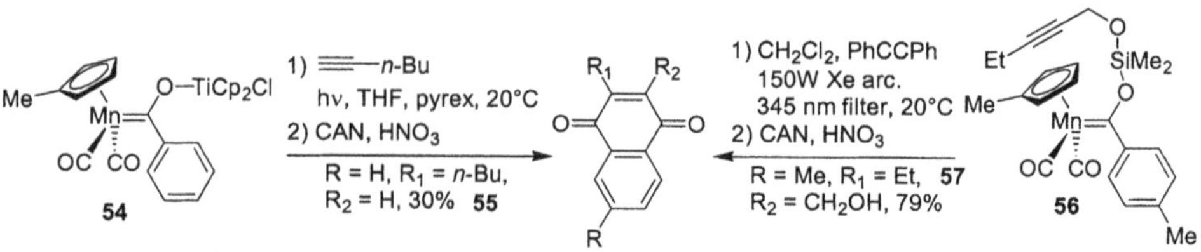

Scheme 34 Manganese carbene complexes in benzannulation reactions

acetylene dicarboxylate [55]. The electron-rich iron(0) carbene complex **58** gave excellent yields of naphthol **59** (Scheme 35). Further studies revealed that the formation of naphthols is restricted to the use of this specific acetylene, as alkyne monoesters give furans.

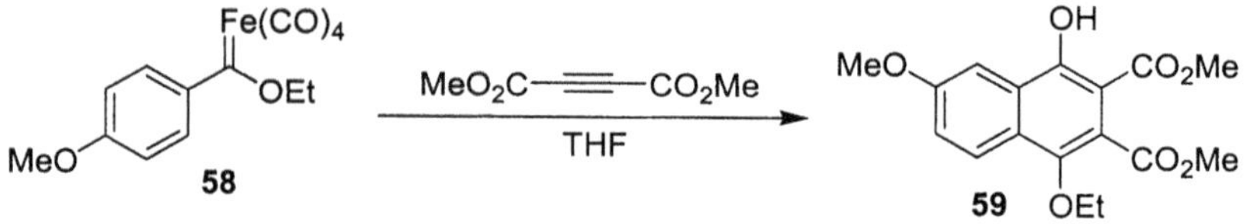

Scheme 35 Iron-mediated benzannulation

A benzannulation reaction yielding the naphthoquinone **61** could also be performed with the ruthenium carborane-stabilised carbene **60** and 1-hexyne [56] (Scheme 36). The ruthenium carbene unit can be regarded as an 18-electron fragment containing a formal Ru(II) centre coordinated to a dianionic six-electron-donor cobaltacarborane ligand.

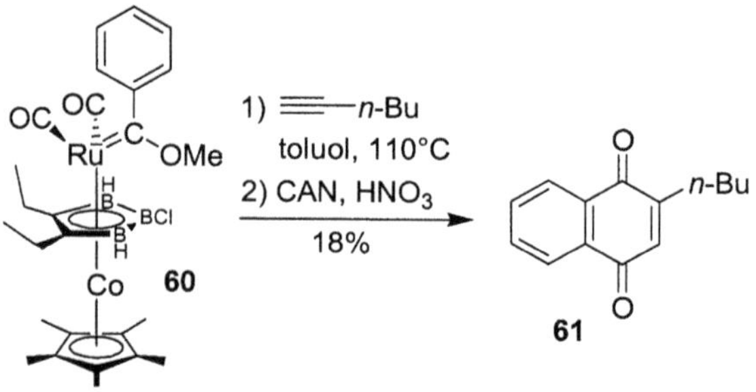

Scheme 36 Ruthenium-mediated benzannulation

A transmetalation of the styrylcarbene chromium complex **62** in the presence of stoichiometric amounts of [Ni(cod)$_2$] to give the nickel carbene intermediate **63** was applied to the synthesis of Cr(CO)$_3$-coordinated cycloheptatriene **64** upon reaction with terminal alkynes [57] (Scheme 37). The formation of pentacarbonyl(acetonitrile)chromium is expected to facilitate the metal exchange.

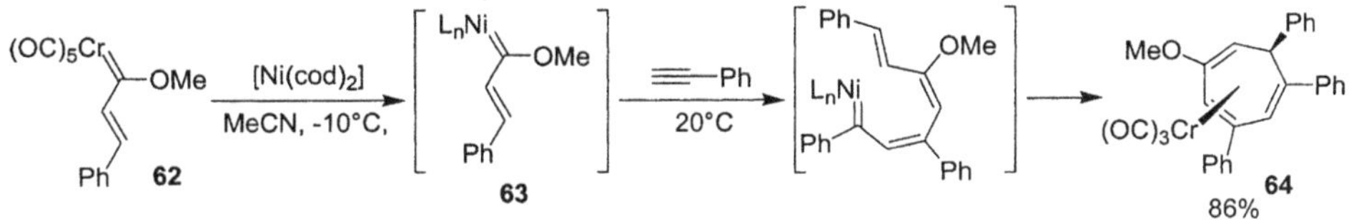

Scheme 37 Transmetalation of chromium to nickel in a metal carbene-mediated cyclisation reaction (L=cod, MeCN, alkyne)

A pathway may be considered which involves a double regioselective alkyne insertion followed by a stereoselective cyclisation to undergo a novel [3+2+2]-cyclisation. These examples illustrate the scope in which the reactivity of Fischer carbene complexes can be tuned in a qualitative manner by transmetalation.

9 Total Synthesis

9.1 Vitamins

The fact that pentacarbonyl carbene complexes react with enynes in a chemoselective and regiospecific way at the alkyne functionality was successfully applied in the total synthesis of vitamins of the K_1 and K_2 series [58]. Oxidation of the intermediate tricarbonyl(dihydrovitamin K) chromium complexes with silver(I) oxide afforded the desired naphthoquinone-based vitamin K compounds **65**. Compared to customary strategies, the benzannulation reaction proved to be superior as it avoids conditions favouring (*E*)/(*Z*)-isomerisation within the allylic side chain. The basic representative vitamin K_3 (menadione) **66** was synthesised in a straightforward manner from pentacarbonyl carbene complex **1** and propyne (Scheme 38).

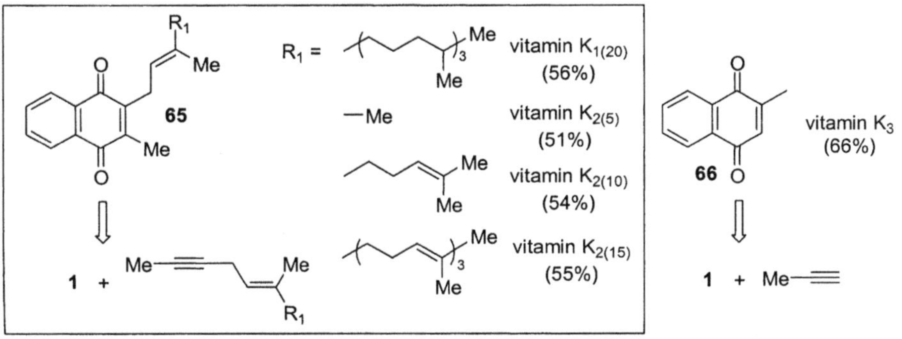

Scheme 38 Metal carbene approach to vitamins K_1, K_2 and K_3

Encouraged by the short synthesis of K vitamins, the chromium-mediated benzannulation was extended to the synthesis of vitamin E **68** [59]. The problem of imperfect regioselectivity of alkyne incorporation – which did not hamper the approach to vitamin K due to the final oxidation to the quinone – was tackled by demethylation of both regioisomeric hydroquinone monomethyl ethers **67** to give the unprotected hydroquinone. Subsequent ring closure yielded α-tocopherol (vitamin E) **68** (Scheme 39).

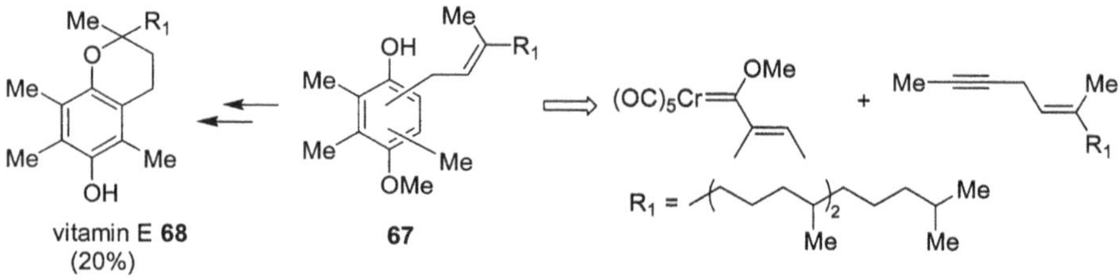

Scheme 39 Metal carbene approach to vitamin E

9.2 Antibiotics

Daunomycinone **72**, one of the clinically important agents in cancer chemotherapy, is a member of the anthracycline familiy of antitumour antibiotics. The 11-deoxy analogue **79** is of current interest due to an improved therapeutic index. The common structure of this family of antibiotics is a linear tetracyclic skeleton containing a quinone C ring attached to a hydroquinone B ring (for daunomycinone) or a phenol B′ ring (for 11-deoxydaunomycinone). Both ring B and ring C can be constructed via benzannulation [60a,b]. The key step of the ring B approach involves the reaction of the ethynyl lactone **69** and the cyclohexenyl(ketal)carbene complex **70** which provides the tetrahydronaphthol **71** in 72–76% yield [60a,b] (Scheme 40).

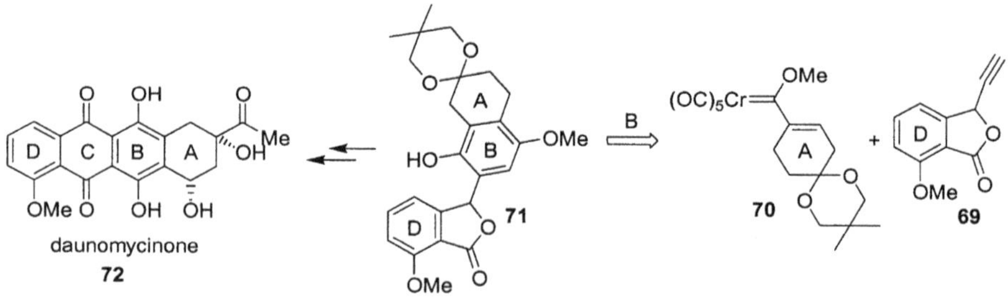

Scheme 40 Benzannulation towards daunomycinone based on ring B formation

Replacing a hydroxy group in daunomycinone by a hydrogen atom leads to 11-deoxydaunomycinone **79**. However, this formally simple transformation affords a fundamental change of the synthetic strategy. Two very similar syntheses with the benzannulation reaction as the key step were developed by Dötz and Wulff [60c–f]. In both cases the carbene complex **77** reacts with a propargylic cyclohexane derivative **73** or **76** to give naphthol **75** and **78**, respectively (Scheme 41). The metal carbene chelate **74** generated by decarbonylation of pentacarbonyl complex **77** readily undergoes opening of the chelate ring and, thus, allows the formation of the alkyne complex intermediate under mild conditions resulting in improved yields. The final B ring closure was effected by Friedel–Crafts cyclisation in an acidic medium.

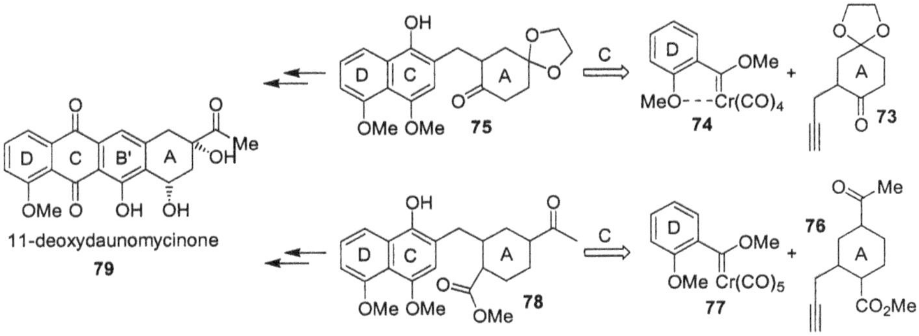

Scheme 41 Ring C benzannulation strategy to 11-deoxydaunomycinone

A similar synthetic strategy was applied in the synthesis of menogaril **83**, another important anthracycline antitumour antibiotic, and to the synthesis of the tricyclic core of olivin **87**, the aglycon of the antitumour antibiotic olivomycin [61, 62]. In both cases a tandem benzannulation/Friedel–Crafts cyclisation sequence yielded the tetracyclic and tricyclic carbon core, respectively (Scheme 42).

Intensive studies towards the total synthesis of fredericamycin A **91** culminated in the enantioselective synthesis of this potent antitumour antibiotic [63].

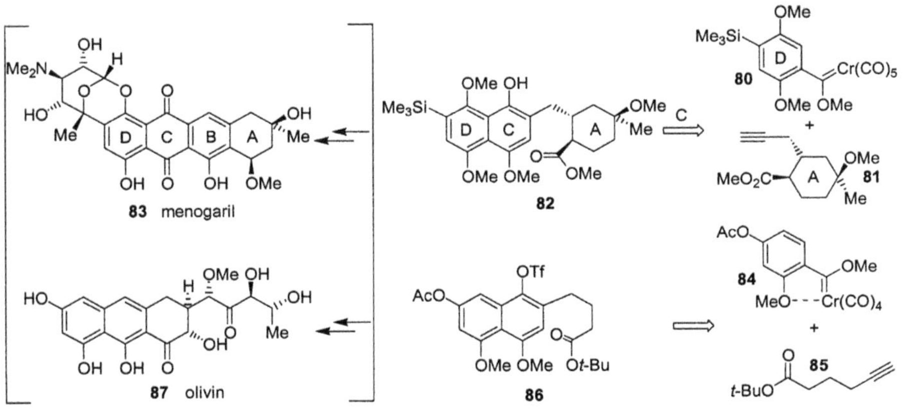

Scheme 42 Studies towards the synthesis of menogaril and olivin

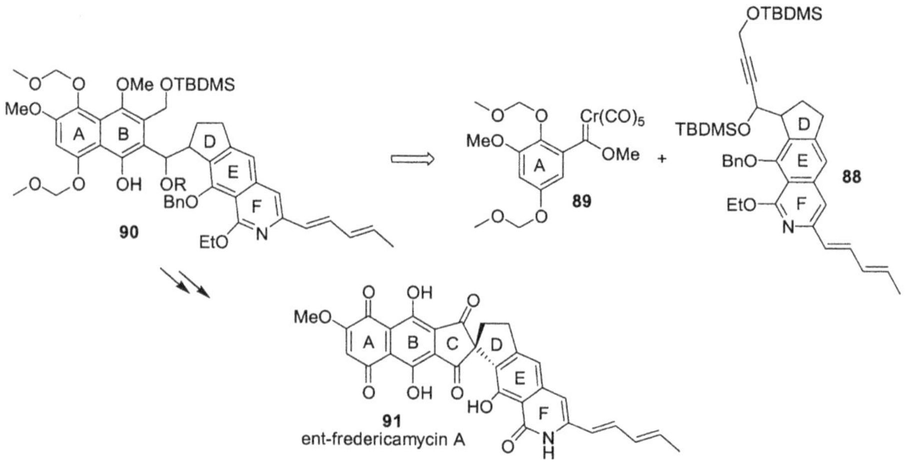

Scheme 43 Total synthesis of fredericamycin A (R=TBDMS)

The convergent approach comprises, among other reaction steps, a regiospecific intermolecular benzannulation reaction between the alkyne **88** and the chromium carbene complex **89** for AB ring construction (Scheme 43). It is noteworthy that the regioselectivity of this reaction is attributed to the bulky TBDMS ether in the alkyne α-substituent, that dictates the incorporation of the large substituent *ortho* to the phenol. Another curiosity is the fact that the reaction failed to provide **90** in the absence of acetic anhydride.

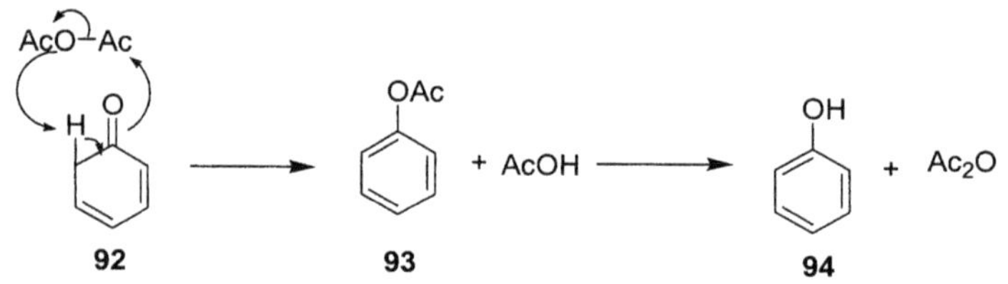

Scheme 44 The role of acetic anhydride in the aromatisation of cyclohexadienone

A plausible pathway is that the aromatisation of the cyclohexadienone **92** by a proton shift is accelerated in the presence of Ac_2O under formation of acetate **93**. The simultaneously generated acetic acid then cleaves the acetate to form the free phenol **94** (Scheme 44). This effect was observed for the first time during studies towards the total synthesis of the lipid-alternating and anti-atherosclerotic furochromone khellin **99** [64]. The furanyl carbene chromium complex **96** was supposed to react with alkoxyalkyne **95** in a benzannulation reaction to give the densely substituted benzofuran derivative **97** (Scheme 45). Upon warming the reaction mixture in tetrahydrofuran to 65 °C the reaction was completed in 4 h, but only a dimerisation product could be isolated. This

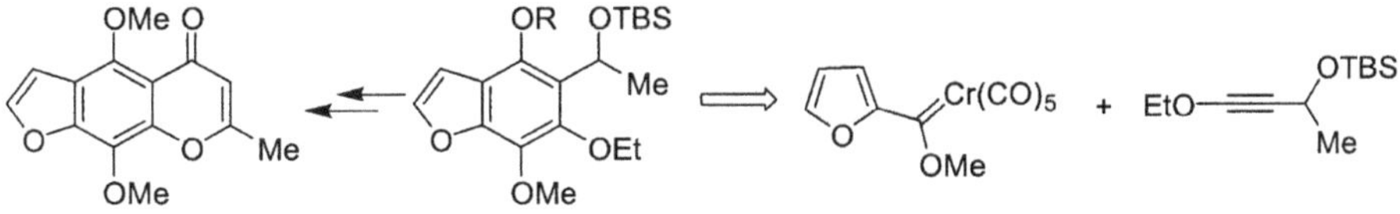

khellin **99** R = H **97** R = Ac **98** **96** **95**

Scheme 45 Total synthesis of khellin

dimer formation was suppressed by in situ protection. When the benzannulation reaction was carried out in the presence of acetic anhydride and triethylamine, the benzofuran acetate **98** was formed in 43% isolated yield; in the absence of triethylamine the unprotected benzofuran **97** was isolated in 36% yield. Triethylamine is not only supposed to deprotonate the phenol intermediate **94**, but also to neutralise the acetic acid formed in an alternative pathway in order to avoid the cleavage of the acetate-protected phenol **93** (Scheme 46).

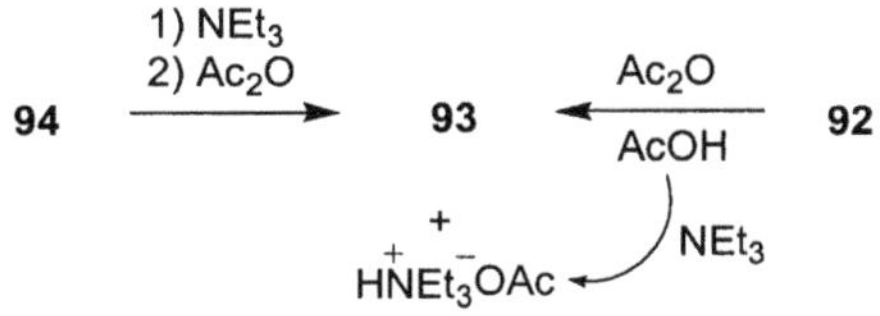

Scheme 46 The role of triethylamine in acetylation and aromatisation

Nanaomycin A **103** and deoxyfrenolicin **108** are members of a group of naphthoquinone antibiotics based on the isochroman skeleton. The therapeutic potential of these natural products has attracted considerable attention, and different approaches towards their synthesis have been reported [65, 66]. The key step in the total synthesis of racemic nanaomycin A **103** is the chemo-and regioselective benzannulation reaction of carbene complex **101** and allylacetylene **100** to give allyl-substituted naphthoquinone **102** after oxidative workup in 52% yield [65] (Scheme 47). The allyl functionality is crucial for a subsequent intramolecular alkoxycarbonylation to build up the isochroman structure. However, modest yields and the long sequence required to introduce the

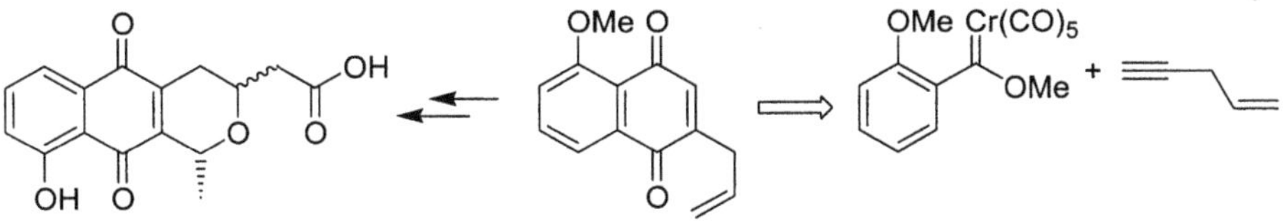

nanaomycin A **103** **102** **101** **100**

Scheme 47 Benzannulation approach to nanaomycin A

hydroxy group in the C3 position, which is the other component in the alkoxycarbonylation, reduces the attractiveness of this benzannulation approach.

For that reason an intramolecular benzannulation was developed, which incorporates all components for the intramolecular alkoxycarbonylation into the naphthoquinone **105** [65]. Based on that strategy a short and convergent pathway for the synthesis of racemic deoxyfrenolicin **108** was accomplished. Xu et al. replaced the allylacetylene **100** in the reaction sequence for nanaomycin A by alkynoate **106**. The benzannulation product **107** was an appropriate precursor for a subsequent tandem oxa-Pictet–Spengler cyclisation/DDQ-induced coupling reaction [66]. Following this strategy the total synthesis of enantiomerically pure deoxyfrenolicin could be accomplished (Scheme 48).

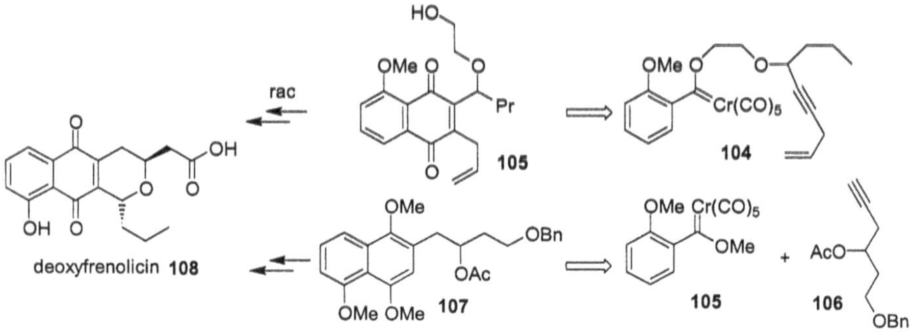

Scheme 48 Benzannulation approach to deoxyfrenolicin

The synthesis of the naphthalene rings found in the gilvocarcin group and in the rubromycin class of natural products via benzannulation was also reported. Both classes show promising antitumour activity [67, 68].

Danishefsky et al. succeeded in preparing the benz[*a*]anthracene core structure **111** of angucycline antibiotics by performing a benzannulation reaction with the cycloalkynone **109** [69]. Deprotonation of the naphthoquinone **110** with DBU yields the desired anthraquinone **111** (Scheme 49).

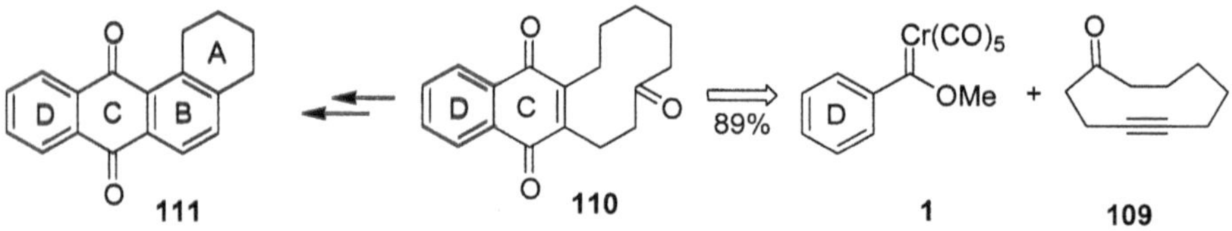

Scheme 49 Synthesis of the benz[*a*]anthracene core structure

9.3 Steroids

The benzannulation reaction with small alkynes such as 1-pentyne may generate a two-alkyne annulation product. In this case the original [3+2+1]-benzannulation is changed to a [2+2+1+1]-benzannulation. After CO dissociation and insertion of the first alkyne, the coordinated α,β-unsaturated moiety in the vinylcarbene complex is supposed to be replaced by the second alkyne. The subsequent reaction steps give the phenol **112** (Scheme 50).

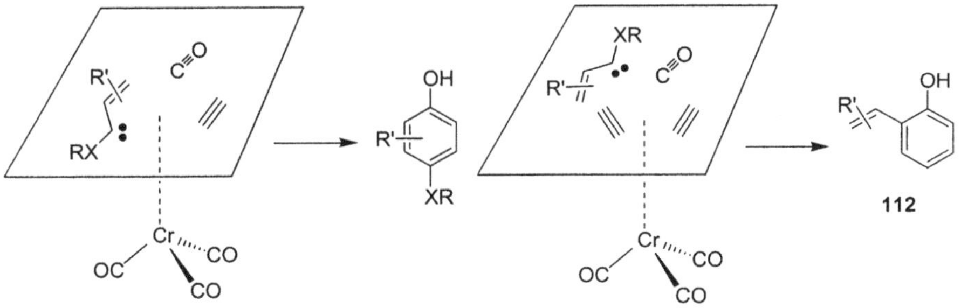

Scheme 50 [3+2+1]- and [2+2+1+1]-benzannulation

The selectivity for two-alkyne annulation can be increased by involving an intramolecular tethering of the carbene complex to both alkynes. This was accomplished by the synthesis of aryl-diynecarbene complexes **115** and **116** from the triynylcarbene complexes **113** and **114**, respectively, and Danishefsky's diene in a Diels–Alder reaction [70a]. The diene adds chemoselectively to the triple bond next to the electrophilic carbene carbon. The thermally induced two-alkyne annulation of the complexes **115** and **116** was performed in benzene and yielded the steroid ring systems **117** and **118** (Scheme 51). This tandem Diels–Alder/two-alkyne annulation, which could also be applied in a one-pot procedure, offers new strategies for steroid synthesis in the class O→ABCD.

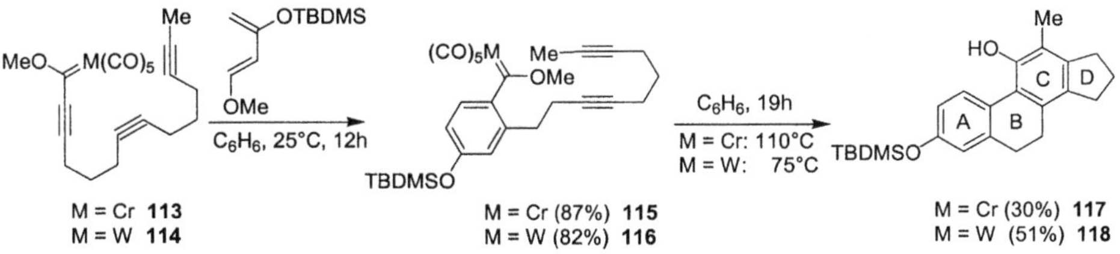

Scheme 51 Synthesis of the steroid skeleton via [2+2+1+1]-benzannulation

The benzannulation reaction of ethynylferrocene **120** with the diterpenoid chromium alkoxycarbene **119** leads to novel diterpenoid ferrocenyl quinones **121** which, due to their electron-transfer properties, are regarded as potential candidates for non-linear optical materials [71] (Scheme 52).

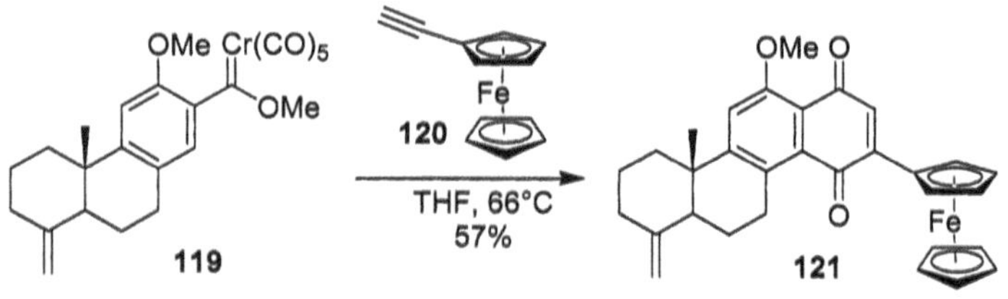

Scheme 52 Synthesis of fused ferrocenyl quinones

9.4 Insecticides

Two very short and elegant syntheses of the antiparasitic agent parvaquone **125** and the insecticide **124** isolated from Scrophulariaceae were developed using the dry-state absorption protocol [72, 73] (Scheme 53).

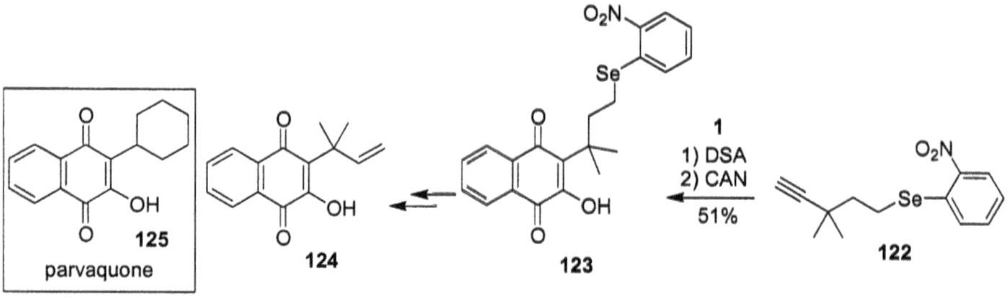

Scheme 53 Total synthesis of insecticides

10 Various Benzannulation Protocols

Merlic et al. were the first to predict that exposing a dienylcarbene complex **126** to photolysis would lead to an *ortho*-substituted phenolic product **129** [74a]. This photochemical benzannulation reaction, which provides products complementary to the classical *para*-substituted phenol as benzannulation product, can be applied to (alkoxy- and aminocarbene)pentacarbonyl complexes [74]. A mechanism proposed for this photochemical reaction is shown in Scheme 54. Photoactivation promotes CO insertion resulting in the chromium ketene in-

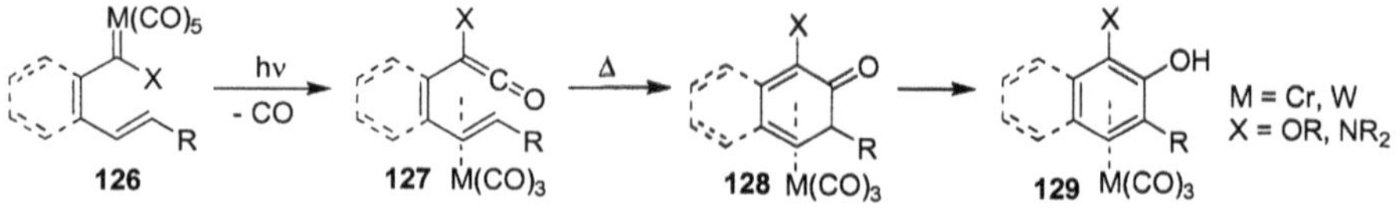

Scheme 54 Complementary benzannulation via photoactivation of carbonyl ligands

termediate **127**. Subsequent thermal electrocyclisation and keto–enol tautomerisim provides *ortho*-(alkoxy or amino)phenols such as **129**.

A similar substitution pattern can be obtained by applying a thermal protocol as well [75]. *Ortho*-methoxyphenol **131** has been synthesised in good yields by warming the cyclobutene-containing 1,3,5-metallatriene **130** in tetrahydrofuran (Scheme 55).

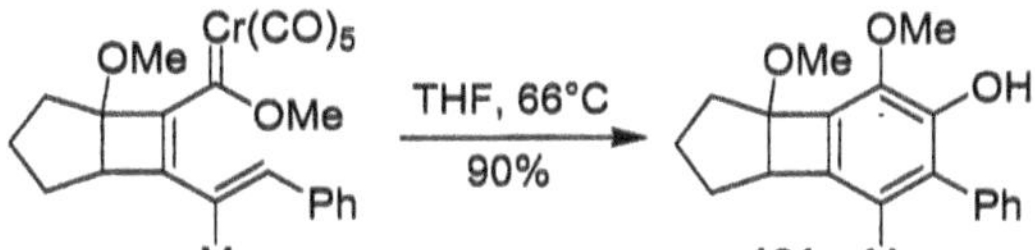

Scheme 55 Thermal benzannulation protocol to *ortho*-alkoxyphenols

Merlic developed a new variation of the thermally induced benzannulation reaction. The dienylcarbene complex **132** was reacted with isonitrile to give an *ortho*-alkoxyaniline derivative **135** [76] (Scheme 56). This annulation product is regiocomplementary to those reported from photochemical reaction of chromium dienyl(amino)carbene complexes. The metathesis of the isocyanide with the dienylcarbene complex **132** generates a chromium-complexed dienylketenimine intermediate **133** which undergoes electrocyclisation. Final tautomerisation and demetalation afford the *ortho*-alkoxyaniline **135**.

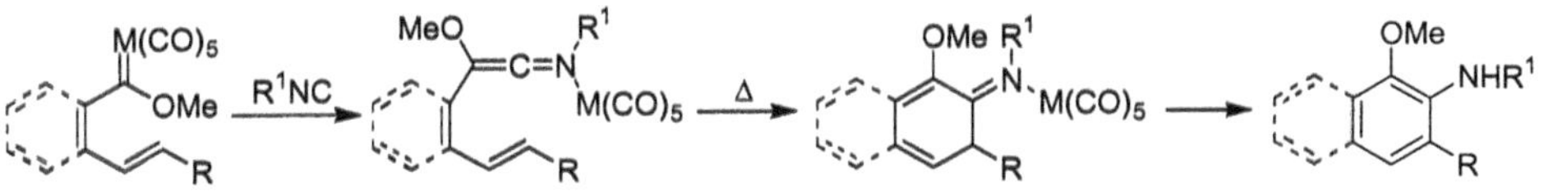

Scheme 56 Thermal benzannulation protocol to *ortho*-alkoxyanilines

Based on this synthetic strategy an efficient method for the synthesis of 2,3-dihydro-1,2-benzisoxazoles **137** and indazoles was elaborated [77a] (Scheme 57).

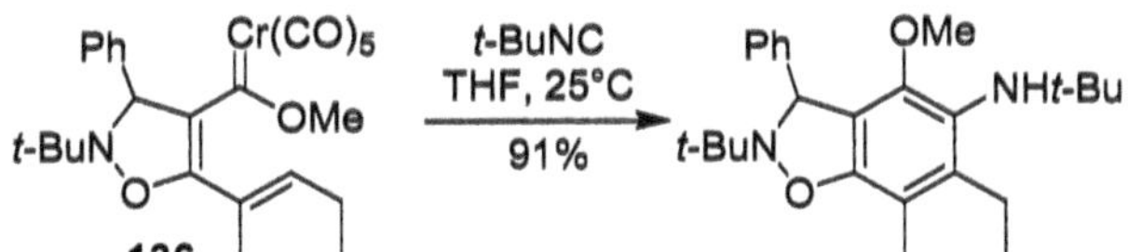

Scheme 57 Synthesis of 2,3-dihydro-1,2-benzisoxazoles

The cyclobutene-containing 1,3,5-metallatriene **130** also reacts with an isocyanide to give the regiocomplementary product [77b], but if the isocyanide is exchanged for a terminal alkyne the course of the reaction is fundamentally changed and a cyclooctatrienone **139** is formed [78] (Scheme 58). The incorporation of the alkyne occurs regioselectively, and a new stereogenic centre is formed during the reaction with high diastereoselectivity. This reaction might be considered as a new variant of the [3+2+1]-benzannulation reaction which involves insertion of both the alkyne and a carbon monoxide ligand. The participation of the additional double bond in the electrocyclic ring closure is responsible for the formation of the eight-membered carbocycle formed in an eight-electron cyclisation.

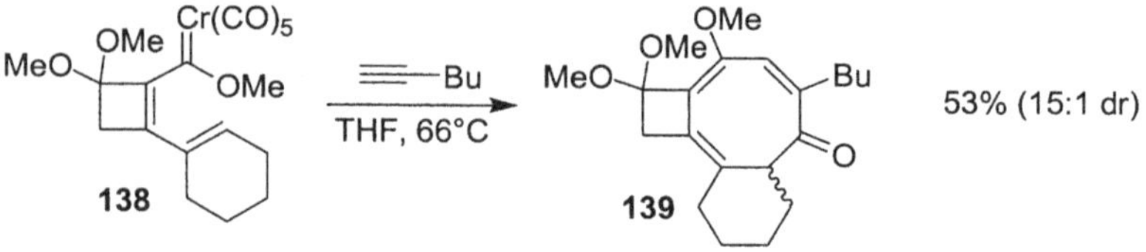

MeO OMe $Cr(CO)_5$ OMe **138** —— ≡—Bu, THF, 66°C ——> MeO OMe OMe Bu O **139** 53% (15:1 dr)

Scheme 58 Formation of eight-membered carbocycles

1-Amino-2-ethoxy-4-phosphinonaphthalene **141** is obtained from the (*E*)-aryl-alkenylcarbene complex **140** and *tert*-butyl isocyanide under mild conditions [79] (Scheme 59). The required substrate **140** is generated from a Michael-type addition of a secondary phosphine to an alkynylcarbene complex.

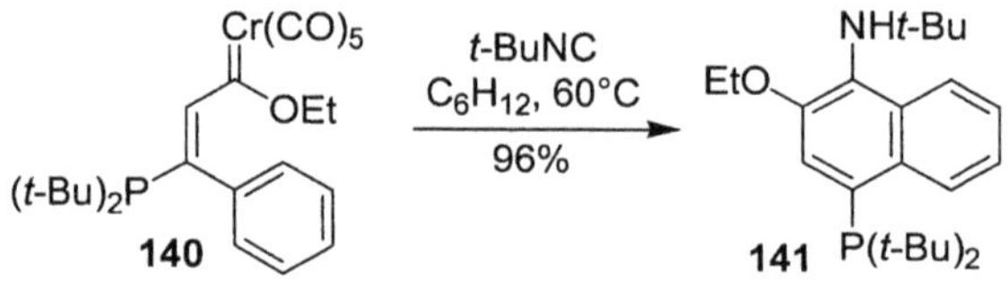

$Cr(CO)_5$ OEt $(t\text{-Bu})_2P$ **140** —— *t*-BuNC, C_6H_{12}, 60°C, 96% ——> NH*t*-Bu EtO $P(t\text{-Bu})_2$ **141**

Scheme 59 Synthesis of phosphinonaphthalenes

These two examples of modified benzannulation reactions were successfully applied to the preparation of analogues of indolocarbazole natural products **143** and the total synthesis of calphostins **146**. The indolocarbazoles **143** have emerged as an important structural class revealing considerable biological activity including antitumour properties. Complementary thermal and photochemical protocols were applied to 2,2′-bisindolyl chromium carbene complexes **142** – which may be regarded as aromatic analogues of dienyl carbene complexes – in order to establish the ABCEF ring system which represents the central core of indolocarbazole alkaloids [80] (Scheme 60).

In the total synthesis of the protein kinase C inhibitors calphostins **146**, the *ortho*-substituted intermediates, which are either obtained from photolysis or from reaction of the dienyl carbene complex **144** with *tert*-butyl isocyanide, were oxidised to yield the 1,2-benzoquinone **145** as a common product [81] (Scheme 61).

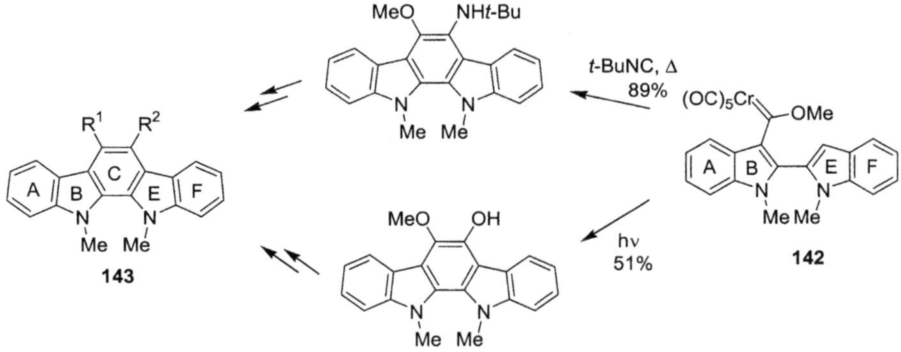

Scheme 60 Synthesis of indolocarbazole alkaloids

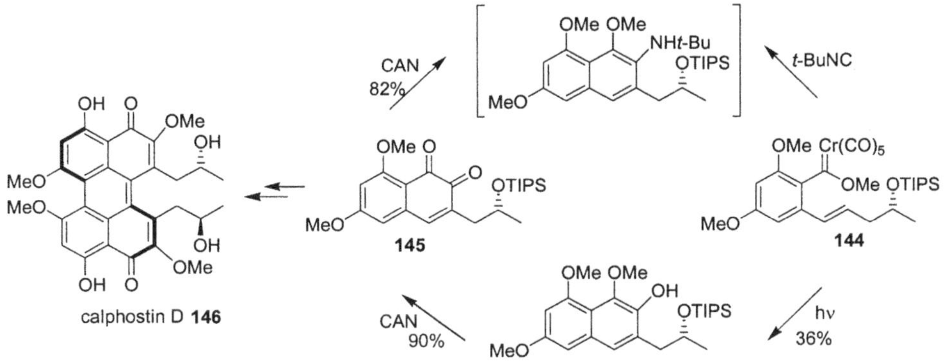

Scheme 61 Complementary thermal and photochemical synthesis of calphostins

11 Final Remarks

The development of the chromium-mediated benzannulation reaction over the past 35 years demonstrates the potential of transition metals in the elaboration of unprecedented reactions. Metals are able to coordinate a variety of organic substrates in a predictable geometry primarily determined by their nature and by their oxidation state. They may act as templates which activate and fix the ligands in an orientation favourable for interligand coupling. The broad and fundamental knowledge of organometallic complexes accumulated over the last half century remains a promising fishing area for the discovery of novel reactions attractive for organic synthesis.

References

1. Dötz KH (1975) Angew Chem Int Ed Engl 14:644
2. Choi YH, Rhee KS, Kim KS, Shin GC, Shin SC (1995) Tetrahedron Lett 36:1871
3. Harrity JPA, Kerr WJ (1993) Tetrahedron 49:5565
4. Hutchinson EJ, Kerr WJ, Magennis EJ (2002) Chem Commun 2262
5. Dötz KH, Fischer H, Mühlemeier J, Märkl R (1982) Chem Ber 115:1355
6. Torrent M, Duran M, Solà M (1998) Organometallics 14:1492
7. Fischer H, Hofmann P (1999) Organometallics 18:2590
8. Barluenga J, Aznar F, Martín A, García-Granda S, Pérez-Carreño E (1994) J Am Chem Soc 116:11191
9. (a) Dötz KH, Schäfer T, Kroll F, Harms K (1992) Angew Chem Int Ed Engl 104:1236; (b) Dötz KH, Siemoneit S, Hohmann F, Nieger M (1997) J Organomet Chem 541:285; (c) Hohmann F, Siemoneit S, Nieger M, Kotila S, Dötz KH (1997) Chem Eur J 3:853
10. Dötz KH (1984) Angew Chem Int Ed Engl 23:587
11. (a) Hofmann P, Hämmerle M (1989) Angew Chem Int Ed Engl 28:908; (b) Hofmann P, Hämmerle M, Unfried G (1991) New J Chem 15:769; (c) Gleichmann MM, Dötz KH, Hess BA (1996) J Am Chem Soc 118:10551
12. (a) Torrent M, Durant M, Solà M (1998) Chem Commun 999; (b) Torrent M, Durant M, Solà M (1999) J Am Chem Soc 121:1309
13. Casey CP (1981) Metal-carbene complexes. In: Jones M Jr, Moss RA (eds) Reactive intermediates. Wiley, New York, p 135
14. Barluenga J, Aznar F, Gutiérrez I, Martín A, García-Granda S, Llorca-Baragano MA (2000) J Am Chem Soc 122:1314
15. (a) Anderson BA, Wulff WD (1990) J Am Chem Soc 112:8615; (b) Chelain E, Parlier A, Rudler H, Daran, JC, Vaissermann J (1991) J Organomet Chem 419:C5; (c) Moser WH, Sun L, Huffman JC (2001) Org Lett 3:3389
16. Wulff WD, Bax BM, Brandvold TA, Chan KS, Gilbert AM, Hsung RS (1994) Organometallics 13:102
17. Waters ML, Brandvold TA, Isaacs L, Wulff WD, Rheingold AL (1998) Organometallics 17:4298
18. Bos ME, Wulff WD, Chamberlin RA, Brandvold TA (1991) J Am Chem Soc 113:9293
19. Wulff WD, Bax BM, Brandvold TA, Chan KS, Gilbert AM, Hsung RP, Mitchell J, Clardy J (1994) Organometallics 13:102
20. Lewis KE, Golden DM, Smith GP (1984) J Am Chem Soc 106:3905
21. Collman JP, Hegedus LS, Norton JR, Finke RG (1987) Principles and applications of organotransition metal chemistry, 2nd edn. University Science Books, Mill Valley, CA
22. (a) Dötz KH, Grotjahn DB, Harms K (1989) Angew Chem Int Ed Engl 28:1384; (b) Grotjahn DB, Dötz KH (1991) Synlett 381; (c) Grotjahn DB, Kroll FEK, Schäfer R, Harms K, Dötz KH (1992) Organometallics 11:298
23. Wulff WD, Gilbert AM, Hsung RP, Rahm A (1995) J Org Chem 60:4566
24. (a) Barluenga J, López LA, Martínez S, Tomás M (1998) J Org Chem 63:7588; (b) Barluenga J, López LA, Martínez S, Tomás M (2000) Tetrahedron 56:4967
25. Gross MF, Finn MG (1994) J Am Chem Soc 116:10921
26. Chamberlin S, Waters ML, Wulff WD (1994) J Am Chem Soc 116:3113
27. Davies MW, Johnson CN, Harrity JPA (2001) J Org Chem 66:3525
28. (a) Wang H, Wulff WD (1998) J Am Chem Soc 120:10573; (b) Dötz KH, Gerhardt A (1999) J Organomet Chem 578:223; (c) Dötz KH, Mittenzwey S (2002) Eur J Org Chem 39
29. (a) Wang H, Wulff WD (2000) J Am Chem Soc 122:9862; (b) Wang H, Huang J, Wulff WD, Rheingold AL (2003) J Am Chem Soc 125:8980

30. (a) Dötz KH, Dietz R (1978) Chem Ber 111:2517; (b) Dötz KH, Stendel J Jr (2002) The chromium-templated carbene benzannulation approach to densely functionalized arenes (Dötz reaction). In: Astruc D (ed) Modern arene chemistry. Wiley-VCH, Weinheim, p 250; (c) Dötz KH, Stendel J Jr (unpublished results)
31. Semmelhack MF, Ho S, Cohen D, Steigerwald M, Lee MC, Lee G, Gilbert AM, Wulff WD, Ball RG (1994) J Am Chem Soc 116:7108
32. (a) Jahr HC, Nieger M, Dötz KH (2002) J Organomet Chem 641:185; (b) Jahr HC, Nieger M, Dötz KH (in preparation)
33. Dötz KH, Schneider J (unpublished results)
34. (a) Fischer EO, Held W, Kreißl FR, Frank A, Huttner G (1977) Chem Ber 110:656; (b) Dötz KH, Dietz R (1977) Chem Ber 110:1555; (c) Dötz KH, Dietz R (1978) Chem Ber 111:2517
35. (a) Dötz KH, Pfeiffer J (1996) Chem Commun 895; (b) Pfeiffer J, Dötz KH (1998) Organometallics 17:4353; (c) Pfeiffer J, Nieger M, Dötz KH (1998) Chem Eur J 4:1843
36. (a) Weyershausen B, Dötz KH (1998) Eur J Org Chem 1739; (b) Weyershausen B, Dötz KH (1999) Synlett 2:231
37. (a) Hsung RP, Wulff WD (1994) J Am Chem Soc 116:6449; (b) Hsung RP, Quinn JF, Weisenberg BA, Wulff WD, Yap GPA, Rheingold AL (1997) Chem Commun 615
38. (a) Pulley SR, Carey JP (1998) J Org Chem 63:5275; (b) Paetsch D, Dötz KH (1999) Tetrahedron Lett 40:487
39. Kretschnik O, Nieger M, Dötz KH (1996) Organometallics 15:3625
40. Bao JB, Wulff WD, Fumo MJ, Grant EB, Heller DP, Whitcomb MC, Yeung SM (1996) J Am Chem Soc 118:2166
41. (a) Dötz KH, Stinner C, Nieger M (1995) J Chem Soc Chem Commun 2535; (b) Dötz KH, Stinner C (1997) Tetrahedron: Asymmetry 8:1751
42. (a) Quinn JF, Powers TS, Wulff WD (1997) Organometallics 16:4945; (b) Hsung RP, Wulff WD, Chamberlin S, Liu Y, Liu RY, Wang H, Quinn JF, Wang SLB, Rheingold AL (2001) Synthesis 8:200
43. (a) Moser E, Fischer EO (1968) J Organomet Chem 13:387; (b) Kreiter CG, Fischer EO (1976) Angew Chem Int Ed Engl 8:761
44. (a) Anderson BA, Wulff WD, Rahm A (1993) J Am Chem Soc 115:4602; (b) Moser R, Fischer EO (1969) J Organomet Chem 16:27
45. Hsung RP, Wulff WD, Challener CA (1996) Synthesis 773
46. (a) King JD, Quayle P (1991) Tetrahedron Lett 32:7759; (b) Beddoes RL, King JD, Quayle P (1995) Tetrahedron Lett 36:3027
47. Neidlein R, Gürtler S, Krieger C (1994) Helv Chim Acta 77:2303
48. Longen A, Nieger M, Airola K, Dötz KH (1998) Organometallics 17:1538
49. Tomuschat P, Kröner L, Steckhan E, Nieger M, Dötz KH (1999) Chem Eur J 5:700
50. (a) Hallett MR, Painter JE, Quayle P, Ricketts D, Patel P (1998) Tetrahedron Lett 39:2851; (b) Dötz KH, Otto F, Nieger M (2001) J Organomet Chem 621:77
51. (a) Vorogushin AV, Wulff WD, Hansen HJ (2001) Org Lett 3:2641; (b) Vorogushin AV, Wulff WD, Hansen HJ (2002) J Am Chem Soc 124:6512
52. Balzer BL, Cazanoue M, Sabat M, Finn MG (1992) Organometallics 11:1759
53. Balzer BL, Cazanoue M, Finn MG (1992) J Am Chem Soc 114:8735
54. (a) Semmelhack MF, Tamura R, Schnatter W, Springer J (1984) J Am Chem Soc 106:5363; (b) Semmelhack MF, Park J (1986) Organometallics 5:2550
55. Rehman AU, Schnatter WFK, Manolache N (1993) J Am Chem Soc 115:9848
56. Stockman KE, Sabat M, Finn MG, Grimes RN (1992) J Am Chem Soc 114:8733
57. Barluenga J, Barrio P, López LA, Tomás M, García-Granda S, Alvarez-Rúa C (2003) Angew Chem Int Ed 42:3008

58. (a) Dötz KH, Pruskil I (1981) J Organomet Chem 209:C4; (b) Dötz KH, Pruskil I, Mühlemeier J (1982) Chem Ber 115:1278
59. Dötz KH, Kuhn W (1983) Angew Chem Int Ed Engl 95:732
60. (a) Wulff WD, Tang PC (1984) J Am Chem Soc 106:434; (b) Wulff WD, Tang PC, Chan KS, McCallum JS, Yang DC, Gilbertson SR (1985) Tetrahedron 41:5813; (c) Dötz KH, Popall M (1987) Angew Chem Int Ed Engl 99:1158; (d) Dötz KH, Popall M, Müller G (1987) J Organomet Chem 334:57; (e) Dötz KH, Popall M (1988) Chem Ber 121:665; (f) Wulff WD, Xu YC (1988) J Am Chem Soc 110:2312
61. Su J, Wulff WD (1998) J Org Chem 63:8440; (b) Wulff WD, Su J, Tang PC, Xu YC (1999) Synthesis 3:415
62. (a) Gilbert AM, Miller R, Wulff WD (1999) Tetrahedron 55:1607; (b) Miller RA, Gilbert AM, Xue S, Wulff WD (1999) Synthesis 1:80; (c) Liptak VP, Wulff WD (2000) Tetrahedron 56:10229
63. (a) Boger DL, Jacobson IC (1989) Tetrahedron Lett 30:2037; (b) Boger DL, Jacobson IC (1990) J Org Chem 55:1919; (c) Boger DL, Jacobson IC (1991) J Org Chem 56:2115; (d) Boger DL, Zhang M (1992) J Org Chem 57:3974; (e) Boger DL, Jacobson IC (1995) J Am Chem Soc 117:11839
64. Yamashita A, Toy A, Scahill TA (1989) J Org Chem 54:3625
65. Semmelhack MF, Bozell JJ, Sato T, Wulff WD, Spiess E, Zask A (1982) J Am Chem Soc 104:5850
66. Xu YC, Kohlman DT, Liang SX, Erikkson C (1999) Org Lett 10:1599
67. Parker KA, Coburn CA (1991) J Org Chem 56:1666
68. Xie X, Kozlowski MC (2001) Org Lett 17:2661
69. Gordon DM, Danishefsky SJ, Schulte GK (1992) J Org Chem 57:7052
70. (a) Bao J, Wulff WD, Dragisich V, Wenglowsky S, Ball RG (1994) J Am Chem Soc 116:7616
71. Woodgate PD, Sutherland HS, Rickard CEF (2001) J Organomet Chem 627:206
72. Harrity JPA, Kerr WJ, Middlemiss D, Scott JS (1997) J Organomet Chem 532:219
73. Caldwell JJ, Colman R, Kerr WJ, Magennis EJ (2001) Synlett 9:1428
74. (a) Merlic CA, Xu D (1991) J Am Chem Soc 113:7419; (b) Merlic CA, Xu D, Gladstone BG (1993) J Org Chem 58:539; (c) Merlic CA, Roberts WM (1993) Tetrahedron Lett 34:7379
75. (a) Barluenga J, Aznar F, Palomero MA, Barluenga S (1999) Org Lett 1:541; (b) Barluenga J, Aznar F, Palomero MA (2003) J Org Chem 68:537
76. (a) Merlic CA, Burns EE, Xu D, Chen SY (1992) J Am Chem Soc 114:8723; (b) Merlic CA, Burns EE (1993) Tetrahedron Lett 34:5401
77. (a) Barluenga J, Aznar F, Palomero MA (2001) Chem Eur J 7:5318; (b) Barluenga J, Aznar F, Palomero MA (2002) Chem Eur J 8:4149
78. Barluenga J, Aznar F, Palomero MA (2000) Angew Chem Int Ed 39:4346
79. Aumann R, Jasper B, Fröhlich R (1995) Organometallics 14:231
80. (a) Merlic CA, McInnes DM, You Y (1997) Tetrahedron Lett 38:6787; (b) Merlic CA, You Y, McInnes DM, Zechman AL, Miller MM, Deng Q (2001) Tetrahedron 57:5199
81. (a) Merlic CA, Aldrich CC, Albaneze-Walker J, Saghatelian A (2000) J Am Chem Soc 122:3224; (b) Merlic CA, Aldrich, CC, Albaneze-Walker J, Saghatelian A, Mammen J (2001) J Org Chem 66:1297

Topics Organomet Chem (2004) 13: 157–201
DOI 10.1007/978-3-540-40910-6

Photoinduced Reactions of Metal Carbenes in Organic Synthesis

Louis S. Hegedus (✉)

Department of Chemistry, Colorado State University, Fort Collins, CO 80523-1872, USA
hegedus@lamar.colostate.edu

Abstract The photoinduced reactions of metal carbene complexes, particularly Group 6 Fischer carbenes, are comprehensively presented in this chapter with a complete listing of published examples. A majority of these processes involve CO insertion to produce species that have ketene-like reactivity. Cycloaddition reactions presented include reaction with imines to form β-lactams, with alkenes to form cyclobutanones, with aldehydes to form β-lactones, and with azoarenes to form diazetidinones. Photoinduced benzannulation processes are included. Reactions involving nucleophilic attack to form esters, amino acids, peptides, allenes, acylated arenes, and aza-Cope rearrangement products are detailed. A number of photoinduced reactions of carbenes do not involve CO insertion. These include reactions with sulfur ylides and sulfilimines, cyclopropanation, 1,3-dipolar cycloadditions, and acyl migrations.

Keywords Metal carbenes · Photochemical reactions · Metal-ketene complexes

Abbreviations

Bn	Benzyl
Cbz	Benzyloxycarbonyl
DMAP	Dimethylaminopyridine
HOMO	Highest occupied molecular orbital
LF	Ligand field
LUMO	Lowest unoccupied molecular orbital
MLCT	Metal-to-ligand charge transfer
PMB	Para-methoxybenzyl
PMP	Para-methoxyphenyl
PPTS	Pyridinium para-toluenesulfonate
tBOC	*t*-Butyloxycarbonyl

1 Introduction

Although many transition metals form carbene complexes, only Group 6 (Cr, Mo, W) heteroatom-stabilized Fischer carbenes of the type

$$(CO)_5M{=}C(Z)(R) \quad (Z = OR', NR_2', \text{etc.}) \qquad \text{(Intro)}$$

have been extensively studied as reagents for organic synthesis (with the obvious exception of olefin metathesis chemistry [1]). Synthetically useful photochemical reactions have largely been restricted to Cr and Mo carbene complexes, thus this chapter will deal primarily with the chemistry of these. Since photochemical reactions involve excited-state chemistry at some stage, the electronic (UV-VIS) spectra of Fischer carbenes are central to a consideration of this chemistry.

The visible spectra of Fischer carbene complexes consist of a very weak band above 500 nm, assigned to a spin-forbidden metal-to-ligand charge transfer band (MLCT), a moderately intense band between 350 and 450 nm assigned as a spin-allowed MLCT, and a weaker band at 300–350 nm assigned as a ligand field (LF) transition [2]. A lower energy LF band is usually masked by the more intense MLCT, although it has been observed in some nonheteroatom-stabilized carbene complexes [3]. The HOMO is metal dπ–pπ centered while the LUMO is carbene-carbon pπ centered [4]. As a result, irradiation into the MLCT band should lead to charge transfer from the metal to the ligand, a formal oxidation of the metal.

In attempts to understand the photochemical reactions of Fischer carbene complexes, several matrix isolation and flash photolysis studies have been conducted using both Cr and W (but not Mo) complexes [5–11]. Although the complexes studied and conditions used varied, several general conclusions were drawn:

1. For chromium alkoxycarbene complexes the MLCT and the lowest energy LF bands overlap. Irradiation at λ>385 nm led to *anti–syn* isomerization

of the OMe group and 30% loss of CO. Loss of CO was even observed at λ>400 nm [9].

2. Tungsten alkoxycarbene complexes underwent similar *anti–syn* rearrangements but were much less prone to undergo CO loss [5–10].
3. No CO-insertion products (metal-ketene complexes) were observed, even when specifically sought [9, 10].

These results suggest that there should be little useful organic chemistry resulting from photoinduced reactions of Fischer carbene complexes. However, this was shown not to be the case.

In studies designed to develop new approaches to β-lactams, Michael McGuire, then a graduate student in the author's research group, discovered that photolysis of a range of Cr Fischer carbene complexes with visible light through Pyrex produced a short-lived species that had ketene-like reactivity [12]. Subsequent studies [13] suggested that irradiation promoted reversible insertion of one of the four *cis*-COs into the metal-carbene-carbon double bond, producing a short-lived metallacyclopropanone-metal-ketene complex (Eq. 1). In the absence of reactive substrates, rapid deinsertion occurred, regenerating the carbene complex.

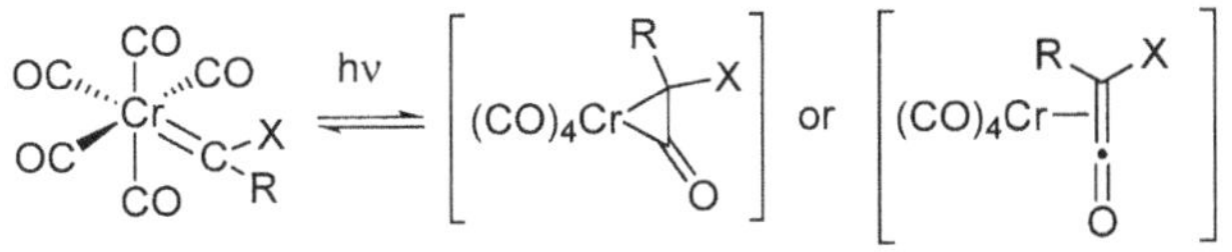

$$(CO)_5Cr{=}C(X)R \; \underset{}{\overset{h\nu}{\rightleftharpoons}} \; \left[(CO)_4Cr\text{–}C(O)\text{–}C(R)(X)\right] \text{ or } \left[(CO)_4Cr\text{–}(R)(X)C{=}C{=}O\right] \qquad (1)$$

Several stable Group 6 metal-ketene complexes are known [14], and photo-driven insertion of CO into a tungsten-carbyne-carbon triple bond has been demonstrated [15]. In addition, *thermal* decomposition of the nonheteroatom-stabilized carbene complexes $(CO)_5M{=}CPh_2$ (M=Cr, W) produces diphenylketene [16]. Thus, the intermediacy of transient metal-ketene complexes in the photodriven reactions of Group 6 Fischer carbenes seems at least possible.

2 Photoinduced Reactions of Fischer Carbene Complexes

2.1 Involving CO Insertion

2.1.1 Cycloaddition Reactions

2.1.1.1 With Imines to Form β-Lactams

The reaction of ketenes (usually formed from treatment of acid chlorides with tertiary amines) with imines is a classic way to form β-lactams [17, 18]. Although widely used, it suffers limitations in scope and efficiency, since free ketenes are highly reactive, and prone to dimerization and multiple incorporations into products. Whatever the nature of the photogenerated species from Fischer carbene complexes, *free* ketenes are not produced, and these by-products are not expected.

Photolysis of chromium alkoxycarbene complexes with a wide range of acyclic imines of aromatic aldehydes produced β-lactams in good to excellent yield (Table 1). The reaction was highly diastereoselective in virtually all cases, giving the relative stereochemistry shown. Cyclic and heterocyclic imines were similarly reactive, again producing single diastereoisomers (Table 2). Of particular note is the clean conversion of protected imidazolines to azapenams. The transformation using ketenes generated from acid chlorides does not take place [27]. Bis-carbene complexes underwent photoreaction with imidazolines to give bis-azapenams as 1:1 mixtures of diastereoisomers. (The *relative* configuration of each azapenam had the two heteroatoms *trans*, as expected, but a 1:1 mixture of [(*R*,*R*)(*S*,*S*)] and (*R*,*S*) diastereomers resulted)

Table 1 Reaction of chromium alkoxycarbenes with acyclic aryl aldimines

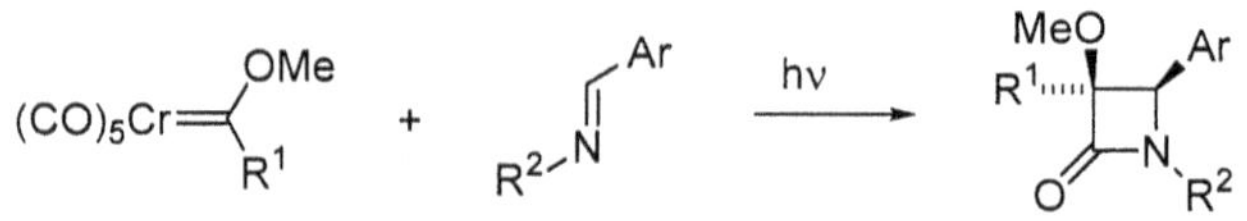

$(CO)_5Cr{=}C(OMe)R^1$ + R^2–N=CH–Ar —hν→ β-lactam (MeO, R^1, Ar, N–R^2)

R^1	R^2	Ar	Yield, %	Ref
Me	Me	Ph	76	[19]
Me	Ph	Ph	52	[19]
Ph	Me	Ph	72	[19]
Ph	Ph	Ph	20	[19]
Me	pMeOPh	Ph	60	[19]
Me	Bn	PhCHCH	45	[19]
Me	$CH_2P(O)(OEt)_2$	Ph	90	[20]
Me	$CH(CO_2Me)P(O)(OEt)_2$	Ph	80	[20]
Me	$CHCH_2$	Ph	41	[20]
Me	Bn	Ph	53	[20]
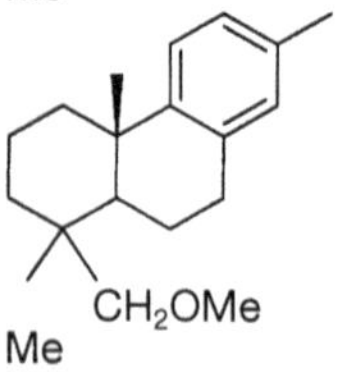 (tricyclic structure with CH_2OMe)	Me	Ph	71	[21]
Me	pMeOPh	Fc[a]	79[b]	[22]
Me	Fc	pMeOPh	85[c]	[22]
Me	Fc	Fc	88[c]	[22]

[a]Fc = ferrocenyl; [b]As a 57:43 cis/trans mixture; [c]The ethoxycarbene was used.

Table 2 Reaction of chromium alkoxycarbenes with cyclic imines

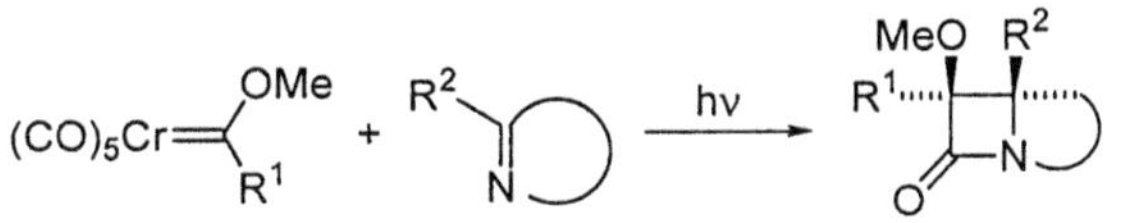

$(CO)_5Cr{=}C(OMe)R^1$ + R^2-imine (cyclic) $\xrightarrow{h\nu}$ β-lactam (MeO, R^1, R^2)

R^1	Imine		Yield, %	Ref.
Me	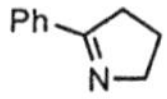 2-Ph-pyrroline		70	[23]
Me	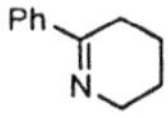 2-Ph-tetrahydropyridine		63	[23]
Me	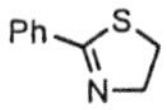 2-Ph-thiazoline		52	[19]
Me	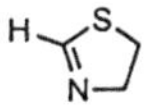 2-H-thiazoline		81	[19]
Me	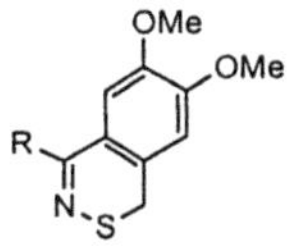 dimethoxy benzothiazine	R = Ph	51	[19]
Me	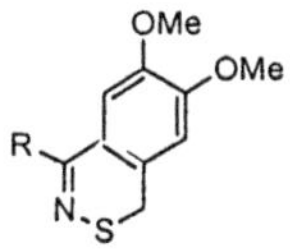 dimethoxy benzothiazine	R = Me	38	[19]
Me		R = H	37	[19]
Ph		R = Ph	27	[19]
Me	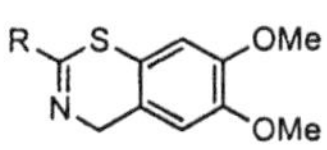 dimethoxy benzothiazine	R = Ph	61	[19]
Ph		R = Ph	25	[19]
Me	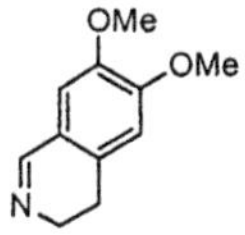 dimethoxy dihydroisoquinoline		43	[19]
Me	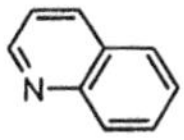 quinoline		38	[19]
Me	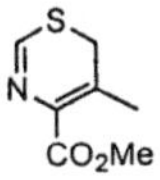 thiazine, CO_2Me		52	[20]
Me	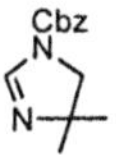 Cbz imidazoline		69	[24]
n-C_6H_{13}	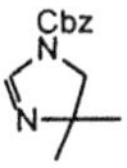 Cbz imidazoline		71	[25]
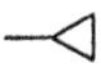 cyclopropyl	"		52	[25]

Table 2 (continued)

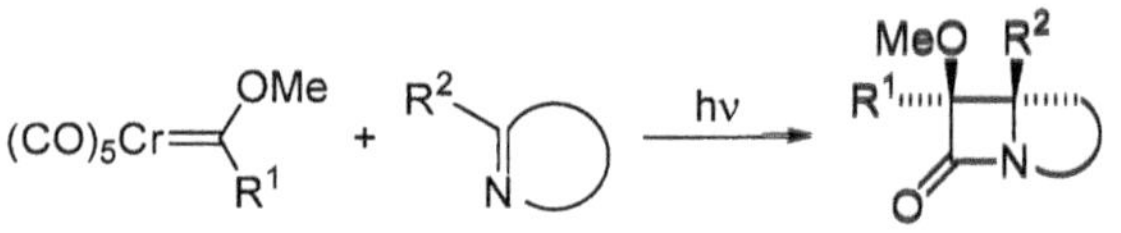

R^1	Imine	Yield, %	Ref.
i-Pr[a]	"	77	[25]
pMeOPh	"	39	[25]
c-hex	"	37	[25]
t-Bu[a]	"	20	[25]
Me[b]	"	74	[26]
Me[c]	"	77	[26]

[a]The ethoxy carbene complex was used. [b]The PMB alkoxycarbene was used. [c]The OCH_2CH_2NHCO quinoxaline carbene was used.

(Eq. 2) [28, 29]. Bis-carbenes linked through the alkyl (rather than alkoxy) groups reacted similarly but in lower yields [30].

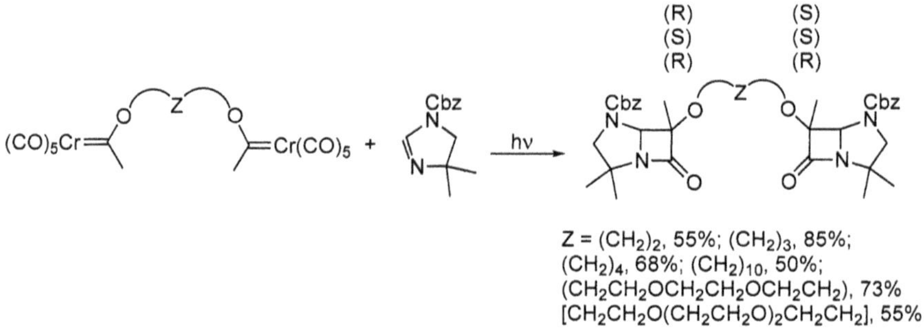

Z = $(CH_2)_2$, 55%; $(CH_2)_3$, 85%; $(CH_2)_4$, 68%; $(CH_2)_{10}$, 50%; $(CH_2CH_2OCH_2CH_2OCH_2CH_2)$, 73% $[CH_2CH_2O(CH_2CH_2O)_2CH_2CH_2]$, 55%

(2)

A few heterocyclic imines reacted poorly if at all with chromium alkoxycarbene complexes. Oxazines required the use of the more reactive (and less stable) molybdenum alkoxycarbenes, producing oxacephams in ≈40% yield. Oxazolines gave low yields (≈12%) of the oxapenam system, along with similar amounts of oxazinone, resulting from incorporation of *two* equivalents of ketene (Eq. 3) [20].

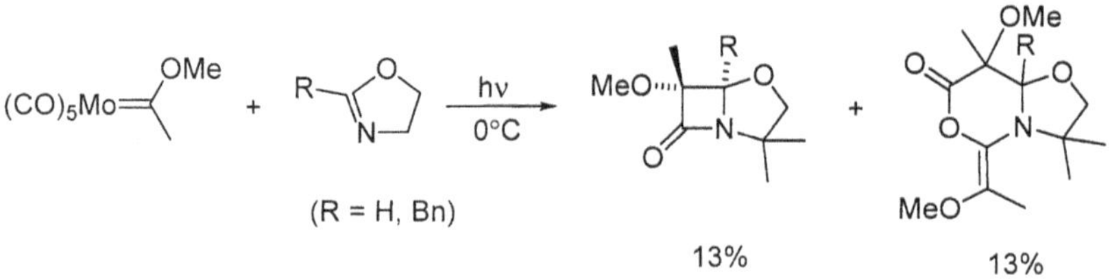

(R = H, Bn) 13% 13% (3)

CBz-protected benzimidazole gave primarily oxazinone [31], while 3*H*-indoles incorporated two equivalents of imine (Eq. 4) [32]. In these cases it appears that the initially formed zwitterionic ketene–imine adduct could not close, and reacted with additional photoactivated carbene or substrate.

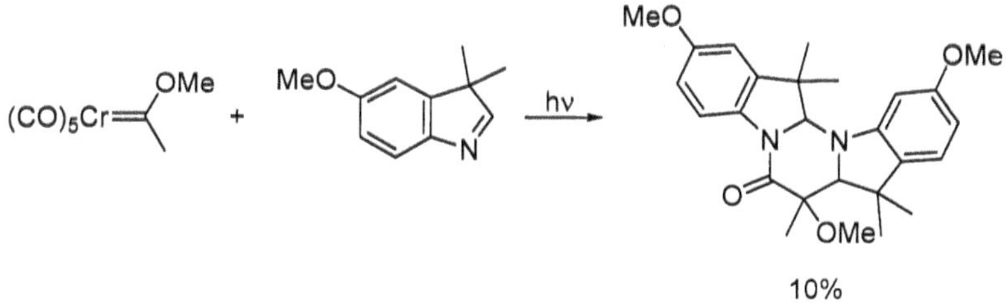

(4)

Other miscellaneous imines that underwent photoreaction with chromium alkoxycarbenes include iminodithiocarbonates [33], the mono-*N*-phenyl imine of benzil and the bis-*N*-phenyl imine of acetoin [20]. By preparing the chromium carbene complex from ^{13}CO-labeled chromium hexacarbonyl, *β*-lactams with two adjacent ^{13}C labels were synthesized [34].

Induction of asymmetry into the *β*-lactam-forming process was inefficient with acyclic imines having chiral groups on the nitrogen [19] but efficient with rigid, cyclic chiral imines (Table 3). One of these was used as a chiral template to produce highly functionalized quaternary systems (Eq. 5) [34].

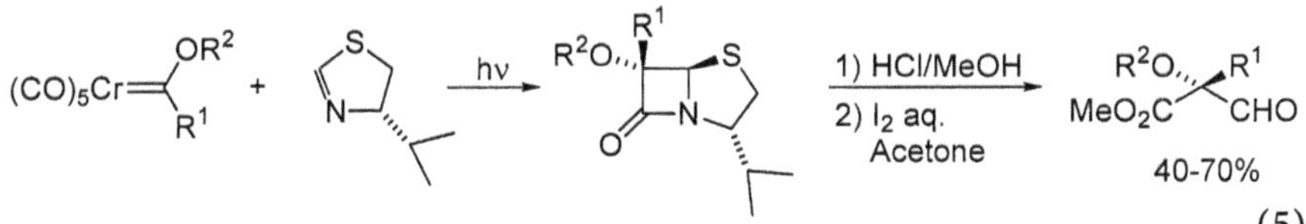

(5)

The mechanism of the classic ketene–imine reaction to form *β*-lactams [17, 18] is thought to involve perpendicular attack of the imine nitrogen on the ketene carbonyl carbon from the side of the sterically smaller of the two groups, followed by conrotatory closure of the zwitterionic intermediate (Eq. 6). This

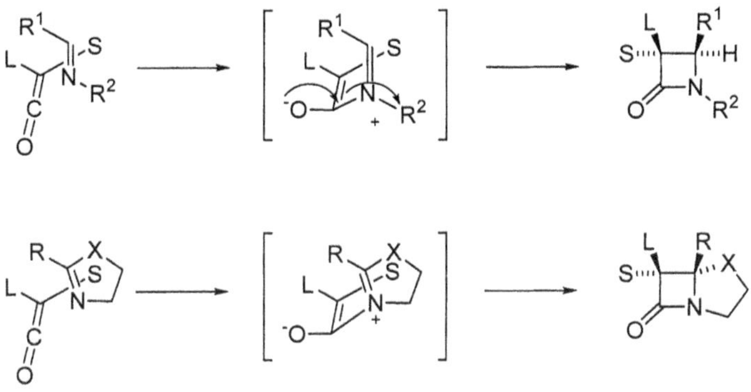

(6)

Table 3 Reaction of chromium alkoxycarbenes with chiral heterocyclic imines

R^1	R^2	R^3	X	Yield, %	ee[a]	Ref.
Me	Bn	*i*-Pr	S	76	>97%	[35]
n-Bu	Bn	*i*-Pr	S	78	>97%	[35]
cyclopropyl	Bn	*i*-Pr	S	39	>97%	[35]
Ph	Bn	*i*-Pr	S	42	>97%	[35]
p-MeOPh	Bn	*i*-Pr	S	42	>97%	[35]
2-furyl			S	29	>97%	[35]
Me	Me	b		42	>97% (X-ray)	[19]
Me	Me	*i*-Pr	NCbz	41	>97% (X-ray)	[24]
Me	Bn	*i*-Pr	NCbz	54	>97%	[24]
—$(CH_2)_3$—		*i*-Pr	NCbz	69	>97%	[24]
Me	Me	Me	NCbz	23	>97%	[36]
Me	Me	Ph	NCbz	35	>97%	[36]
Me	Me	c	NCbz	75	>97%	[36]

[a]Absolute configuration determined by conversions to compounds of known absolute configuration, and by correlation to closely related compounds for which X-ray structures are available. [b]The substrate was [structure: thiazoline with gem-dimethyl and CO_2Me]

[c]The substrate was [structure: N-Cbz imidazoline with gem-dimethyl, methyl and CO_2Me]

places the large substituent of the ketene *cis* to the *anti*-substituent of the imine. In *all* the cases cited above, the observed stereochemistry was exactly opposite that expected on these steric grounds. Initially, this difference was thought to be due to the presence of the metal during the cycloadditions, biasing the process to produce the contrasteric product. However, subsequent considerations [37], supported by theoretical calculations relating the closure step of β-lactam formation to the electronic bias observed ("torquoselectivity") in the ring opening of cyclobutenes bearing heteroatom substituents [38], suggested that the observed stereoselectivity was due to the presence of the donor methoxy group on the ketene. This greatly lowers the energy for closure of the zwitterion resulting from attack over the large R group, from the face opposite the donor groups, leading to the contrasteric product (Eq. 7).

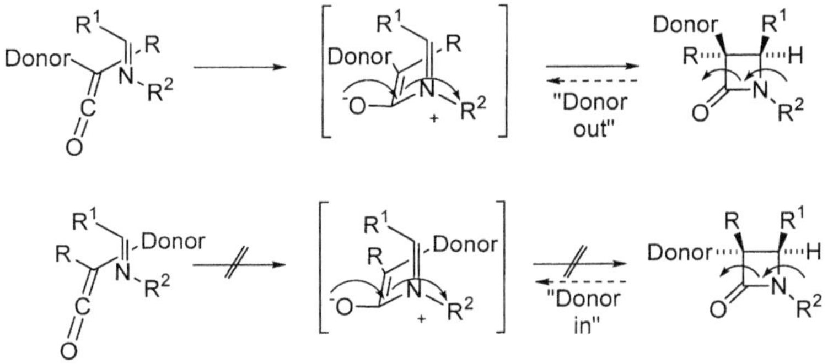

(7)

Chromium aminocarbenes [39] are readily available from the reaction of $K_2Cr(CO)_5$ with iminium chlorides [40] or amides and trimethylsilyl chloride [41]. Those from *formamides* (H on carbene carbon) readily underwent photoreaction with a variety of imines to produce β-lactams, while those having R-groups (e.g., Me) on the carbene carbon produced little or no β-lactam products [13]. The dibenzylaminocarbene complex underwent reaction with high diastereoselectivity (Table 4). As previously observed, cyclic, optically active imines produced β-lactams with high enantioselectivity, while acyclic, optically active imines induced little asymmetry. An intramolecular version produced an unusual anti-Bredt lactam rather than the expected β-lactam (Eq. 8) [44].

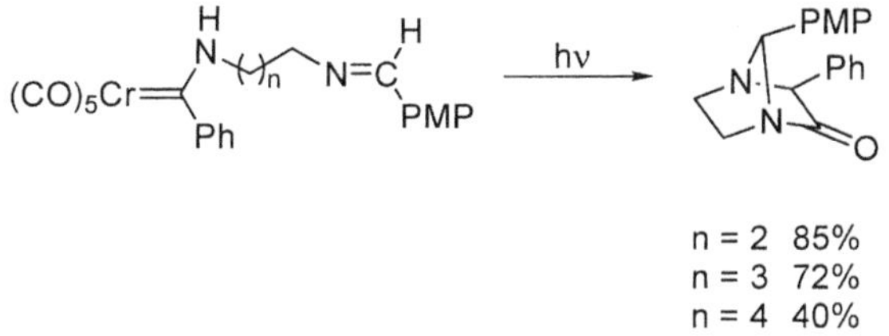

(8)

n = 2 85%
n = 3 72%
n = 4 40%

With optically active formamide-derived aminocarbene complexes high enantioselectivity was observed in most cases (Table 5). This chemistry was used in the synthesis of 1-carbacephalathin and 3-ANA precursors (Eq. 9) [48], as well as the synthesis of α,α'-disubstituted amino acids (Scheme 1) [49].

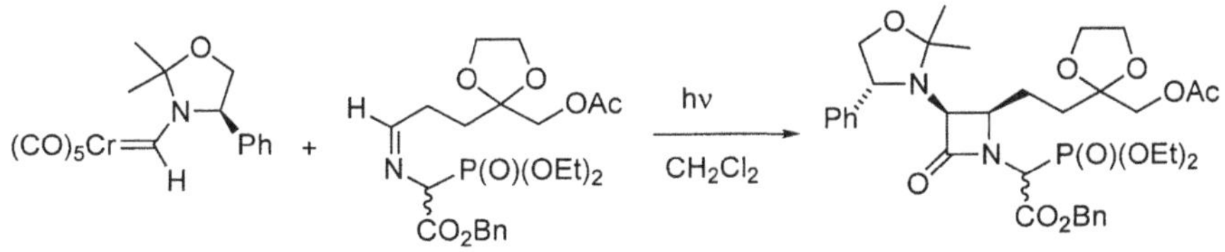

32% (9)

Although the photodriven reactions of chromium carbene complexes with imines superficially resemble those of free ketenes, there are major differences. The optically active oxazolidine carbene (Table 5) gave excellent yields and high ee values when allowed to react with imidates, oxazines, thiazines, and

Table 4 Reaction of chromium aminocarbenes with imines

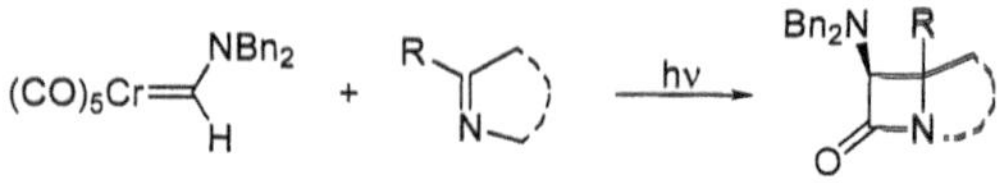

Imine	Product	Yield, %	Ref.
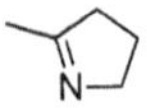	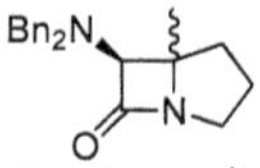	51[a]	[40]
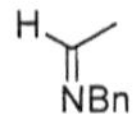	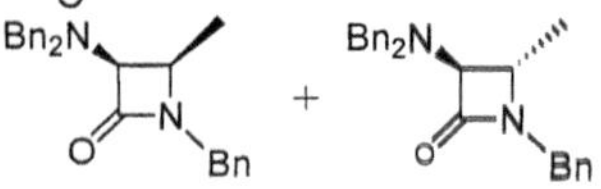	74 (7:4)	[40]
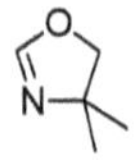	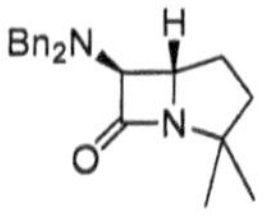	32	[40]
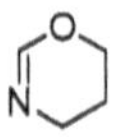	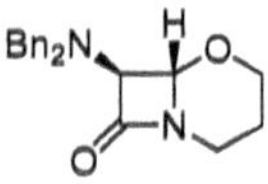	81	[40]
	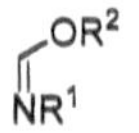 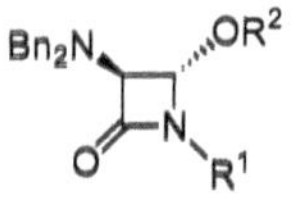 R^1 = Me, R^2 = Bn	79	[40]
	R^1 = Et, R^2 = Ph	76	[40]
	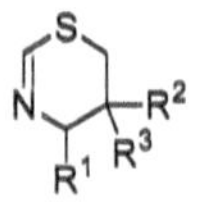 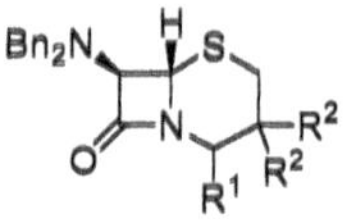 R^1 = R^2 = R^3 = H	73	[40]
	R^1 = CO_2Et, R^2 = OH, R^3 = Me	76	[40]
	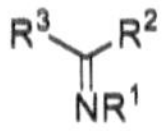 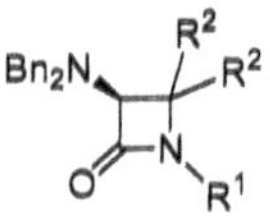 R^1 = Bu^t, R^2, R^3 = H	72	[42]
	R^1 = Bn, R^2, R^3 = Me	56	[42]
	R^1 = Bn, R^2, R^3 = H	72	[42]
	R^1 = *p*-$MeOC_6H_4$, R^2, R^3 = H	44	[42]
	 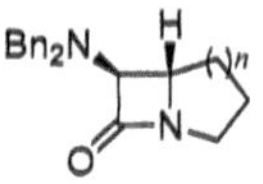 n = 1	54	[42]
	n = 2	85	[42]
	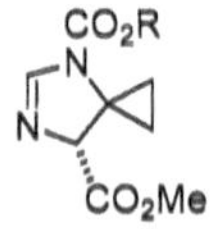 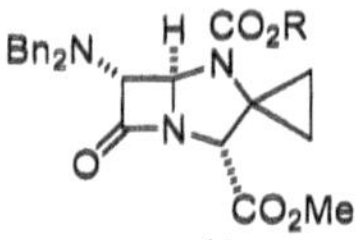 R = Menthyl	80	
	R = Bn	50	[43]
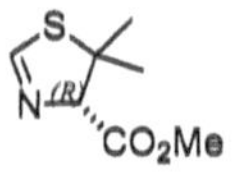	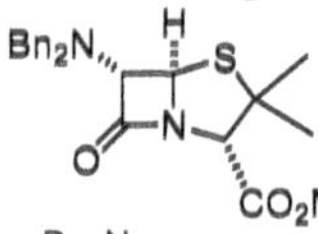	93[b]	[40]
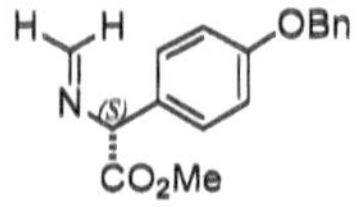	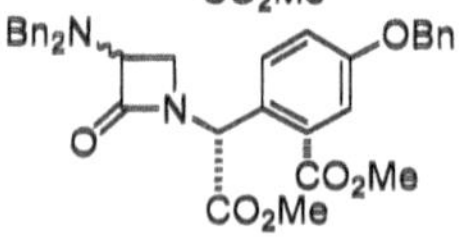	46 (1:1)	[42]

[a]Single isomer. [b]Single enantiomer.

Table 5 Reaction of optically active aminocarbenes with imines

$(CO)_5Cr$=C(H)(oxazolidine, Z) + $R^3C(R^2)=NR^1$ $\xrightarrow{h\nu}$ β-lactam

Z	R¹	R²	R³	Yield (%)	de (%)	ref
(S) iPr	Bn	Me	Me	59	70 (S,S)	[45]
(S) iPr	Bn	Me	H *(trans)*	40	≥97 (S,S)	[45]
(S) iPr	Bn	H	Me *(cis)*	14	≥97 (S,S)	[45]
(S) iPr	$-(CH_2)_4-$		H	55	≥97 (S,S)	[45]
(S) iPr	Bn	OMe	H	76	≥97 (S,S)	[45]
(S) iPr	$-(CH_2)_3O-$		H	70	≥97 (S,S)	[45]
(R)Ph	Bn	H	H	74	70 (R,R)	[45]
(R)Ph	Bn	Me	Me	79	70 (R,R)	[45]
(R)Ph	Bn	Me	H *(trans)*	41	≥97 (R,R)	[45]
(R)Ph	Bn	H	Me *(cis)*	20	≥97 (R,R)	[45]
(R)Ph	$-(CH_2)_3-$		H	75	≥97 (R,R)	[45]
(R)Ph	$-(CH_2)_4-$		H	91	≥97 (R,R)	[45]
(R)Ph	Bn	OMe	H	91	≥97 (R,R)	[45]
(R)Ph	$-(CH_2)_3O-$		H	95	≥97 (R,R)	[45]
(S)Ph	Bn	SMe	SMe	61	60 (S,S)	[46]
(S)Ph	PMP	SMe	SMe	60	60 (S,S)	[46]
(S)Ph	(imidazoline, N-P)		p = tBoc	78	>97 (S,S)	[47]
			p = pTs	93	96 (S,S)	[47]
			p = $BnSO_2$	86	80 (S,S)	[47]
			p = Cbz	80	80 (S,S)	[47]

aliphatic imines, but modest yields of mixtures of *cis* and *trans* isomers with aryl or α,β-unsaturated imines [50]. In contrast, the corresponding *oxazolidinone* ketene (from the acid chloride [51]) gave excellent yields and ee values with aryl and α,β-unsaturated imines but very low yields of β-lactams with other imines. Clearly chromium is influencing the outcome of the process.

Pyrrolocarbenes produced low yields of β-lactams in photodriven reactions with imines [52], while *o*-acylimidatocarbene complexes gave a mixture of compounds with β-lactams being minor components [53].

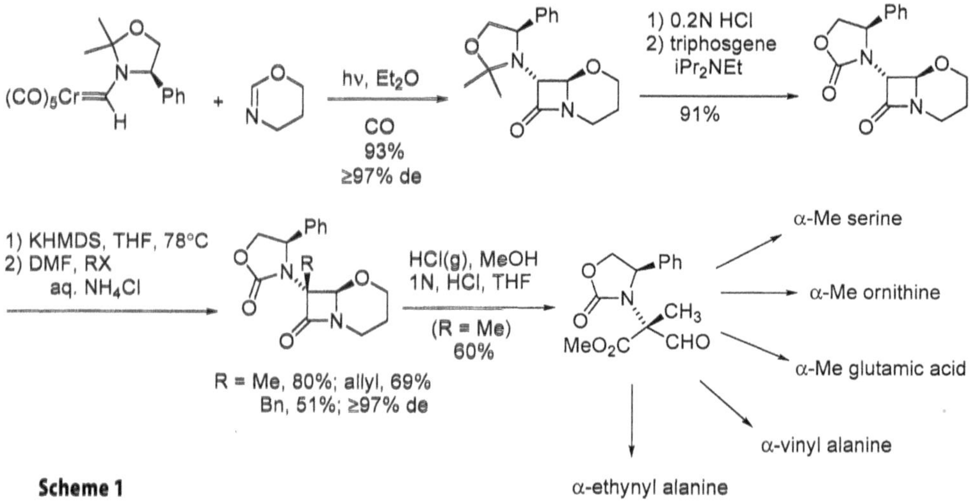

Scheme 1

2.1.1.2
With Olefins to Give Cyclobutanones

The first report of the reaction of a chromium alkoxycarbene with an alkene to give a cyclobutanone came in 1974 [54], when it was reported that treatment of the (phenyl)(methoxy) chromium carbene complex with *N*-vinyl pyrrolidinone under 150 atm of CO pressure produced the corresponding cyclobutanone, presumably via the ketene or ketene complex produced by pressure-driven insertion of CO into the metal-carbene-carbon bond. It wasn't until 1989 that the photodriven version of this process was reported [55]. Monosubstituted, electron-rich alkenes underwent photochemical reaction with chromium alkoxycarbenes to produce cyclobutanones in fair to good yield and with high stereoselectivity for the *more hindered* cyclobutanone (Table 6) [56], the same selectivity as that observed with free ketenes [57]. Di- and trisubstituted alkenes were somewhat less efficient (Table 7) [56], while dienes underwent cycloaddition to one of the two alkenes (Table 8) [56]. Intramolecular versions were also efficient to form five- and six-membered rings, but larger rings failed to form and tethered alkynes gave complex mixtures of unidentified products (Table 9) [56].

Alkoxycarbene complexes with unsaturation in the alkyl side chain rather than the alkoxy chain underwent similar intramolecular photoreactions (Eqs. 10 and 11) [60]. Cyclopropyl carbene complexes underwent a facile vinylcyclopropane rearrangement, presumably from the metal-bound ketene intermediate (Eqs. 12 and 13) [61]. A cycloheptatriene carbene complex underwent a related [6+2] cycloaddition (Eq. 14) [62].

$(CO)_5Cr$=C(OMe)– ... hν → 78% (10)

(cis → complex mixture)

Table 6 Reaction of alkoxycarbenes with monosubstituted alkenes

$(CO)_5Cr$=C(OMe)Me + CH2=CH–Z —hν→ MeO / Z / O (cyclobutanone)

Z	Yield (%)	ratio	ref
Ph	44	10:1	[56]
CH_2OAc	31	10:1	[56]
OEt	87	6:1	[56]
OAc	19	20:1	[56]
NHAc	96	8:1	[56]
N-pyrrolidinone (N, O)	78	7:1	[56]
nBu	55	>20:1	[58]
$TMSCH_2$	78	>20:1	[58]
$-(CH_2)_3-$[a]	84	>20:1	[58]
$-CH=CH-CH_2$[b]	80	>20:1	[58]
p-HOPh	51	>20:1	[58]
pMeOPh	59	>20:1	[58]
PhS	58	>20:1	[58]
$-(CH_2)_3O-$	80[c]	>20:1	[58]
tBuO	51	>20:1	[58]
BnO	82	>20:1	[58]
(±) Ph–CH(CH3)–O	68	1:1	[58]
2,4,6-iPrPhCH$_2$O	72	1:1	[58]
(±) 2,4,6-iPrPhCHO(CH$_3$)	67	1:1	[58]

[a]cyclopentene; [b]cyclopentadiene; [c]dihydropyran.

Table 7 Reaction of alkoxycarbenes with polysubstituted alkenes

$(CO)_5Cr=C(OMe)Me$ + $R^2R^1C=CHR^3$ —hν→ (cyclobutanone product: MeO, R^1, R^2, R^3)

R^1	R^2	R^3	ratio	yield	ref
Me	H	Me	5:1	51	[56]
H	Me	Me	5:1	13	[56]
$-(CH_2)_5-$		H	---	45	[56]
Me	Me	Me	>95:5	77	[56]
OAc	Me	H	1:1	16	[56]
$R^1R^3 = O(CH_2)_4$		$R^2 = H$	11:1	80	[56]
	(1,4-dioxene)		---	47	[59]

Table 8 Reaction of alkoxycarbenes with cyclic dienes

$(CO)_5Cr=C(OR^1)R$ + cyclic diene $(CH_2)_n$ —hν→ (bicyclic cyclobutanone product: R, R^1O, n)

R	R^1	n	ratio	yield	ref
Me	Me	1	10:1	67	[56]
Me	Me	2	14:1	52	[56]
cyclopropyl	Me	2	13:1	46	[56]
Me	Et	2	10:1	85	[56]
Ph	Me	2	20:1	75	[56]
nBu	Et	2	14:1	96	[56]
Me	Bn	2	9:1	70	[56]
Me	TMS	2	5:1	37	[56]
PhCHCH	Me	2	---	0	[56]
R=R′=OMe	(1-methoxy-1,3-cyclohexadiene, OMe)	---	15:1	73	[59]
R=R′=OMe	(1,2-dimethylenecyclopentane)	---	3:1	67	[59]

Table 9 Intramolecular cyclobutanone-forming reactions

R^1	R^2	R^3	R^4	R^5	n	yield (%)	ref
Me	H	H	H	H	0	88	[56]
Me	H	Me	H	H	0	62	[56]
Me	H	H	Et	H	0	47	[56]
Me	H	H	H	Et	0	29[a]	[56]
Me	Me	H	H	H	0	62[b]	[56]
Me	H	H	Me	Me	0	80[c]	[56]
Me	H	H	H	H	1	73	[56]
cyclopropyl	H	Me	H	H	0	65	[56]
Bn	H	Me	H	H	0	95	[56]
Ph	H	H	Et	H	0	42	[59][d]
Ph	H	H	H	Et	0	72	[59][e]
						97[f]	[56]

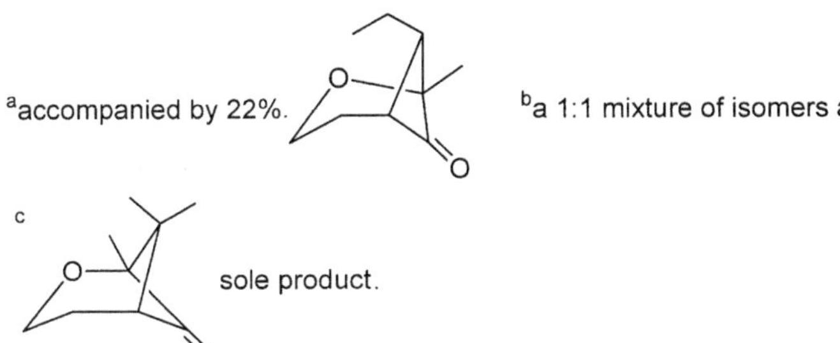

[a]accompanied by 22%. [b]a 1:1 mixture of isomers at R^2.

[c] sole product.

[d]2:1 mixture of bridged bicyclic to fused bicyclic ketone.
[e]1:1 mixture of bridged bicyclic to fused bicyclic ketone.
[f]2:1 mixture of trans isomers at 6,6 ring junction.

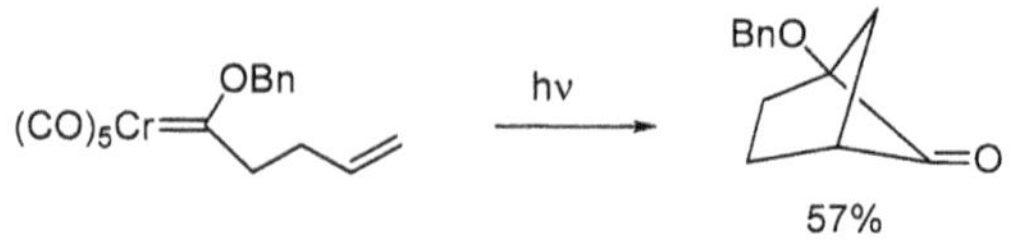

(11)

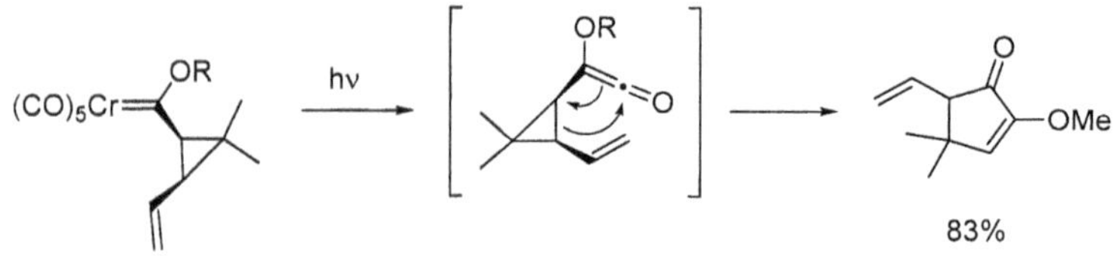

(12)

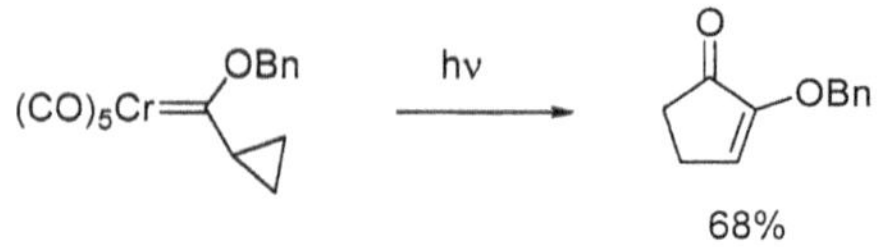

(13)

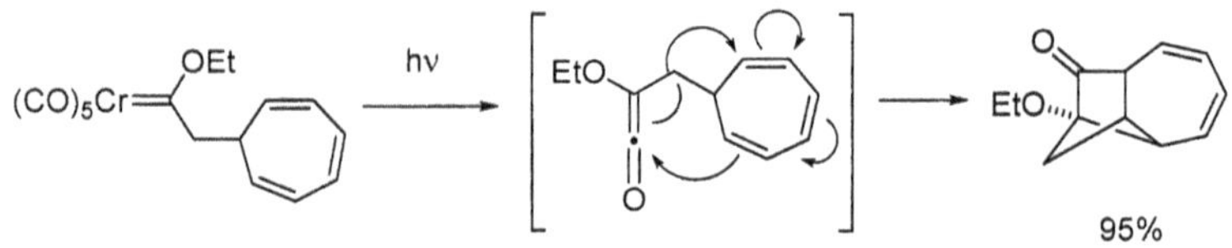

(14)

In contrast to alkoxycarbene complexes, most aminocarbene complexes appear too electron-rich to undergo photodriven reaction with olefins. By replacing aliphatic amino groups with the substantially less basic aryl amino groups, modest yields of cyclobutanones were achieved (Table 10) [63], (Table 11) [64]. Both reacted with dihydropyran to give modest yields of cyclobutanone. Thiocarbene complexes appeared to enjoy reactivity similar to that of alkoxycarbenes (Eq. 15) [59].

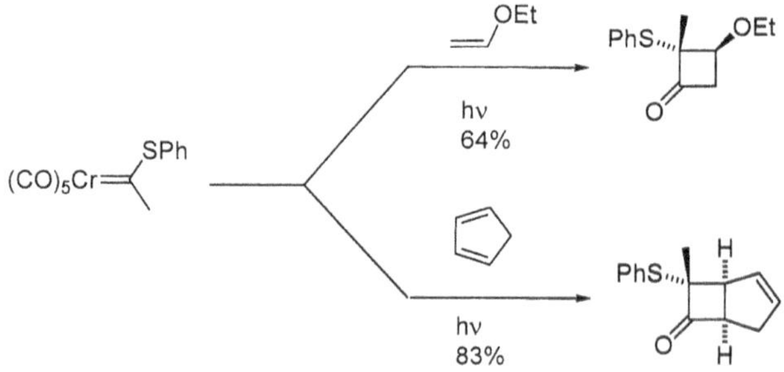

(15)

Of perhaps greater use for organic synthesis was the observation that photodriven reactions of alkoxycarbenes with unsubstituted optically active ene carbamates [65] produced aminocyclobutanones in fair yield with high diastereoselectivity (Table 12) [66]. In contrast, with a gem-disubstituted ene carbamate, the *syn–anti* selectivity was low but high asymmetric induction α to nitrogen was observed (Eq. 16). *Trans*-monosubstituted ene carbamates failed to react, as did α,β-unsaturated chromium carbene complexes.

Table 10 Reaction of arylaminocarbenes with alkenes

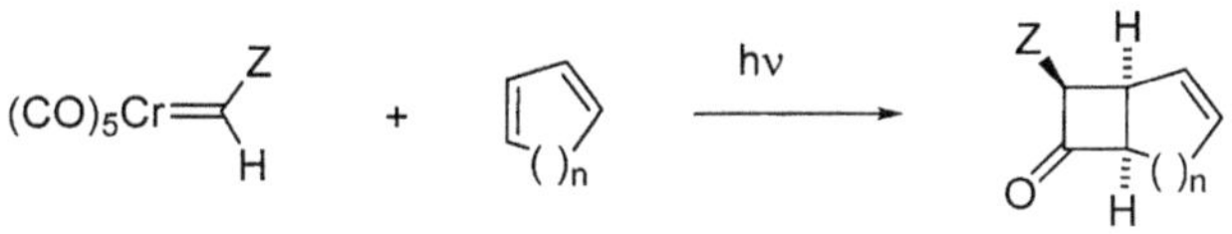

Z	n	yield (%)	ref
PhNMe	1	44	[63]
	2	37	[63]
pMeOPhNMe	1	37	[63]
Ph_2N	1	45	[63]
2,6-Me_2PhNMe	1	38	[63]
tetrahydroquinolyl	1	44	[63]
3-Me indolyl	1	38	[63]

Table 11 Reaction of chromium pyrrolocarbene complexes with alkenes

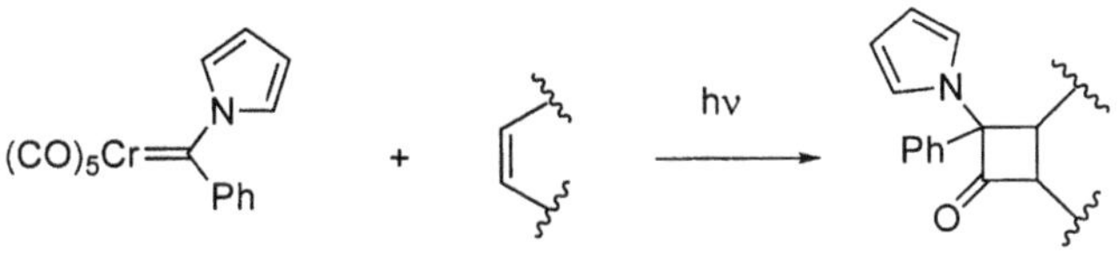

Alkene	Product	yield (%)	ref
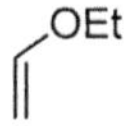	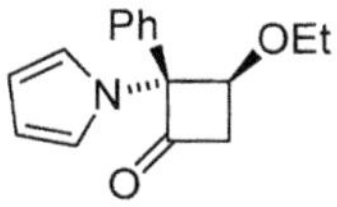	35	[64]
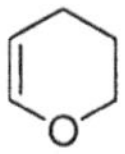	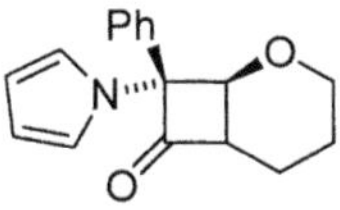	49	[64]
	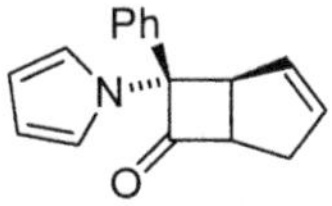	58	[64]
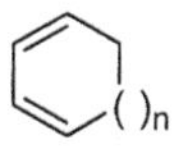	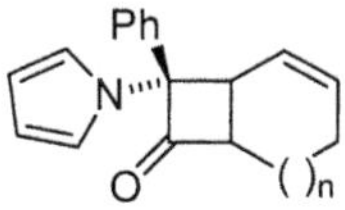 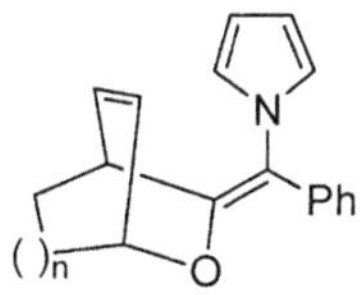		[64]
		n = 1 24, 18	[64]
		n = 2 12, 12	[64]

Table 12 Reaction of alkoxycarbenes with optically active ene carbamates

R^1	R^2	yield (%)	de (%)	ref
Me	Me	61	94	[66]
Me	Bn	56	86	[66]
Ph	Me	67	≥97	[66]
cyclopropyl	Me	59	≥97	[66]
MeCH=CH	Me	---		
$-(CH_2)_4-$		50	≥97	[66]

R^1 = Me; R^2 = Me	58%	29%
R^1 = Me; R^2 = Bn	45%	33%
R^1, R^2 = $(CH_2)_3$	60%	

(16)

Although optically active functionalized cyclobutanones themselves are of little intrinsic interest, they are highly reactive and have been converted to a number of synthetically useful intermediates. Further functionalization was achieved at both the α-position using enolate chemistry, and by nucleophilic attack at the ketone carbonyl [67]. They underwent facile Baeyer–Villiger ring expansion and elimination of the oxazolidinone group to produce optically active butenolides, which were subjected to a number of 1,4-addition reactions and 1,3-dipolar cycloaddition reactions [68]. This facile approach to optically active butenolides was used to prepare several biologically active systems, including two butenolides isolated from *Plakortis lita* [a, b, Eq. 17], tetrahydrocerulenin [69], and cerulenin (Scheme 2) [70]. By using cyclic alkoxycarbene complexes, optically active spiroketals were synthesized (Eq. 18) [71]. Optically active cyclobutanones produced as in Table 12 have also been used as precursors for palladium-catalyzed ring expansion to cyclopentenones [72], and for the study of the effect of adjacent chiral tertiary and quaternary centers on metal-catalyzed allylic substitutions [73].

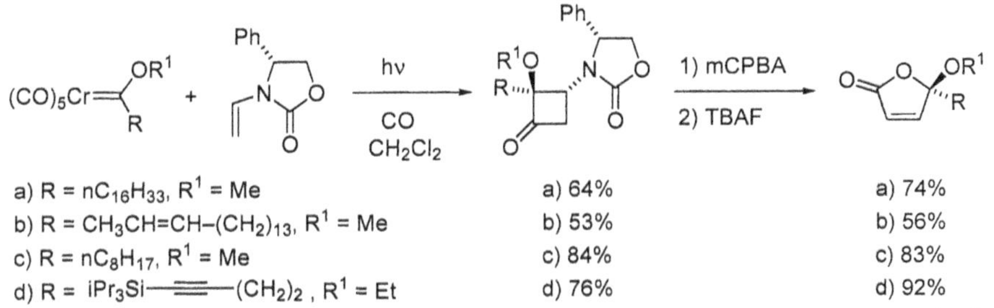

hν
CO
CH2Cl2
1) mCPBA
2) TBAF
a) R = nC16H33, R1 = Me
b) R = CH3CH=CH–(CH2)13, R1 = Me
c) R = nC8H17, R1 = Me
d) R = iPr3Si–≡–(CH2)2 , R1 = Et
a) 64%
b) 53%
c) 84%
d) 76%
a) 74%
b) 56%
c) 83%
d) 92%

(17)

NaOCl
DMF/Et2O
30%
NH3
Et2O
99%
(+)-tetrahydrocerulenin
(+)-cerulenin

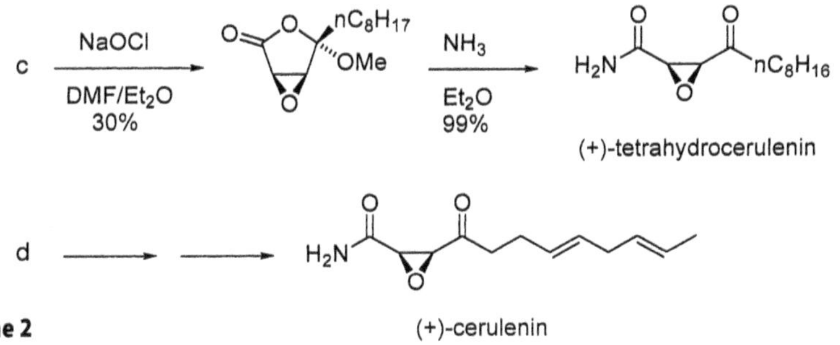

Scheme 2

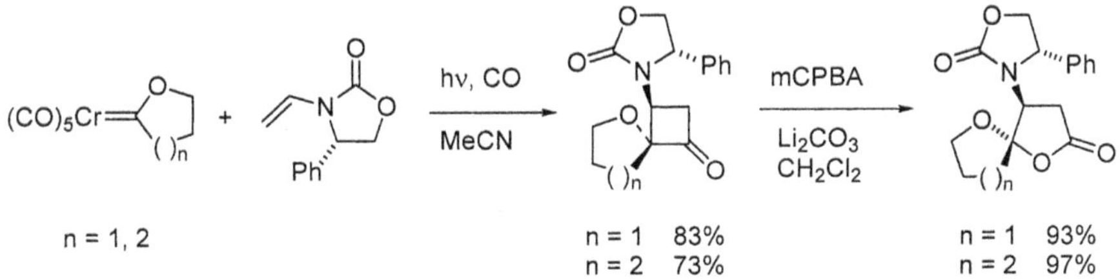

hν, CO
MeCN
mCPBA
Li2CO3
CH2Cl2
n = 1, 2
n = 1 83%
n = 2 73%
n = 1 93%
n = 2 97%

(18)

The optically active cyclobutanone from the (benzyloxymethyl)(ethoxy) carbene complex has been developed as a template for the synthesis of 4′-substituted nucleoside analogs (Schemes 3 and 4) [74]. Photochemical ring expansion in acetic acid directly produced the acylated ketal. Treatment with a Lewis acid and a silylated nucleophile produced 4′-disubstituted deoxyribo analogs [75]. Baeyer–Villiger oxidation followed by oxazolidinone elimination gave the (benzyloxymethyl)(ethoxy)butenolide. Carbonyl reduction and acylation gave a 1:1 mixture of epimeric allyl acetates, which were subjected to Vörbruggen coupling to give an epimeric mixture of 4,4′-disubstituted didehydrodideoxy ribonucleoside derivatives [76]. Palladium-catalyzed allylic amination with common nucleoside bases in the presence of chiral phosphines resulted in kinetic resolution to give a single β-epimer (Scheme 3) [77]. This same optically active cyclobutanone intermediate was the starting point for the synthesis of (−)-cyclobut-A, (±)-3′-epi-cyclobut-A [78], carbovir and aristeromycin [79], and (+)-neplanocin A [80], as well as aminocyclopentitols [81] and, from the (methoxy)(methyl) analog, 6-deoxy-4-aminohexoses (Scheme 4) [82].

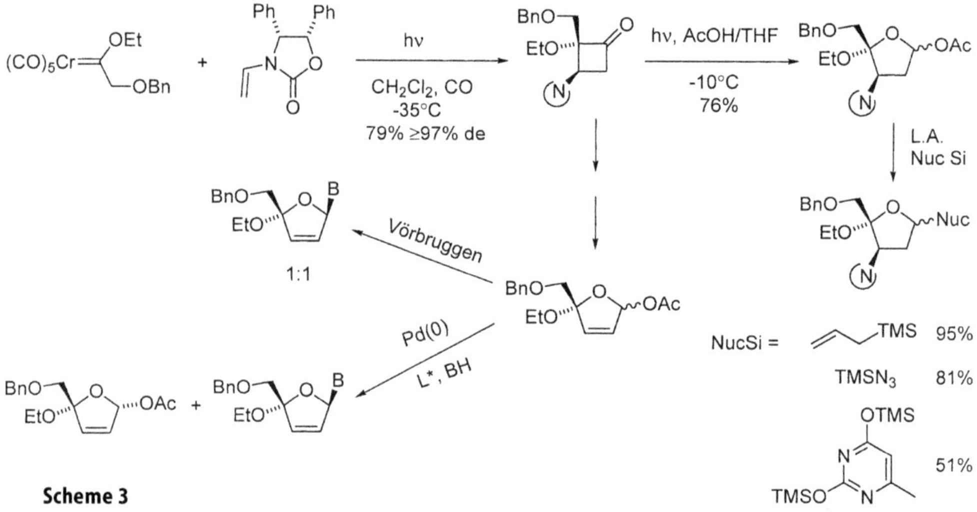

Scheme 3

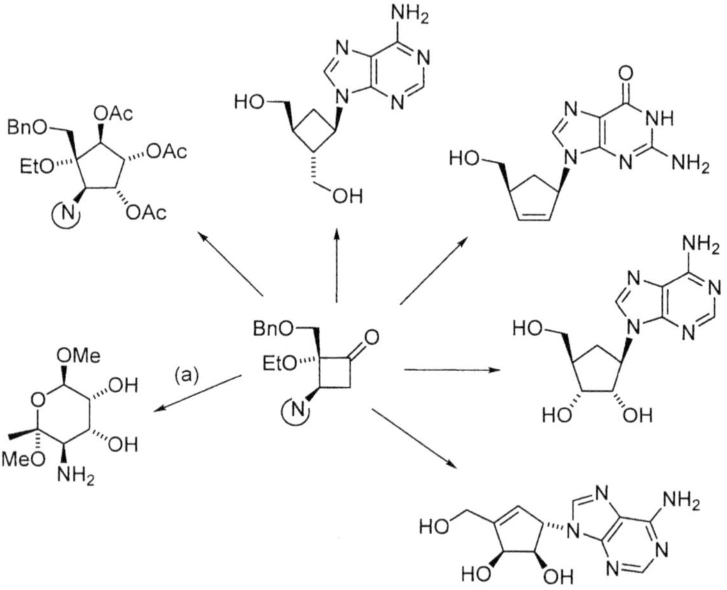

Scheme 4 [a]From the (methoxy)(methyl) analog of the cyclobutanone

2.1.1.3
With Aldehydes to Give *β*-Lactones

Photolysis of chromium alkoxycarbene complexes with aldehydes in the presence of Lewis acids produced *β*-lactones [83]. Intermolecular reactions were slow, low-yielding, and nonstereoselective, while intramolecular reactions were more efficient (Eqs. 19 and 20). Subsequent studies showed that amines, particularly DMAP, could also catalyze this process (Table 13) [84], resulting in reasonable yields and diastereoselectivity in intermolecular cases.

$(CO)_5Cr$ … CHO; CH_3 — hν, CO / $ZnCl_2$, CH_2Cl_2 → H_3C … H (19)

n = 2 50%
n = 3 59%
n = 4 8%

$(CO)_5Cr$ … OMe … CHO — hν / CO THF $ZnCl_2$ → MeO … H (20)

72%

Table 13 Reaction of alkoxycarbenes with aldehydes

$(CO)_5Cr$=C(OMe)Ph + RCHO — hν / DMAP CO → (*syn*) + (*anti*)

R	R	yield (%), (*syn/anti*)
C_2H_5	C_2H_5	53 (15/1)
CH_3	CH_3	55 (8/1)
i-Pr	i-Pr	35 (4/1)
n-Pr	n-Pr	43 (23/1)
Ph	Ph	33 (15/1)
(dioxolanyl, H)	(dioxolanyl, H)	(1.3/1 *syn* diast.) 65 ≥99/1

2.1.1.4
With Azoarenes to Give Diazetidinones

Photolysis of chromium alkoxycarbenes with azoarenes produced 1,2- and 1,3-diazetidinones, along with imidates from formal azo metathesis (Eq. 21) [85, 86]. Elegant mechanistic studies [87–89] indicated the primary photoprocess was *trans*-to-*cis* isomerization of the azoarene followed by subsequent thermal reaction with the carbene complex. Because of the low yields and mixtures obtained the process is of little synthetic use.

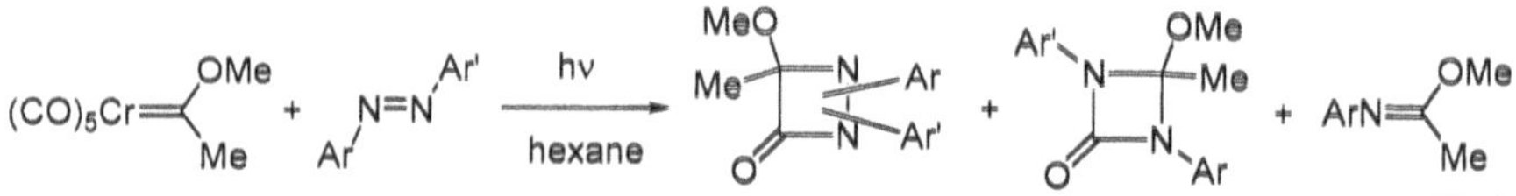

(21)

2.1.1.5
Photochemical Benzannulation Reactions

The *thermal* benzannulation of Group 6 carbene complexes with alkynes (the Dötz reaction) is highly developed and has been used extensively in synthesis [90, 91]. It is thought to proceed through a chromium vinylketene intermediate generated by sequential insertion of the alkyne followed by carbon monoxide into the chromium-carbene-carbon double bond [92]. The realization that photodriven CO insertion into *Z*-dienylcarbene complexes should generate the same vinylketene intermediate led to the development of a photochemical variant of the Dötz reaction (Table 14).

Table 14 Photo-driven Benzannulation Reactions

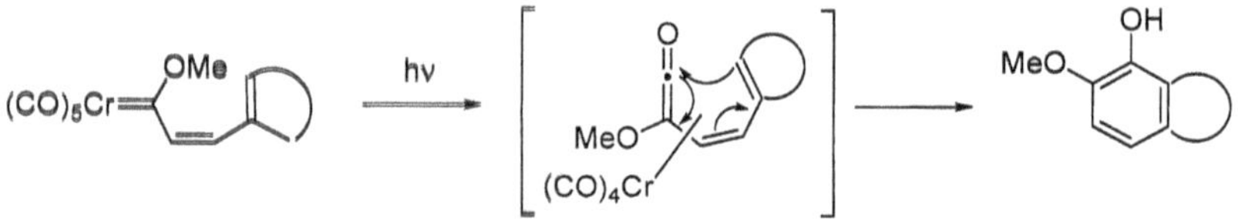

Carbene	Product		yield (%)	ref
		X = H	90	[93]
		X = OH	78	[93]
		R = Ph	87	[93]
		R = Me	90	[93]

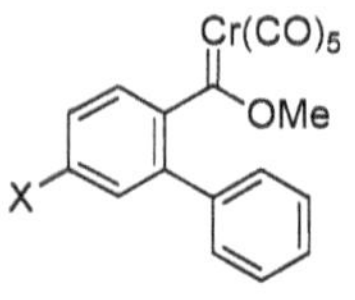

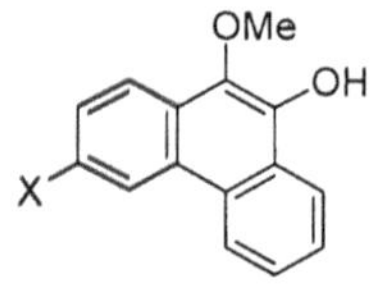

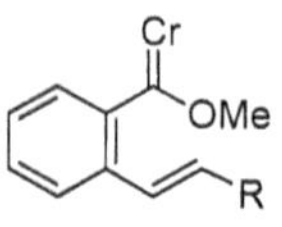

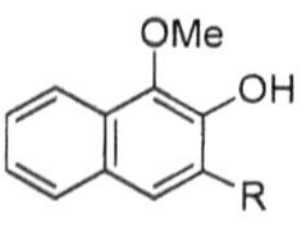

Table 14 (continued)

Carbene	Product		yield (%)	ref
		X = H	90	[93]
		X = OH	78	[93]
		R = Ph	87	[93]
		R = Me	90	[93]
			50	[93]
		X = OMe	93	[93]
		X = NMe_2	75	[94]
			62	[94]
			81	[94]

Table 14 (continued)

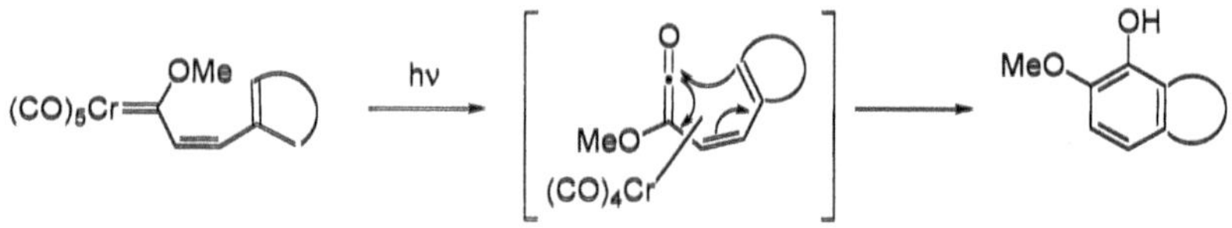

$(CO)_5Cr$=C(OMe)– … –hν→ [MeO, $(CO)_4Cr$ ketene intermediate] → MeO, OH product

Carbene	Product	yield (%)	ref
Cr–O, N–Me, OBu^t, Ph	Bu^{t}O, O, N–Me, OH	83	[94]
Cr–O, N, Bz, OBu^t	Bu^{t}O, O, NBz, OH	28	[94]
Cr–O, N, Bz, OBu^t, Ph	Bu^{t}O, O, NBz, OH, Ph	33	[94]
$Cr(CO)_5$, OMe, O	OMe, OH, O	65	[95]
$Cr(CO)_5$, OMe	OMe, OH; OMe, OH	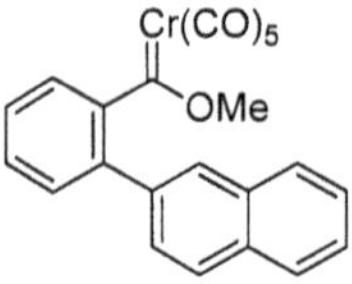 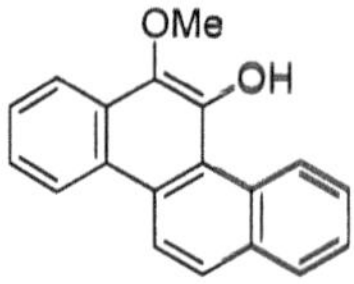 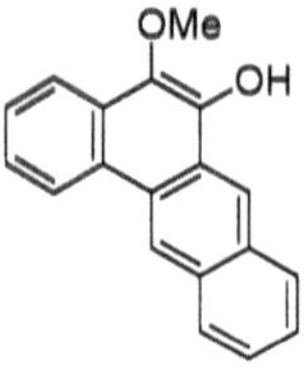 83 (11:1)	[95]
$Cr(CO)_5$, OMe, X	OMe, OH, X + OMe, OH, X		[95]

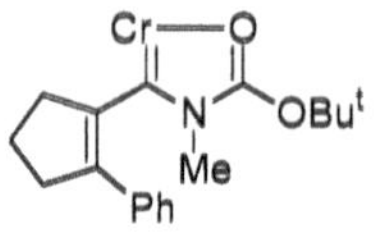

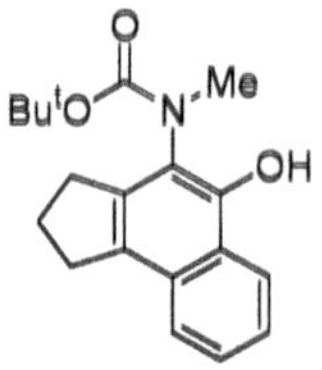

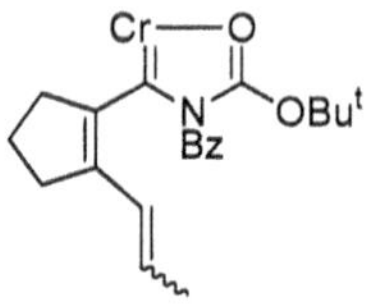

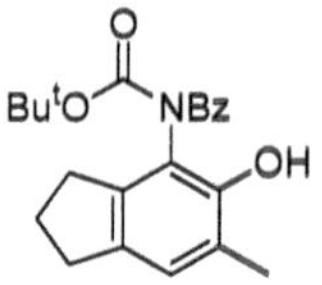

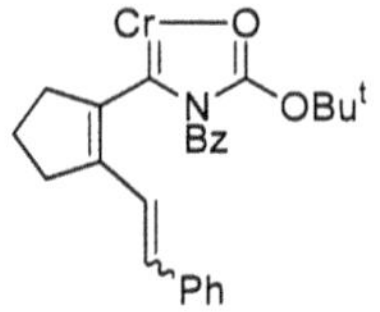

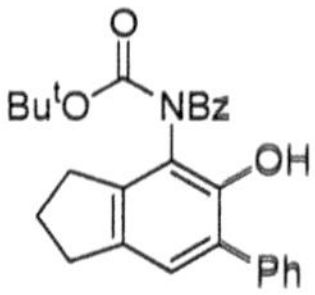

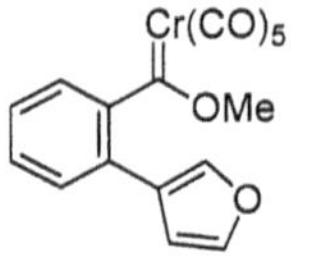

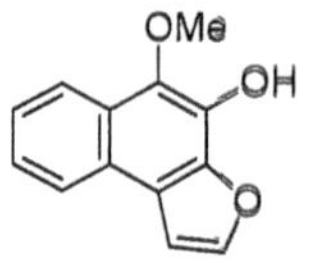

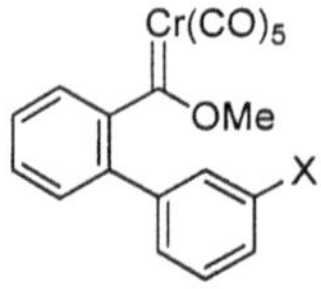

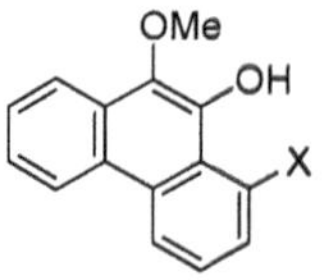

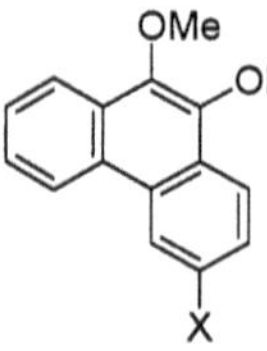

Table 14 (continued)

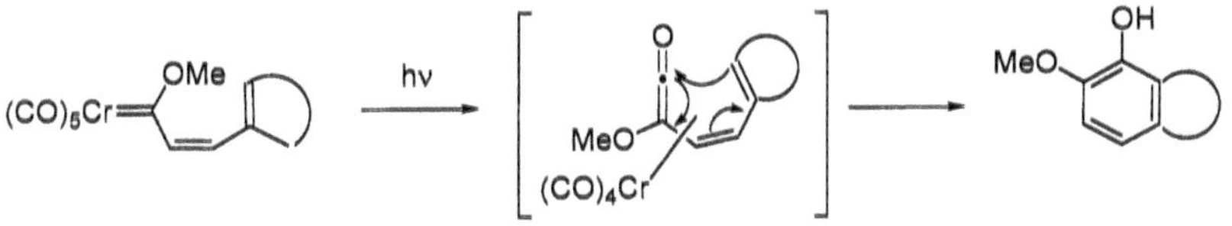

Carbene	Product	yield (%)	ref
X = OH[a]		65 (1.2:1)	
X = Me		71 (1.2:1)	
X = CHO		71 (1:1.3)	[95]
X = $CH(OMe)_2$		74 (1:1.5)	
X = F		66 (1:5)	
X = OMe		70 (1:7)	
X = Cl		89 (1:14)	
X = CF_3		69 (<1:>25)	
X = OH[b]		75 (5.4:1)	
X = Me		96 (1:12)	
X = CHO		88 (1:3)	
C = $CH(OMe)_2$		92 (<1:>25)	
X = F		58 (<1:>25)	
X = Cl		89 (<1:>25)	

[a]THF solvent. [b]toluene solvent.

This photodriven benzannulation was used in the synthesis of indolocarbazoles (Eq. 22) [96] and calphostins (Eq. 23) [97]. The thermal insertion of isonitriles into these same classes of carbenes provided a complementary approach to similar benzannulations [98–100]. Manganese alkoxycarbene complexes underwent both inter- [101] and intramolecular [102] photodriven benzannulation reactions with alkynes (Eqs. 24 and 25).

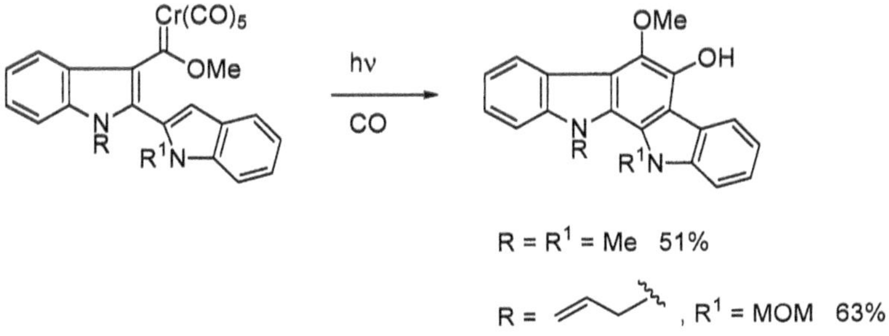

(22)

R = R¹ = Me 51%

R = allyl, R¹ = MOM 63%

OMe $Cr(CO)_5$ OMe OTIPS MeO — hν, THF or $PhCH_3$ → OMe OMe OH OTIPS MeO 36% (23)

O-$TiCp_2Cl$ OC Mn OC R + — hν, Ce(IV)/HNO_3 → R O O (24)

R = Me 28%
R = H 30%

R^2 R^3 O ( )n $SiMe_2$ O OC Mn OC R^1 — hν, oxidation 30-79% → O ( )n OH R^1 R^3 R^2 O (25)

R^1 = H, Me, OMe n = 0, 1
R^2 = H, Et
R^3 = H, Me, Et, Ph, TMS

2.1.2 Nucleophilic Attack

2.1.2.1 By Alcohols to Give Esters

Photodriven reactions of Fischer carbenes with alcohols produces esters, the expected product from nucleophilic addition to ketenes. Hydroxycarbene complexes, generated in situ by protonation of the corresponding "ate" complex, produced α-hydroxyesters in modest yield (Table 15) [103]. Ketals, presumably formed by thermal decomposition of the carbenes, were major by-products. The discovery that amides were readily converted to aminocarbene complexes [104] resulted in an efficient approach to α-amino acids by photodriven reaction of these aminocarbenes with alcohols (Table 16) [105, 106]. α-Alkylation of the (methyl)(dibenzylamino)carbene complex followed by photolysis produced a range of *racemic* alanine derivatives (Eq. 26). With chiral oxazolidine carbene complexes optically active amino acid derivatives were available (Eq. 27). Since both enantiomers of the optically active chromium aminocarbene are equally available, both the natural *S* and unnatural *R* amino acid derivatives are equally

Table 15 Photo-driven reactions of hydroxycarbene complexes with alcohols

$(CO)_5Cr{=}C(ONMe_4)R$ → 1) HCl; 2) hν R^1OH, 6 at CO, -50°C → $R^1O_2C{-}CH(OH){-}R$

R	R^1	yield, %
Ph	Me	60
Ph	Et	19
Me	Me	10
Me	Bn	---[a]
nBu	Bn	21
nBu	Me	62
sec Bu	Me	28
$PhCH_2CH_2$	Me	22
cyclopropyl	Bn	4
cyclopropyl	Me	19

[a] the dibenzyl ketal of acetaldehyde was the sole product isolated (30%).

available. Even α-deuteroglycine (Eq. 28) [107] and ^{13}C-labeled, α-deuterated amino acids [34] were available by this methodology (Eq. 29). 2,6-Imino-D-allonates were prepared using this chemistry (Eq. 30) [108].

$(CO)_5Cr{=}C(NBn_2)CH_3$ → 1) B^-; 2) RX; 3) hν, MeOH/CO → $RCH_2CH(NBn_2)CO_2Me$ (26)

R = nBu, 48%; Bn, 72%; allyl, 72%
R = $pMeOPhCH_2$, 73%; EtO_2CCH_2, 66%

$(CO)_5Cr{=}C(CH_3)$–N[(S)-4-Ph-2,2-dimethyloxazolidine] → 1) BuLi/THF; 2) R^1X; 3) hν, R^2OH → R^1CH_2–(R)CH(CO_2R^2)–N[oxazolidine, Ph] (27)

R^1 = Bn, R^2 = t-Bu, 42% yield, ≥93% de
R^1 = t-BuO_2CCH_2, R^2 = Me, 52% yield, ≥93% de

Table 16 Photo-driven reactions of aminocarbenes to produce α amino acids

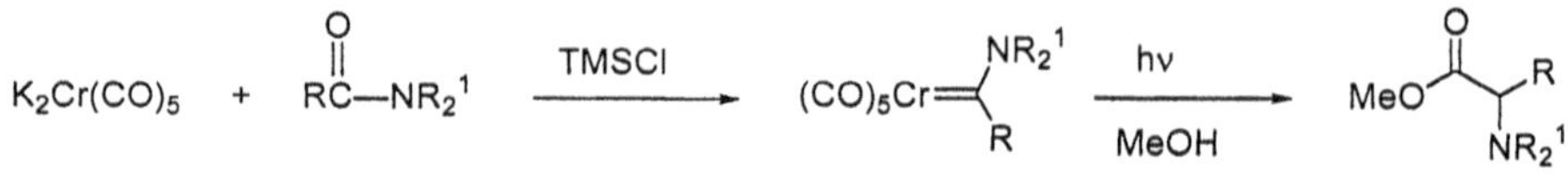

$K_2Cr(CO)_5$ + $RC(O)—NR_2^1$ —TMSCl→ $(CO)_5Cr{=}C(NR_2^1)R$ —hν, MeOH→ $MeO_2C—CH(R)—NR_2^1$

R	R^1	Complex yield %	Amino acid yield %
Me	Bn	44	87
Me	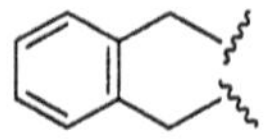	57	84
Me	 O	78	98
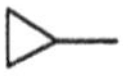	Me	43	85
$BnOCH_2$	Me	50	82
Ph	Et	91	98
pCF_3Ph	Me	63	88
3-furyl	Me	96	76
Ph	H	96	84
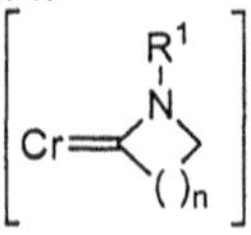 R^1 N Cr ( )n, n = 1	Bn	35	45
n = 2	Bn	49	---[a]
n = 2	Me	98	---[a]
n = 3	Me	97	90
n = 3	Bn	32	90
n = 4	Bn	29	58
n = 10	Bn	36	96

[a]only slow photochemical decomposition was observed. The same was true for styryl-, phenylacetylidiyl, and o-chlorophenyl carbenes

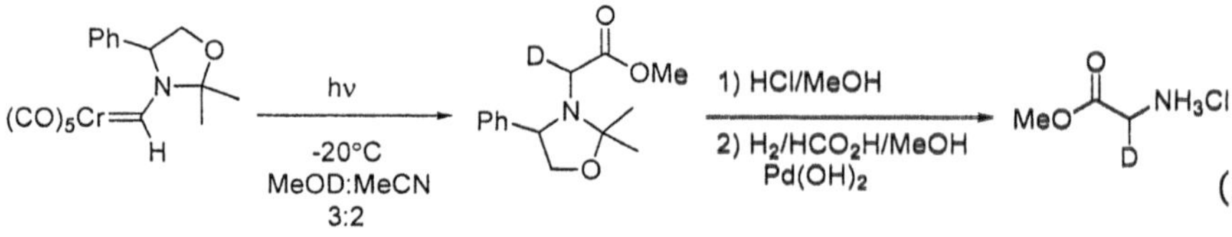

Ph O N $(CO)_5Cr$ H —hν, -20°C, MeOD:MeCN 3:2→ D OMe O Ph N O —1) HCl/MeOH, 2) $H_2/HCO_2H/MeOH$, $Pd(OH)_2$→ MeO O NH_3Cl D (28)

(R) *(S)*

(R,R) 93% yield
(S,S) ≥97% D 86% de

(R) 82%
(S) 82%

(29)

R = Me, 78%
R = t-BuO_2CCH_2, 97%
R = allyl, 57%
R = Bn, 70%
>95% de *(R)*

(30)

Activated esters for use in peptide-coupling reactions were produced by photolysis of optically active chromium aminocarbenes with alcohols which are good leaving groups, such as phenol, pentafluorophenol, 2,4,5-trichlorophenol, and *N*-hydroxysuccinimide (Table 17) [109]. Since arylcarbenes bearing the op-

Table 17 Synthesis of optically active activated amino esters

R	R[1]	Product yield, % (config.)	de
Me	Ph	55 *(S,R)*	86
Me	2,4,5-Cl_3Ph	52 *(S,R)*	90
Me	2,4,5-Cl_3Ph	57 *(R,S)*	91
Me	C_6F_5	38 *(S,R)*	80
allyl	2,4,5-Cl_3Ph	60 *(R,S)*	95
Bn	2,4,5-Cl_3Ph	51 *(S,R)*	92
$tBuO_2CCH_2$	2,4,5-Cl_3Ph	48 *(S,R)*	85
nC_8H_{17}	2,4,5-Cl_3Ph	40 *(R,S)*	a
MeO_2CCH_2	2,4,5-Cl_3Ph	29 *(R,S)*	86
3-oxocyclohexyl	2,4,5-Cl_3Ph	36 *(R,S)*	a

[a] de could not be determined

tically active oxazolidine auxilliary are difficult to synthesize and often unstable, the above chemistry does not afford an effective approach to aryl glycines. In contrast, a wide range of arylcarbenes having an optically active diphenylamino alcohol as a chiral auxilliary were readily synthesized [110]. Photolysis of these resulted in intramolecular trapping of the ketene-like intermediate, producing aryl-substituted oxazinones in good yield. Although diastereoselectivity for the process was only fair, diastereoisomers were readily separated to provide reasonable yields of optically pure aryl glycines (Table 18). Intramolecular trapping by a pendant OH group from aldol reactions at the α-carbon produced 2-aminobutyrolactones (Table 19) [111]. These were converted to homoserines and were used in the total synthesis of (+)-bulgecinine.

Table 18 Synthesis of aryl glycines

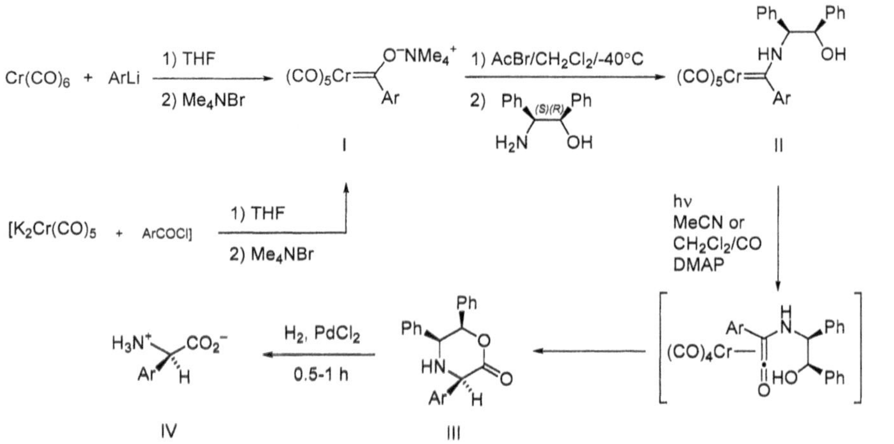

I (% yield)[a]	II (% yield)[a]	III (% yield)[a]	de[b] (ratio)	III (% yield)[c]	IV (% yield)[a,d]	ee (Mosher amide)[e]
Ph, 62	79	81	64 (82:18)	67	69 (23)[d]	94
p-meOPh, 78	81	82	76 (88:12)	72	91 (41)	96
p-ClPh, 64	73	73	60 (80:20)	59	81 (22)	84
p-FPh, 62	77	77	72 (86:14)	69	80 (26)	98
p-CF_3Ph, 80	84	84	60 (80:20)	73	86 (42)	98
o-MeOPh, 81	86	78	76 (88:12)	61	24[f] (11)	64[g]
2,6-F_2Ph, 63	78	62	64 (82:18)	63	95 (29)	56
1-Naphth, 75	51	76	62 (81:19)	66	40[h] (10)	91
3-Thienyl, 80	55	75	66 (83:17)	62	[h,i]	

[a]Reported yields are for isolated, purified materials. [b]Determined by integration of appropriate signals in the ^{1}H NMR spectrum in the crude reaction mixture. [c]Yield of isolated, pure, single (major) diastereoisomer based on ArLi. [d]Overall yield of amino acid, from ArLi + $Cr(CO)_6$. [e]Determined by integration of appropriate peaks in ^{1}H NMR spectra of the Mosher's amide. [f]As the hydrochloride salt. The free amine unstable. [g]The amino acid decomposed during conversion to Mosher's amide, complicating the ee determination. [g]Oxidative removal of the chiral auxiliary was used. [i]Deprotection was unsuccessful.

Table 19 Photo-driven synthesis of α-aminobutyrolactones

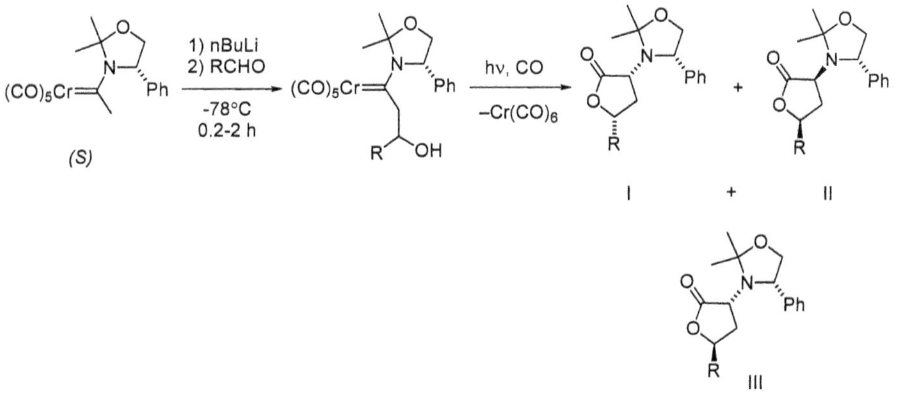

R	yield (%)[a]	I	II	III
C_6H_5	78	92	8	---
pMeOPh	39[b]	100	---	---
2-furyl	44	100	---	---
tBu	32 (50)[c]	100	---	---
Me	61	22	67	11
iPr	59	22	68	10[d]
$CH=CH_2$	37[e]	41	41	18

[a]Combined yield of separated, isolatled, pure diastereoisomers. [b]Aldol product was purified before photolysis; yield for the aldol reaction 53%, yield for the photolysis 74%. [c]The aldol reaction was stirred at 0°C for 4 h under artificial light. After recooling to -78°C and quenching, only the lactone and starting carbene were isolated. [d]Ratio determined by GC. [e]Aldol reaction was carried out at -100°C.

2.1.2.2
By Amino Acids to Give Peptides

A major justification for the synthesis of unnatural amino acids is to incorporate them into peptides to alter their biological activity/stability. By using an amino acid ester as the nucleophile both the peptide bond and the new stereogenic center, the absolute configuration of which is controlled by the carbene not the amino acid, were generated in the coupling step (Table 20) [112]. The reaction was efficient for a range of amino acid esters, including those having side chain functionality [113], and a modest range of carbene complexes. This system experienced modest "double diastereoselection" with (*R*)(*S*) or (*S*)(*R*) being the "matched" pair and (*S*)(*S*) or (*R*)(*R*) being the mismatched pair. This effect only slightly eroded diastereoselectivity. Even sterically hindered α,α-dialkyl amino acid esters and *N*-alkyl amino acid esters coupled reasonably well, because the species photogenerated from the carbene complex was highly reactive [113]. This chemistry worked well on Merrifield resin-supported sys-

Table 20 Synthesis of dipeptides from aminocarbenes and α-aminoesters

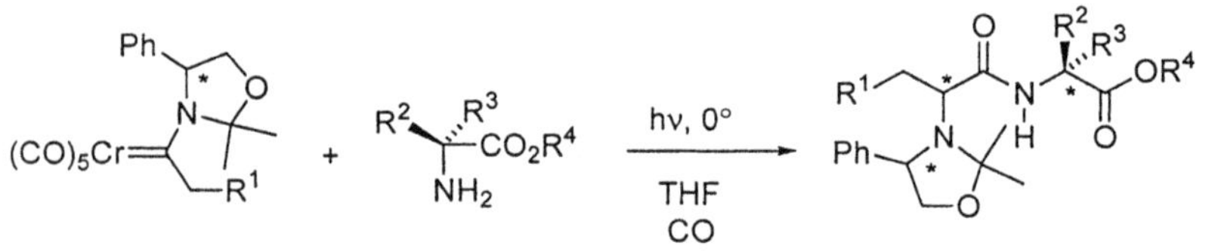

R^1	R^2	R^3	R^4	dr	yield, %[a]	***	ref
H	Me	H	t-Bu	98:2	88	(R)(S)(S)	[112]
H	Me	H	t-Bu	90:10	68	(S)(R)(S)	[112]
H	H	H	t-Bu	94:6	68	(R)(S)(–)	[112]
H	Bn	H	t-Bu	97:3	72	(R)(S)(S)	[112]
H	Bn	H	t-Bu	92:8	65	(S)(R)(S)	[112]
H	Ph	H	t-Bu	98:2	86	(R)(S)(S)	[112]
H	Ph	H	t-Bu	90:10	70	(S)(R)(S)	[112]
H	$CH_3CH(OH)$	H	Me	95:5	56	(R)(S)(S)	[112]
H	2-indolyl	H	Me	90:10	60	(R)(S)(S)	[112]
H	(proline)	H	t-Bu	---	61	(R)(S)(S)	[112]
H	(proline)	H	Me	74:26	60	(R)(S)(S)	[113]
H	CH_2OH	H	Me	98:2	61	(R)(S)(S)	[113]
H	CH_2SH	H	Me	92:8	37	(R)(S)(S)	[113]
H	$(CH_2)_2SH$	H	Me	96:4	68	(R)(S)(S)	[113]
H	$pHOPhCH_2$	H	Me	94:6	64	(R)(S)(S)	[113]
H	$(CH_2)_2CO_2Me$	H	Me	98:2	75	(R)(S)(S)	[113]
H	CH_2CO_2Me	H	Me	95:5	64	(R)(S)(S)	[113]
CH_2OCO_2tBu	Me	H	t-Bu	93:7	77	(R)(S)(S)	[112]
Bn	Me	H	t-Bu	87:13	82	(R)(S)(S)	[112]
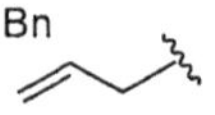	Me	H	t-Bu	77:23	57	(R)(S)(S)	[112]
H	Me	Me	Me	98:2	78	(R)(S)(–)	[113]
H	Ph	Ph	Me	98:2	82	(R)(S)(–)	[113]
H	Bn	Me	Me	98:2	84	(R)(S)(S)	[113]
H	nPr	Me	Me	97:3	67	(R)(S)(S)	[113]
H	Bn	nPr	Me	92:8	68	(R)(S)(S)	[113]
H	CH_2OH	Me	Me	93:7	68	(R)(S)(S)	[113]
iPr	Me	Me	Me	98:2	76	(R)(S)(–)	[113]
H	Bn	Me	Me	90:10	76	(S)(R)(S)	[113]
H	nPr	Me	Me	95:5	71	(S)(R)(S)	[113]
H	Bn	nPr	Me	91:9	53	(S)(R)(S)	[113]

[a]Yield of pure, isolated single diastereoisomer.

tems and was used to synthesize an octapeptide having three unnatural, chromium carbene-derived residues in the middle [114]. Soluble poly(ethylene glycol) (PEG)-supported systems also coupled effectively [115]. Photolysis of *o*-silylcarbenes in the presence of *p*-anisidine gave *N*-*p*-anisyl-α-hydroxyamides in fair yield [116].

2.1.2.3
By Stabilized Ylides to Produce Captodative Allenes

Photolysis of alkoxycarbene complexes in the presence of stabilized ylides produced allenes having a donating group at one terminus and an accepting group at the other. These were highly reactive and rearranged to 1,3-dienes under mildly acidic conditions and hydrolyzed to γ-keto-α,β-unsaturated esters (Eq. 31) [117].

$(CO)_5Cr=C(OR^2)R^1$ + Ph_3PCHX — hν, PhH / CO → [allene] — Et_2O / HCl → (31)

$X = CO_2R$, SO_2tolyl

PPTs, CH_2Cl_2

2.1.2.4
By Arenes: Intramolecular Friedel–Crafts Arene Acylation

Chromium carbene complexes having electron-rich arenes tethered to the carbene oxygen or carbon underwent photodriven intramolecular Friedel–Crafts acylation in the presence of zinc chloride (Eqs. 32 and 33) [118]. The process was highly regioselective, undergoing acylation exclusively *para* to the activating group.

$(CO)_5Cr=C(R)O(CH_2)_n$–Ar(X)(Y) — hν / CH_2Cl_2, $ZnCl_2$, CO, 0° → (32)

Substituents	Yield
R = Me, n = 1, X = OMe, Y = H	69%
R = Me, n = 1, X = OH, Y = H	59%
R = Me, n = 1, X = Y = OMe	38%
R = Me, n = 2, X = OMe, Y = H	54%
R = cyclopropyl, n = 1, X = OMe, Y = H	43%
R = Ph, n = 1, X = OMe, Y = H	15%

$(CO)_5Cr$=C(OMe)–CH₂–(CH(m-MeOC₆H₄))ₙ ... $\xrightarrow[\text{ZnCl}_2\ \text{CO}\ 0°]{h\nu,\ CH_2Cl_2}$ MeO-tetralone (OMe, Z) (33)

n = 1, Z = H 60%
n = 2, Z = $mMeOPhCH_2$ 62%

2.1.2.5
By Tertiary Allylic Amines: Zwitterionic Aza-Cope Rearrangement

Ketenes react with tertiary allylic amines in the presence of Lewis acids to give zwitterionic intermediates which undergo [3,3]-sigmatropic rearrangement [119]. Photolysis of chromium carbene complexes in the presence of tertiary amines results in similar chemistry [120]. Cyclic (Table 21) and strained allylic amines (Eq. 34) work best, while acylic amines are less reactive (Eq. 35).

Table 21 Zwitterionic aza-Cope reaction

$(CO)_5Cr$=C(X)(R¹) + N-Bn 2-vinyl cyclic amine (( )ₙ, R²) $\xrightarrow[\text{CO, L.A., THF}]{h\nu}$ N-Bn lactam (R², X, R¹)

R^1	X	n	R^2	L.A.	yield, %
Me	OMe	1	H	$ZnCl_2$	71
Me	OBn	1	H	$ZnCl_2$	66
$-(CH_2)_3O-$		1	H	Me_2AlCl	22
H	Me_2N	1	H	Me_2AlCl	9
H	(4-Ph-2,2-dimethyloxazolidin-3-yl)	1	H	$ZnCl_2$	19
Me	OMe	1	Me	$ZnCl_2$	20
Me	OBn	1	Me	$ZnCl_2$	40
Me	OMe	2	H	Me_2AlCl	33

$(CO)_5Cr{=}C(X)R$ + bicyclic alkene–NBn $\xrightarrow[\text{CO, THF}]{h\nu}$ bicyclic lactam (NBn, O, H, H, X, R) (34)

R = Me. X = OMe	60% (9:1)
R = Me, X = OBn	51%
R,X = $(CH_2)_3O$–	10%
R = H, X = oxazolidinyl (N, O, Ph)	50% (3:1)

$(CO)_5Cr{=}C(X)R$ + Bn_2N-allyl + Me_2AlCl $\xrightarrow[\text{CO, THF}]{h\nu}$ Bn_2N–C(O)–C(X)(R)–allyl (35)

R = Me, X = OMe	45%
R = Me, X = OBn	68%
R = X = –$(CH_2)_3O$–	14%

2.2
Photodriven Reactions of Fischer Carbenes Not Involving CO Insertion

2.2.1
Nucleophilic Addition/Elimination at the Carbene Carbon

Sulfur-stabilized ylides underwent photodriven reaction with chromium alkoxy-carbenes to produce 2-acyl vinyl ethers as *E*/*Z* mixtures with the *E* isomer predominating (Table 22) [121–123]. The reaction is thought to proceed by nucleophilic attack of the ylide carbon at the chromium carbene carbon followed by elimination of $(CO)_5CrSMe_2$. The same reaction occurred thermally, but at a reduced rate. Sulfilimines underwent a similar addition/elimination process to produce imidates or their hydrolysis products (Table 23) [124, 125]. Again the reaction also proceeded thermally but much more slowly. Less basic sulfil-imines having acyl or sulfonyl groups on nitrogen failed to react.

A narrow range of 2-phenyl-1-azirines underwent photodriven reactions with alkoxycarbenes to give *N*-vinylimidates, in a process probably related to the above reactions (Table 24) [126].

Table 22 Photo-driven reaction of sulfur-stabilized ylides with alkoxycarbenes

R	R^1	R^2	E/Z	yield, % E	yield, % Z
Me	Me	OMe	2.3:1	65	
Me	Me	Ot-Bu	4:1	73	
Me	Me	Ph	100:0	81	
Me	Bn	OMe	5.6:1	60	
Ph	Me	OMe	4:1	70	
Ph	Me	Ph	4.6:1	90	
Me		OMe	2.3:1	41	29
Me		OMe	3.4:1	68	22
Me		OMe	4:1	51	15
Me		Ph	4:1	48	
Me		OMe	1.6:1	34	23
Bu	Me	OMe	3.5:1	52	20
	Me	OMe	2.4:1	23	30
	Me	OMe	2.4:1	50	12
	Me	OMe	1:1	19	41

2.2.2 Cyclopropanation and Other Cycloadditions

One of the earliest reported *thermal* reactions of Fischer carbene complexes was the reaction with olefins to give cyclopropanes [127]. More recently it has been shown that photolysis accelerates intermolecular cyclopropanation of electron-poor alkenes [128]. Photolysis of Group 6 imine carbenes with alkenes

Table 23 Photo-driven reactions of sulfilimines with alkoxycarbenes

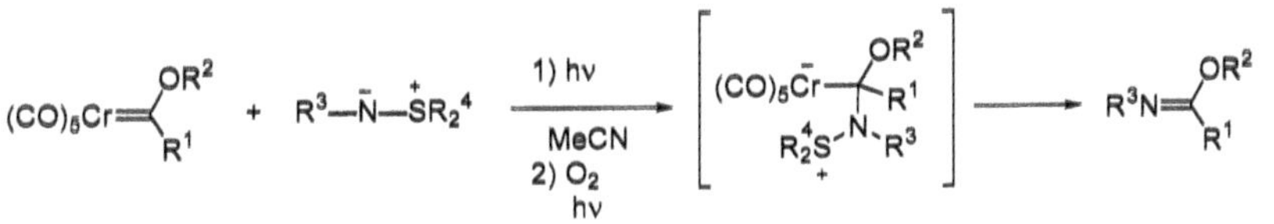

$(CO)_5Cr{=}C(OR^2)R^1 + R^3{-}\bar{N}{-}\overset{+}{S}R^4_2 \xrightarrow[\text{MeCN; 2) } O_2,\ h\nu]{\text{1) } h\nu} [(CO)_5\bar{C}r{-}C(OR^2)(R^1){-}N(R^3){-}\overset{+}{S}R^4_2] \longrightarrow R^3N{=}C(OR^2)R^1$

R^1	R^2	R^3	R^4	yield, %
Me	Me	2-Py	Me	80
Me	Et	2-Thiazolyl	Me	65
Me	Me	2-pyrimidinyl	Me	45
Me	Me	pO_2NPh	Me	52
Me	Me	Ph	Me	70
Me	Me	pMeOPh	Me	90
Ph	Me	phth	Me	60
Me	Me	$EtO(CH_2)_2$	Ph	50[a]
Ph	Me	$MeO_2C(CH_2)_2$	Ph	50[a]
Ph	Me	$EtO_2C(CH_2)_2$	Ph	55[a]
Ph	Me	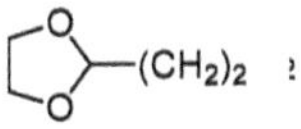 1,3-dioxolan-2-yl–$(CH_2)_2$	Ph	98[a]
Ph	Me	$PhSO_2(CH_2)_2$	Ph	80
Ph	Me	$NC(CH_2)_2$	Ph	62
Me	Me	pMeOPh	Me	90
Ph	Me	pMeOPh	Me	71
Me	Bn	pMeOPh	Me	85
Me	allyl	pMeOPh	Me	54
Me	H–C≡C–$(CH_2)_2$	pMeOPh	Me	50
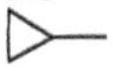 cyclopropyl	Me	pMeOPh	Me	63
PhCH=CH	Me	pMeOPh	Me	85
TMS–C≡C	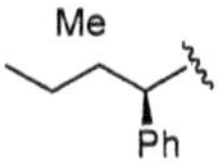 Me	pMeOPh	Me	59
Me	(structure, Ph)	pMeOPh	Me	70
Me	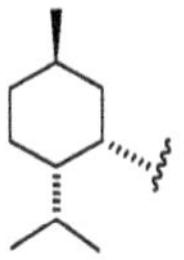 (structure)	pMeOPh	Me	98
Me	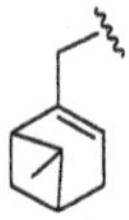 (structure)	pMeOPh	Me	40
Me	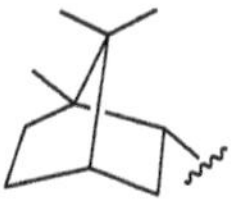 (structure)	pMeOPh	Me	66

[a]Directly hydrolyzed to the amide.

Table 24 Photo-driven reaction of 2-phenyl-1-azirines with alkoxycarbenes

R^1	R^2	R^3	yield, %
Me	Me	Me	77
Ph	Me	Me	74
Me	Me	H	33
Ph	Ph	H	55
Me	Ph	H	61

produced 1-pyrrolines [129, 130]. Although this was initially thought to involve a [3+2] cycloaddition it was subsequently shown to involve *two* photochemical steps, initial cyclopropanation with photolysis simply acting to eject a CO from the carbene to allow cyclopropanation, followed by photodriven rearrangement of the cyclopropyl ketimine to the 1-pyrroline [131]. A wide range of electron-poor alkenes were reactive, as were several imine carbenes (Table 25). A related photodriven reaction of Group 6 imine carbenes with alkynes produced 2*H*-pyrroles. However, this process was thought to proceed via a six-membered aza-metallacycle rather than via cyclopropene intermediates (Table 26) [132].

Finally, chromium imine carbenes underwent photoreaction with imines to give azadienes (metathesis) (Eq. 36), with azobenzene to give both metathesis and cycloaddition products (Eq. 37), and with ketones to give oxazolines (Eq. 38) [133].

(36)

R^1 = Ph, Et 25-28%
R^2 = H, Ph

(37)

22% 35%

Table 25 Photo-driven reaction of Group 6 imine carbenes with alkenes

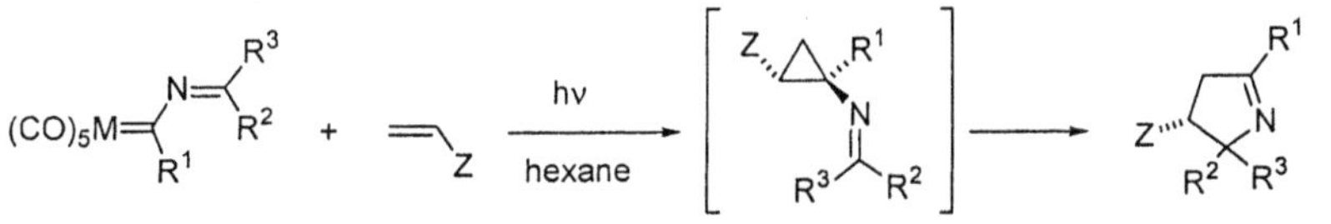

M	R^1	R^2	R^3	Z	yield, %
Cr	Me	Ph	Ph	CO_2Et	54
Cr	Me	Ph	Ph	COMe	76
Cr	Me	Ph	Ph	Ph	24
Cr	Ph	Ph	Ph	COMe	61
Mo	Me	Ph	Ph	COMe	54
W	Me	Ph	Ph	CO_2Et	43
W	Me	Ph	Ph	COMe	50
Cr	Me	Ph	Ph	CO_2tBu	35
Cr	Me	Ph	Ph	$CO_2C_{10}H_{21}$	37
Cr	Me	Ph	Ph	CO_2nBu	47
Cr	Me	Ph	Ph	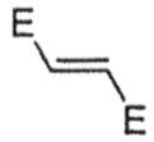 E–CH=CH–E	58
Cr	Me	Ph	Et	COMe	81[a]
Cr	Me	Et	Et	COMe	54
Cr	Me	Ph	H	COMe	43
Cr	Ph	Ph	H	COMe	43[b]

[a] 1.6/1 mixture of diastereoisomer. [b] 1:1 mixture of diastereoisomers.

Table 26 Photo-driven reactions of Group 6 imine carbenes with alkynes

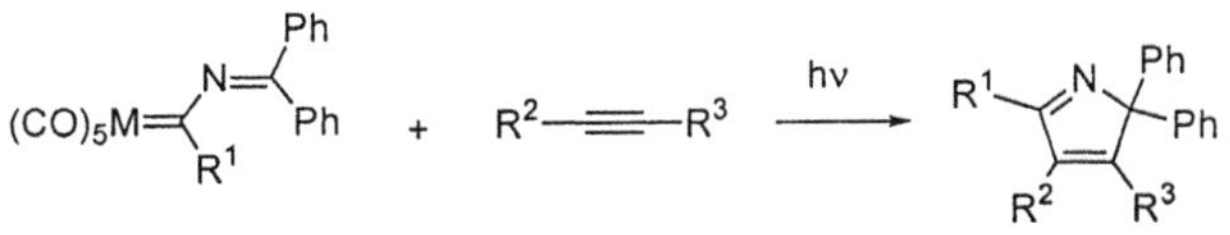

M	R^1	R^2	R^3	yield, %
Cr	Me	Ph	Ph	51
Cr	Ph	Ph	Ph	49
Mo	Me	Ph	Ph	21
Cr	Me	Ph	H	43
Cr	Me	CO_2Me	CO_2Me	5
Cr	Me	OEt	H	73
W	Me	OEt	H	63

R^1 = Ph, 12%; Me, 15%; $(CH_2)_4$, 14%

(38)

1,3-Dipolar cycloadditions to alkynylcarbenes followed by photolysis led to β-enamino ketoaldehydes (Eq. 39) [134]. Photolysis of *N*-acylamino carbene complexes produced munchnones, which were trapped with alkynes to give pyrroles (Table 27) [135]. This same reaction occurred in the dark under 30 psi carbon monoxide pressure. Tungsten carbonyl cyclized *N*-(*o*-alkynylphenyl)-imines into indoles via a photodriven process proceeding through a tungsten-carbene-containing azomethine ylide (Table 28) [136]. With internal alkynes 1,2-*R* migration occurred (Eq. 40).

Ar = Ph, 32%; pMePh, 31%
$pMeO_2CPh$, 37%;
$p\text{-}Me_2NPh$, 43%

(39)

Table 27 Cycloaddition of alkynesto acylaminocarbenes

R^1	R^2	yield, %
MeO_2C	CO_2Me	90
Bu	CO_2Me	36
H	CO_2Me	80
Ph	Ph	12
H	Bu	27
Ph	$(CO)_5Cr{=}C(OMe)$–	65

Table 28 Photo-driven tungsten carbonyl- assisted 1,3-dipolar cycloadditions

R^1	R^2	R^3	R^4	yield, %
H	Ph	H	OtBu	52
H	Ph	Me	OMe	86
H	Ph	4-(cyclopent-1-enyl)morpholine		67
H	Ph	2-methoxyfuran (OMe)		74
H	OEt	OTMS	OEt	74
OMe	Me	OTMS	OEt	55

$$\text{(40)}$$

R = nPr, 76%; Me, 55%; Ph, 61%

2.2.3 Acyl Migration

Photolysis of (2-acyloxyethenyl)carbene complexes produced 2-butene-1,4-diones (Eq. 41) [137].

$$\text{(41)}$$

60-68%

R = Ph, tBu, iPr

R^1 = Ph, iPr, $iC_7H_7CH_2$, ,tBu, iPr

3 Conclusions

Despite the unpromising UV-visible spectra and flash photolysis studies, the carbene complexes presented in this chapter have a rich photochemistry at wavelengths exceeding 300 nm. A wide range of synthetically useful transformations has been developed, and continued studies are likely to reveal more.

References

1. Fürstner A (1998) Alkene metathesis in organic synthesis (Topics in organometallic chemistry I). Springer, Berlin Heidelberg New York
2. Foley HC, Strubinger LM, Targos TS, Geoffroy GL (1983) J Am Chem Soc 105:3064
3. Fong LK, Cooper NJ (1984) J Am Chem Soc 106:2595
4. Block TJ, Fenske RF, Casey CP (1976) J Am Chem Soc 98:441; Nakatsuji H, Uskio J, Yonezawa T (1983) J Am Chem Soc 105:426
5. Bell SEJ, Gordon KC, McGarvey JJ (1988) J Am Chem Soc 110:3107
6. Servaas PC, Stufkens DJ, Oskam A (1990) J Organometal Chem 390:61
7. Rooney AD, McGarvey JJ, Gordon KC, McNicholl RA, Schubert U, Hepp W (1993) Organometallics 12:1277
8. Rooney AD, McGarvey JJ, Gordon KC (1995) Organometallics 14:107
9. Gallagher ML, Green JB, Rooney AD (1997) Organometallics 16:5260
10. Doyle KO, Gallagher ML, Pryce MT, Rooney AD (2001) J Organomet Chem 617–618:269
11. For a review on chromium carbene complex photochemistry in organic syntheses see: Hegedus LS (1997) Tetrahedron 53:4105
12. McGuire MA, Hegedus LS (1982) J Am Chem Soc 104:5538
13. Hegedus LS, deWeck G, D'Andrea S (1988) J Am Chem Soc 110:2122
14. For a review on metal-ketene complexes see: Geoffroy GL, Bassner SL (1988) Adv Organomet Chem 28:1
15. Sheriden JB, Geoffroy GL, Rheingold AL (1986) Organometallics 5:1514
16. Fischer H (1983) Angew Chem 95:913
17. Tidwell TT (1995) Ketenes. Wiley, New York
18. Georg GI (ed) (1993) The organic chemistry of β-lactams. VCH, New York
19. Hegedus LS, McGuire MA, Schultze LM, Yijun C, Anderson OP (1984) J Am Chem Soc 106:2680
20. Hegedus LS, Schultze LM, Toro J, Yijun C (1985) Tetrahedron 41:5833
21. Woodgate PD, Sutherland PS, Richard CEF (2001) J Organomet Chem 629:114
22. Sierra MA, Mancheno MJ, Vicente R, Gomez-Gallego M (2001) J Org Chem 66:8920
23. Borel C, Hegedus LS, Krebs J, Satoh Y (1987) J Am Chem Soc 109:1101
24. Betschart C, Hegedus LS (1992) J Am Chem Soc 114:5010
25. Hegedus LS, Moser WH (1994) J Org Chem 59:7779
26. Hegedus LS, Greenberg MM, Wendling JJ, Bullock JP (2003) J Org Chem 68:4179
27. Bose AK, Kapin JE, Fabey JL, Mankas MS (1973) J Org Chem 38:3437
28. Dumas S, Lastra E, Hegedus LS (1995) J Am Chem Soc 117:3368
29. Puntener K, Hellman MP, Kuester E, Hegedus LS (2000) J Org Chem 65:8301
30. Kuester E, Hegedus LS (1999) Organometallics 18:5318
31. Brugel TA, Hegedus LS (2003) J Org Chem 68:8409

32. Giordani C, Licandro E, Maiorana S, Papagni A, Slawin AM, Williams DJ (1990) J Organomet Chem 393:227
33. Alcaide B, Dominguez G, Plumet J, Sierra MA (1992) J Org Chem 57:447
34. Lastra E, Hegedus LS (1993) J Am Chem Soc 115:87
35. Thompson DK, Suzuki N, Hegedus LS, Satoh Y (1992) J Org Chem 57:1461
36. Hsiao Y, Hegedus LS (1997) J Org Chem 62:3586
37. Dumas S, Hegedus LS (1994) J Org Chem 59:4967
38. Lopez R, Sordo TL, Sordo JA, Gonzalez J (1993) J Org Chem 58:7036; Cossio F, Ugalde JM, Lopez X, Lecea B, Palomo C (1993) J Am Chem Soc 115:995; Cossio FP, Arrieta A, Lecea B, Ugalde JM (1994) J Am Chem Soc 116:2085; Gerrieta A, Lecea B, Cossio FP (1998) J Org Chem 63:5869. For a theoretical treatment of the photoreaction of chromium carbene complexes with imines see: Arrieta A, Cossio FP, Fernandez I, Gomez-Gallego M, Lecea B, Mancheño MJ, Sierra MA (2000) J Am Chem Soc 122:11509
39. For reviews on chromium aminocarbenes see: Schwindt MA, Miller JR, Hegedus LS (1991) J Organometal Chem 413:143; Grotjahn DB, Dötz KH (1991) Synlett 381
40. Borel C, Hegedus LS, Krebs J, Satoh Y (1987) J Am Chem Soc 109:1101
41. Imwinkelried R, Hegedus LS (1988) Organometallics 7:702; Schwindt MA, Lejon T, Hegedus LS (1990) Organometallics 9:2814
42. Hegedus LS, D'Andrea S (1988) J Org Chem 53:3113
43. Es-Sayed M, Heiner T, deMeijere A (1993) Synlett 57
44. Alcaide B, Casarrubios L, Dominguez G, Sierra MA, Monge N (1995) J Am Chem Soc 117:5604
45. Hegedus LS, Imwinkelried R, Alarid-Sargent M, Dvorak D, Satoh Y (1990) J Am Chem Soc 112:1109
46. Alcaide B, Casarrubios L, Dominguez G, Sierra MA (1994) 59:7934
47. Ronan B, Hegedus LS (1993) Tetrahedron 49:5549
48. Narukawa Y, Juneau KN, Snustad D, Miller DB, Hegedus LS (1992) J Org Chem 57:5453.
49. Colson PJ, Hegedus LS (1993) J Org Chem 58:5918
50. Hegedus LS, Montgomery J, Narukawa Y, Snustad DC (1991) J Am Chem Soc 113:5784
51. Evans DA, Sjogren EB (1985) Tetrahedron Lett 26:3783; 3787
52. Merino I, Hegedus LS (1995) Organometallics 14:2522
53. Hegedus LS, Schultze LM, Montgomery J (1989) Organometallics 8:2189
54. Dorrer B, Fischer EO (1974) Chem Ber 107:2683
55. Sierra MA, Hegedus LS (1989) J Am Chem Soc 111:2335
56. Soderberg B, Hegedus LS, Sierra MA J Am Chem Soc 112:4364
57. Valente E, Pericas MA, Moyano A (1990) J Org Chem 55:3582 and references therein
58. Reeder LM, Hegedus LS (1999) J Org Chem 64:3306
59. Koebbing S, Mattay J (1992) Tetrahedron Lett 33:927
60. Soderberg BC, Hegedus LS (1990) Organometallics 9:3113
61. Moser WH, Hegedus LS (1996) J Am Chem Soc 118:7873
62. Aumann R, Kruger C, Goddard R (1992) Chem Ber 125:1627
63. Soderberg BC, Hegedus LS (1991) J Org Chem 56:2209
64. Merino I, Hegedus LS (1995) Organometallics 14:2522
65. Montgomery J, Wieber GM, Hegedus LS (1990) J Am Chem Soc 112:6255
66. Hegedus LS, Bates RW, Soderberg BC (1991) J Am Chem Soc 113:923
67. Riches AG, Wernersbach LA, Hegedus LS (1998) J Org Chem 63:4691
68. Reed AD, Hegedus LS (1995) J Org Chem 60:3787
69. Miller M, Hegedus LS (1993) J Org Chem 58:6779
70. Kedar TE, Miller MW, Hegedus LS (1996) J Org Chem 61:6121
71. Bueno A, Hegedus LS (1998) J Org Chem 63:684

72. Hegedus LS, Ranslow PB (2000) Synthesis 953
73. Sebahar HL, Yoshida K, Hegedus LS (2002) J Org Chem 67:3788
74. Reed AD, Hegedus LS (1997) Organometallics 16:2313
75. Umbricht G, Hellman MD, Hegedus LS (1998) J Org Chem 53:5173
76. Hegedus LS, Geisler L, Riches AG, Salman SS, Umbricht G (2002) J Org Chem 67:7649
77. Hegedus LS, Hervert KL, Matsui S (2002) J Org Chem 67:4076
78. Brown B, Hegedus LS (1998) J Org Chem 63:8012
79. Brown B, Hegedus LS (2000) J Org Chem 65:1865
80. Hegedus LS, Geisler L (2000) J Org Chem 65:4200
81. Wen X, Norling H, Hegedus LS (2000) J Org Chem 65:2096
82. Heileman MJ, Hegedus LS (2001) Synthesis 1356
83. Colson PJ, Hegedus LS (1994) J Org Chem 59:4972
84. Merlic CA, Doroh BC (2003) J Org Chem 68:6056
85. Hegedus LS, Kramer A (1984) Organometallics 3:1263
86. Hegedus LS, Lundmark BR (1989) J Am Chem Soc 111:9194
87. Sleiman HF, McElwee-White L (1988) J Am Chem Soc 110:8700
88. Arndtsen BA, Sleiman HF, Chang AK, McElwee-White L (1991) J Am Chem Soc 113:4871
89. Maxey CT, Sleiman HF, Massey ST, McElwee-White L (1992) J Am Chem Soc 114:5153
90. Wulff WD (1995) In: Abel EW, Stone FGA, Wilkinson G (eds) Comprehensive organometallic chemistry II, vol 12. Pergamon, Oxford, p 470
91. Wulff WD (1991) In: Trost BM, Fleming D (eds) Comprehensive organic synthesis, vol 5. Pergamon, Oxford, p 1065
92. Fischer H, Muhlemeier J, Märkl R, Dötz KH (1982) Chem Ber 115:1355
93. Merlic A, Xu D (1991) J Am Chem Soc 113:7418. For photoaccelerated classical Dötz benzannulations see: Choi YH, Rhee KS, Shin GP, Shin SC (1995) Tetrahedron Lett 36:1871; Weyershausen B, Dötz KH (1999) Synlett 231
94. Merlic CA, Xu D, Gladstone BG (1993) J Org Chem 58:538
95. Merlic CA, Roberts WM (1993) Tetrahedron Lett 34:7379
96. Merlic CA, Mcinnes DM, You Y (1997) Tetrahedron Lett 38:6787
97. Merlic CA, Aldrich CC, Albaneze-Walker J, Saghatelian A, Mammen J (2001) J Org Chem 66:1297
98. Merlic CA, Burns EE, Xu D, Chen SY (1992) J Am Chem Soc 114:8722
99. Merlic CA, Burns EE (1993) Tetrahedron Lett 34:5401
100. Merlic CA, Aldrich CC, Albaneze-Walker J, Saghatelian A (2000) J Am Chem Soc 122:3224
101. Balzer BL, Cazanone M, Sabat M, Finn MG (1992) Organometallics 11:1759
102. Balzer BL, Cazanone M, Finn MG (1992) J Am Chem Soc 114:8735
103. Soderberg BC, Odens HH (1996) Organometallics 15:5080
104. Imwinkelried R, Hegedus LS (1988) Organometallics 7:702
105. Hegedus LS, Schwindt MA, DeLombaert S, Imwinkelried R (1990) J Am Chem Soc 112:2264; Schwindt MA, Lejon T, Hegedus LS (1990) 9:2814
106. For a review see: Hegedus LS (1995) Acc Chem Res 28:299
107. Hegedus LS, Lastra E, Narukawa Y, Snustad DC (1992) J Am Chem Soc 114:2991
108. Klumpe M, Dötz KH (1998) Tetrahedron Lett 39:3683
109. Zhu J, Deur C, Hegedus LS (1997) J Org Chem 62:7704
110. Vernier J-M, Hegedus LS, Miller DB (1992) J Org Chem 57:6914
111. Schmeck C, Hegedus LS (1994) J Am Chem Soc 116:9927
112. Miller JR, Pulley SR, Hegedus LS, DeLombaert S (1992) J Am Chem Soc 114:5602
113. Dubuisson C, Fukumoto Y, Hegedus LS (1995) J Am Chem Soc 117:3697

114. Pulley SR, Hegedus LS (1993) J Am Chem Soc 115:9037
115. Zhu J, Hegedus LS (1995) J Org Chem 60:5831
116. Alcaide B, Casarrubios L, Dominguez G, Sierra MA (1994) Inorg Chim Acta 222:261
117. Sestrick MR, Miller M, Hegedus LS (1992) J Am Chem Soc 114:4079
118. Bueno AB, Moser WH, Hegedus LS (1998) J Org Chem 63:1462
119. Edstrom E (1991) J Am Chem Soc 113:6690
120. Deur CJ, Miller MW, Hegedus LS (1996) J Org Chem 61:2871
121. Alcaide B, Dominguez G, Rodriguez-Lopez J, Sierra MA (1992) Organometallics 11:1979
122. Alcaide B, Cassarubios L, Dominguez G, Sierra MA (1996) Organometallics 15:4612
123. For a review on reactions of Group 6 metal carbenes with ylides and related dipolar species see: Alcaide B, Cassarubios L, Dominguez G, Sierra MA (1998) Curr Org Chem 2:551
124. Alcaide B, Dominguez G, Plumet J, Sierra MA (1991) Organometallics 10:11
125. Alcaide B, Cassarubios L, Dominguez G, Sierra MA (1993) J Org Chem 58:3886.
126. Hegedus LS, Kramer A, Chen Y (1985) Organometallics 4:1747
127. Doyle MP (1995) In: Abel EW, Stone FGA, Wilkinson G (eds) Comprehensive organometallic chemistry II, vol 12. Pergamon, Oxford, p 387
128. Sierra MA, del Amo JC, Mancheño MJ, Gomez-Gallegos M (2001) Tetrahedron Lett 42:5435
129. Campos PJ, Sampedro D, Rodriguez MA (2002) Organometallics 21:4076
130. Campos PJ, Sampedro D, Rodriguez MA (2000) Organometallics 19:3802
131. Campos PJ, Sampedro D, Rodriguez MA (2001) Org Lett 3:4087
132. Campos PJ, Sampedro D, Rodriguez MA (2003) J Org Chem 68:4674
133. Campos PJ, Sampedro D, Rodriguez MA (2002) Tetrahedron Lett 43:73
134. Barluenga J, Fernandez-Mari F, Gonzalez R, Aguilar E, Revelli GA, Viado AL, Fañanas FJ, Olano B (2000) Eur J Org Chem 1773
135. Merlic A, Baur A, Aldrich CC (2000) J Am Chem Soc 122:7398
136. Kusama H, Takaya J, Iwasawa N (2002) J Am Chem Soc 124:11592
137. Aumann R, Jasper B (1995) Organometallics 14:1461

Topics Organomet Chem (2004) 13: 203–222
DOI 10.1007/978-3-540-40910-6

Metal Carbene Reactions from Dirhodium(II) Catalysts

Michael P. Doyle (✉)

Department of Chemistry, University of Arizona, 1306 E. University, Tucson, AZ 85721, USA
mdoyle@u.arizona.edu
Present address: Department of Chemistry and Biochemistry, University of Maryland, College Park, MD 20742, USA
mdoyle3@umd.edu

Abstract The dirhodium(II) core is a template onto which both achiral and chiral ligands are placed so that four exist in a paddle wheel fashion around the core. The resulting structures are effective electrophilic catalysts for diazo decomposition in reactions that involve metal carbene intermediates. High selectivities are achieved in transformations ranging from addition to insertion and association. The syntheses of natural products and compounds of biological interest have employed these catalysts and methods with increasing frequency.

Keywords Cyclopropanation · Insertion · Ylide reactions · Asymmetric catalysis · Synthesis

1 Introduction

Few methodologies have either the diversity of synthetic transformations or the high level of product selectivity as catalytic reactions with the intermediate involvement of metal carbenes [1–5]. They provide synthetic opportunities that are clearly demonstrated in the preparation of the antidepressant sertraline (**1**)

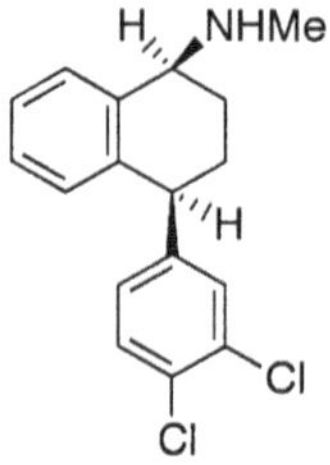

sertraline (**1**) 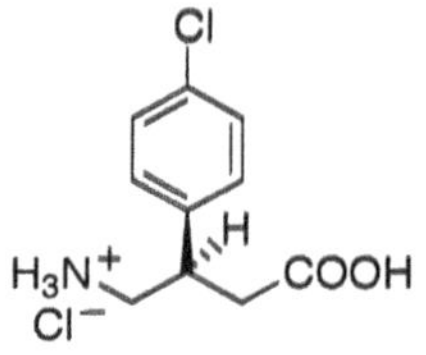 (*R*)-(-)-baclofen (**2**) 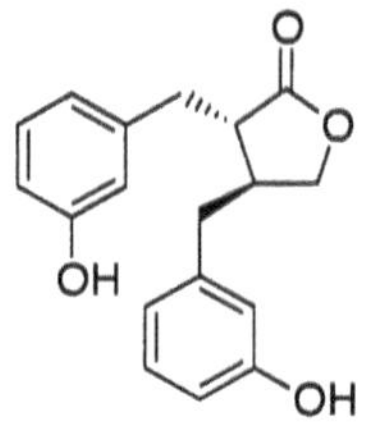 (-)-enterolactone (**3**)

[6], the GABA receptor agonist (*R*)-baclofen (**2**) [7], the lignan lactone (–)-enterolactone (**3**) [8], the metabolite presqualene alcohol (**4**) [9], and the cyclopropane-NMDA receptor antagonist milnacipran (**5**) [10], where the key step in each synthesis is a catalytic reaction of a diazocarbonyl compound.

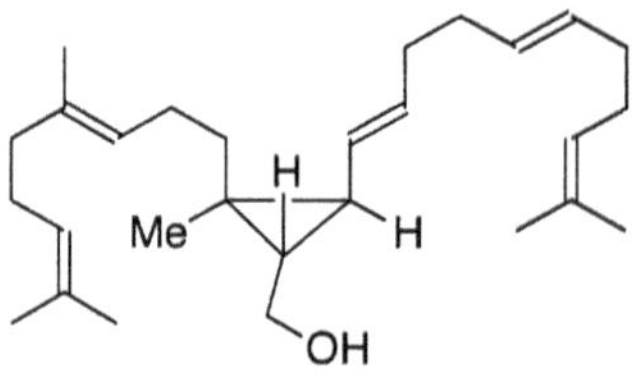

presqualene alcohol (**4**) 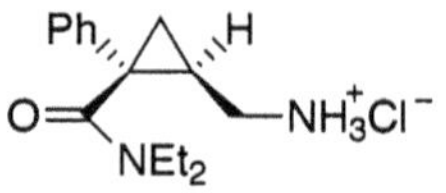 (-)-milnacipran (**5**)

Diazocarbonyl compounds are especially useful in these reactions because of their ease of formation, relative stability, and controlled reactivity in catalytic reactions [1, 11]. As outlined in Scheme 1, a wide diversity of methodologies are available for this synthesis, with access dependent on the nature of Z. Vinyl- and aryldiazoacetates are accessible by other pathways [2]. The order of reactivity toward diazo decomposition has diazoketones and diazoacetates much more reactive than diazoacetoacetates or diazomalonates. However, the influence of electronic effects on reactivities is more pronounced with phenyl- and vinyldiazoacetates than with diazoacetoacetates and, especially, diazoacetates [12].

The mechanism through which catalytic metal carbene reactions occur is outlined in Scheme 2. With dirhodium(II) catalysts the open axial coordination site on each rhodium serves as the Lewis acid center that undergoes electrophilic addition to the diazo compound. Lewis bases that can occupy the axial coor-

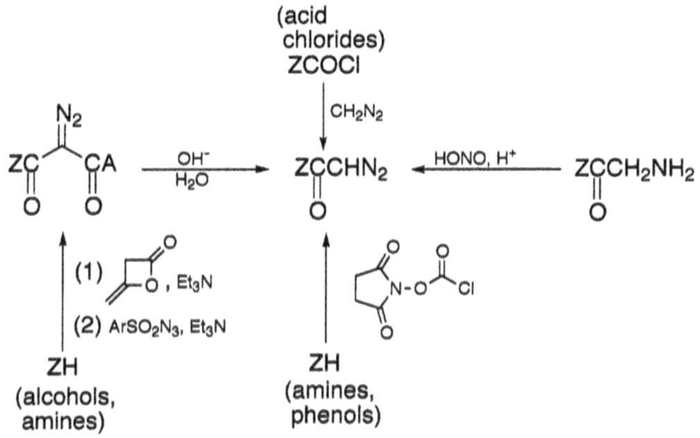

Scheme 1

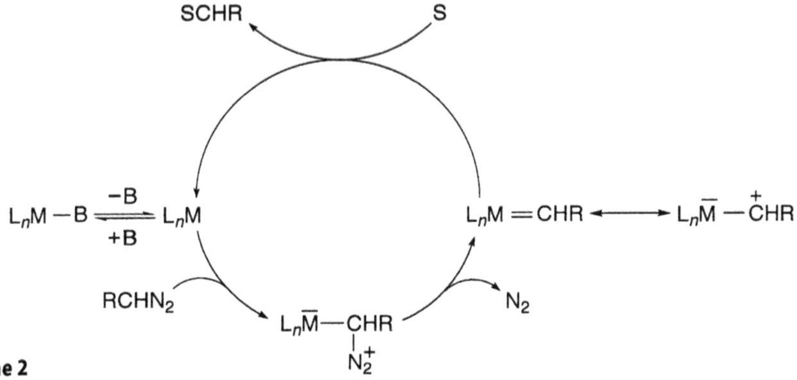

Scheme 2

dination site, which include, but are not limited to, amines, nitriles, and ketones, inhibit reaction with diazo compounds [13]. Loss of dinitrogen from the metal-associated diazo compound forms the electrophilic metal carbene intermediate that then transfers the carbene to a substrate to regenerate the catalytically active species. The highly electrophilic nature of the carbene transfer step suggests that the intermediate metal carbene may be better represented as a metal-stabilized carbocation than the traditional metal carbene.

2 The Catalysts

The use of dirhodium(II) catalysts for catalytic reactions with diazo compounds was initiated by Ph. Teyssie [14] in the 1970s and rapidly spread to other laboratories [1]. The first uses were with dirhodium(II) tetraacetate and the more soluble tetraoctanoate, $Rh_2(oct)_4$ [15]. Rhodium acetate, revealed to have the paddle wheel structure and exist with a Rh–Rh single bond [16], was conveniently prepared from rhodium(III) chloride trihydrate [17]. The more commonly employed achiral dirhodium(II) carboxylates – those without chiral carboxylate ligands – are listed in Scheme 3 (oct=octanoate, tfa=trifluoroacetate, pfb=perfluorobutyrate, tpa=triphenylacetate).

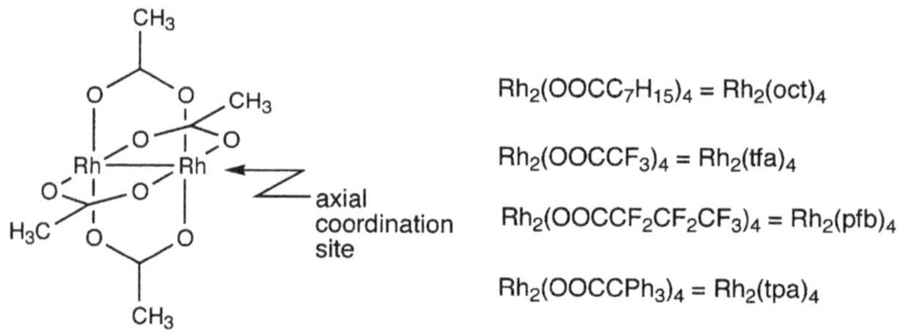

axial coordination site

$Rh_2(OOCC_7H_{15})_4 = Rh_2(oct)_4$

$Rh_2(OOCCF_3)_4 = Rh_2(tfa)_4$

$Rh_2(OOCCF_2CF_2CF_3)_4 = Rh_2(pfb)_4$

$Rh_2(OOCCPh_3)_4 = Rh_2(tpa)_4$

Scheme 3 $Rh_2(OAc)_4$

Based on their unique stereochemistry in which two nitrogens and two oxygens are bound to each rhodium in a *cis*-2,2 fashion, dirhodium(II) carboxamidates, exemplified by dirhodium(II) acetamidate [$Rh_2(acam)_4$],

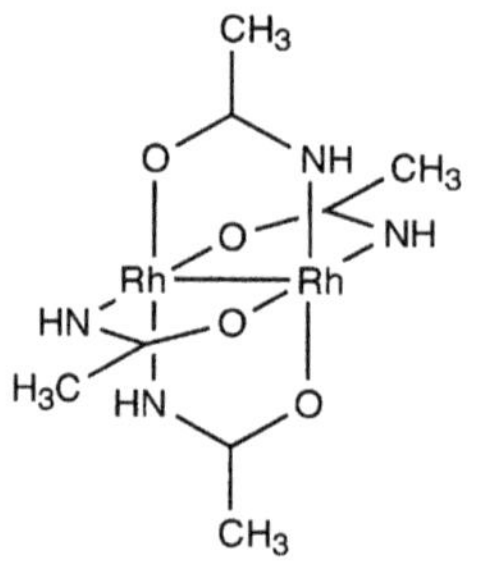

$Rh_2(acam)_4$

were extensively investigated in the 1980s [18]. However, not until a practical methodology was developed for the synthesis of these materials [19] was it possible to utilize them effectively as catalysts. As a class they are less reactive and more selective than rhodium(II) carboxylates in their reactions with diazocarbonyl compounds.

By the late 1980s efforts were begun to prepare dirhodium(II) carboxylates [20, 21] and carboxamidates [22] that possess chiral ligands, and these efforts are ongoing [23–25]. Among the carboxylates, those designed with prolinate (**6**) and phenylalanate or *tert*-lucinate (**7**) ligands have proven to be the most effective. An even broader range of chiral carboxamidate ligated dirhodium(II) catalysts (**8**, **9**, **10**, and **11**) have been prepared. Their structures are like those of $Rh_2(acam)_4$ with two oxygen and two nitrogen atoms bound to the dirhodium(II)

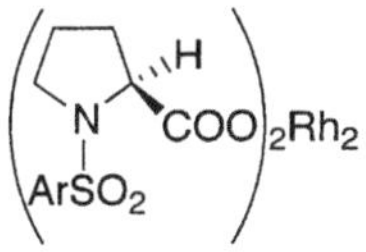

6a : Ar = *p*-*t*-BuC_6H_4, Rh_2(*S*-TBSP)$_4$ [26]
6b : Ar = *p*-$C_{12}H_{25}C_6H_4$, Rh_2(*S*-DOSP)$_4$ [27]

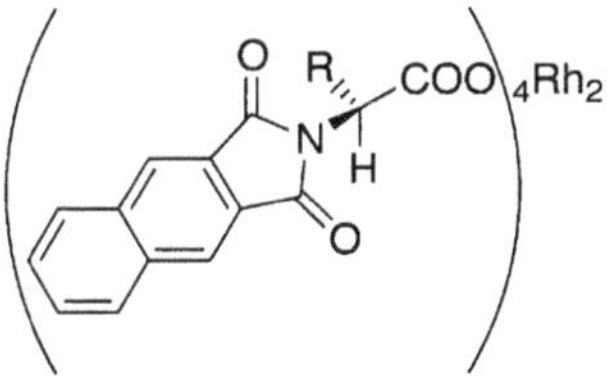

7a : R = Bn, Rh_2(*S*-BPTPA)$_4$ [28]
7b : R = *t*-Bu, Rh_2(*S*-BPTTL)$_4$ [28]

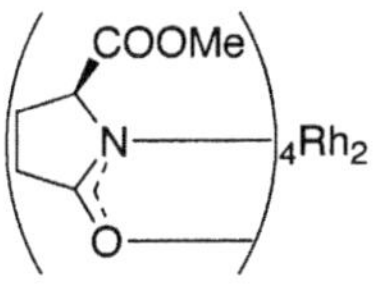

8 : Rh_2(5*S*-MEPY)$_4$ [29]

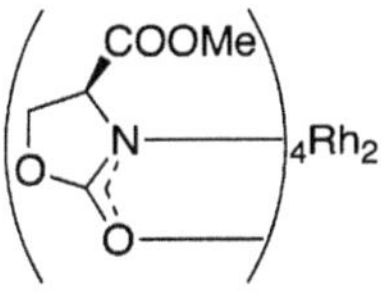

9 : Rh_2(4*S*-MEOX)$_4$ [30]

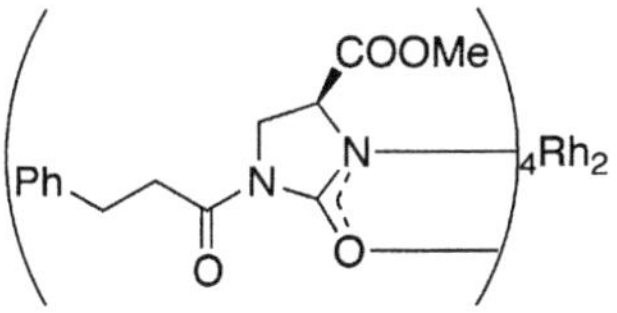

10 : Rh_2(4*S*-MPPIM)$_4$ [31]

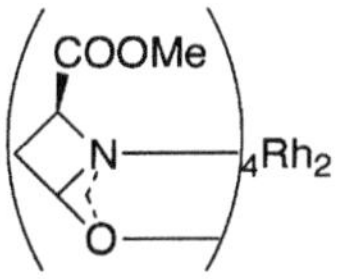

11 : Rh_2(4*S*-MEAZ)$_4$ [32]

core in a *cis*-2,2 arrangement [33]. Their reactivities toward diazo decomposition are also lower than those of the chiral dirhodium carboxylates. However, azetidinone-ligated catalysts such as **11** have enhanced reactivity relative to their five-membered ring counterparts, owing to their longer Rh–Rh bond length produced by the wider OCN bite angle of the azetidinone ring [34].

The design of dirhodium(II) catalysts offers unique geometries to enhance selectivities [33]. The flexibility of chiral carboxylates such as **6** or **7** affords potential arrays (Scheme 4) that can be and are influenced by solvent effects [35, 36]. According to Davies, the catalyst conformation that offers the highest level of selectivity in catalytic reactions is the one in which all of the chiral groups are aligned in the same direction (**12**) [35]. In contrast, the carboxamidates have a rigid structure that, as exemplified by the crystal structure

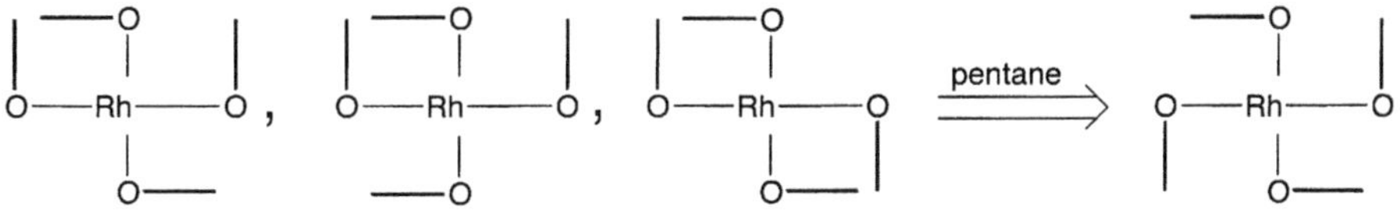

pentane

most selective (**12**)

Scheme 4

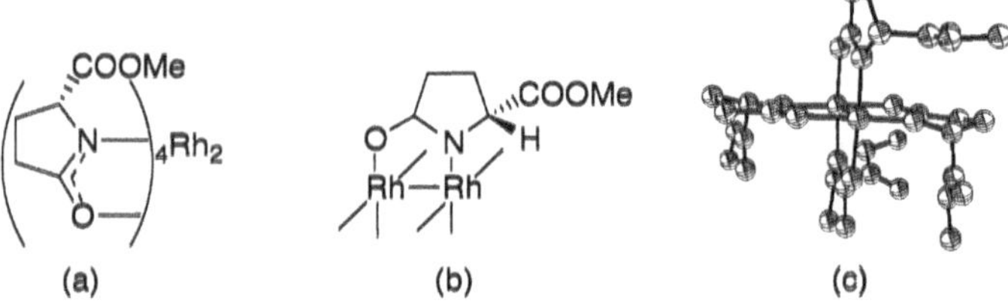

Fig. 1 Crystal structure of $Rh_2(5R\text{-MEPY})_4$

of $Rh_2(5R\text{-MEPY})_4$ (Fig. 1), offers greater access to the carbene center in catalytic reactions because of the absence of substituents in two adjacent quadrants of the catalyst surface. Only two carboxylate substituents, both closer to the carbene center than those of the dirhodium(II) carboxylates, are present to define the directional influence of the reacting system. The net result of these structural distinctions is that chiral dirhodium(II) carboxamidates are best suited for intramolecular reactions, and chiral dirhodium(II) carboxylates work well with aryl- and vinyldiazoacetates in intermolecular reactions [38].

3 Cyclopropanation

The best known of metal carbene reactions, cyclopropanation reactions, have been used since the earliest days of diazo chemistry for addition reactions to the carbon–carbon double bond. Electron-donating groups (EDG) on the carbon–carbon double bond facilitate this catalytic reaction [37], whereas electron-withdrawing groups (EWG) inhibit addition while facilitating noncatalytic dipolar cycloaddition of the diazo compound [39] (Scheme 5). There are several reviews that describe the earlier synthetic approaches [1, 2, 4, 5, 40–43], and these will not be duplicated here. Focus will be given in this review to control of stereoselectivity.

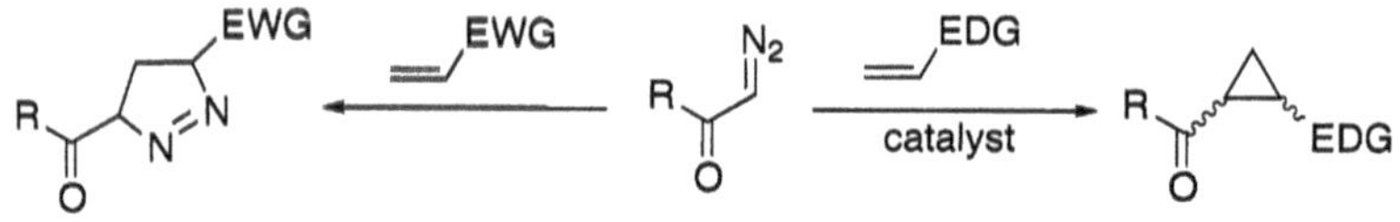

EWG EWG N_2 EDG
R N=N R catalyst R EDG
O O O

Scheme 5

3.1
Intramolecular Allylic and Homoallylic Cyclopropanation

Chiral dirhodium(II) carboxamidate catalysts are, by far, the most effective for reactions of allylic diazoacetates [44, 45] and allylic diazoacetamides [46]. Product yields are high, catalyst loading is low (less than 1 mol%), and enantioselectivities are exceptional (Scheme 6). The catalysts of choice are the two

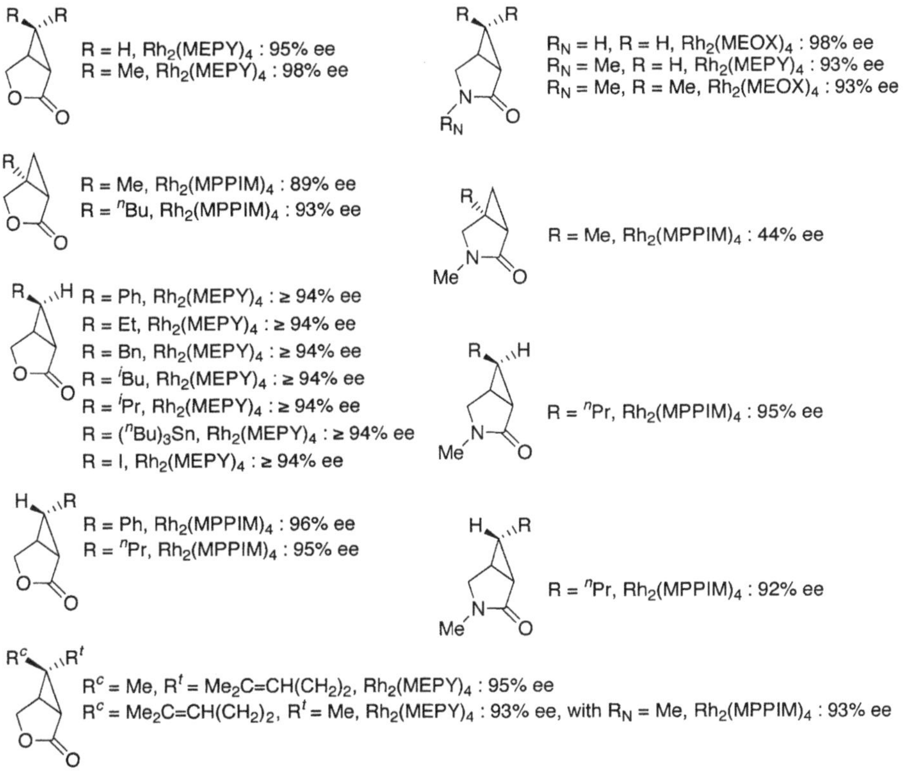

R = H, $Rh_2(MEPY)_4$: 95% ee
R = Me, $Rh_2(MEPY)_4$: 98% ee

R_N = H, R = H, $Rh_2(MEOX)_4$: 98% ee
R_N = Me, R = H, $Rh_2(MEPY)_4$: 93% ee
R_N = Me, R = Me, $Rh_2(MEOX)_4$: 93% ee

R = Me, $Rh_2(MPPIM)_4$: 89% ee
R = ^{n}Bu, $Rh_2(MPPIM)_4$: 93% ee

R = Me, $Rh_2(MPPIM)_4$: 44% ee

R = Ph, $Rh_2(MEPY)_4$: ≥ 94% ee
R = Et, $Rh_2(MEPY)_4$: ≥ 94% ee
R = Bn, $Rh_2(MEPY)_4$: ≥ 94% ee
R = ^{i}Bu, $Rh_2(MEPY)_4$: ≥ 94% ee
R = ^{i}Pr, $Rh_2(MEPY)_4$: ≥ 94% ee
R = $(^nBu)_3Sn$, $Rh_2(MEPY)_4$: ≥ 94% ee
R = I, $Rh_2(MEPY)_4$: ≥ 94% ee

R = ^{n}Pr, $Rh_2(MPPIM)_4$: 95% ee

R = Ph, $Rh_2(MPPIM)_4$: 96% ee
R = ^{n}Pr, $Rh_2(MPPIM)_4$: 95% ee

R = ^{n}Pr, $Rh_2(MPPIM)_4$: 92% ee

R^c = Me, R^t = $Me_2C{=}CH(CH_2)_2$, $Rh_2(MEPY)_4$: 95% ee
R^c = $Me_2C{=}CH(CH_2)_2$, R^t = Me, $Rh_2(MEPY)_4$: 93% ee, with R_N = Me, $Rh_2(MPPIM)_4$: 93% ee

Scheme 6

$Rh_2(MEPY)_4$ enantiomers, but for methallyl diazoacetate and *trans*-substituted allylic diazoacetates the highest levels of enantiocontrol are achieved with $Rh_2(MPPIM)_4$ catalysts [47]. With homoallylic systems enantioselectivities are lower by 10–20% ee with these catalysts (Scheme 7) [44, 48, 49], and further extensions in ring size give even lower ee percentage values but also a predominance of carbon–hydrogen insertion products [50]. Lower enantioselectivities are also observed for allylic diazopropionates [51], in which the size of the methyl group (relative to hydrogen in diazoacetates) causes a decrease in the energy difference between the diastereomeric transition states.

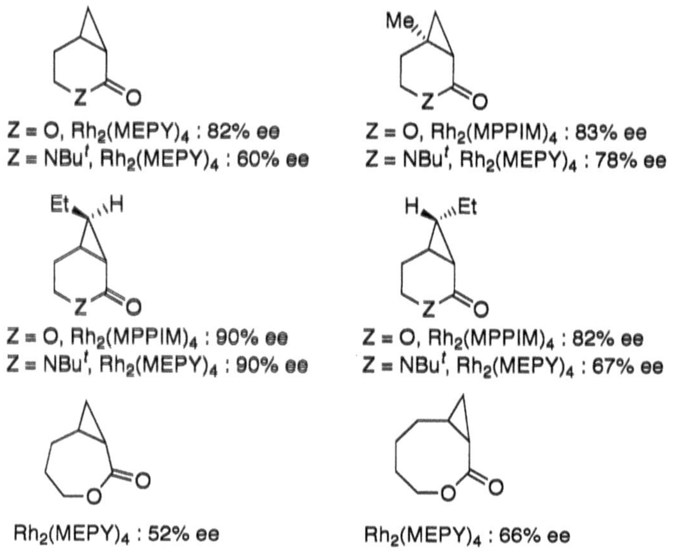

Z = O, $Rh_2(MEPY)_4$: 82% ee
Z = NBu^t, $Rh_2(MEPY)_4$: 60% ee

Z = O, $Rh_2(MPPIM)_4$: 83% ee
Z = NBu^t, $Rh_2(MEPY)_4$: 78% ee

Z = O, $Rh_2(MPPIM)_4$: 90% ee
Z = NBu^t, $Rh_2(MEPY)_4$: 90% ee

Z = O, $Rh_2(MPPIM)_4$: 82% ee
Z = NBu^t, $Rh_2(MEPY)_4$: 67% ee

Scheme 7 $Rh_2(MEPY)_4$: 52% ee $Rh_2(MEPY)_4$: 66% ee

The use of chiral dirhodium carboxamidates has made possible the highly enantioselective synthesis of presqualene alcohol (**4**) from farnesyl diazoacetate (**14**) through cyclopropane **15** [9] (Eq. 1). Highly enantiomerically en-

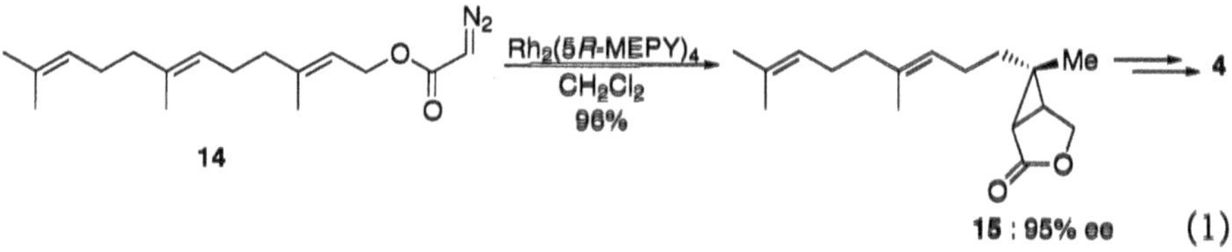

14 $\xrightarrow[\text{CH}_2\text{Cl}_2,\ 96\%]{\text{Rh}_2(5R\text{-MEPY})_4}$ **15** : 95% ee $\longrightarrow$ **4** (1)

riched 1,2,3-trisubstituted cyclopropanes have been used by Martin and coworkers to prepare conformationally restricted peptide isosteres for renin (**16**) [52], HIV-1 protease [53], matrix metalloproteinases [54], and Ras farnesyltransferase inhibitors [55], among others [56, 57]; the cyclopropane ring replaces two adjacent atoms in the peptide backbone of the native dipeptide, orienting both the peptide backbone and the amino acid side chain by varying the stereochemistry (R^1 or R^2) in the cyclopropane ring [57].

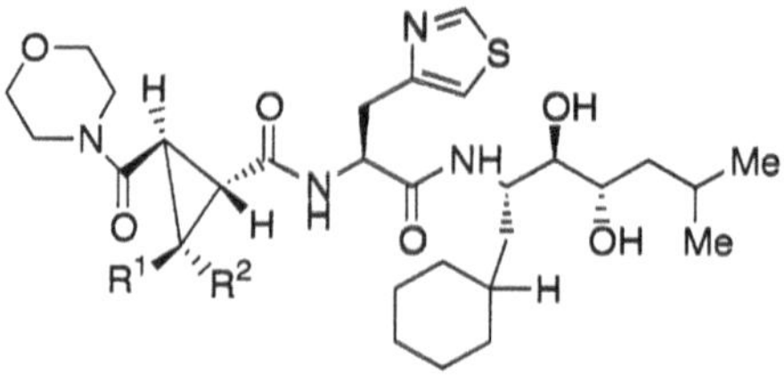

16 : a renin inhibitor

Although dirhodium(II) carboxamidates are less reactive toward diazo decomposition than are dirhodium carboxylates, and this has limited their uses with diazomalonates and phenyldiazoacetates, the azetidinone-ligated catalysts **11** cause rapid diazo decomposition, and this methodology has been used for the synthesis of the cyclopropane-NMDA receptor antagonist milnacipran (**17**) and its analogs (Eq. 2) [10, 58]. In the case of R=Me the turnover number with $Rh_2(4S\text{-MEAZ})_4$ was 10,000 with a stereochemical outcome of 95% ee.

down to 0.01 mol % $Rh_2(4S\text{-MEAZ})_4$, CH_2Cl_2, 0° C

18 R = Me, 95% ee

17 R = H, milnacipran

(2)

3.2 Intermolecular Cyclopropanation

The $Rh_2(DOSP)_4$ catalysts (**6b**) of Davies have proven to be remarkably effective for highly enantioselective cyclopropanation reactions of aryl- and vinyldiazoacetates [2]. The discovery that enantiocontrol could be enhanced when reactions were performed in pentane [35] added advantages that could be attributed to the solvent-directed orientation of chiral attachments of the ligand carboxylates [59]. In addition to the synthesis of (+)-sertraline (**1**) [6], the uses of this methodology have been extended to the construction of cyclopropane amino acids (Eq. 3) [35], the synthesis of tricyclic systems such as **22** (Eq. 4) [60], and, as an example of tandem cyclopropanation–Cope rearrangement, an efficient asymmetric synthesis of epi-tremulane **23** (Eq. 5) [61].

$Rh_2(S\text{-DOSP})_4$, pentane

19 : 90% ee

20

(3)

$Rh_2(S\text{-DOSP})_4$, pentane; Et_2AlCl, 79%

21: 86% ee

22

(4)

$Rh_2(S\text{-DOSP})_4$, pentane, -78°C, 93% ee; 140°C

23

(5)

Although dirhodium(II) carboxamidates normally do not give high diastereoselectivities in intermolecular cyclopropanation reactions [62], there are other advantages for their use. Surprisingly, the azetidinone-ligated catalysts such as **26** provide high selectivities for *cis*-disubstituted cyclopropanes [63] and, as exemplified by the synthesis of urea-PETT analog **25** that is an HIV-1 reverse transcriptase inhibitor [64], such applications benefit from the use of these catalysts (Scheme 8).

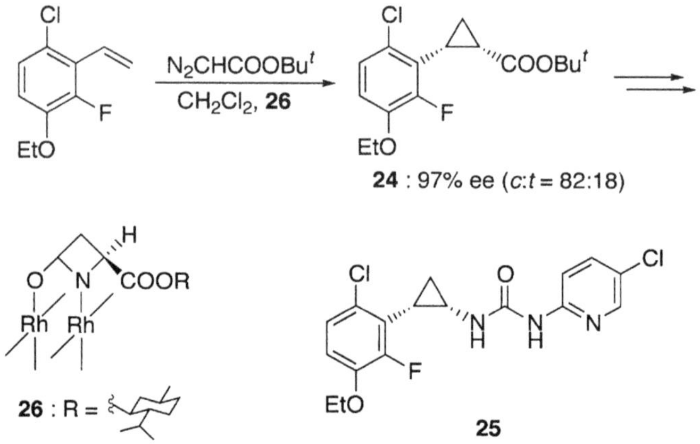

$N_2CHCOOBu^t$ / CH_2Cl_2, **26**

24 : 97% ee (*c*:*t* = 82:18)

26 : R =

25

Scheme 8 $Rh_2(4S,R\text{-MenthAZ})_4$

3.3 Macrocyclization

The search for the racemic form of **15**, prepared by allylic cyclopropanation of farnesyl diazoacetate **14**, prompted the use of $Rh_2(OAc)_4$ for this process. But, instead of **15**, addition occurred to the terminal double bond exclusively and in high yield (Eq. 6) [65]. This example initiated studies that have demonstrated the generality of the process [66–68] and its suitability for asymmetric cyclopropanation [69]. Since carbon–hydrogen insertion is in competition with addition, only the most reactive carboxamidate-ligated catalysts effect macrocyclic cyclopropanation [70] (Eq. 7), and $CuPF_6$/bis-oxazoline **28** generally produces the highest level of enantiocontrol.

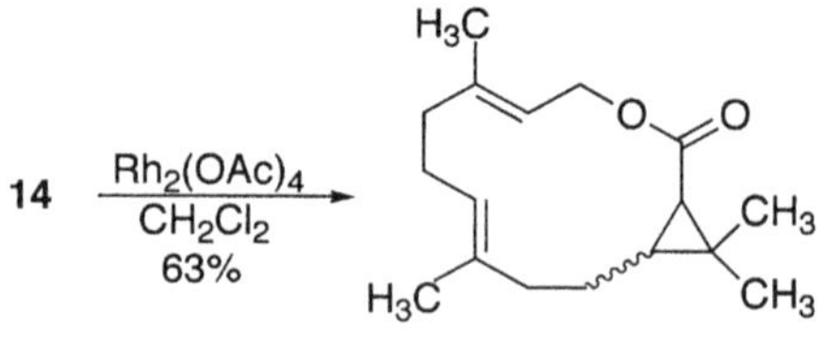

14 $\xrightarrow[\text{63\%}]{Rh_2(OAc)_4,\ CH_2Cl_2}$ (6)

26 : *t*:*c* = 86:14

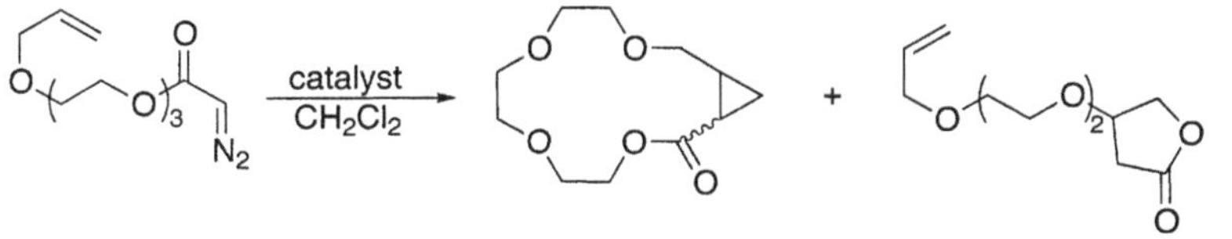

(7)

	29 (Z:E)	30
$Rh_2(OAc)_4$	97 (64:36)	3
Rh_2(dFIBAZ)$_4$ (**27a**)	83 (65:35)	13
Rh_2(MEOX)$_4$ (**9**)	7 (64:36)	93
$CuPF_6$/ **28**	83 (40:60)	13

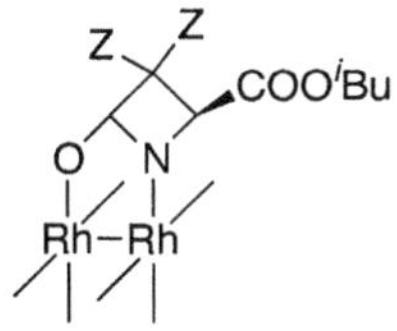

27a : Z = F
27b : Z = H

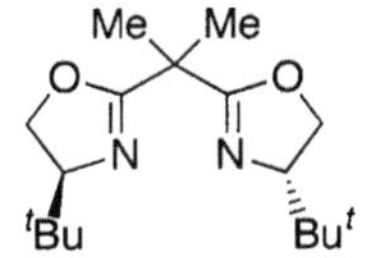

28

These reactions serve as a link in understanding selectivity differences between inter- and intramolecular cyclopropanation reactions, and they have been useful in defining the mechanism of addition as a function of catalyst [50,69,70].

4 Cyclopropenation

Addition to a carbon–carbon triple bond is even more facile than addition to a carbon–carbon double bond, and there are now several reports of intermolecular [71] and intramolecular reactions [72–74] that produce stable cyclopropene products with moderate to high enantioselectivities. One of the most revealing examples is that shown in Scheme 9 [72] where the allylic cyclopropanation product (**30**) is formed by the less reactive Rh_2(MEPY)$_4$ catalyst, but macrocyclization is favored by the more reactive Rh_2(TBSP)$_4$ and Rh_2(IBAZ)$_4$ catalysts and, as expected, the highest enantioselectivities are derived from the use of chiral dirhodium(II) carboxamidate catalysts.

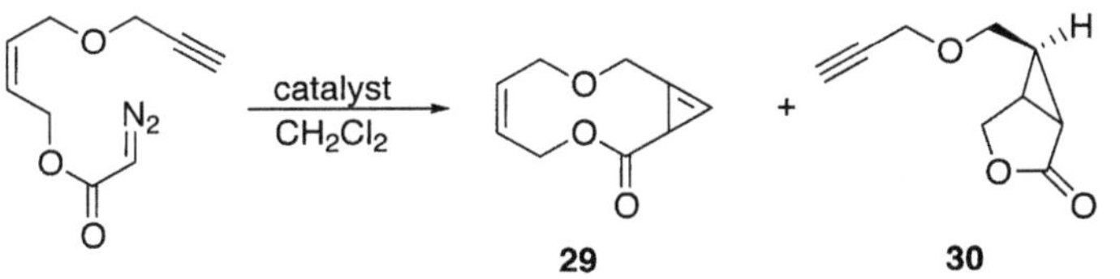

	29	30
Rh_2(4*S*-IBAZ)$_4$ (**27b**)	84 (97% ee)	16 (88% ee)
Rh_2(5*S*-MEPY)$_4$ (**8**)	4	96 (96% ee)
Rh_2(*S*-TBSP)$_4$ (**6a**)	96 (12% ee)	4 (20% ee)
$CuPF_6$/ **28**	31 (75% ee)	69 (46% ee)

Scheme 9

5 Insertion Reactions

The insertion of a carbene into a Z–H bond, where Z=C, Si, is generally referred to as an insertion reaction, whereas those occurring from Z=O, N are based on ylide chemistry [75]. These processes are unique to carbene chemistry and are facilitated by dirhodium(II) catalysts in preference to all others [1, 3, 4]. The mechanism of this reaction involves simultaneous Z–H bond breaking, Z–carbene C and carbene C–H bond formation, and the dissociation of the rhodium catalyst from the original carbene center [1].

5.1 Intramolecular Carbon–Hydrogen Insertion

The most useful of the insertion processes is the intramolecular reactions that occur with high selectivity for the formation of five-membered ring products. The electrophilic nature of the process is suggested by C–H bond reactivity in competitive experiments (3°>2°⩾1°) [76, 77]. Asymmetric catalysis with $Rh_2(MPPIM)_4$ has been used to prepare a wide variety of lignans that include (–)-enterolactone (**3**) [8], as well as (*R*)-(–)-baclofen (**2**) [7], 2-deoxyxylolactone (**31**) [80, 81], and (*S*)-(+)-imperanane (**32**) [82]. Enantioselectivities are 91–96%

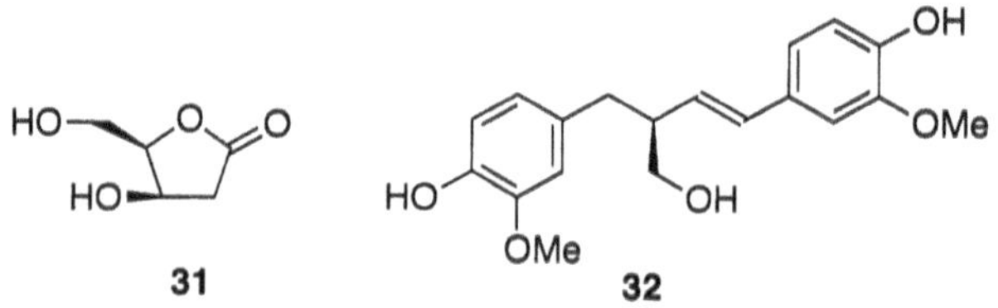

HO HO O O **31**

HO OMe OH OH OMe **32**

ee for the broad range of applications. Where there is more than one equivalent site for insertion with these diazoacetates, the formation of the *cis* product (e.g., **31** and Eq. 8) is preferred [49, 83]. The same advantages are obtained in the synthesis of pyrrolizidine bases (**36** and **37**) from pyrrolidine diazoamides [84], and the thermodynamically less stable isomer is preferred with high stereocontrol. Furthermore, *β*-lactam products such as **38** can be formed from selected diazoacetamides with exceptionally high enantioselectivities [79].

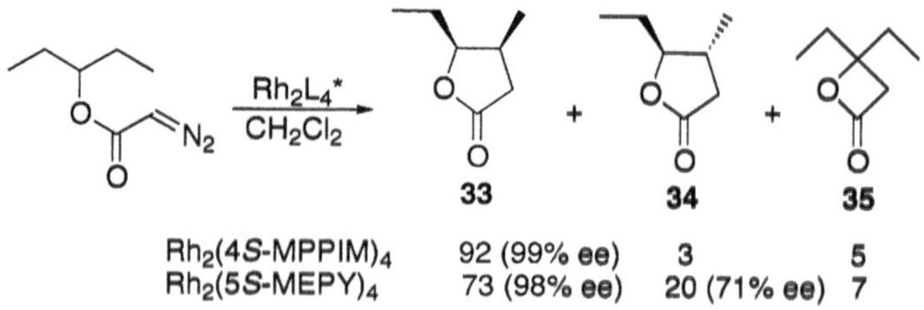

N_2 —$Rh_2L_4^*$, CH_2Cl_2→ **33** + **34** + **35** (8)

	33	34	35
$Rh_2(4S\text{-}MPPIM)_4$	92 (99% ee)	3	5
$Rh_2(5S\text{-}MEPY)_4$	73 (98% ee)	20 (71% ee)	7

36 37 38

With nonracemic chiral diazoacetates the insertion process occurs with evident match/mismatch characteristics. This has been demonstrated in reactions of optically pure 2-methylcyclohexyl diazoacetates (Eq. 9) [85] and in carbon–hydrogen insertion reactions of steroidal diazoacetates (Eq. 10) [86], as well as with the synthesis of pyrrolizidines **36** and **37** [84]. The mechanistic preference for formation of a β-lactone in Eq. 10 over insertion into the 4-position is not clear, but there are other examples of β-lactone formation [87]. In these and related examples, selectivities in match/mismatch examples are high, and future investigations are anticipated to show even greater applicability.

40 ← $Rh_2(5S\text{-MEPY})_4$, CH_2Cl_2, 94% — $Rh_2(4R\text{-MPPIM})_4$, CH_2Cl_2, 98% → 39 (9)

Rh_4L_4, CH_2Cl_2 → 41 + 42

	41	42
$Rh_2(4R\text{-MEOX})_4$	89	11
$Rh_2(4S\text{-MEOX})_4$	10	90

(10)

Diastereoselective carbon–hydrogen insertion with diazoesters leading to cyclopentane rings remains a challenge. As with intramolecular cyclopropanation reactions of diazoketones [88], diazoacetoacetates or diazomalonates [1, 2], control of selectivity is more difficult to achieve than with diazoacetates. Among recent examples are the preparation of a *cis*-isoprostane synthon **43** (Eq. 11) [89] and a key step in the construction of the marine secosteroid (–)-astrogorgiadiol **44** (Eq. 12) [90].

$Rh_2(TPA)_4$, CH_2Cl_2, 91% (11)

TPA = Ph_3CCOO 13 : 87 (**43**)

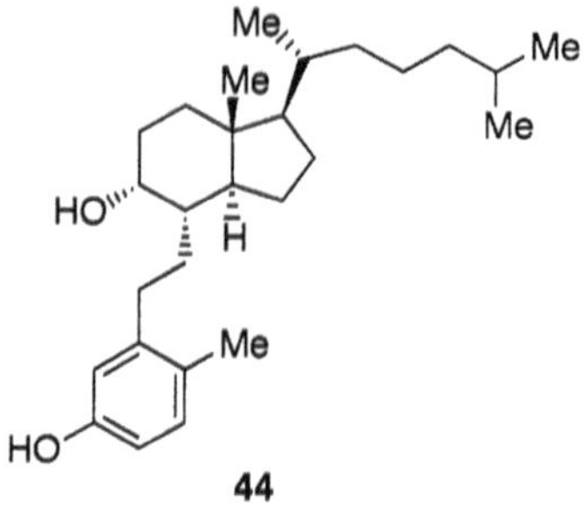

44

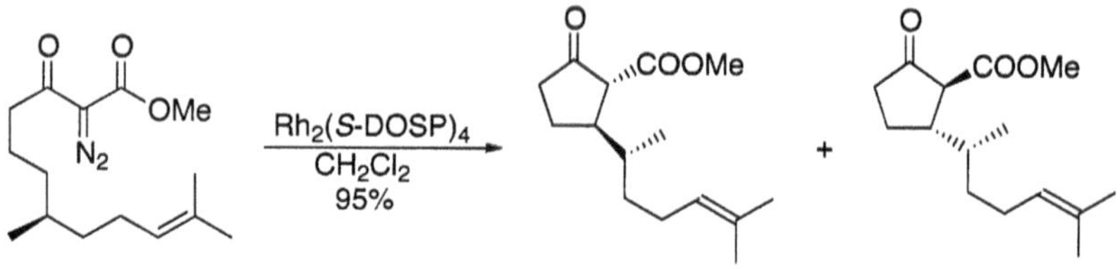

(12)

Rh$_2$(*S*-DOSP)$_4$
CH$_2$Cl$_2$
95%

5.2 Intermolecular Carbon–Hydrogen Insertion

One of the most dramatic recent developments in metal carbene chemistry catalyzed by dirhodium(II) has been demonstration of the feasibility and usefulness of intermolecular carbon–hydrogen insertion reactions [38, 91]. These were made possible by recognition of the unusual reactivity and selectivity of aryl- and vinyldiazoacetates [12] and the high level of electronic control that is possible in their reactions. Some of the products that have been formed in these reactions, and their selectivities with catalysis by Rh$_2$(*S*-DOSP)$_4$, are reported in Scheme 10.

This methodology has provided an alternative, highly enantioselective route to sertraline **1** [94]. Insertion into the oxygen-activated CH$_2$ position of allyl ethers yields *syn*-aldol products with high stereocontrol (Eq. 13) [97], and

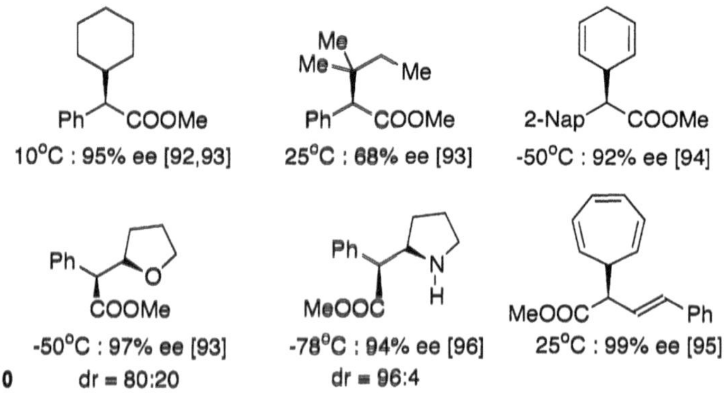

10°C : 95% ee [92,93]
25°C : 68% ee [93]
-50°C : 92% ee [94]
-50°C : 97% ee [93]
dr = 80:20
-78°C : 94% ee [96]
dr = 96:4
25°C : 99% ee [95]

Scheme 10

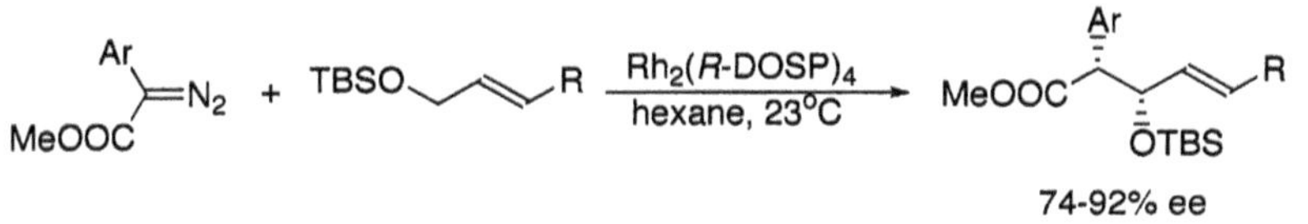

$$\text{ArC(=N}_2\text{)COOMe} + \text{TBSO-CH}_2\text{CH=CHR} \xrightarrow[\text{hexane, 23°C}]{Rh_2(R\text{-DOSP})_4} \text{MeOOC-CH(Ar)-CH(OTBS)-CH=CH-R} \tag{13}$$

74-92% ee

efficient benzylic C–H insertion has also been achieved [98]. This is a promising area for further development.

5.3 Silicon–Hydrogen Insertion

Early work by Landais and coworkers [99, 100] established the viability of aryl- and vinyldiazoacetates for silicon–hydrogen insertion which, like C–H insertion, occurs in rhodium(II)-catalyzed reactions in a concerted fashion. Subsequently, Doyle, Moody, and Davies showed that chiral dirhodium(II) catalysts could be used to effect asymmetric induction [101, 102]. Not surprisingly, the highest enantiomeric excess achieved at room temperatures or in refluxing CH_2Cl_2 was with the $Rh_2(MEPY)_4$ catalysts [101] (Eq. 14); however, these reactions were sluggish and generally impractical. Work by Davies showed that $Rh_2(S\text{-DOSP})_4$, operating at –78 °C in pentane for 48 h, gave **45** in 50% yield with 85% ee [102]; even higher selectivities could be obtained with vinyldiazoacetates.

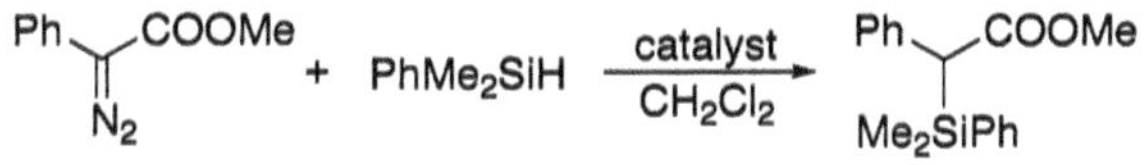

$$\text{PhC(=N}_2\text{)COOMe} + \text{PhMe}_2\text{SiH} \xrightarrow[CH_2Cl_2]{\text{catalyst}} \text{PhCH(SiMe}_2\text{Ph)COOMe} \tag{14}$$

45

6 Ylide Generation and Reactions

The use of dirhodium(II) catalysts to generate ylides that, in turn, undergo a vast array of chemical transformations is one of the major achievements in metal carbene chemistry [1, 103]. Several recent reviews have presented a wealth of information on these transformations [1, 103–106], and recent efforts have been primarily directed to establishing asymmetric induction, which arises when the chiral catalyst remains bound to the intermediate ylide during bond formation (Scheme 11).

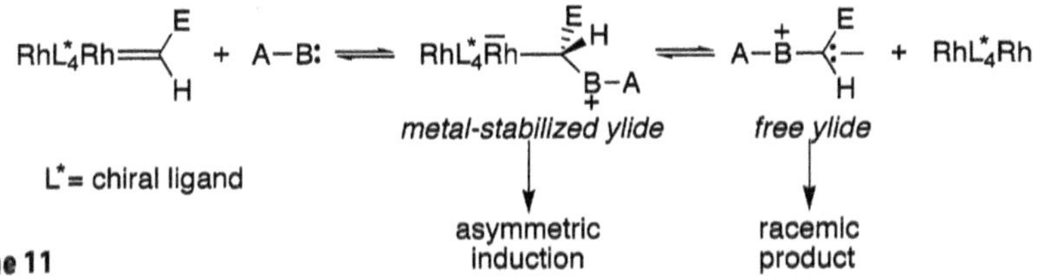

$$\text{RhL}^*_4\text{Rh=C(E)H} + \text{A–B:} \rightleftharpoons \text{RhL}^*_4\bar{\text{Rh}}\text{–C(E)(H)–}\overset{+}{\text{B}}\text{–A} \rightleftharpoons \text{A–}\overset{+}{\text{B}}\text{–}\bar{\text{C}}\text{(E)H} + \text{RhL}^*_4\text{Rh}$$

metal-stabilized ylide → asymmetric induction

free ylide → racemic product

L* = chiral ligand

Scheme 11

The premier example of this process in an ylide transformation designed for [2,3]-sigmatropic rearrangement is reported in Eq. 15 [107]. The *threo* product **47** is dominant with the use of the chiral $Rh_2(MEOX)_4$ catalysts but is the minor product with $Rh_2(OAc)_4$. That this process occurs through the metal-stabilized ylide rather than a chiral "free ylide" was shown from asymmetric induction using allyl iodide and ethyl diazoacetate [107]. Somewhat lower enantioselectivities have been observed in other systems [108].

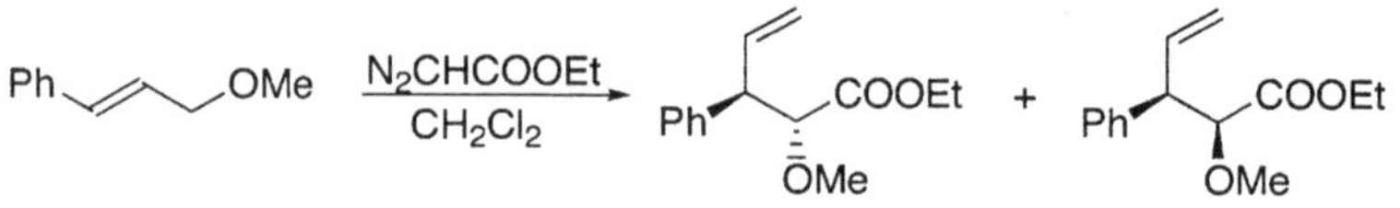

Ph–CH=CH–CH$_2$OMe $\xrightarrow[CH_2Cl_2]{N_2CHCOOEt}$ **46** + **47** (15)

	46	**47**
$Rh_2(OAc)_4$	83	17
$Rh_2(4S\text{-}MEOX)_4$	15 (95% ee)	85 (98% ee)
$Rh_2(4R\text{-}MEOX)_4$	15 (94% ee)	85 (98% ee)

6.1 Carbonyl Ylides

Carbonyl ylides continue to be targets of opportunity because of their suitability for trapping by dipolar addition. High enantiocontrol has been achieved in the process described by Eq. 16 [109], but such high enantioselectivity is not general [110] and is dependent on those factors suggested by Scheme 11. Using achiral dirhodium(II) catalysts, Padwa and coworkers have developed a broad selection of tandem reactions of which that in Eq. 17 is illustrative [111]; these

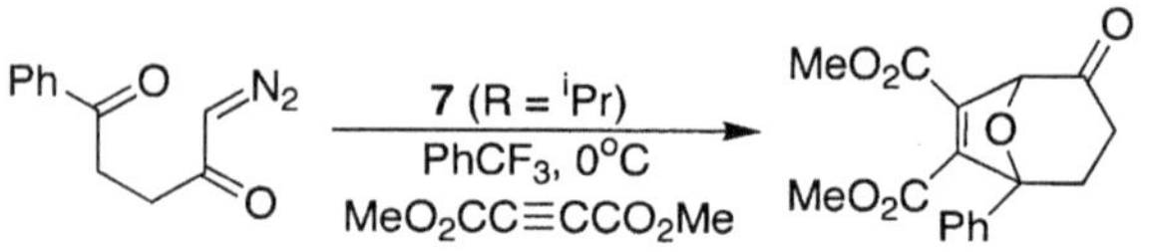

7 (R = ^{i}Pr), $PhCF_3$, 0°C, $MeO_2CC{\equiv}CCO_2Me$ → 90% ee (16)

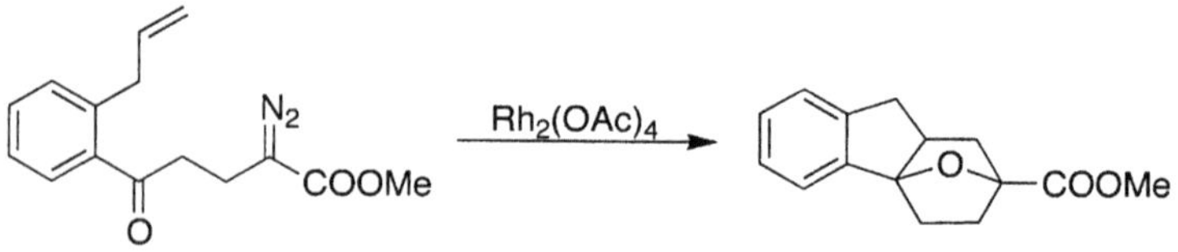

$Rh_2(OAc)_4$ (17)

intramolecular reactions indicate the multiplicity of processes catalyzed by dirhodium(II) compounds that can be used for the synthesis of complex organic compounds.

More recently carbonyl ylides and the corresponding imino ylides generated from aryl- and vinyldiazoacetates have been shown to undergo a variety of processes not previously encountered (Scheme 12) [112, 113]. The difference in

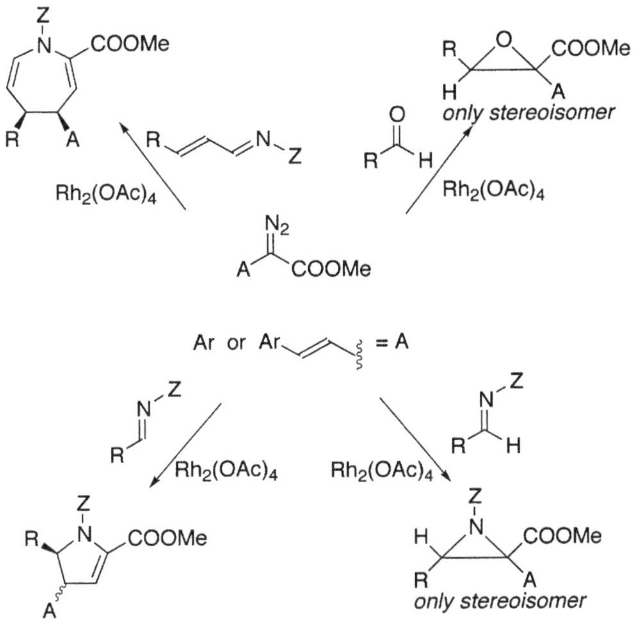

Scheme 12

these results from those obtained with the use of diazoacetates [114] is due to differences in the internal stabilities of the intermediate onium ylides, and one can anticipate a spectrum of outcomes that may result with variously constituted diazo compounds.

References

1. Doyle MP, McKervey MA, Ye T (1998) Modern catalytic methods for organic synthesis with diazo compounds. Wiley, New York
2. Davies HML, Antoulinakis EG (2001) Org React (NY) 57:1
3. Taber DF (1995) In: Helmchen G (ed) Houben-Weyl: Methods of organic chemistry, vol E21a. Thieme, Stuttgart, chap 1.2
4. Doyle MP (2000) In: Ojima I (ed) Catalytic asymmetric synthesis. Wiley-VCH, New York, chap 5
5. Burke SD, Grieco PA (1979) Org React (NY) 26:361
6. Corey EJ, Grant TG (1994) Tetrahedron Lett 35:5373
7. Doyle MP, Hu W (2002) Chirality 14:169
8. Bode JW, Doyle MP, Protopopova MN, Zhou QL (1996) J Org Chem 61:9146
9. Rogers DH, Yi EC, Poulter CD (1995) J Org Chem 60:941
10. Doyle MP, Hu W (2001) Adv Synth Catal 343:299
11. Regitz M, Maas G (1986) Diazo compounds: properties and syntheses. Academic, New York
12. Davies HML, Panaro SA (2000) Tetrahedron 56:4871
13. Pirrung MC, Liu H, Morehead AT Jr (2002) J Am Chem Soc 124:1014
14. Paulissen R, Reinhinger H, Hayez E, Hubert AJ, Teyssie P (1973) Tetrahedron Lett 2233

15. Doyle MP (1986) Chem Rev 86:919
16. Cotton FA, Walton RA (eds) (1993) Multiple bonds between metal atoms. Oxford University Press, Oxford
17. Rampel GA, Legzdino P, Smith H, Wilkinson G (1972) Inorg Synth 13:90
18. Ahsam MQ, Bernal I, Bear JL (1986) Inorg Chem 25:260
19. Doyle MP, Bagheri V, Wandless TJ, Harn NK, Brinker DA, Eagle CT, Loh KL (1990) J Am Chem Soc 112:1906
20. Brunner H, Kluschanzoff H, Wutz K (1989) Bull Soc Chem Belg 98:63
21. Kennedy M, McKervey MA, Maguire AR, Roos GHP (1990) J Chem Soc Chem Commun 361
22. Doyle MP, Brandes BD, Kazala AP, Pieters RJ, Jarstfer MB, Watkins LM, Eagle CT (1990) Tetrahedron Lett 31:6613
23. Davies HML, Walji AM (2003) Org Lett 5:479
24. Doyle MP, Yan M, Gau HM, Blossey EC (2003) Org Lett 5:561
25. Taber DF, Malcolm SC, Bieger K, Lahuerta P, Sanau M, Stiriba SE, Perez-Prieto J, Monge MA (1999) J Am Chem Soc 121:860
26. Davies HML, Peng ZQ, Houser JH (1994) Tetrahedron Lett 35:8939
27. Davies HML, Bruzinski PR, Fall MJ (1996) Tetrahedron Lett 37:4133
28. Kitagaki S, Anada M, Kataoka O, Matsuno K, Umeda C, Watanabe N, Hashimoto S (1999) J Am Chem Soc 121:1417
29. Doyle MP, Winchester WR, Hoorn JAA, Lynch V, Simonsen SH, Ghosh R (1993) J Am Chem Soc 115:9968
30. Doyle MP, Dyatkin AB, Protopopova MN, Yang CI, Miertschin CS, Winchester WR, Simonsen SH, Lynch V, Ghosh R (1995) Recl Trav Chim Pays Bas 114:163
31. Doyle MP, Zhou QL, Raab CE, Roos GHP, Simonsen SH, Lynch V (1996) Inorg Chem 35:6064
32. Doyle MP, Davies SB, Hu W (2000) Org Lett 2:1145
33. Doyle MP, Ren T (2001) Prog Inorg Chem 49:113
34. Doyle MP, Zhou QL, Simonsen SH, Lynch V (1996) Synlett 697
35. Davies HML, Bruzinski P, Hutcheson DK, Kong N, Fall MJ (1996) J Am Chem Soc 118:6897
36. Kitagaki S, Matsuda H, Watanabe N, Hashimoto S (1997) Synlett 1171
37. Doyle MP, Griffin JH, Bagheri V, Dorow RL (1984) Organometallics 3:53
38. Davies HML (1999) Eur J Org Chem 2459
39. Doyle MP, Doro RL, Tamblyn WH (1982) J Org Chem 47:4059
40. Ye T, McKervey MA (1994) Chem Rev 94:1091
41. Padwa A, Krumpe KE (1992) Tetrahedron 48:5385
42. Nefedov OM, Shapiro EA, Dyatkin AB (1992) In: Patai S (ed) The chemistry of acid derivatives, vol 2. Wiley, London, chap 25
43. Khlebnikov AF, Novikov MS, Kostikov RR (1996) Adv Heterocycl Chem 65:93
44. Doyle MP, Austin RE, Bailey AS, Dwyer MP, Dyatkin AB, Kalimin AV, Kwan MMY, Liras S, Oalmann CJ, Pieters RJ, Protopopova MN, Raab CE, Roos GHP, Zhou QL, Martin SF (1995) J Am Chem Soc 117:5763
45. Doyle MP, Peterson CS, Zhou QL, Nishiyama H (1997) J Chem Soc Chem Commun 211
46. Doyle MP, Kalinin AV (1996) J Org Chem 61:2179
47. Doyle MP, Protopopova MN (1998) Tetrahedron 54:7919
48. Doyle MP, Eismont MY, Protopopova MN, Kwan MMY (1994) Tetrahedron 50:1665
49. Doyle MP, Zhou QL, Dyatkin AB, Ruppar DA (1995) Tetrahedron Lett 36:7579
50. Doyle MP, Phillips IM (2001) Tetrahedron Lett 42:3155
51. Doyle MP, Zhou QL (1995) Tetrahedron Asymmetry 6:2157

52. Martin, SF, Austin RE, Oalmann CJ, Baker WR, Condon SL, deLara E, Rosenberg SH, Spina KP, Stein HH, Cohen J, Kleinert HD (1992) J Med Chem 35:1710
53. Martin SF, Dorsey GO, Game T, Hillier M, Kessler H, Baur M, Mathä B, Erickson JW, Bhat TN, Munshi S, Gulnik SV, Topal IA (1998) J Med Chem 41:1581
54. Martin SF, Dwyer MP, Hartmann B, Knight KS (2000) J Org Chem 65:1305
55. Hillier MC, Davidson JP, Martin SF (2001) J Org Chem 66:1657
56. Martin SF, Oalmann CJ, Liras S (1993) Tetrahedron Lett 49:3521
57. Davidson JP, Lubman O, Rose T, Waksman G, Martin SF (2002) J Am Chem Soc 124:205
58. Doyle MP, Hu W, Weathers TM Jr (2003) Chirality 15:369
59. Doyle MP, Zhou QL, Charnsangavej C, Longoria MA, McKervey MA, Garcia CF (1996) Tetrahedron Lett 37:4129
60. Davies HML, Kong N, Churchill MR (1998) J Org Chem 63:4129
61. Davies HML, Doan BD (1996) Tetrahedron Lett 37:3967
62. Müller P, Baud C, Ené D, Motallebi S, Doyle MP, Brandes BD, Dyatkin AB, See MM (1995) Helv Chim Acta 78:459
63. Doyle MP, Davies SB, Hu W (2000) J Chem Soc Chem Commun 867
64. Hu W, Timmons DJ, Doyle MP (2002) Org Lett 4:901
65. Doyle MP, Protopopova MN, Poulter CD, Rogers DH (1995) J Am Chem Soc 117:7281
66. Doyle MP, Peterson CS, Protopopova MN, Marnett AB, Parker DL Jr, Ene DG, Lynch V (1997) J Am Chem Soc 119:8826
67. Doyle MP, Peterson CS, Parker DL Jr (1996) Angew Chem Int Ed Engl 35:1334
68. Doyle MP, Hu W (2001) Synlett 1364
69. Doyle MP, Hu W, Chapman B, Marnett AB, Peterson CS, Vitale JP, Stanley SA (2000) J Am Chem Soc 122:5718
70. Doyle MP, Hu W (2000) J Org Chem 65:8839
71. Doyle MP, Protopopova MN, Müller P, Ene DG, Shapiro E (1994) J Am Chem Soc 116:8492
72. Doyle MP, Ene DG, Peterson CS, Lynch V (1999) Angew Chem Int Ed Engl 38:700
73. Doyle MP, Ene DG, Forbes DC, Pillow TH (1999) J Chem Soc Chem Commun 1691
74. Doyle MP, Hu W (2000) Tetrahedron Lett 41:6265
75. Moody CJ, Miller DJ (1998) Tetrahedron 54:2257
76. Taber DF, Ruckle RE Jr (1986) J Am Chem Soc 108:7686
77. Doyle MP, Westrum LJ, Wolthius WNE, See MM, Boone WP, Bagheri V, Pearson MM (1993) J Am Chem Soc 115:958
78. Wang P, Adams J (1994) J Am Chem Soc 116:3296
79. Doyle MP, Kalinin AV (1995) Synlett 1075
80. Doyle MP, Dyatkin AB, Tedrow JS (1994) Tetrahedron Lett 35:3853
81. Doyle MP, Tedrow JS, Dyatkin AB, Spaans CJ, Ene DG (1999) J Org Chem 64:8907
82. Doyle MP, Hu W, Valenzuela MV (2002) J Org Chem 67:2954
83. Doyle MP, Dyatkin AB, Roos GHP, Cañas F, Pierson DA, van Basten A, Müller P, Polleux P (1994) J Am Chem Soc 116:4507
84. Doyle MP, Kalinin AV (1996) Tetrahedron Lett 37:1371
85. Doyle MP, Kalinin AV, Ene DG (1996) J Am Chem Soc 118:8837
86. Doyle MP, Davies SB, May EJ (2001) J Org Chem 66:8112
87. Doyle MP, May EJ (2001) Synlett 967
88. Taber DF, Kanai K (1999) J Org Chem 64:7983
89. Taber DF, Green JH, Zhang W, Song R (2000) J Org Chem 65:5436
90. Taber DF, Malcolm SC (2001) J Org Chem 66:944
91. Davies HML, Antoulinakis EG (2001) J Organometal Chem 617–618:47
92. Davies HML, Hansen T (1997) J Am Chem Soc 119:9075

93. Davies HML, Hansen T, Churchill MR (2000) J Am Chem Soc 122:3063
94. Davies HML, Stafford DG, Hansen T (1999) Org Lett 1:233
95. Davies HML, Stafford DG, Hansen T, Churchill MR, Keil KM (2000) Tetrahedron Lett 41:2035
96. Davies HML, Hansen T, Hopper D, Panaro SA (1999) J Am Chem Soc 121:6509
97. Davies HML, Antoulinakis EG, Hansen T (1999) Org Lett 1:383
98. Davies HML, Jin Q, Ren P, Kovalevsky AY (2002) J Org Chem 67:4165
99. Landais Y, Planchenault D (1994) Tetrahedron Lett 35:4565
100. Bulugahapitiya P, Landais Y, Parra-Rapado L, Planchenault D, Weber V (1997) J Org Chem 62:1630
101. Buck RT, Doyle MP, Drysdale MJ, Ferris L, Forbes DC, Haigh D, Moody CJ, Pearson ND, Zhou QL (1996) Tetrahedron Lett 37:7631
102. Davies HML, Hansen T, Rutberg J, Bruzinski P (1997) Tetrahedron Lett 38:1741
103. Doyle MP, Forbes DC (1998) Chem Rev 98:911
104. Padwa A, Hornbuckle SF (1991) Chem Rev 91:263
105. Li AH, Dai LX, Aggarwal VK (1997) Chem Rev 97:2341
106. Hodgson DM, Pierard FYTM, Stupple PA (2001) Chem Soc Rev 30:50
107. Doyle MP, Forbes DC, Vasbinder MM, Peterson CS (1998) J Am Chem Soc 120:7653
108. Kitagaki S, Yanamoto Y, Tsutsui H, Anada M, Nakajima M, Hashimoto S (2001) Tetrahedron Lett 42:6361
109. Kitagaki S, Anada M, Kataoka O, Matsumo K, Umeda C, Watanabe N, Hashimoto S (1999) J Am Chem Soc 121:1417
110. Hodgson DM, Glen R, Grant GH, Redgrave AJ (2003) J Org Chem 68:581
111. Padwa A, Zhang ZJ, Zhi L (2000) J Org Chem 65:5223
112. Doyle MP, Hu W, Timmons DJ (2001) Org Lett 3:933
113. Doyle MP, Hu W, Timmons DJ (2001) Org Lett 3:3741
114. Doyle MP, Forbes DC, Protopopova MN, Stanley SA, Vasbinder MM, Xavier KR (1997) J Org Chem 62:7210

Topics Organomet Chem (2004) 13: 223–267
DOI 10.1007/978-3-540-40910-6

Olefin Metathesis Directed to Organic Synthesis: Principles and Applications

Bernd Schmidt · Jolanda Hermanns

Universität Dortmund, Fachbereich Chemie, Organische Chemie, Otto-Hahn-Strasse 6, 44227 Dortmund, Germany
bschmidt@chemie.uni-dortmund.de

Abstract For many years after its discovery, olefin metathesis was hardly used as a synthetic tool. This situation changed when well-defined and stable carbene complexes of molybdenum and ruthenium were discovered as efficient precatalysts in the early 1990s. In particular, the high activity and selectivity in ring-closure reactions stimulated further research in this area and led to numerous applications in organic synthesis. Today, olefin metathesis is one of the most dynamic areas in organic synthesis and homogeneous catalysis. This contribution will focus on the development of ruthenium-based metathesis precatalysts with extended scope and their application in organic synthesis. Applications in polymer chemistry and in the total synthesis of complex natural products are not covered.

Keywords Metathesis · Alkenes · Catalysis · Ruthenium · Metal carbene complexes

1 Introduction

Olefin metathesis is the transition-metal-catalyzed inter- or intramolecular exchange of alkylidene units of alkenes. The metathesis of propene is the most simple example: in the presence of a suitable catalyst, an equilibrium mixture of ethene, 2-butene, and unreacted propene is obtained (Eq. 1). This example illustrates one of the most important features of olefin metathesis: its reversibility. The metathesis of propene was the first technical process exploiting the olefin metathesis reaction. It is known as the Phillips triolefin process and was run from 1966 till 1972 for the production of 2-butene (feedstock: propene) and from 1985 for the production of propene (feedstock: ethene and 2-butene, which is nowadays obtained by dimerization of ethene). Typical catalysts are oxides of tungsten, molybdenum or rhenium supported on silica or alumina [1].

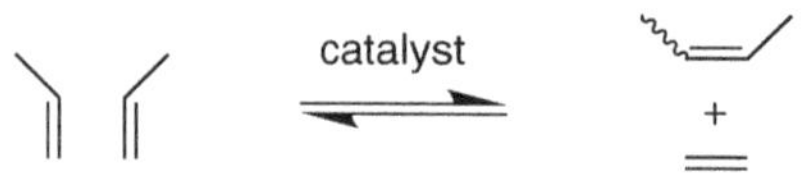

catalyst (1)

Olefin metathesis was originally discovered during research directed to the development of catalysts for olefin polymerization. Typically, these catalysts are generated in situ from a transition metal halide and a main group organometallic compound [2]. It was rapidly discovered that certain catalyst systems behave significantly different toward alkenes. For example, Truett et al. reported in 1960 that treatment of norbornene with a catalyst generated in situ from lithium aluminum tetraalkyl and titanium tetrachloride did not give the expected polyolefin, but a highly unsaturated polymer that could be degraded to *cis*-1,3-cyclopentanedicarboxylic acid (Eq. 2) [3].

A mechanism for olefin metathesis reactions, which is now generally accepted, was first proposed in 1970 by Hérisson and Chauvin [4]. It is outlined

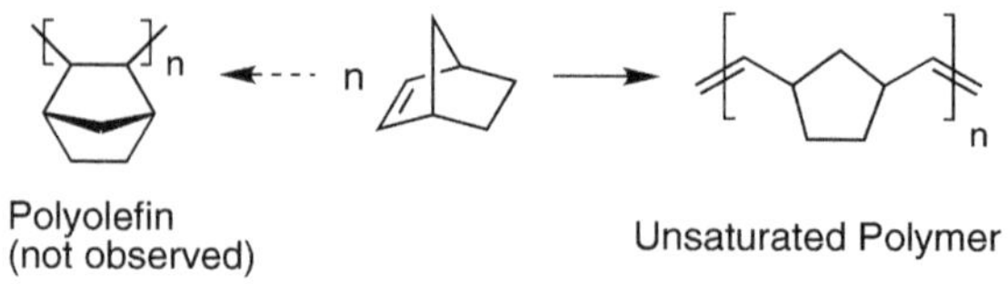

Polyolefin (not observed) Unsaturated Polymer (2)

in Scheme 1 for the case of an intramolecular olefin metathesis, generally considered as ring-closing metathesis (RCM). The components of the catalyst system react to a metal carbene **A**, which is the catalytically active species. **A** and the substrate, in this case an α,ω-diene **B**, undergo a [2+2]-cycloaddition to a metallacyclobutane **C**. Cycloreversion of **C** gives a new carbene species **D** and ethylene. Now the sequence of [2+2]-cycloaddition and cycloreversion is repeated: intramolecular cycloaddition leads to a metallacyclobutane **E**; cycloreversion of **E** regenerates the catalytically active species **A** and gives the metathesis product, a cycloalkene **F**.

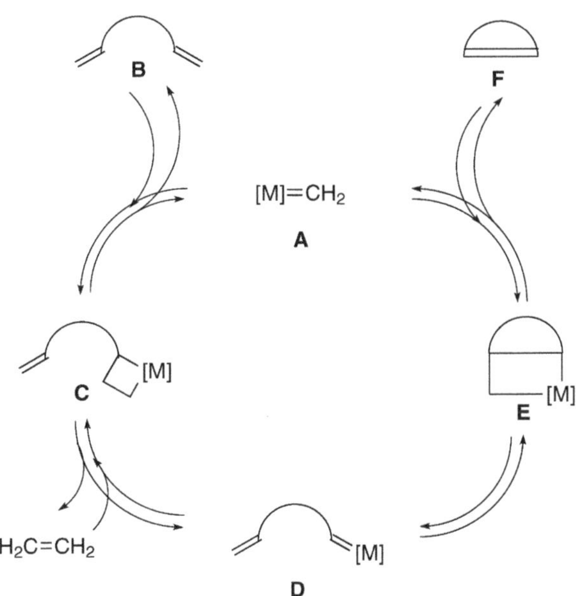

Scheme 1 Chauvin mechanism for ring-closing metathesis [4]

As stated above, olefin metathesis is in principle reversible, because all steps of the catalytic cycle are reversible. In preparatively useful transformations, the equilibrium is shifted to one side. This is most commonly achieved by removal of a volatile alkene, mostly ethene, from the reaction mixture. An obvious and well-established way to classify olefin metathesis reactions is depicted in Scheme 2. Depending on the structure of the olefin, metathesis may occur either inter- or intramolecularly. Intermolecular metathesis of two alkenes is called cross metathesis (CM) (if the two alkenes are identical, as in the case of the Phillips triolefin process, the term self metathesis is sometimes used). The intermolecular metathesis of an α,ω-diene leads to polymeric structures and ethene; this mode of metathesis is called acyclic diene metathesis (ADMET). Intramolecular metathesis of these substrates gives cycloalkenes and ethene (ring-closing metathesis, RCM); the reverse reaction is the cleavage of a cyclo-

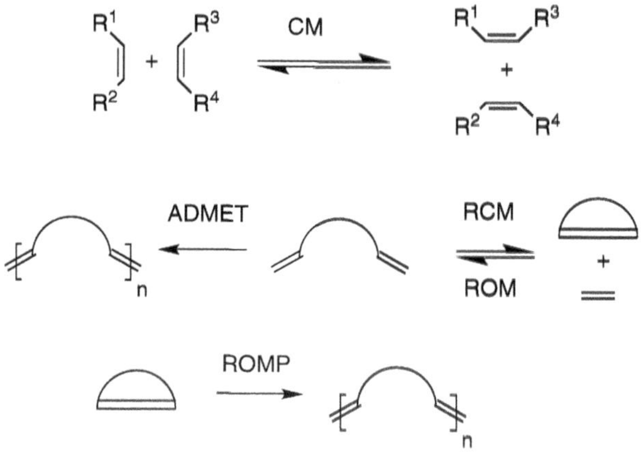

Scheme 2 Different modes of the olefin metathesis reaction: cross metathesis (CM), ring-closing metathesis (RCM), ring-opening metathesis (ROM), acyclic diene metathesis polymerization (ADMET), and ring-opening metathesis polymerization (ROMP)

alkene with ethene or another alkene (ring-opening metathesis, ROM). The latter reaction mode is particularly important for strained bicyclic alkenes, such as norbornene. In the absence of an acyclic alkene, a ring-opening metathesis may be the initiating step for a polymerization reaction, a metathesis type which is described as ring-opening metathesis polymerization (ROMP). This process is very important from an industrial point of view. For example, the ring-opening metathesis polymerization of cyclooctene is exploited for the production of Vestenamer, an elastomeric additive for rubbers [5].

Although olefin metathesis had soon after its discovery attracted considerable interest in industrial chemistry, polymer chemistry and, due to the fact that transition metal carbene species are involved, organometallic chemistry, the reaction was hardly used in organic synthesis for many years. This situation changed when the first structurally defined and stable carbene complexes with high activity in olefin metathesis reactions were described in the late 1980s and early 1990s. A selection of precatalysts discovered in this period and representative applications are summarized in Table 1.

Molybdenum complex **5** [7a] and ruthenium complex **7** [7b] turned out to be "lead structures" for the development of metathesis precatalysts for organic synthesis. The overwhelming majority of studies dealing with olefin metathesis directed to organic synthesis rely on precatalysts derived from either **5** or **7**. The advent of these reagents caused a revolution in organic synthesis because synthetic strategies can now be realized that would have been utopian 15 years ago. Not surprisingly, olefin metathesis became one of the most dynamically developing areas in contemporary chemistry. The publication of numerous excellent reviews over the past few years reflects the importance of this reaction for organic synthesis [7]. Due to limitations in space it is impossible to cover all facets of olefin metathesis in this contribution.

Table 1 Examples of defined metathesis precatalysts and selected applications

Precatalyst	Application	Reference
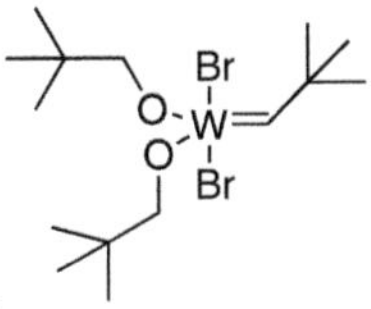 1	ROMP of norbornene	[6a-c]
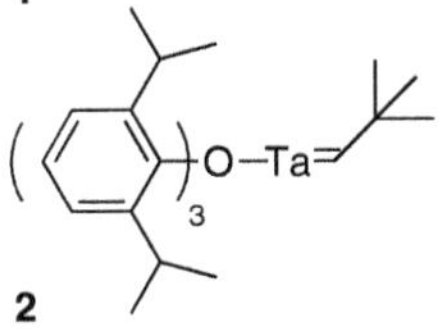 2	Self metathesis of *Z*-2-Pentene	[6d]
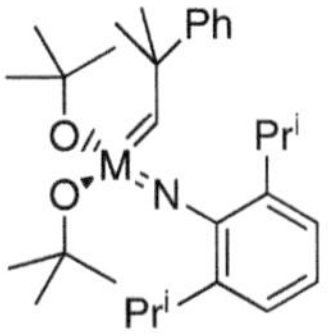 **3a**: M = Mo **3b**: M = W	ROMP of norbornene	[6e]
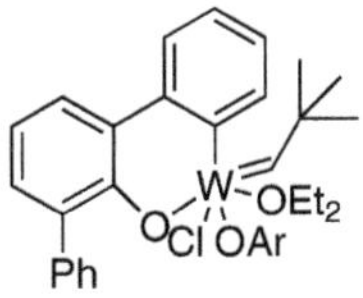 4	Self metathesis of ethyl oleate	[6f]
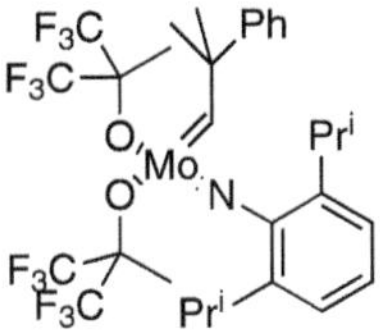 5	Self metathesis of *Z*-2-Pentene	[6g]
	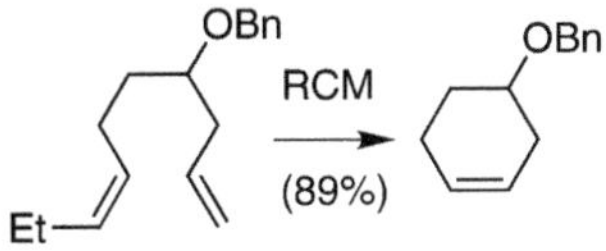 RCM (89%)	[6h]
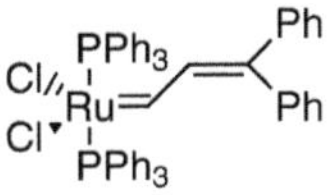 6	ROMP of norbornene	[6i]
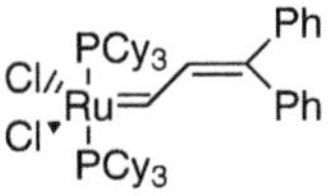 7 (Cy = cyclohexyl)	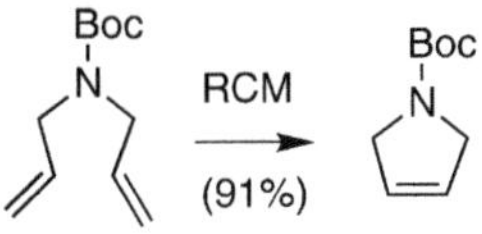 RCM (91%)	[6j] [6k]

We will focus on the development of ruthenium-based metathesis precatalysts with enhanced activity and applications to the metathesis of alkenes with nonstandard electronic properties. In the class of molybdenum complexes [7a,g,h] recent research was mainly directed to the development of homochiral precatalysts for enantioselective olefin metathesis. This aspect has recently been covered by Schrock and Hoveyda in a short review and will not be discussed here [8h]. In addition, several important special topics have recently been addressed by excellent reviews, e.g., the synthesis of medium-sized rings by RCM [8a], applications of olefin metathesis to carbohydrate chemistry [8b], cross metathesis [8c,d], enyne metathesis [8e,f], ring-rearrangement metathesis [8g], enantioselective metathesis [8h], and applications of metathesis in polymer chemistry (ADMET, ROMP) [8i,j]. Application of olefin metathesis to the total synthesis of complex natural products is covered in the contribution by Mulzer et al. in this volume.

2
Ruthenium-Based Olefin Metathesis Catalysts

Soon after their first description in the literature, some significant differences in reactivity between Schrock's molybdenum complex **5** and Grubbs' ruthenium complex **7** were discovered. Complex **5** was found to be less tolerant toward air and moisture and also significantly less tolerant toward polar functional groups, such as alcohols, than **7**. On the other hand, **5** was found to be much more reactive in the formation of tri- or even tetrasubstituted double bonds, a situation where precatalyst **7** is significantly less reactive and sometimes even inactive. The issue of functional group compatibility of **5** and **7** has been addressed in a review by Armstrong where numerous examples and references are given [7d]. While the original molybdenum complex **5** is still used in organic synthesis, **7** was replaced by more conveniently available or more active derivatives. The evolution of ruthenium-based olefin metathesis catalysts is described in this chapter.

2.1
A Survey of Methods for Introduction of the Carbene Ligand: First-Generation Metathesis Catalysts

Synthetic routes to active ruthenium metathesis catalysts are classified according to the ruthenium precursor used.

2.1.1
... From $[RuCl_2(PPh_3)_3]$ or $[RuCl_2(PPh_3)_4]$

The ruthenium complexes $[RuCl_2(PPh_3)_3]$ and $[RuCl_2(PPh_3)_4]$ are conveniently available from $RuCl_3 \cdot 3H_2O$ and triphenylphosphine [9]. Reaction with 3,3-diphenylcyclopropene gives **6**, which catalyzes the living polymerization of norbornene [6i], but shows no activity in the metathesis of acyclic dienes [6k]. Precatalyst **7** with significantly enhanced activity was obtained by exchanging the triphenylphosphine ligands in **6** against the more nucleophilic tricyclohexylphosphine (PCy_3) or triisopropylphosphine (PPr^i_3) [6k]. Complex **7** was found to be an efficient catalyst for the self metathesis of *Z*-2-pentene as well as for various RCM reactions; an example is given in Scheme 3 [6j].

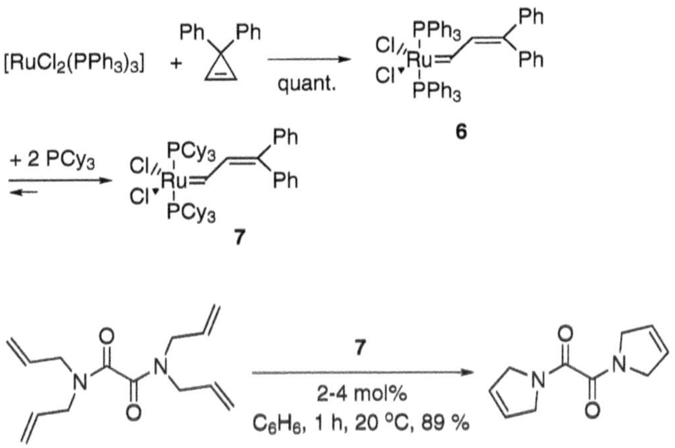

Scheme 3 Grubbs' route to Ru carbene complex **7** [6j,k]

Various attempts were made to develop more convenient approaches to ruthenium precatalysts derived from **7** by choosing carbene sources that are more easily available than cyclopropenes. Grubbs et al. discovered an improved approach starting from $[RuCl_2(PPh_3)_3]$, phenyldiazomethane (**8**), and PCy_3 [10]. The resulting ruthenium carbene complex **9** rapidly became one of the standard catalysts for olefin metathesis and is now commercially available. More recently, Milstein et al. devised an alternative route to **9** by using a sulfur ylide **10**, generated in situ from benzyldiphenylsulfonium tetrafluoroborate and a base [11]. This route appears to be general and was also applied to the synthesis of carbene complexes of rhodium and osmium (Scheme 4).

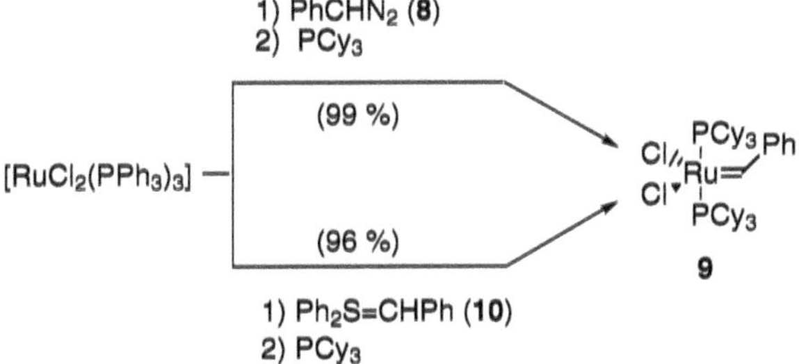

Scheme 4 Grubbs' [10] and Milstein's [11] synthesis of **9**

By treating [$RuCl_2(PPh_3)_3$] with an isopropoxy-substituted phenyldiazomethane **11**, Hoveyda et al. obtained the carbene complex **12**, which upon treatment with excess PCy_3 undergoes a ligand exchange to yield the monophosphine complex **13** (Scheme 5) [12]. While **12** was found to be inactive, **13** is an efficient precatalyst for olefin metathesis reactions. Kinetic studies revealed that **13** initiates the ROMP of cyclooctene 30 times slower than **9**, but propagation is four times faster. Mechanistic implications of this observation will be discussed later in this chapter.

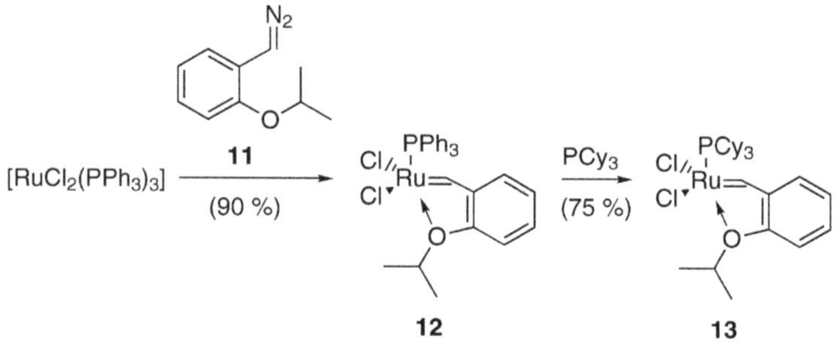

Scheme 5 Hoveyda's monophosphine precatalyst [12]

Propargylic alcohols were found to be a source for the carbene ligand in ruthenium-based metathesis catalysts by Hill et al. Reaction of [$RuCl_2(PPh_3)_3$] with alcohol **14** gives a carbene complex **16**. Originally, the structure of an allenylidene complex **15** was assigned to the reaction product [13a] but it is now assumed that **15** rearranges to a carbene complex **16** [13b]. Ligand exchange with PCy_3 gives a metathesis-active precatalyst **17** [13c]. The synthesis of **17** and its application in the RCM of **18** to the azacyclic fragment **19** of the alkaloid balanol is outlined in Scheme 6 [13d].

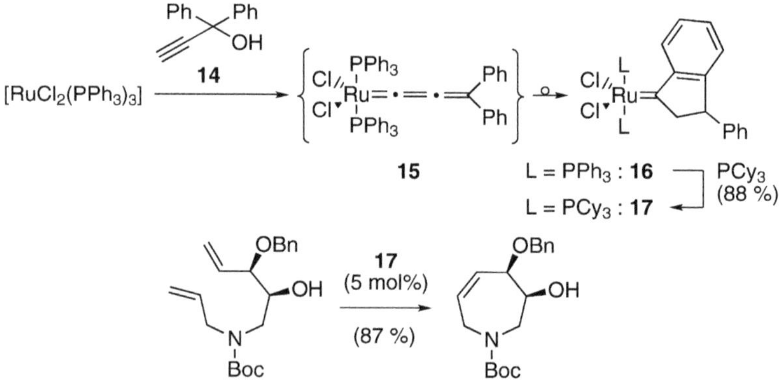

Scheme 6 Ru carbenes from propargyl alcohols and application (Hill, Fürstner [13])

2.1.2
... From [(*p*-cymene)$RuCl_2$]$_2$

The title compound, which is commercially available, may serve as an entry point to a variety of cationic olefin metathesis precatalysts, as was demonstrated by Fürstner and Dixneuf (Scheme 7). Cleaving the dimer **20** with a phosphine ligand yields an 18-electron complex **21**, which upon treatment with the propargylic alcohol **14** (Scheme 6) in the presence of halide scavengers gives cationic allenylidene complex **22** [14a]. As exemplified by the conversion of **23** to **24** (Scheme 7), these allenylidene complexes catalyze RCM reactions [14b]. Interestingly, the counterion has a strong effect on the efficiency of the metathesis reaction. In certain cases, a cycloisomerization might even be preferred over metathesis reactions [14c]. Generation of a metathesis catalyst in situ from dimer **20** and PCy_3 was described by Fürstner [14d]. Irradiation of the mixture with neon light is essential for the success of the reaction. Conditions B in Scheme 7 describe the application of these conditions to the RCM of **23**.

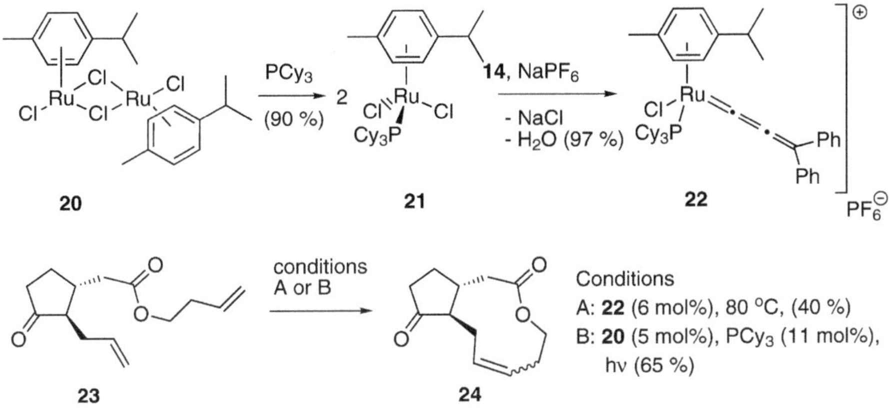

PCy3 (90 %) 2 · 14, NaPF6 - NaCl - H2O (97 %) · 20 · 21 · 22 · PF6⊖

conditions A or B · 23 · 24

Conditions
A: **22** (6 mol%), 80 °C, (40 %)
B: **20** (5 mol%), PCy3 (11 mol%), hν (65 %)

Scheme 7 A cationic precatalyst (Dixneuf, Fürstner [14])

2.1.3
... From Ruthenium Hydride Complexes

It was demonstrated by Grubbs et al. that the complex [Ru(H)(H_2)Cl(PCy_3)$_2$] (**25**) [15] upon reaction with propargylic chlorides gives ruthenium carbene complexes **26** [16a]. Precursor **25** was obtained by reduction of the polymeric [$RuCl_2$(COD)]$_x$ in the presence of a base and PCy_3 under an atmosphere of hydrogen (Scheme 8). The same carbene complex **26** was more recently synthesized from [RuHCl(PPh_3)$_3$] and the same propargyl chloride [16b,c].

Complex **25** was also used in an efficient one-pot synthesis of ruthenium-based precatalysts developed by Werner et al. [17a,b]; it is generated in situ from $RuCl_3$, H_2, PCy_3, magnesium, and 1,2-dichlorethane. Upon reaction of **25** with

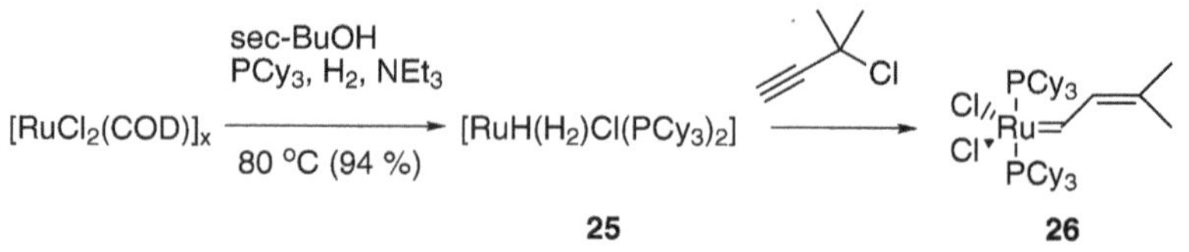

Scheme 8 Ruthenium carbene complexes from propargyl halides [16]

acetylene an allenylidene complex **27** results which, under the reaction conditions used, reacts to form the carbene complex **28**. Complex **28** has been used in a synthesis of the unsaturated proline analogue **31**. RCM of **29** was achieved in the presence of as little as 0.1 mol% **28** in nearly quantitative yield. The standard precatalyst **9** showed comparable activity in this example (Scheme 9) [17c].

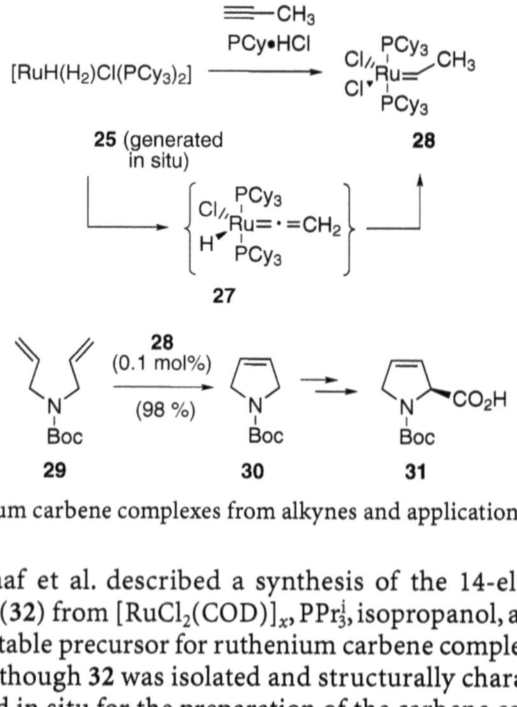

Scheme 9 Ruthenium carbene complexes from alkynes and application [17]

Van der Schaaf et al. described a synthesis of the 14-electron complex $[RuHCl(PPr^{i}_{3})_2]$ (**32**) from $[RuCl_2(COD)]_x$, PPr^{i}_{3}, isopropanol, and a base. Compound **32** is a suitable precursor for ruthenium carbene complex **33**, as outlined in Scheme 10. Although **32** was isolated and structurally characterized, it may also be generated in situ for the preparation of the carbene complex **33** [18].

Ruthenium hydride complexes, e.g., the dimer **34**, have been used by Hofmann et al. for the preparation of ruthenium carbene complexes [19]. Reaction of **34** with two equivalents of propargyl chloride **35** gives carbene complex **36** with a chelating diphosphane ligand (Eq. 3). Complex **36** is a remarkable example because its phosphine ligands are, in contrast to the other ruthenium carbene complexes described so far, arranged in a fixed *cis* stereochemistry. Although **36** was found to be less active than conventional metathesis catalysts, it catalyzes the ROMP of norbornene or cyclopentene.

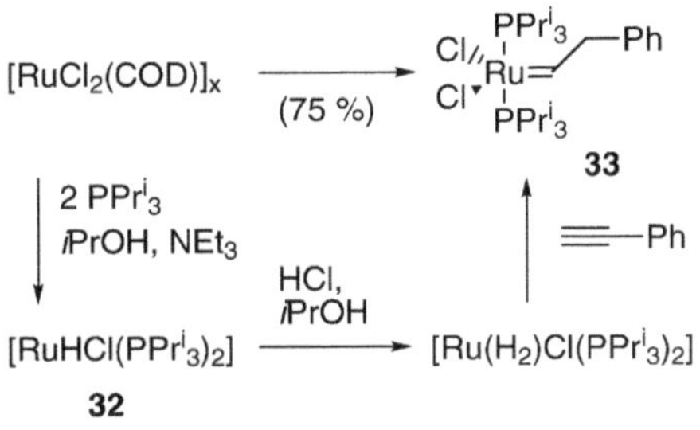

$$[RuCl_2(COD)]_x \xrightarrow{(75\,\%)} \mathbf{33}$$

[RuHCl(PPr$^i_3$)$_2$] (32) prepared with 2 PPr$^i_3$, iPrOH, NEt$_3$; HCl, iPrOH → [Ru(H$_2$)Cl(PPr$^i_3$)$_2$]; ≡—Ph → 33

Scheme 10 Ruthenium carbene complexes from ruthenium hydride species prepared in situ [18]

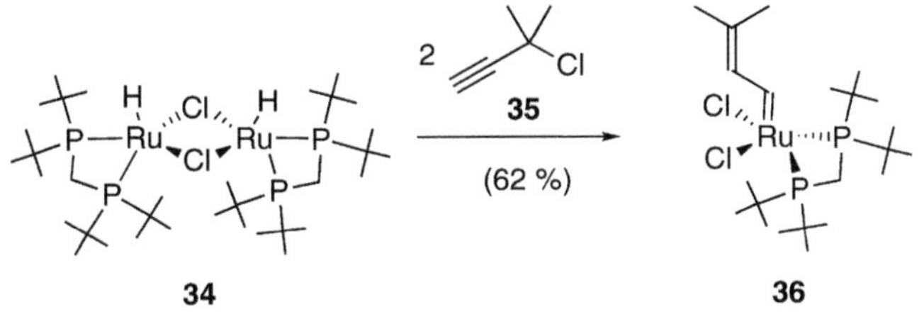

$$\mathbf{34} \xrightarrow[(62\,\%)]{2\ \mathbf{35}} \mathbf{36} \quad (3)$$

As a final example in this section, a contribution by Grubbs et al. is discussed. The chloride-free ruthenium hydride complex $[RuH_2(H_2)_2(PCy_3)_2]$ (**37**) is believed to react, in the presence of alkenes, to form an unidentified ruthenium(0) species which undergoes oxidative additions with dihalo compounds, e.g., **38**, to give the corresponding ruthenium carbene complex **9** (Eq. 4) [20].

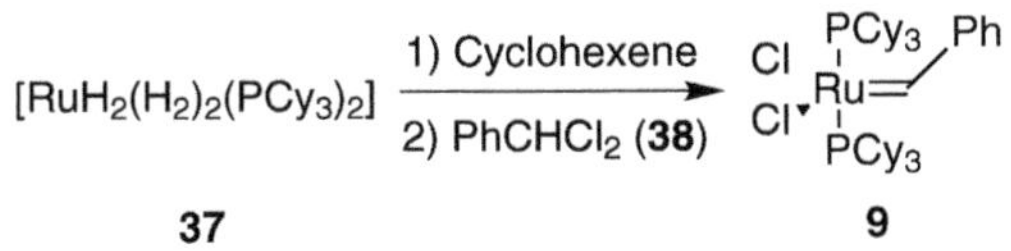

$$\underset{\mathbf{37}}{[RuH_2(H_2)_2(PCy_3)_2]} \xrightarrow[\text{2) } PhCHCl_2\ (\mathbf{38})]{\text{1) Cyclohexene}} \underset{\mathbf{9}}{Cl_2(PCy_3)_2Ru{=}CHPh} \quad (4)$$

Other sources of Ru(0) can also be used for this synthesis. For example, it was recently demonstrated that [Ru(arene)(diene)] complexes such as **39** undergo double oxidative addition of heterosubstituted dihalo compounds **40** in the presence of phosphine ligands (Eq. 5) [21].

The resulting carbene complex **41b** bears a hetero substituent and shows activity in the ring-opening/cross metathesis of strained bicyclic alkenes and

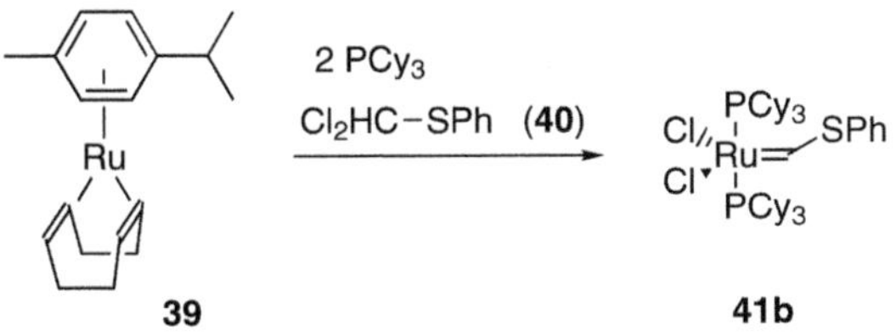

$$\mathbf{39} \xrightarrow[Cl_2HC{-}SPh\ (\mathbf{40})]{2\ PCy_3} \underset{\mathbf{41b}}{Cl_2(PCy_3)_2Ru{=}CHSPh} \quad (5)$$

42 + 43 → 44 (41b (2 mol%), (85 %)) (6)

vinyl sulfur compounds. The transformation **42**+**43**→**44** depicted in Eq. 6 is quite remarkable due to the high chemoselectivity observed: the sulfur-substitued C=C double bond is exclusively attacked [21].

2.1.4
... From Other Carbene Complexes by Exchange of the Carbene Ligand

Grubbs et al. demonstrated, in the original contribution on the synthesis of the now widely used ruthenium carbene complex **9**, that exchange of the carbene ligand by reacting **9** with terminal olefins is an option for the synthesis of other derivatives **45–50** (Table 2) [10b]. Formation of the new carbene complexes upon reaction of excess alkene with **9** is fast. If, however, these complexes are not immediately isolated, further reaction with the alkene occurs which finally leads to the methylidene compound **45**. Hoveyda's monophosphine complex **13** can also be prepared via this method [12]. Reaction of vinylferrocene with **9** gives the bimetallic complex **51** [22], and reaction with heteroalkenes yields Fischer-type carbene complexes **41a–d** [23]. The latter aspect has been systematically investigated by Grubbs et al. There appears to be a general trend that the thermal stability of complexes $[(PCy_3)_2Cl_2Ru{=}CHER]$ decreases in the order E=N>C>S>O, while their activity decreases in the order E=C>N>S>O [24]. Some examples of carbene complexes prepared by exchange of the carbene ligand from **9** are listed in Table 2.

2.2
Mechanistic Studies

Soon after the potential of ruthenium-catalyzed olefin metathesis for organic synthesis was discovered, efforts were made to get more insight into the mechanism. One intention of these studies was to understand the influence of the phosphine ligand on the catalytic activity and to apply information obtained from mechanistic studies to catalyst tuning. In a pioneering paper by Grubbs et al., the catalytic activity of precatalysts with the general formula $[X_2(L_2)Ru{=}CHCPh_2]$ was found to correlate with the ligand sphere in the following way: catalyst activities decrease in the order X=Cl>Br>I and $L{=}PCy_3{>}PPr^i_3{>}PCy_2Ph{\gg}PPh_3$ [25]. Thus, the original precatalyst **7** gives the best results. Furthermore, addition of CuCl leads to an enhancement of catalytic

Table 2 Ru- carbene complexes by exchange of the carbene ligand in **9**

Alkene	Complex		Yield	δ(^{31}P)	Ref.
C_2H_4	$Cl_2(PCy_3)_2Ru{=}CH_2$	(**45**)	100 %	43.7	[10b]
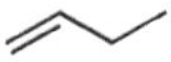	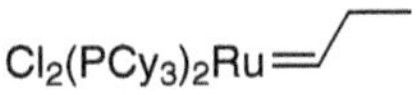 $Cl_2(PCy_3)_2Ru{=}$	(**46**)	97 %	36.4	[10b]
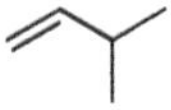	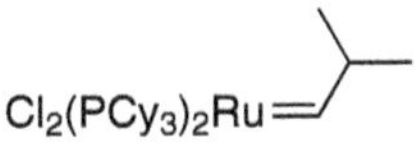 $Cl_2(PCy_3)_2Ru{=}$	(**47**)	not isolated		[10b]
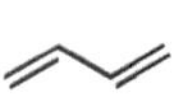	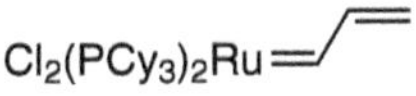 $Cl_2(PCy_3)_2Ru{=}$	(**48**)	95 %	36.2	[10b]
	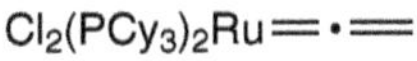 $Cl_2(PCy_3)_2Ru{=}\bullet{=}$	(**49**)	98 %	35.4	[10b]
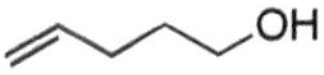 OH	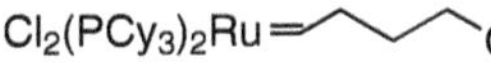 $Cl_2(PCy_3)_2Ru{=}$ OH	(**50**)	76 %	37.1	[10b]
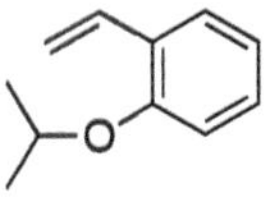 O	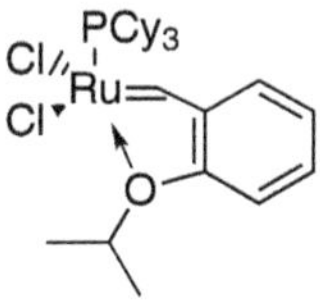 PCy_3, Cl, Cl, Ru=, O	(**13**)	67 %	59.2	[12]
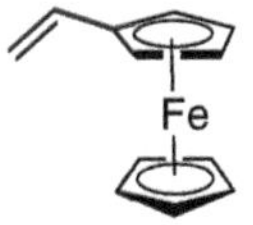 Fe	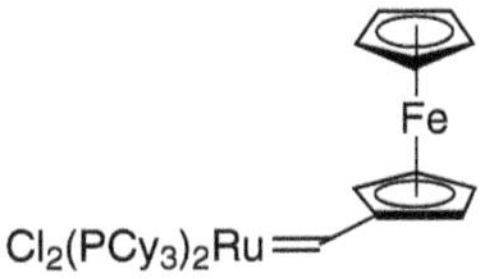 Fe, $Cl_2(PCy_3)_2Ru{=}$	(**51**)	87 %	34.4	[22]
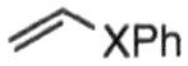 XPh	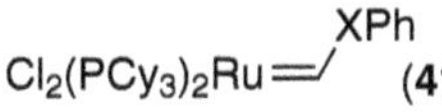 $Cl_2(PCy_3)_2Ru{=}$ XPh	(**41a**) X = O	72 %	35.1	[23]
		(**41b**) X = S	68 %	30.9	
		(**41c**) X = Se	77 %	29.5	
		(**41d**) X = Te	61 %	26.3	
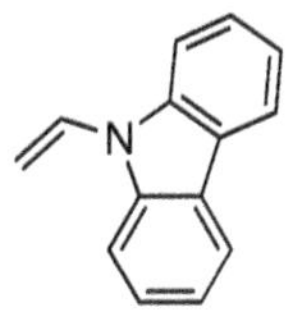 N	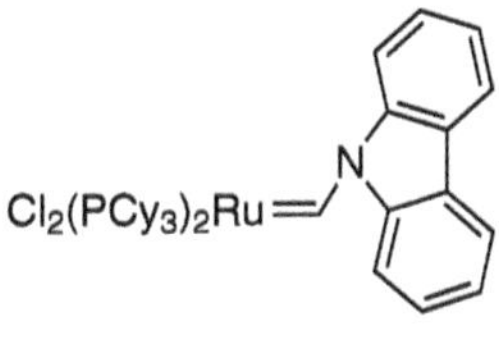 $Cl_2(PCy_3)_2Ru{=}$ N	(**52**)	77 %	41.3	[24]

activity. An investigation of the kinetics of an olefin metathesis reaction led to a proposal that favors a "dissociative" mechanistic pathway over an "associative" one (Scheme 11). In this context, associative means that in the initiating step an alkene coordinates to the 16-electron complex **7** to give an 18-electron complex **11-A**, which then undergoes the sequence of [2+2]-cycloadditions and -reversions. Dissociative means that actually a ligand substitution occurs where one phosphine ligand is replaced by the alkene to give another 16-electron species **11-B** (Scheme 11). The latter pathway accounts for approximately 95% of the turnover and can explain why bulky and electron-rich phosphines are favorable in ruthenium-based metathesis catalysts. Furthermore, the observed rate enhancement upon addition of CuCl can also be explained along these lines, as CuCl is known to bind phosphines.

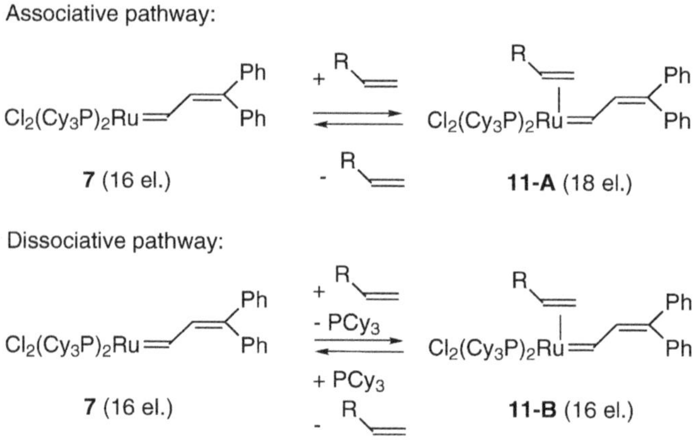

Scheme 11 "Associative" and "dissociative" mechanistic pathways [25]

More recently, Grubbs et al. obtained a refined mechanistic picture of the initiating step by conducting a ^{31}P NMR spectroscopic study of the phosphine exchange in precatalysts **12-A**. These investigations revealed that substitution of the phosphine proceeds via a dissociative–associative mechanism, i.e., a 14-electron species **12-B** is involved that coordinates the alkene to give a 16-electron species **12-C** (Scheme 12) [26a]. Increased initiation rates are observed if the substituents R′ and the phosphine ligands PR_3 in precatalysts

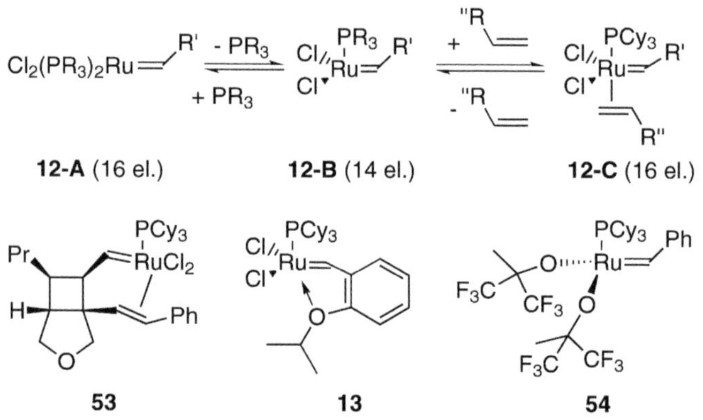

Scheme 12 Fourteen-electron species as active catalysts [26a,b] (**53** [27], **13** [12], **54** [28])

12-A are electron donating and bulky, because this situation facilitates the formation of the 14-electron species **12-B** [26b]. However, a recent study by Werner et al. revealed that the PCy_3 ligand obviously represents the optimum, as the sterically even more demanding tricyclooctylphosphine ligand ($PCoc_3$) leads to a strongly reduced activity [26c]. Evidence for the importance of 14-electron species as active catalysts is provided by a number of experimental results:

a) Snapper et al. described "a ruthenium catalyst caught in the act." The complex **53** (Scheme 12) was isolated from a ring-opening metathesis reaction of a cyclobutene with precatalyst **9** and characterized by X-ray crystallography [27]. It was found to be an active metathesis catalyst. Although the initiation of ROMP of cyclooctene proceeds four times slower compared to **9**, propagation is eight times faster, which is attributed to the absence of a second phosphine ligand.
b) Hoveyda et al. made the same observations for precatalyst **13** using the same test reaction. Compound **13** initiates the ROMP of cyclooctene 30 times slower than **9**, but propagation is four times faster [12].
c) In order to probe the mechanistic assumption outlined in Scheme 12, Grubbs et al. synthesized tetracoordinate 14-electron complex **54** (Scheme 12) and derivatives [28]. While **54** is inactive at ambient temperature, it is converted to a highly active species by addition of two equivalents of HCl. It is believed that acid-mediated exchange of the alkoxy ligands by chloride occurs under these conditions generating the corresponding 14-electron species **12-B** (R'=Ph).
d) Chen et al. have investigated olefin metathesis reactions using the electrospray ionization tandem mass spectrometry technique [29]. They were able to detect 14-electron species resulting from dissociation of one phosphine ligand [29a]. In ROMP the monophosphine complex is the resting state of the catalyst, as the free coordination site is blocked by the neighbouring alkene moiety of the growing polymer chain [29c]. The experimental technique employed by Chen et al. has also potential for catalyst screening [29d–f].
e) Various computational studies on the mechanism of the olefin metathesis reaction were performed [30]. Thiel et al. calculated a comparatively low $\Delta G^{\#}_{298}$ value of 5–10 kcal/mol for the dissociation of one phosphine ligand in a model system [30b], and other calculations also suggest that dissociation of one phosphine ligand is the initiating step [30a–d]. In contrast, Bottoni et al. discovered that at least one probable reaction pathway might proceed via a bisphosphine complex [30e].

Apart from the question whether the 14-electron species **12-B** is a relevant intermediate, computational studies have been conducted in order to shed light on other aspects of the mechanism. Stereochemical issues, for instance, have not yet been investigated by experiment. DFT calculations suggest that attack of the alkene to **12-B** occurs *trans*, because *cis* attack is associated with a rather high barrier [30b].

The mechanistic investigations presented in this section have stimulated research directed to the development of advanced ruthenium precatalysts for olefin metathesis. It was pointed out by Grubbs et al. that "the utility of a catalyst is determined by the ratio of catalysis to the rate of decomposition" [31]. The decomposition of ruthenium methylidene complexes, which attribute to approximately 95% of the turnover, proceeds monomolecularly, which explains the commonly observed problem that slowly reacting substrates require high catalyst loadings [31]. This problem has been addressed by the development of a novel class of ruthenium precatalysts, the so-called second-generation catalysts.

2.3 Ruthenium Precatalysts with *N*-Heterocyclic Carbene Ligands

It was rapidly discovered that the ruthenium precatalysts described so far have some limitations: for example, they are significantly less active than molybdenum- or tungsten-based catalysts, they do not react with highly substituted alkenes, and they are only moderately stable to air and elevated temperatures in solution [32b]. Attempts to improve the performance of ruthenium-based metathesis catalysts focused on the ligand sphere. For example, heterobimetallic derivatives were prepared and tested [32a], a trispyrazolylborate was used to substitute one PCy_3 and one chloride [32b], and bidentate Schiff-base ligands were also investigated [32c]. The breakthrough was achieved when ruthenium alkylidene complexes with *N*-heterocyclic nucleophilic carbene (NHC) ligands [33] were synthesized and investigated for their catalytic activity. *N*-Heterocyclic carbenes had previously been recognized as good σ-donor and poor π-acceptor ligands, similarly to electron-rich phosphines. They can be expected to bind tightly to a metal fragment and stabilize the catalytically active species [34]. Based on these assumptions, it was expected that introducing one basic, sterically demanding *N*-heterocyclic carbene ligand to a ruthenium alkylidene complex should significantly reduce the rate of decomposition and facilitate dissociation of a phosphine ligand, leading to an overall enhanced catalytic activity.

2.3.1 Synthesis of NHC-Ligated Ruthenium Carbene Complexes

Ruthenium carbene complexes bearing one (**56**) or two (**57**) NHC ligands are generally prepared by exchange of one or two phosphine ligands against the *N*-heterocyclic carbene **55** (Eq. 7). Whether one phosphine ligand is replaced or both depends on the stoichiometry, the bulkiness of the NHC ligand, and the phosphine which is substituted. Nucleophilic carbenes are normally obtained by deprotonation of the corresponding imidazolium salt. Their isolation can be avoided by generating the carbene in situ [36c, 40]. A variety of examples of NHC-ligated ruthenium-based carbene complexes are depicted in Table 3.

$$Cl_2(PR_3)_2Ru{=}CHR' \xrightarrow{\mathbf{55}} \mathbf{56} \text{ and/or } \mathbf{57} \quad (7)$$

Table 3 Ruthenium alkylidene complexes with *N*-heterocyclic carbene ligands

Complex	Ref.	Complex	Ref.
57a	[35a]	**57b**	[38]
56a	[35b,c]	**56e**	[38]
56b (R = Ph) **56c** (R = Cy)	[36]	**56f**	[39]
56d	[36c, 37, 40]	**56g**	[40]

The bis-NHC complex **57a** was obtained from [$Cl_2(PPh_3)_2Ru{=}CHPh$] or [$Cl_2(PCy_3)_2Ru{=}CHPh$] (**9**) in comparable yields [35a,c]. At low temperatures and with the appropriate stoichiometry, selective substitution of one phosphine was achieved to obtain **56a** [35b,c]. In contrast, Nolan et al. obtained **56c** selectively from **9** even in the presence of a tenfold excess of the sterically more demanding bismesityl-NHC ligand [36a]. For the saturated system **56d** it was recently demonstrated that substitution of the second phosphine is not possible [40]. Interestingly, replacement of a PPh_3 ligand in a mono-NHC complex **56e** is obviously much more facile, as recently illustrated by Fogg et al. for the synthesis of bis-NHC complex **57b** [38]. For several complexes structural information obtained from X-ray crystallographic studies is available. These data reveal the significant difference between the two types of Ru–C bonds: representative values for the Ru–NHC bond are 211 pm (**57a** [35a]), 207 pm (**56c** [36a]), and 209 pm (**56d** [40]). In contrast, the Ru–alkylidene bonds are much shorter: 182 pm (**57a** [35a]), 184 pm (**56c** [36a]), and 183.5 pm (**56d** [40]).

2.3.2 Comparative Investigations of Catalytic Activities in RCM and CM

Grubbs et al. compared the activity and stability of **9**, **57a**, and **56c** for the test reaction **58**→**59** at 55 °C (Eq. 8) [31]. With the first-generation catalyst **9** a yield of 80% was obtained after 8 h, when the reaction progress ceased. As expected, **56c** is significantly more reactive giving a quantitative yield after 4 h with the catalyst still being active. The bis-NHC complex **57a** is only moderately active, with a yield of 33% after 3 h. Given the mechanism postulated by Grubbs (Sect. 2.2), it is surprising that a bis-NHC complex shows any catalytic activity at all, because according to this mechanistic picture dissociation of one NHC ligand is required, which appears to be unlikely if DFT calculations on this issue are taken into account [35b]. Nevertheless, Grubbs et al. recently presented evidence that in bis-NHC complexes one ligand can be replaced by PCy_3 via a dissociative–associative mechanism [40].

MeO_2C CO_2Me [Ru] → MeO_2C CO_2Me (8)

58 **59**

Comparative studies [41] focused on the activity of molybdenum precatalyst 5 (Table 1) and the ruthenium precatalysts **9** (Scheme 4), **56c**, and **56d** (Table 3) in RCM and cross-metathesis reactions. Table 4 lists representative examples and references. In the formation of tetrasubstituted double bonds (e.g., **60**), **5** appears to be superior [37, 41a]. In other systems that are difficult to cyclize (e.g., **61**), probably for conformational reasons, the performance of **56c,d** is comparable to 5 or significantly better, as in the case of **62** [41c]. The utility of NHC catalysts in cross metathesis, illustrated for **63** [41d], is especially remarkable, as this metathesis variant has long been considered extremely difficult and unselective [8c,d]. Formation of a triple-substituted double bond in macrocycle **64** fails with **9**, while the second-generation catalysts **56c,d** both give preparatively useful yields [39]. The metathesis of electron-deficient double bonds, which is an important issue in natural product synthesis, is often difficult to achieve with the first-generation catalyst. Addition of a Lewis acid is sometimes beneficial [42] but high catalyst loadings are normally required. The second-generation catalysts **56c** or **56d** can solve the problem in most cases [41f]; this point is illustrated with the formation of **65** [41e] and will be discussed for further examples in Sect. 3.1.

Table 4 Comparative investigations: efficiency of some metathesis catalysts

Product	Yield obtained with				Ref.
	5 (mol%)	**9** (mol%)	**56c** (mol%)	**56d** (mol%)	
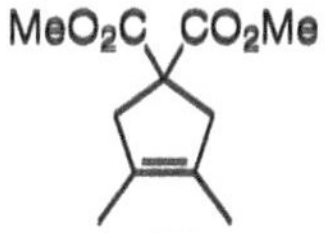 MeO_2C CO_2Me **60**	93 % (5)	N. R.[a]	31 % (5)	40 % (5)	[37,41a]
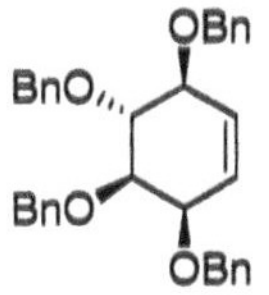 OBn, BnO, BnO, OBn **61**	92 % (5,1 h)	32 % (5, 60 h)	89 % (5, 2 h)	–	[41c]
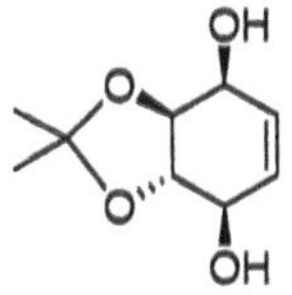 OH, O, O, OH **62**	N. R.[a]	N. R.[a] (20, 20 h)	65 % (1.5, 2 h)	–	[41c]
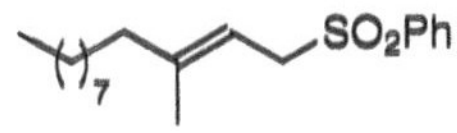 SO_2Ph **63**	–	N. R.[a]	–	87 %	[41d]
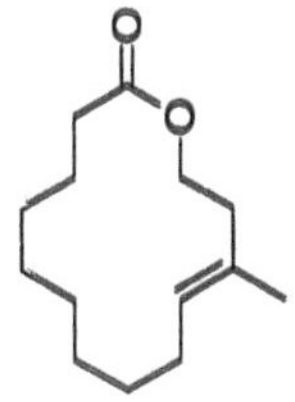 O, O **64**	–	N. R.[b] (10, 17 h)	65 % (10, 38 h)	57 % (6, 40 h)	[39]
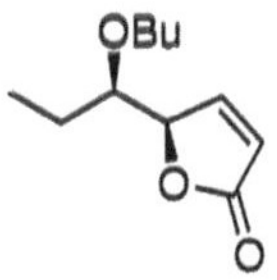 OBu, O, O **65**	–	45 % (50, 48 h)[c]	–	80 % (5, 24 h)	[41e]

[a] N.R. = no reaction [b] only self metathesis.
[c] addition of $Ti(OPr^i)_4$ (1 equ.) required.

In summary, the order of reactivity for the most commonly used ruthenium-based metathesis catalysts was found to be **56d**>**56c**>**9**≅**7**. This order of reactivity is based on IR thermography [39], determination of relative rate constants for the test reaction **58**→**59** (Eq. 8) [40], and determination of turnover numbers for the self metathesis of methyl-10-undecenoate [43].

2.3.3
E/*Z* Selectivity in RCM Leading to Macrocycles (Macro-RCM)

The *E*/*Z* selectivity problem is restricted to cross metathesis and RCM leading to macrocycles (macro-RCM). Both aspects have recently been covered in reviews by Blechert et al. [8d] and by Prunet [44]. *E*/*Z* selectivity can be influenced by reaction temperature, solvent or substitution pattern of the substrate. Here, we will only discuss the influence of the precatalyst.

That low *E*/*Z* selectivity in macrocycle formation can be a problem has early been recognized in various studies directed to the total synthesis of epothilones. Epothilones are 16-membered macrocyclic lactones bearing a *cis*-epoxide attached to the macrocycle. Several synthetic strategies investigated in the late 1990s involve RCM and subsequent epoxidation of the newly formed double bond. The configuration of the epoxide moiety is important for the biological activity: the diastereomer resulting from epoxidation of the *Z*-alkene is the naturally occurring and more potent derivative. Unfortunately, most RCM approaches give mixtures of *E*- and *Z*-isomers with ratios close to 1:1. All attempts to improve the *E*/*Z* ratio in favor of the *Z*-isomer, e.g., by variation of the substitution pattern, led to a stronger preference for the undesired *E*-isomer [45]. A more recent approach to the *Z*-selective formation of macrocycloalkenes is based on ring-closing alkyne metathesis and subsequent stereoselective hydrogenation of the resulting cycloalkyne to the *Z*-cycloalkene. This strategy was developed by Fürstner et al. and has recently been applied to the synthesis of epothilones [46a,b].

The influence of the precatalyst on the *E*/*Z* ratio of a given macro-RCM has been systematically investigated by Grubbs et al. for 14-membered lactones [46c]. Thus, a significantly higher amount of *E*-isomer was observed if the more reactive second-generation catalyst **56d** was used rather than the first-generation catalyst **9**. A closer look at the kinetics revealed that for catalyst **9** the *E*/*Z* ratio is nearly independent of conversion, whereas for **56d** the amount of *E*-isomer dramatically increases with increasing conversion. Furthermore, addition of precatalyst **56d** to a mixture of *E*- and *Z*-macrolactones leads to a higher *E*/*Z* ratio. A similar observation had previously been made by Kalesse et al. [46d], who described the formation of a 1:12 mixture of *E*- and *Z*-**66** by RCM. Isolation of the minor product *E*-**66** and treatment with precatalyst **9** causes an isomerization to *Z*-**66** (Eq. 9). Computational studies revealed that in the case of **66** the *Z*-isomer is thermodynamically more stable.

Interconversion of *E*- and *Z*-macrocycles in metathesis reactions proceeds via a ring-opening metathesis/ring-closing metathesis sequence; thus, more

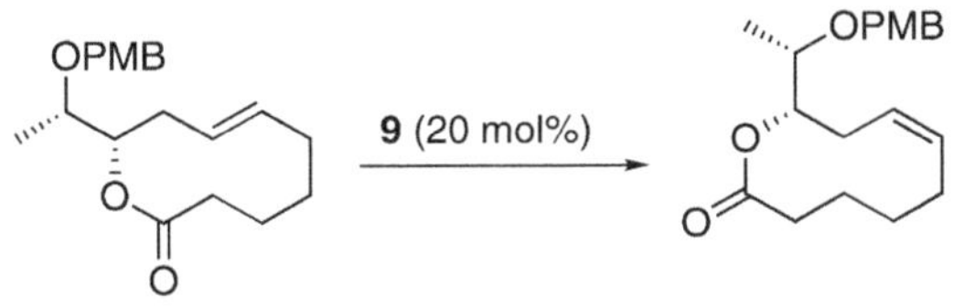

OPMB — **9** (20 mol%) → OPMB

E-**66** *Z*-**66** (9)

reactive second-generation catalysts will more easily induce the ring-opening reaction and finally produce the thermodynamically preferred product. If, however, the kinetically preferred product is required, less reactive first-generation catalysts are the better choice. This point is illustrated for a few examples from natural product synthesis in Scheme 13 [46e–g]. For **67** [46e], the *E*-isomer is apparently more stable, while for **68** [46f] and **69** [46g] the *Z*-isomer is preferred. For the latter two examples further evidence for the greater stability of the *Z*-isomer is provided by computational analysis.

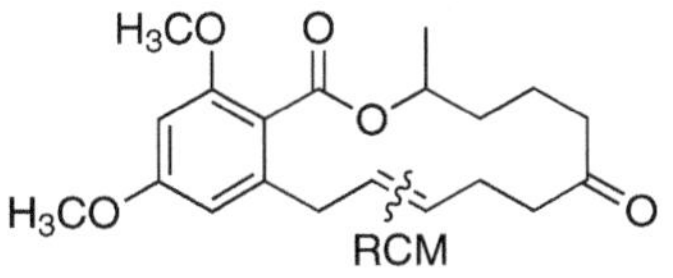

H_3CO O
H_3CO RCM

67

9 (19 mol%): *E*:*Z* = 2.4:1 (73 %)
56c (5 mol%): *E*:*Z* = 8:1 (85 %)

Ref [46e]

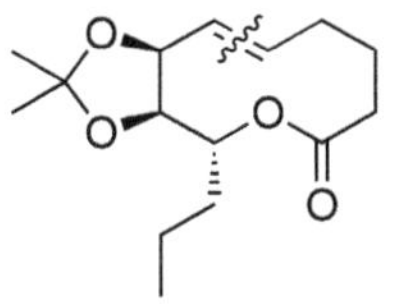

68

17 (10 mol%): *E*:*Z* = 7.9:1 (77 %)
56c (10 mol%): only *Z* (86 %)

Ref [46f]

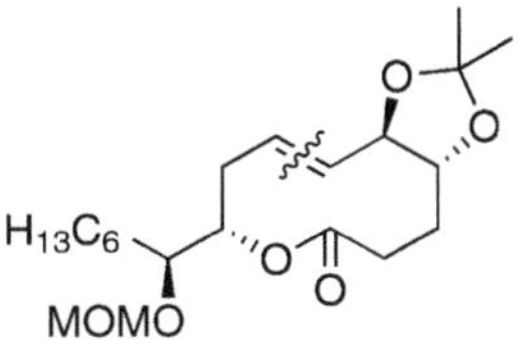

$H_{13}C_6$ MOMO

69

9 (20 mol%): *E*:*Z* = 2:1 (67 %)
56d only *Z*

Ref [46g]

Scheme 13 Influence of the Ru catalyst on *E*/*Z* ratio in macrocyclization (**68** [46e], **68** [46f], **69** [46g])

2.3.4 Origins of Enhanced Reactivity of Second-Generation Metathesis Catalysts

The discovery of second-generation metathesis catalysts was based on the hypothesis that bulky and electron-rich NHC ligands would facilitate dissociation of the remaining phosphine ligand, hence leading to an improved overall activity. While the improved metathesis activity was indeed observed, the hypothesis turned out to be wrong. Detailed kinetic studies by Grubbs et al. revealed that the rate constant for loss of phosphine (k_1 in Scheme 14) is two orders of magnitude greater for the first-generation catalyst **9** than for the second-generation catalyst **56d.** However, the 14-electron intermediate **14-A** is rapidly removed from the catalytic cycle by trapping with free phosphine. In contrast, the NHC-substituted analog **14-B**, albeit formed much slower than **14-A**, remains longer in the catalytic cycle because recoordination of free phosphine is less favorable. This observation is described by the ratio k_{-1}/k_2 (Scheme 14), which is four orders of magnitude greater for **14-A** than for **14-B.** Thus, the second-generation catalyst initiates much slower than the first-generation catalyst, but propagation is much faster, leading to an overall increased metathesis activity [26a,b].

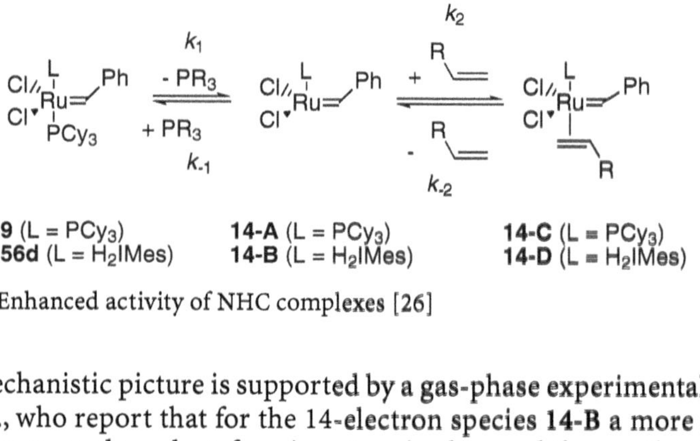

9 (L = PCy_3) **56d** (L = H_2IMes) — **14-A** (L = PCy_3) **14-B** (L = H_2IMes) — **14-C** (L = PCy_3) **14-D** (L = H_2IMes)

Scheme 14 Enhanced activity of NHC complexes [26]

This mechanistic picture is supported by a gas-phase experimental study by Chen et al., who report that for the 14-electron species **14-B** a more favorable partitioning toward product-forming steps is observed than with **14-A** [47a]. Computational studies by Cavallo suggest that upon dissociation of the phosphine ligand the Ru–L bond in **14-B** becomes significantly shorter, leading to enhanced steric interaction between the NHC ligand and the other ligands of the 14-electron fragment. As a result, **14-B** is destabilized, which is partly compensated by coordination of the olefin and subsequent metallacyclobutane formation. The latter step is associated with release of steric pressure and should be facilitated [30c]. Based on theoretical studies, Adlhart and Chen suggest that in the course of the metathesis reaction with first-generation catalysts an unfavorable rotation around the Ru–L bond occurs, which can be avoided if L has twofold rather than threefold symmetry. This unfavorable rotation results in higher barriers for product-forming steps [47b].

Consequently, further tuning of ruthenium-based metathesis catalysts requires improved rates of initiation while maintaining the rates of propagation observed for the second-generation catalysts. One approach toward this goal is the exchange of the PCy_3 ligand in mono-NHC complexes by less basic phosphine ligands. Nolan et al. reported that the PPh_3 complex **56b** (Table 3) reacts significantly faster in RCM reactions than the PCy_3 analog **56c** [36a]. More recently, this concept was systematically evaluated by Grubbs et al. who prepared a variety of triaryl phosphine analogs of mono-NHC complex **56d** (Scheme 15) [48a]. As direct substitution of PCy_3 by less basic phosphine ligands is not possible, a two-step procedure was developed. The bispyridine complex **70** reacts with a variety of triaryl phosphines PAr_3 to give complexes **56h–m**. These precatalysts show significantly enhanced metathesis activity, with the electron-poor phosphine in **56h** being the most active derivative (Scheme 15) [48b].

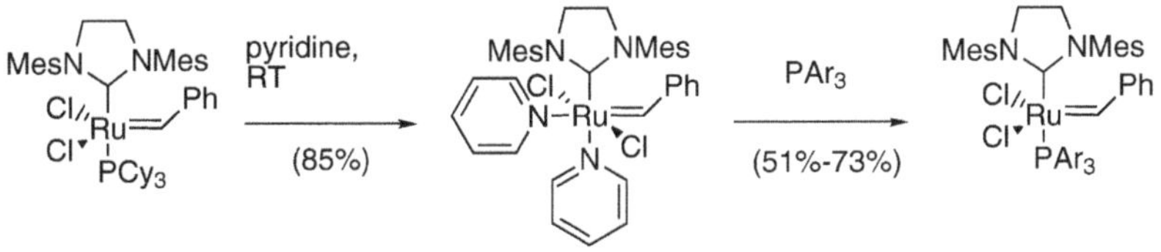

56d → pyridine, RT (85%) → **70** → PAr_3 (51%-73%) → **56h-m**

56	PAr_3	Yield	k_{rel} [a]
56h	$P(p\text{-}CF_3C_6H_4)_3$	51%	340
56i	$P(p\text{-}ClC_6H_4)_3$	70%	95
56j	$P(p\text{-}FC_6H_4)_3$	72%	50
56k	PPh_3	73%	50
56l	$P(p\text{-}CH_3C_6H_4)_3$	64%	20
56m	$P(p\text{-}CH_3OC_6H_4)_3$	63%	10

[a] rate constant for ROMP of COD relative to **56d**

Scheme 15 Two-step route to triarylphosphine-NHC complexes (Grubbs [48])

2.4 Phosphine-Free Ruthenium Precatalysts with One NHC Ligand

The search for even more active and recyclable ruthenium-based metathesis catalysts has recently led to the development of phosphine-free complexes by combining the concept of ligation with *N*-heterocyclic carbenes and benzylidenes bearing a coordinating isopropoxy ligand. The latter was exemplified for Hoveyda's monophosphine complex **13** in Scheme 5 [12]. Pioneering studies in this field have been conducted by the groups of Hoveyda [49a] and Blechert [49b], who described the phosphine-free precatalyst **71a**. Compound **71a** is prepared either from **56d** [49a] or from **13** [49b], as illustrated in Scheme 16.

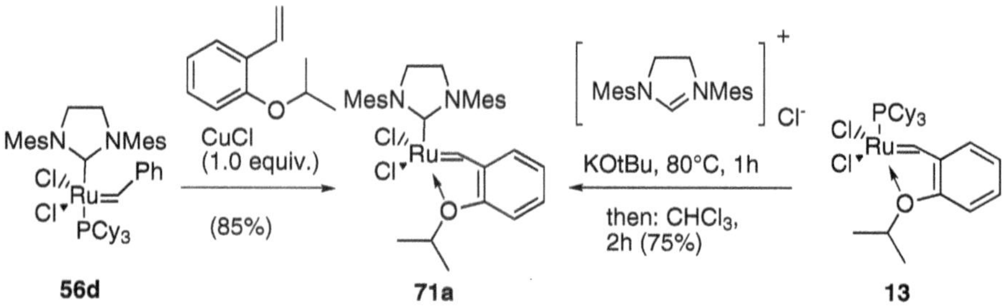

Scheme 16 Hoveyda's (*left*, [49a]) and Blechert's (*right*, [49b]) synthesis of phosphine-free complex **71a**

A first evaluation of complex **71a** by Blechert et al. revealed that its catalytic activity differs significantly from that of the monophosphine complex **56d** [49b]. In particular, **71a** appears to have a much stronger tendency to promote cross metathesis rather than RCM. Follow-up studies by the same group demonstrate that **71a** allows the cross metathesis of electron-deficient alkenes with excellent yields and chemoselectivities [50]. For instance, alkene **72** undergoes selective cross metathesis with 3,3,3-trifluoropropene to give **73** in excellent yield and selectivity. Precatalyst **56d**, under identical conditions, furnishes a mixture of **73** and the homodimer of **72** (Scheme 17) [50a]. While **56d** was found to be active in the cross metathesis involving acrylates, it failed with acrylonitrile [51]. With **71a**, this problem can be overcome, as illustrated for the conversion of **72**→**74** (Scheme 17) [50b].

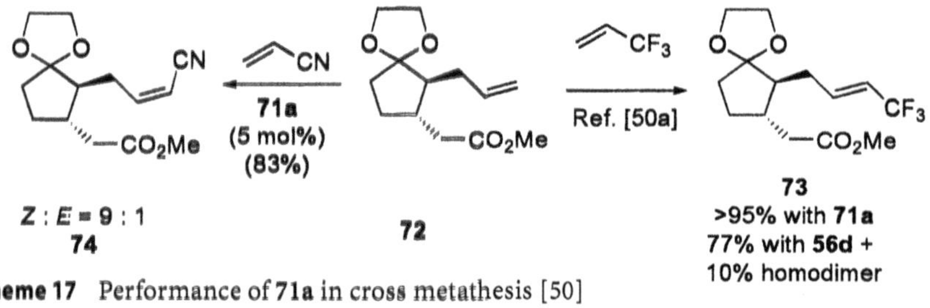

Scheme 17 Performance of **71a** in cross metathesis [50]

Reports in the literature for cross-metathesis reactions involving vinylsulfones are somewhat contradictory: while Grubbs et al. state that a mixed NHC-PCy_3 precatalyst fails in such reactions [51], Grela et al. reported the successful cross metathesis of phenylvinylsulfone using **56d** for a variety of examples [52]. Comparable results were obtained with the phosphine-free precatalyst **71a** by the same group [53], whereas Blechert et al. report a significant improvement of the yields when switching from **56d** to **71a** [8d]. The utility of phosphine-free precatalyst **71a** for selective cross-metathesis reactions involving electron-deficient alkenes has also been demonstrated by Cossy et al. [54]. This aspect is

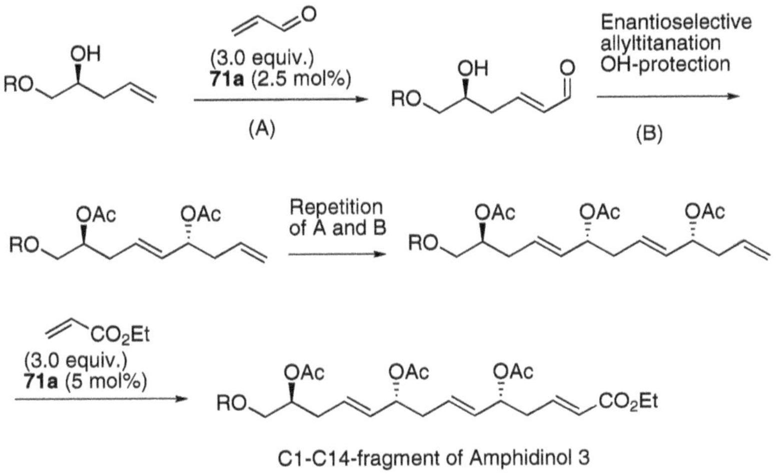

(3.0 equiv.)
71a (2.5 mol%)
(A)
Enantioselective allyltitanation OH-protection
(B)
Repetition of A and B
CO_2Et
(3.0 equiv.)
71a (5 mol%)
C1-C14-fragment of Amphidinol 3

Scheme 18 Application of **71a** in target molecule synthesis [54b]

illustrated by the elegant synthesis of the C1-C14 fragment of amphidinol 3, which is based on the iterative application of selective cross metathesis with acrolein and highly enantioselective allyltitanation (Scheme 18) [54b].

The following order of initiation rate constants was found by Grubbs et al. for **71a** and some precatalysts containing one phosphine ligand: **56d** $\ll$ **56k** $\cong$ **71a** $<$ **56h** (cf. Scheme 15 for structures of **56d,h,k**) [48b, 55]. Thus, **71a** shows a rate of initiation comparable to that of **56k** but three orders of magnitude higher than that of **56d**. Nevertheless, **56d** appears to be more reactive in RCM reactions than **71a** [56]. Wakamatsu and Blechert were the first to report that the activity of precatalysts related to **71a** can be dramatically enhanced by modification of the benzylidene unit [56]. For example, RCM of **75** using 1 mol% of BINOL-derived complex **71b** yields the azacyclic product **76** in quantitative yield within 20 min (Eq. 10), whereas with **56d** only 4% of **76** was obtained under these conditions [56].

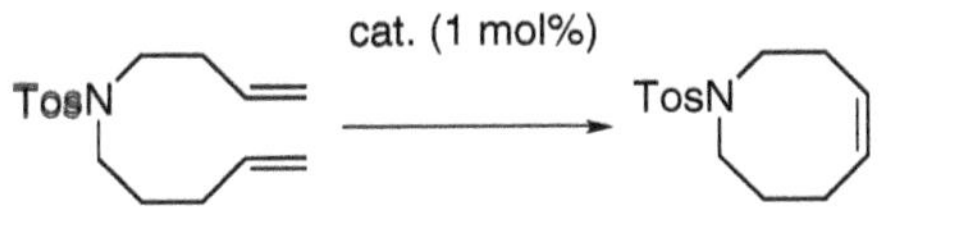

cat. (1 mol%)
TosN — TosN (10)

75　**76** (99% with **71b**, 4% with **56d**)

Other precatalysts that are structurally related to **71a** have recently been described. Structures and references are given in Table 5. Complex **71c** is obviously even more reactive than **71b**. The variation in these complexes compared to the parent compound **71a** appears to be mainly steric. In contrast, complexes **71d** and **71e** differ significantly in the electronic properties of the aromatic system.

Table 5 Phosphine-free ruthenium metathesis catalysts

Complex	Ref.	Complex	Ref.
71b	[56]	**71d** (X = Br) **71e** (X = NO_2)	[58]
71c	[57]	**71f**	[59]
		71g	[60]

Thus, electron-poor **71e** was found to be much more reactive than the unsubstituted **71a.** In contrast **71d,** bearing a donor atom on the aromatic moiety, is significantly less reactive than **71a.** From these results, it might be concluded that sterically demanding and/or electron-withdrawing substituents facilitate initial ligand dissociation and lead to enhanced catalytic activities. Interestingly, the ester-substituted complex **71g,** synthesized by Fürstner et al. in the course of a study concerning chelation effects in metathesis reactions, is significantly less reactive [60].

Several approaches toward immobilization of phosphine-free ruthenium-based metathesis catalysts bearing a coordinating ether group have been made over the past 3 years [61]. This aspect has been covered in a recently published review by Blechert and Connon [8d] and will therefore not be discussed here.

An alternative approach to phosphine-free ruthenium precatalysts is based on pyridine complex **70** [48], which has been established by Grubbs et al. as a valuable precursor for other mixed NHC-phosphine complexes (cf. Scheme 15). Complex **70** is only moderately active in the cross metathesis of allylbenzene and acrylonitrile, but its initiation rate constant is five orders of magnitude higher than the value obtained for the standard second-generation catalyst **56d** [55]. Replacement of the pyridine ligand in **70** by 3-bromopyridine yields a highly active metathesis catalyst that not only displays an even higher rate of initiation, but also catalyzes the cross metathesis with high efficiency [55]. This novel catalyst was recently used to promote selective cross metathesis of conjugated enynes. Standard catalysts fail in this particular reaction [62]. Other pyridine-containing precatalysts for olefin metathesis have recently been described by Herrmann et al. [63].

3 Olefin Metathesis of "Nonstandard" Double Bonds

3.1 α,β-Unsaturated C–C Double Bonds

α,β-Unsaturated γ- and δ-lactones are important structural patterns in a large variety of target molecules. It was demonstrated by Ghosh et al. that olefin metathesis of acrylates using the first-generation Grubbs' catalyst **9** opens up a pathway to these structures [64]. Unfortunately, reactivity of acrylates toward **9** is rather low, which is in part attributed to the formation of chelates. Addition of a Lewis acid [42] often leads to enhanced reactivity [64, 65]. Nevertheless, comparatively high catalyst loadings are normally required if **9** is used [66]. As was briefly mentioned in Sect. 2.3.2, significant improvements are often achieved when NHC-ligated ruthenium precatalysts are used [41f]. This point will be illustrated for the examples depicted in Scheme 19. Compound **77** was synthesized by Cossy et al. as a key building block in the synthesis of methynolide [67]. It was obtained in near quantitative yield using 15 mol% of **56c**, while the saturated **56d** gave a significantly lower yield. Oxacyclic ketone **78** is not formed with the first-generation catalyst **9**, even in the presence of a Lewis acid, whereas a yield of 58% was obtained with 15 mol% of **56d** [68]. It is possible to reduce the catalyst loading significantly with only slight reduction of the yield; however, comparatively high dilution (10^{-2}–10^{-3} M) is required to suppress undesired intermolecular metathesis. Interestingly, Eustache et al. observed that the second-generation catalyst **56d** fails in the synthesis of cyclohexenone **79**

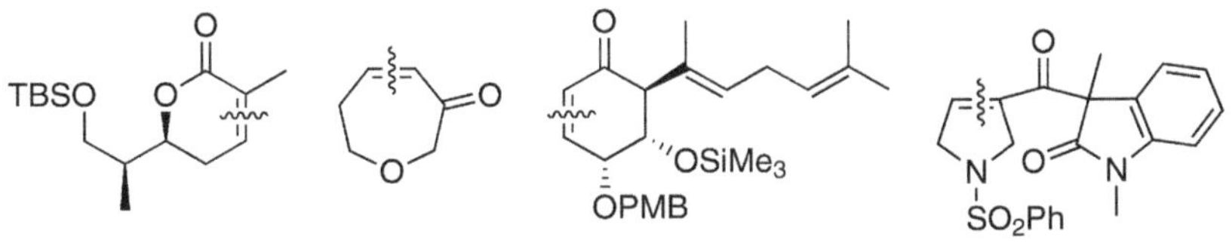

TBSO OSiMe3 OPMB SO2Ph

77 **78** **79** **80**

Scheme 19 Examples of RCM of electron-deficient alkenes (**77** [67], **78** [68], **79** [69], **80** [70])

(a key intermediate for the natural product fumagillol), whereas a 53% yield was obtained with the first-generation catalyst **9** and Ti(Oi-Pr)$_4$ as a Lewis acid [69]. Grigg et al. described the use of **56d** in the synthesis of acylated heterocycles such as **80** [70]. Acceleration of the reaction by microwave irradiation was also observed for a set of examples [70b].

The cross metathesis of acrylic amides [71] and the self metathesis of two-electron-deficient alkenes [72] is possible using the precatalyst **56d.** The performance of the three second-generation catalysts **56c,d** (Table 3) and **71a** (Scheme 16) in a domino RCM/CM of enynes and acrylates was recently compared by Grimaud et al. [73]. Enyne metathesis of **81** in the presence of methyl acrylate gives the desired product **82** only with phosphine-free **71a** as a precatalyst. With monophosphine complexes **56c,d** only the primary enyne metathesis product **83** is obtained under these conditions (Eq. 11).

OTBS **56c, d** or **71a** CO$_2$Me 3 equ. TBSO CO$_2$Me TBSO **81** **82** + **83** (11)

3.2 Enol Ethers and Enamines

Cyclic enol ethers are attractive building blocks for polyether natural products and carbohydrate chemistry. Approaches based on RCM were investigated, but the first-generation ruthenium catalyst **9** was found to be not very reactive, making rather high catalyst loadings and high dilution necessary [74a]. Although examples of the successful utilization of **9** in enol ether metathesis have been published [74b,c], the much more reactive molybdenum complex **5** was normally employed in this reaction [75]. Recently, it was demonstrated that second-generation ruthenium catalysts give comparable results in many examples [76]. For instance, RCM of **84** to the trans-fused polyether **85** (Eq. 12) is achieved in excellent yield with both **5** and **56d.** Not unexpectedly, **9** fails completely in this particular transformation [76a].

An alternative approach to cyclic enol ethers that avoids the metathesis of vinyl ethers has recently been developed by Snapper et al. [77a] and by Schmidt

OBn BnO BnO O O cat BnO BnO BnO H O H **84** **85** (12)

[77b]. It is based on the RCM of homoallyl-allyl ethers, which is an extraordinarily facile process, and subsequent regioselective double bond isomerization of the metathesis product to the cyclic enol ether. The isomerization is achieved by converting the metathesis catalyst to a ruthenium-hydride species in situ, either by the action of molecular hydrogen [77a] or by the action of inorganic hydrides [77b]. The sequence is outlined in Scheme 20.

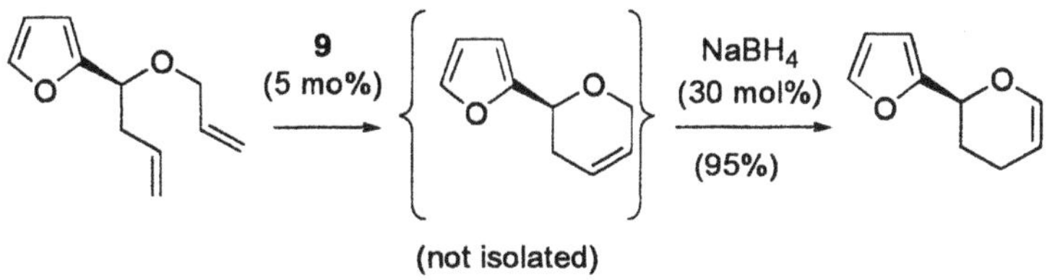

9 (5 mo%) → (not isolated) → $NaBH_4$ (30 mol%) (95%)

Scheme 20 RCM-isomerization sequence for the synthesis of cyclic enol ethers [77b]

RCM of enol ethers has been exploited for the synthesis of masked carbonyl compounds, especially silyl enol ethers [78a–d]. Very recently, a report describing the metathesis of an enol phosphate **94** (Table 6, entry 5) was published [78e]. Compound **91** (Table 6, entry 3) is the product resulting from ene–yne metathesis of an acetylenic ether **90** after desilylation [78c]. The influence of the silyl protecting group was investigated by Aggarwal et al. Surprisingly, RCM of the sterically more demanding TBS ether **92** was found to be more efficient than for the corresponding TMS analog. A competing double bond isomerization, which turned out to be a problem in some cases, can be suppressed by addition of activated molecular sieves (Table 6, entry 4) [78d]. This strategy for the synthesis of masked enolates might be complementary to established routes, e.g., enolization of cyclic ketones and subsequent trapping with enolates.

Metathesis of *N*-tosylated ene-amides and yne-amides has been less extensively investigated. An example of the RCM of ene-amides is a new indole synthesis developed by Nishida [79]: metathesis precursor **96** (prepared by ruthenium-catalyzed isomerization of the corresponding allyl amide) is cyclized to indole **97** in the presence of **56d** (Eq. 13).

The metathesis of ene-ynamides has been investigated by Mori et al. and Hsung et al. [80]. Second-generation ruthenium catalysts and elevated temperatures were required to obtain preparatively useful yields. Witulski et al. published a highly regioselective cyclotrimerization of 1,6-diynes such as **98** and terminal alkynes using the first-generation ruthenium metathesis catalyst **9**

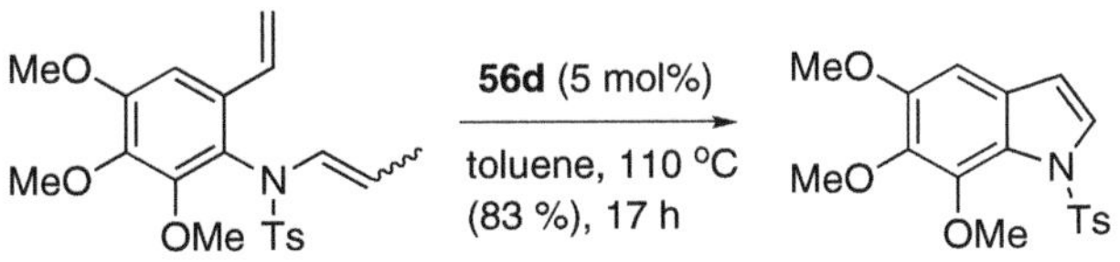

MeO, MeO, OMe, N, Ts **96** → **56d** (5 mol%), toluene, 110 °C (83 %), 17 h → MeO, MeO, OMe, N, Ts **97** (13)

Table 6 RCM products as masked enolates

entry	starting material	product	cat. (mol%)	yield	Ref.
1	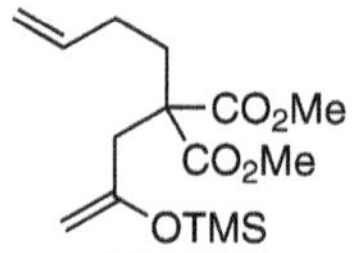 **86**	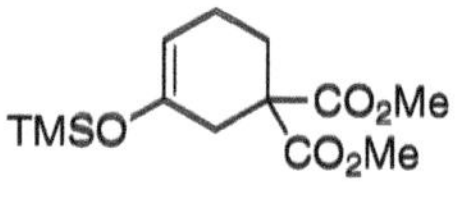 **87**	**56d** (7 mol%)	92 %	[78a]
2	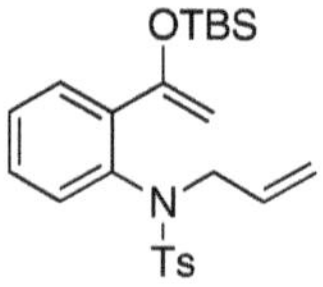 **88**	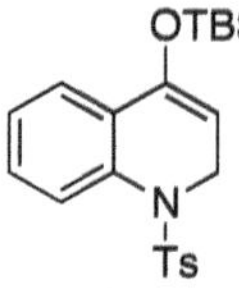 **89**	**9** (5 mol%) **56d** (5 mol%)	0 % 95 %	[78b]
3	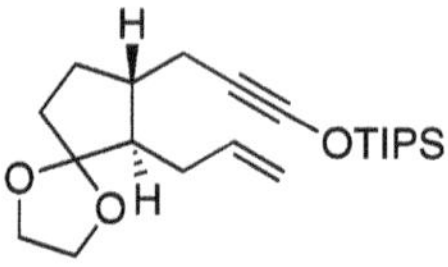 **90**	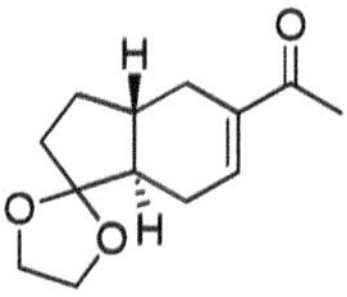 **91**	**56d** (5 mol%)	68 %	[78c]
4	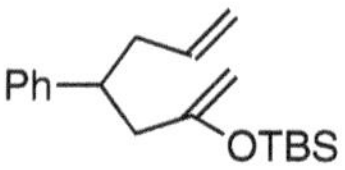 **92**	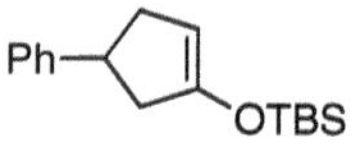 **93**	**56d** (20 mol%)	60 %	[78d]
5	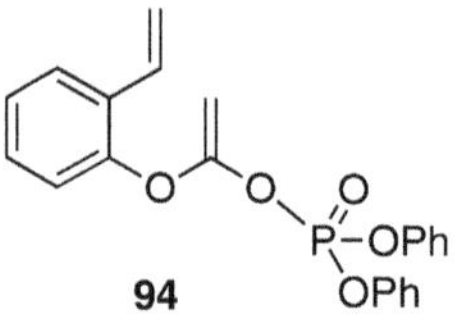 **94**	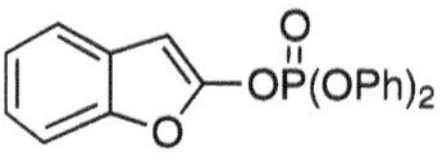 **95**	**56d** (5 mol%)	99 %	[78e]

(Eq. 14) [81]. Although this transformation does not appear to be a metathesis reaction, it is thought to proceed via the formation of ruthenium carbene species and not via classical [2+2+2]-cycloaddition pathways. A rationale for the strong preference of the *meta* isomer **99** was provided on the basis of a metathesis-type mechanism.

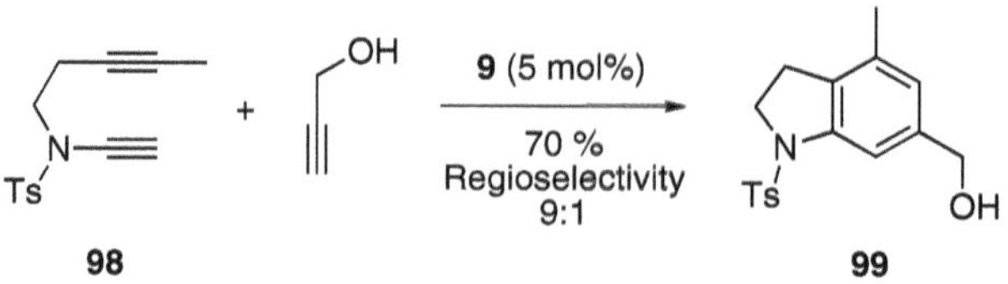

$$\mathbf{98} + \text{HC}\equiv\text{C–CH}_2\text{CH}_2\text{OH} \xrightarrow[\text{70 \%, Regioselectivity 9:1}]{\mathbf{9}\text{ (5 mol\%)}} \mathbf{99} \tag{14}$$

3.3 Substrates Containing Other Heteroatoms

In this section, some examples of metathesis reactions of substrates containing less common heteroatoms will be described.

3.3.1 Organosilicon Compounds

Numerous examples have been published of cross-metathesis reactions of allyltrimethylsilane [8d]. While these reactions appear to be facile and general, the RCM of diallylsilanes and allyl-homoallylsilanes to the corresponding silacyclopentenes and silacyclohexenes gives slightly lower yields compared to those obtained for oxa- or azacycle formation. Furthermore, the five-membered silacycles require rather high catalyst loadings if the first-generation catalyst **9** is used [82]. Comparatively few examples of the RCM of vinylsilanes have been published. These transformations require the more reactive second-generation ruthenium catalysts [83, 84] or the molybdenum complex **5** (Table 1) [85]. Molybdenum-catalyzed RCM of vinylsilanes in combination with Pd-catalyzed cross-coupling reactions has been exploited by Denmark et al. for the total synthesis of the natural product brasilenyne; the RCM step **100**→**101** is depicted in Eq. 15 [85c].

Si O O I OPMB **100** —— **5** (5 mol%), RT, 1 h, (92 %) → Si O O I OPMB **101** (15)

The cross metathesis of vinylsilanes is catalyzed by the first-generation ruthenium catalyst **9**. This transformation has been extensively investigated from both preparative and mechanistic points of view by Marciniec et al. [86]. Interestingly, the same vinylsilanes obtained from cross metathesis may also result from a ruthenium-hydride-catalyzed silylative coupling and there might be some interference of metathesis and nonmetathesis mechanisms [87].

3.3.2 Organohalides

Few reports describe the cross metathesis of allyl halides [88]. First-generation catalyst **9** does not seem to be sufficiently reactive to promote this reaction in preparatively useful yields and acceptable catalyst loadings, but second-generation catalyst **56d** gives good results for allyl chloride. Cross-metathesis reactions involving allyl bromide and allyl iodide give significantly lower yields, probably due to coordination of the halide to the ruthenium atom [88b]. 1,4-Dihalo-2-butenes also undergo cross-metathesis reactions in the presence of 2 mol% of **56d**. In contrast to the previously cited study, bromo and chloro compounds give similar yields and *E*/*Z* selectivities [88c]. Both first- and second-generation ruthenium catalysts fail in cyclization reactions involving vinyl bromides [89], whereas the corresponding vinyl chlorides are converted to cycloalkenes in good to excellent yield in the presence of 10 mol% of **56d** at elevated temperatures [89b]. The method has potential for combination with cross-coupling reactions; this point is illustrated in Scheme 21 for the sequence **102**→**103**→**104**. RCM of vinyl fluorides has very recently been published by Brown et al. using the same precatalyst under similar conditions [89c].

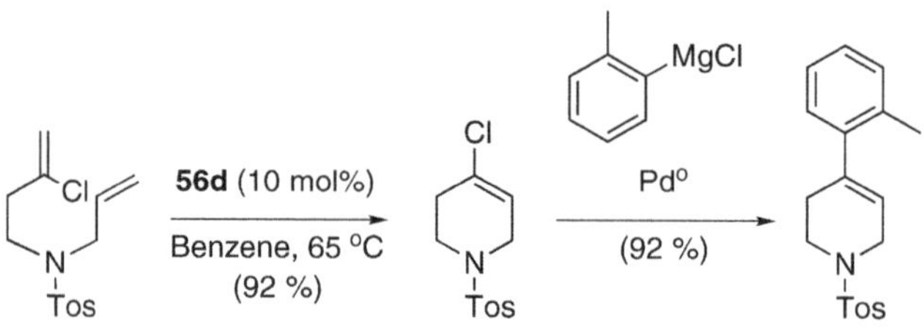

Scheme 21 RCM of vinyl chlorides and subsequent cross coupling [89b]

3.3.3 Organophosphorus and -sulfur Compounds

Free phosphines and thioethers were long believed to be incompatible with ruthenium-based metathesis catalysts. For instance, no cyclization product is obtained from diallyl sulfide upon reaction with **7** (Table 1), whereas quantitative conversion to the corresponding thiacycle is observed with Schrock's catalyst **5** [90]. Similarly, diallylphenylphosphine undergoes RCM in the presence of tungsten carbene complex **4** (Table 1) [91], whereas no conversion was observed with ruthenium catalyst **9** [92]. The low reactivity of substrates containing these heteroatoms may be attributed to inhibition of the catalyst by coordination. The observation that **9** catalyzes the RCM of allyl sulfides bearing sterically demanding substituents is in line with this assumption [93]. Formation of unsubstituted or less substituted thiacycles requires the use of either molybdenum catalyst **5** or second-generation catalysts **56c,d** [94]. It has also been described that **56d** is an efficient catalyst for the intermolecular enyne metathesis of propargyl sulfides and ethene [95]. Obviously, the NHC-ligated metathesis catalysts are less sensitive to coordination by thioethers.

Another way to reduce the nucleophilicity of the sulfur atom is oxidation to the corresponding sulfones, sulfonamides or sultones. Thus, diallyl sulfone undergoes RCM to the corresponding sulfolene smoothly in the presence of first-generation catalyst **9**. Second-generation catalyst **56d** is only required for the synthesis of sulfolenes containing tri- or tetrasubstituted double bonds [96]. Several contributions have appeared dealing with the metathesis-based synthesis of cyclic sulfonamides. Normally, the first-generation catalyst **9** is sufficiently reactive to obtain preparatively useful yields [97]. Cyclic sultones are also available by RCM; however, **56d** was found to be clearly superior over **9** in this case [98].

Similar to sulfur compounds, organophosphorus compounds can also be substrates for olefin metathesis when the coordinating ability of the phosphorus atom is significantly reduced. As mentioned above, diallylphenylphosphine does not undergo olefin metathesis in the presence of ruthenium-based precatalysts; if, however, the phosphorus atom is complexed by borane, RCM is possible and affords phosphacycles in good yields [92]. In most examples where olefin metathesis has been applied to organophosphorus compounds, the phosphorus is present in a higher oxidation state, e.g., as a phosphonate. Several contributions dealing with RCM of organophosphorus compounds were published by Hanson et al. [99]. Enyne metathesis of an acetylenic phosphonate [100] and various cross-metathesis reactions of vinylphosphonates [101] have also been reported in the literature. One example of the latter class of reactions is outlined in Eq. 16: the nucleotides **105** and **106** are coupled in the presence of second-generation catalyst **56d** to the nucleotide dimer **107** in fair yield and very high levels of *E* selectivity. Rather high catalyst loadings of up to 20% are required for this particular example, probably because of steric hindrance [101a].

TBSO … MeO … + … OTBDPS … **56d** (20 mol%) (58 %) … TBSO … MeO … OTBDPS

105 **106** **107** (16)

3.3.4 Organoboranes

Olefin metathesis of vinylboronates [102] and allylboronates [103, 104] has been investigated over the past few years because organoboranes are versatile intermediates for organic synthesis. Cross metathesis of vinylboronate **108** and 2-butene **109**, for example, yields the boronate **110**, which can be converted to the corresponding vinyl bromide **111** with high *Z* selectivity. Vinyl iodides can be obtained analogously. It should be noted that vinyl bromides and vinyl

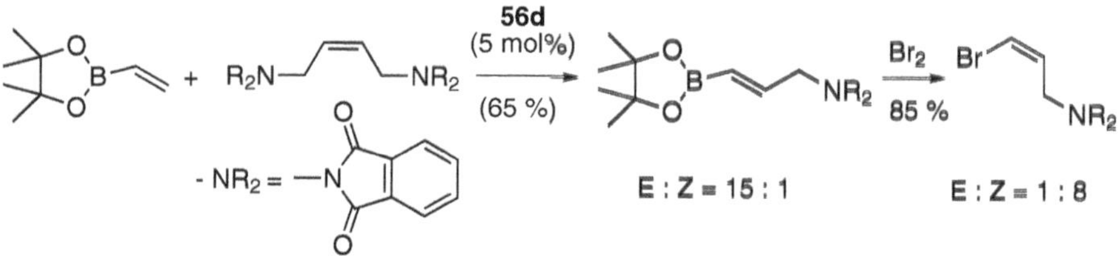

Scheme 22 CM of vinylboronates [102a]

iodides are not directly accessible by cross metathesis with bromoethene or iodoethene (Scheme 22) [102a].

Allylboronates are attractive reagents for the highly diastereoselective allylation of carbonyl compounds. A sequential cross-metathesis–allylation reaction has recently been developed by Grubbs et al. [88c] and by Miyaura et al. [103]. The sequence is illustrated in Scheme 23 for the formation of homoallylic alcohol **114** from allylboronate **112**, acetal **113**, and benzaldehyde [88c].

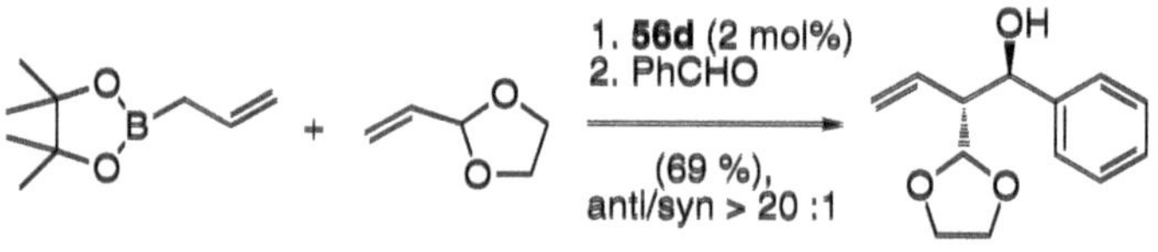

Scheme 23 CM of allylboronates [88c]

Enyne metathesis starting either from acetylenic boronates and homoallylic alcohols [104a,c] or from propargyl alcohols and allylboronates [104b] has recently been described. The resulting boronated dienes can be converted to allenes or cycloaddition products. The cross metathesis of vinylcyclopropylboronates directed toward the total synthesis of natural products has very recently been investigated by Pietruszka et al. [104d].

3.4 Conjugated and Cumulated Dienes in Olefin Metathesis

Grubbs et al. reported that the ruthenium-catalyzed RCM of a conjugated diene proceeds in such a way that the less hindered olefin moieties participate in the reaction. Consequently, RCM of **115** gives *exo*-methylene compound **116**, and not *exo*-vinyl compound **117** (Scheme 24) [105]. This regioselectivity is complementary to that observed for enyne metathesis of **118**, which gives exclusively **117** (Scheme 24) [106a].

Olefin metathesis of a conjugated diene has recently been exploited by Hiemstra et al. for the synthesis of bicyclic analogs of **116**. Second-generation

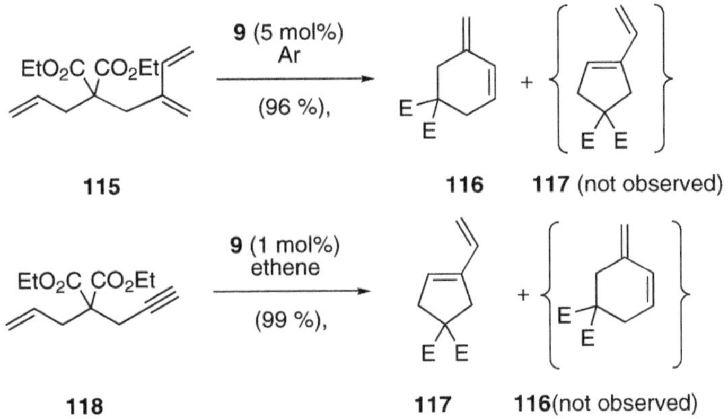

Scheme 24 Complementary results for ene–diene metathesis [105] and enyne metathesis [106a]

catalyst **56d** gives significantly better yields for medium-sized rings but leads to the formation of the same regioisomer as **9** [107]. A study on the metathesis-based synthesis of macrolides by Wagner et al. revealed that this is not necessarily the case: while **119** is cyclized by 15 mol% of first-generation catalyst **9** to the conjugated diene **120**, second-generation catalyst **56c** gives the *E*-cycloalkene **121** under these conditions by selective attack of the internal double bond (Scheme 25) [108].

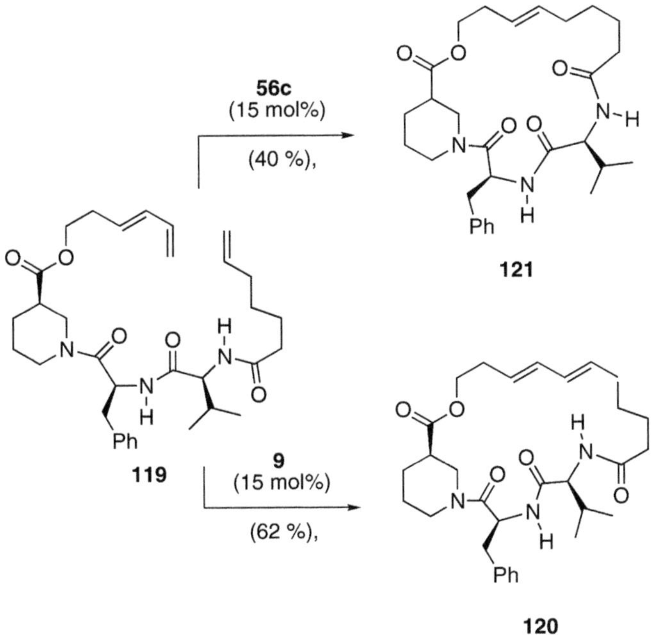

Scheme 25 Differing regioselectivity of **9** and **56c** in ene–diene macrocyclization [108]

This selectivity is not general for macrocyclization reactions involving conjugated dienes. For example, Danishefsky et al. reported the selective formation of the larger ring in an olefin metathesis-based synthesis of the natural products radicicol and monocillin I using the second-generation precatalyst **56d** [109]. The effects of substrate and precatalyst on the ring size in ene–diene RCM reactions were systematically investigated by Paquette et al. by analysis of product distribution and molecular mechanics calculations [110].

The stronger tendency of second-generation catalyst **56c** to attack the internal double bond of a conjugated diene has recently been employed in an elegant ring-contraction metathesis: the 16-membered cyclodiene antibiotic josamycin is converted in the presence of a terminal alkene and **56c** to a 14-membered cycloalkene. The reaction proceeds via a ring-opening metathesis/ring-closing metathesis [111]. Comparatively little work has been done on the field of olefin metathesis with cumulated double bonds. A study by Barrett et al. describes the self metathesis of a variety of monosubstituted allenes [112]. Polymerization appears to be a major problem in many cases and small changes in the substrate structure have a significant influence on the amount of metathesis product formed.

3.5 Olefin Metathesis in the Ligand Sphere of Metal Complexes

Although the number of applications of olefin metathesis to transition metal complexes is small compared to the number of applications in organic synthesis, this field is becoming increasingly important. Spectacular examples are the double RCM reactions of copper phenanthroline complexes as a synthetic route to catenanes [113] or a recently reported approach to steric shielding of rhenium complex terminated sp-carbon chains [114].

Transition metals have been used to complex Lewis-basic centers in metathesis substrates and to arrange the reacting olefins in such a way that cyclization is facilitated. Olefin metathesis of **122**, for example, proceeds with good yield to the bispyridine macrocycle **123** (Eq. 17) [115].

Cl–Pd–Cl (**122**) —[**9** (5 mol%), RT; (80 %)]→ Cl–Pd–Cl (**123**) (17)

Pyridine complexes of Pd- and Pt-pincer ligands are also suitable substrates for olefin metathesis [116]. The first-generation catalyst **9** efficiently mediates the RCM of diallylphosphines and diallyl sulfide when the heteroatom is complexed by a cationic $[C_5H_5(NO)(PPh_3)Re]$ moiety [117]. This principle has been exploited in the same study for tungsten, rhodium, and platinum complexes.

In a variety of studies, olefin metathesis has been applied to transition metal π-complexes such as η^4-tricarbonyliron complexes [118], η^6-arenechromiumtricarbonyl complexes [119], and η^5-cyclopentadienyl complexes of iron [120] and group 4 metals [120c, 121]. An interesting example of the kinetic resolution of diastereomers was described by Richards et al. [120b] and by Ogasawara et al. [120c] for the synthesis of ferrocenophanes bearing substituents at the bridge. As depicted in Eq. 18, a mixture of diastereomeric metathesis precursors **124** is converted to the ferrocenophanes **125** with a strong preference for *meso*-**125**. The *rac*-**124** obviously reacts significantly slower and is not cyclized if the substituents R are bulky (e.g., R=Ph) [120b] or if lower catalyst loadings are applied [120c]. While these transformations are catalyzed by the first-generation ruthenium catalyst **9**, analogous phosphaferrocene derivatives require molybdenum catalyst **5**, presumably due to inhibition of ruthenium catalysts by coordination to the Lewis-basic phosphorus atom present in the substrate [120d].

9 (10 mol%), toluene, 130 °C, R = Ph (18)

meso : rac = 3.5 : 1 **124** → *meso*-**125** + *rac*-**125**

Bent *ansa*-metallocenes of early transition metals (especially Ti, Zr, Hf) have attracted considerable interest due to their catalytic activity in the polymerization of α-olefins. Ruthenium-catalyzed olefin metathesis has been used to connect two Cp substituents coordinated to the same metal [120c, 121a] by RCM or to connect two bent metallocenes by cross metathesis [121b]. A remarkable influence of the catalyst on *E*/*Z* selectivity was described for the latter case: while first-generation catalyst **9** yields a 1:1 mixture of *E*- and *Z*-dimer **127**, *E*-**127** is the only product formed with **56d** (Eq. 19).

It has been demonstrated that group 6 Fischer-type metal carbene complexes can in principle undergo carbene transfer reactions in the presence of suitable transition metals [122]. It was therefore interesting to test the compatibility of ruthenium-based metathesis catalysts and electrophilic metal carbene functionalities. A series of examples of the formation of oxacyclic carbene complexes by metathesis (e.g., **128**, **129**, Scheme 26) was published by Dötz et al. [123]. These include substrates where double bonds conjugated to the pentacarbonyl metal moiety participate in the metathesis reaction. Evidence is

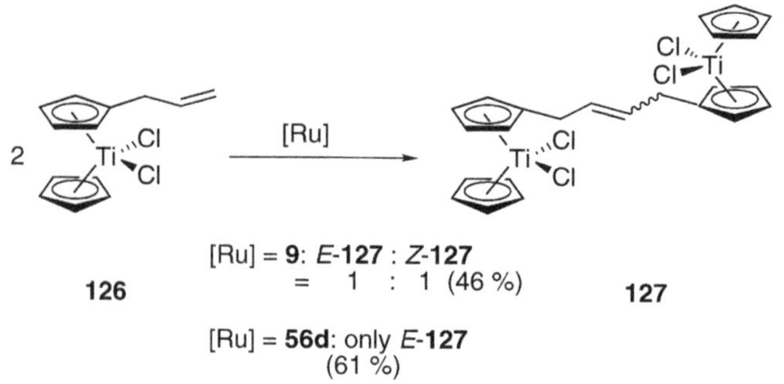

2 126 —[Ru]→ 127 (19)

[Ru] = **9**: *E*-**127** : *Z*-**127** = 1 : 1 (46 %)

[Ru] = **56d**: only *E*-**127** (61 %)

provided that a pentacarbonyl metal group promotes cyclization, whereas for the corresponding ester analog intermolecular metathesis is preferred. The observation is discussed as an "organometallic variant of the Thorpe–Ingold effect." Hydrazine derivatives of penta- and tetracarbonylchromium carbene complexes (e.g., **130**, **131**, Scheme 26) were investigated as substrates for metathesis reactions by Licandro et al. [124]. The 7-, 8-, and 9-membered heterocycles were synthesized in good to moderate yields. A metal-free analog did not undergo ring closure, presumably due to an unfavorable configuration of the substrate.

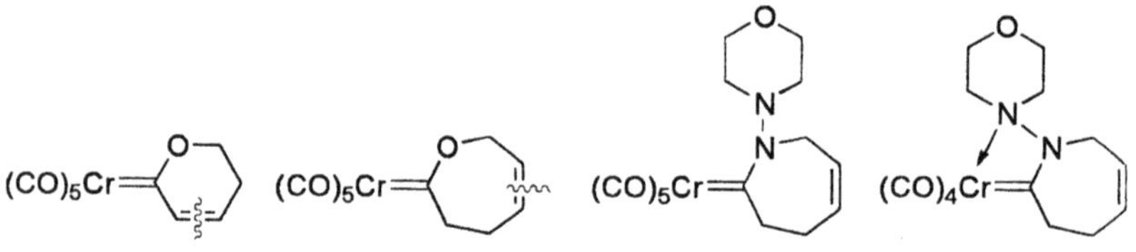

128 **129** **130** **131**

Scheme 26 RCM of substrates with Fischer carbene functionality (**128**, **129** [123]; **130**, **131** [124])

Hexacarbonyldicobalt complexes of alkynes have served as substrates in a variety of olefin metathesis reactions. There are several reasons for complexing an alkyne functionality prior to the metathesis step [125]: (a) the alkyne may chelate the ruthenium center, leading to inhibition of the catalytically active species [125d]; (b) the alkyne may participate in the metathesis reaction, giving undesired enyne metathesis products [125f]; (c) the linear structure of the alkyne may prevent cyclization reactions due to steric reasons [125a–d]; and (d) the hexacarbonylcobalt moiety can be used for further transformations [125c,f].

RCM of **132** to the medium-sized enyne **135**, for example, appears to be highly unlikely. This transformation was achieved by conversion of **132** to the cobalt complex **133**, which is cyclized to the protected cycloenyne **134**. Deprotection yields **135**, and a subsequent Pauson–Khand reaction yields the interesting tricyclic structure **136** (Scheme 27) [125c].

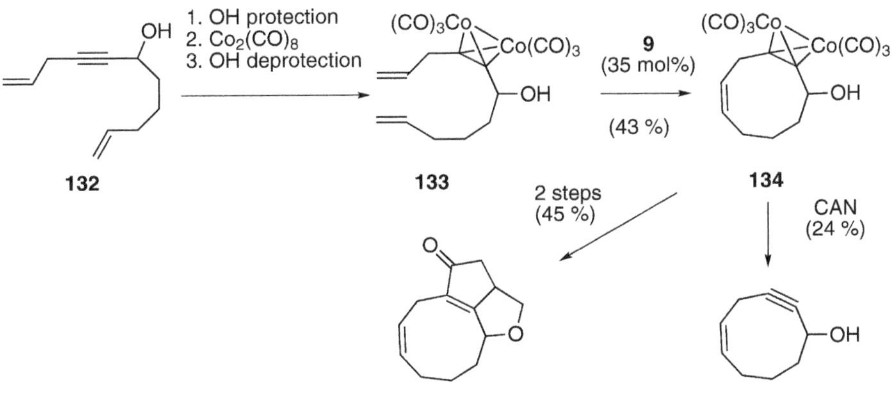

Scheme 27 RCM of substrates bearing alkyne cobalt complexes [125c]

4
Conclusions and Perspective

The discovery of Schrock's molybdenum catalyst and Grubbs' ruthenium catalysts made olefin metathesis into one of the most important C–C bond-forming reactions in synthetic organic chemistry. Over the past 10 years, significant progress has been made in the development of novel precatalysts derived from the complexes originally introduced by Schrock and Grubbs: while molybdenum precatalysts have been further developed to obtain homochiral catalysts (an aspect not covered in this contribution), research in the field of ruthenium-based precatalysts focused on the development of more convenient methods of preparation and on the development of more active precatalysts. The combination of Ru carbene complexes with one *N*-heterocyclic carbene ligand turned out to be extraordinarily valuable because these precatalysts have led to a significant extension of the scope of Ru-catalyzed olefin metathesis reactions. The development of phosphine-free Ru carbene complexes will further broaden the scope of olefin metathesis. This is especially important for the field of cross metathesis, which has long been neglected but offers exciting opportunities for organic synthesis.

Acknowledgements We thank Andrea Bokelmann for assistance in the preparation of this manuscript and Dr. Bernd Plietker for helpful discussions.

References

1. Obenaus F, Droste W, Neumeister J (2001) Butenes. In: Ullmann's encyclopedia of industrial chemistry, electronic release, 6th edn. Wiley-VCH, Weinheim, chap 3; Weissermel K, Arpe H-J (1994) Industrielle organische chemie, 4th edn. VCH, Weinheim
2. Kaminsky W, Arndt M (1996) Reactions of unsaturated compounds. In: Cornils B, Herrmann WA (eds) Applied homogeneous catalysis with organometallic compounds, vol 1. VCH, Weinheim, p 220
3. Truett WL, Johnsen DR, Robinson IM, Montague BA (1960) J Am Chem Soc 82:2337
4. Hérisson J-L, Chauvin Y (1976) Makromol Chem 141:161
5. Degussa AG (2003) Vestenamer – the rubber with unique properties. Company publication, available viahttp://www.degussa-hpp.de/e/index.html
6. (a) Kress J, Osborn JA, Amir-Ebrahimi V, Ivin KJ, Rooney JJ (1988) Chem Commun 1164; (b) Kress J, Osborn JA, Ivin KJ (1989) Chem Commun 1234; (c) Ivin KJ, Kress J, Osborn JA (1988) J Mol Catal 46:351; (d) Wallace KC, Liu AH, Dewan JC, Schrock RR (1988) J Am Chem Soc 110:4964; (e) Schrock RR (1990) Acc Chem Res 23:158; (f) Couturier J-L, Paillet C, Leconte M, Basset JM, Weiss K (1992) Angew Chem Int Ed 31:628; (g) Schrock RR, Murdzek JS, Bazan GC, Robbins J, DiMare M, O'Regan M (1990) J Am Chem Soc 112:3875; (h) Fu GC, Grubbs RH (1993) J Am Chem Soc 115:3800; (i) Nguyen ST, Johnson LK, Grubbs RH (1992) J Am Chem Soc 114:3974; (j) Fu GC, Nguyen ST, Grubbs RH, Ziller JW (1993) J Am Chem Soc 115:9856; (k) Nguyen ST, Grubbs RH, Ziller JW (1993) J Am Chem Soc 115:9858
7. (a) Schrock RR (1998) Top Organomet Chem 1:1; (b) Fürstner A (1998) Top Organomet Chem 1:37; (c) Schuster M, Blechert S (1997) Angew Chem Int Ed 36:2036; (d) Armstrong SK (1998) J Chem Soc Perkin Trans I 371; (e) Grubbs RH, Chang S (1998) Tetrahedron 54:4413; (f) Fürstner A (2000) Angew Chem Int Ed 39:3013; (g) Schrock RR (1999) Tetrahedron 55:8141; (h) Grubbs RH (ed) (2003) Handbook of metathesis. Wiley-VCH, Weinheim
8. (a) Maier ME (2000) Angew Chem Int Ed 39:2073; (b) Roy R, Das SK (2000) Chem Commun 519; (c) Gibson SE, Keen SP (1998) Top Organomet Chem 1:155; (d) Connon SJ, Blechert S (2003) Angew Chem Int Ed 42:1900; (e) Mori M (1998) Top Organomet Chem 1:133; (f) Poulsen CS, Madsen R (2003) Synthesis 1; (g) Arjona O, Csaky AG, Plumet J (2003) Eur J Org Chem 611; (h) Hoveyda AH, Schrock RR (2001) Chem Eur J 7:945; (i) Tindall D, Pawlow JH, Wagener KB (1998) Top Organomet Chem 1:183; (j) Kiessling LL, Strong LE (1998) Top Organomet Chem 1:199
9. Hallman PS, Stephenson TA, Wilkinson G (1977) Inorg Synth 12:238
10. (a) Schwab P, France MB, Ziller W, Grubbs RH (1995) Angew Chem Int Ed 34:2039; (b) Schwab P, Grubbs RH, Ziller JW (1996) J Am Chem Soc 118:100
11. Gandelman M, Rybtchinski B, Ashkenazi N, Gauvin RM, Milstein D (2001) J Am Chem Soc 123:5372
12. Kingsbury JS, Harrity JPA, Bonitatebus PJ, Hoveyda AH (1999) J Am Chem Soc 121:791
13. (a) Harlow KJ, Hill AF, Wilton-Ely JDET (1999) J Chem Soc Dalton Trans 285; (b) Fürstner A, Grabowski J, Lehmann CW (1999) J Org Chem 64:8275; (c) Fürstner A, Hill AF, Liebl M, Wilton-Ely JDET (1999) Chem Commun 601; (d) Fürstner A, Thiel OR (2000) J Org Chem 65:1738; (e) Fürstner A, Guth O, Düffels A, Seidel G, Liebl M, Gabor B, Mynott R (2001) Chem Eur J 7:4811
14. (a) Fürstner A, Picquet M, Bruneau C, Dixneuf PH (1998) Chem Commun 1315; (b) Fürstner A, Liebl M, Lehmann CW, Picquet M, Kunz R, Bruneau C, Touchard D, Dixneuf PH (2000) Chem Eur J 6:1847; (c) Picquet M, Touchard D, Bruneau C, Dixneuf P (1999) New J Chem 141; (d) Fürstner A, Ackermann L (1999) Chem Commun 95

15. Christ ML, Sabo-Etienne S, Chaudret B (1994) Organometallics 13:3800
16. (a) Wilhelm TE, Belderrain TR, Brown SN, Grubbs RH (1997) Organometallics 16:3867; (b) Amoroso D, Snelgrove JL, Conrad JC, Drouin SD, Yap GPA, Fogg DE (2002) Adv Synth Catal 344:757; (c) Volland MAO, Rominger F, Eisenträger F, Hofmann P (2002) J Organomet Chem 641:220
17. (a) Wolf J, Stüer W, Grünwald C, Werner H, Schwab P, Schulz M (1998) Angew Chem Int Ed 37:1124; (b) Wolf J, Stüer W, Grünwald C, Gevert O, Laubender M, Werner H (1998) Eur J Inorg Chem 1827; (c) Stürmer R, Schäfer B, Wolfart V, Stahr H, Kazmaier U, Helmchen G (2001) Synthesis 46
18. van der Schaaf PA, Kolly R, Hafner A (2000) Chem Commun 1045
19. (a) Hansen SM, Rominger F, Metz M, Hofmann P (1999) Chem Eur J 5:557; (b) Hansen SM, Volland MAO, Rominger F, Eisenträger F, Hofmann P (1999) Angew Chem Int Ed 38:1273
20. Belderrain TR, Grubbs RH (1997) Organometallics 16:4001
21. Katayama H, Nagao M, Ozawa F (2003) Organometallics 22:586
22. Maishal TK, Sarkar A (2002) Synlett 1925
23. Katayama H, Uruschima H, Nishioka T, Wada C, Nagao M, Ozawa F (2000) Angew Chem Int Ed 39:4513
24. Louie J, Grubbs RH (2002) Organometallics 21:2153
25. Dias EL, Nguyen ST, Grubbs RH (1997) J Am Chem Soc 119:3887
26. (a) Sanford MS, Ulman M, Grubbs RH (2001) J Am Chem Soc 123:749; (b) Sanford MS, Love JA, Grubbs RH (2001) J Am Chem Soc 123:6543; (c) Stüer W, Wolf J, Werner H (2002) J Organomet Chem 641:203
27. Tallarico JA, Bonitatebus PJ, Snapper ML (1997) J Am Chem Soc 119:7157
28. Sanford MS, Henling LM, Day MW, Grubbs RH (2000) Angew Chem Int Ed 39:3451
29. (a) Hinderling C, Adlhart C, Chen P (1998) Angew Chem Int Ed 37:2685; (b) Adlhart C, Hinderling C, Baumann H, Chen P (2000) J Am Chem Soc 122:8204; (c) Adlhart C, Chen P (2000) Helv Chim Acta 83:2192; (d) Adlhart C, Volland MAO, Hofmann P, Chen P (2000) Helv Chim Acta 83:3306; (e) Volland MAO, Adlhart C, Kiener CA, Chen P, Hofmann P (2001) Chem Eur J 7:4621. For an approach to metathesis catalyst screening based on IR thermography, see: (f) Reetz MT, Becker MH, Liebl M, Fürstner A (2000) Angew Chem Int Ed 39:1234
30. (a) Aagaard OM, Meier RJ, Buda F (1998) J Am Chem Soc 120:7174; (b) Vyboishchikov SF, Bühl M, Thiel W (2002) Chem Eur J 8:3962; (c) Cavallo L (2002) J Am Chem Soc 124:8965; (d) Fomine S, Vargas SM, Tlenkopatchev MA (2003) Organometallics 22:93; (e) Bernardi F, Bottoni A, Miscione GP (2003) Organometallics 22:940
31. Ulman M, Grubbs RH (1999) J Org Chem 64:7202
32. (a) Diaz EL, Grubbs RH (1998) Organometallics 17:2758; (b) Sanford MS, Henling LM, Grubbs RH (1998) Organometallics 17:5384; (c) Chang S, Jones LI, Wang C-C, Henling LM, Grubbs RH (1998) Organometallics 17:3460
33. For a review on *N*-heterocyclic carbenes, see: Herrmann WA, Köcher C (1997) Angew Chem Int Ed 36:2162
34. For a general account of transition-metal-NHC complexes in homogeneous catalysis, see: Yong BS, Nolan SP (2003) Chemtracts 16:205
35. (a) Weskamp T, Schattenmann WC, Spiegler M, Herrmann WA (1998) Angew Chem Int Ed 37:2490; (b) Weskamp T, Kohl FJ, Hieringer W, Gleich D, Herrmann WA (1999) Angew Chem Int Ed 38:2416; (c) Weskamp T, Kohl FJ, Herrmann WA (1999) J Organomet Chem 582:362
36. (a) Huang J, Stevens ED, Nolan SP, Petersen JL (1999) J Am Chem Soc 121:2674; (b) Huang J, Schanz H-J, Stevens ED, Nolan SP (1999) Organometallics 18:5375; (c) Jafarpour L, Hillier AC, Nolan SP (2002) Organometallics 21:442

37. Scholl M, Ding S, Lee CW, Grubbs RH (1999) Org Lett 1:953
38. Conrad JC, Yap GPA, Fogg DE (2003) Organometallics 22:1986
39. Fürstner A, Ackermann L, Gabor B, Goddard R, Lehmann CW, Mynott R, Stelzer F, Thiel OR (2001) Chem Eur J 7:3236
40. Trnka TM, Morgan JP, Sanford MS, Wilhelm TE, Scholl M, Choi TL, Ding S, Day MW, Grubbs RH (2003) J Am Chem Soc 125:2546
41. (a) Scholl M, Trnka TM, Morgan JP, Grubbs RH (1999) Tetrahedron Lett 40:2247; (b) Ackermann L, Fürstner A, Weskamp T, Kohl FJ, Herrmann WA (1999) Tetrahedron Lett 40:4787; (c) Ackermann L, El Tom D, Fürstner A (2000) Tetrahedron 56:2195; (d) Chatterjee AK, Grubbs RH (1999) Org Lett 1:1751; (e) Carda M, Rodríguez S, González F, Castillo E, Villanueva A, Marco JA (2002) Eur J Org Chem 2649; (f) Fürstner A, Thiel OR, Ackermann L, Schanz H-J, Nolan SP (2000) J Org Chem 65:2204
42. (a) Fürstner A, Langemann K (1997) J Am Chem Soc 119:9130; (b) Fürstner A, Langemann K (1997) Synthesis 792
43. (a) Dinger MB, Mol JC (2002) Adv Synth Catal 344:671; (b) Mol JC (2002) Green Chem 4:5
44. Prunet J (2003) Angew Chem Int Ed 42:2826
45. Review: Nicolaou KC, King NP, He Y (1998) Top Organomet Chem 1:73
46. (a) Fürstner A, Mathes C, Grela K (2001) Chem Commun 1057; (b) Fürstner A, Mathes C, Lehmann CW (2001) Chem Eur J 7:5299; (c) Lee CW, Grubbs RH (2000) Org Lett 2:2145; (d) Kalesse M, Quitschalle M, Claus E, Gerlach K, Pahl A, Meyer HH (1999) Eur J Org Chem 2817; (e) Fürstner A, Thiel OR, Kindler N, Bartkowska B (2000) J Org Chem 65:7990; (f) Fürstner A, Radkowski K, Wirtz C, Goddard R, Lehmann CW, Mynott R (2002) J Am Chem Soc 124:7061; (g) Murga J, Falomir E, Garcia-Fortanet J, Carda M, Marco JA (2002) Org Lett 4:3447
47. (a) Adlhart C, Chen P (2003) Helv Chim Acta 86:941; (b) Adlhart C, Chen P (2002) Angew Chem Int Ed 41:4484
48. (a) Sanford MS, Love JA, Grubbs RH (2001) Organometallics 20:5314; (b) Love JA, Sanford MS, Day MW, Grubbs RH (2003) J Am Chem Soc 125:10103
49. (a) Garber SB, Kingsbury JS, Gray BL, Hoveyda AH (2000) J Am Chem Soc 122:8168; (b) Gessler S, Randl S, Blechert S (2000) Tetrahedron Lett 41:9973
50. (a) Imhof S, Randl S, Blechert S (2001) Chem Commun 1692; (b) Randl S, Gessler S, Wakamatsu H, Blechert S (2001) Synlett 430; (c) Randl S, Connon SJ, Blechert S (2001) Chem Commun 1796
51. Chatterjee AK, Morgan JP, Scholl M, Grubbs RH (2000) J Am Chem Soc 122:3783
52. Grela K, Bieniek M (2001) Tetrahedron Lett 42:6425
53. Michrowska A, Bieniek M, Kim M, Klajn R, Grela K (2003) Tetrahedron 59:4525
54. (a) Cossy J, BouzBouz S, Hoveyda AH (2001) J Organomet Chem 624:327; (b) BouzBouz S, Cossy J (2001) Org Lett 3:1451; (c) BouzBouz S, De Lemos E, Cossy J (2002) Adv Synth Catal 344:627
55. Love JA, Morgan JP, Trnka TM, Grubbs RH (2002) Angew Chem Int Ed 41:4035
56. Wakamatsu H, Blechert S (2002) Angew Chem Int Ed 41:794
57. Wakamatsu H, Blechert S (2002) Angew Chem Int Ed 41:2403
58. Grela K, Harutyunyan S, Michrowska A (2002) Angew Chem Int Ed 41:4038
59. Grela K, Kim M (2003) Eur J Org Chem 963
60. Fürstner A, Thiel OR, Lehmann CW (2002) Organometallics 21:331
61. (a) Kingsbury JS, Garber SB, Giftos JM, Gray BL, Okamoto MM, Farrer RA, Fourkas JT, Hoveyda AH (2001) Angew Chem Int Ed 40:4251; (b) Dowden J, Savovic J (2001) Chem Commun 37; (c) Randl S, Buschmann N, Connon SJ, Blechert S (2001) Synlett 1547; (d) Grela K, Tryznowski M, Bieniek M (2002) Tetrahedron Lett 43:9055; (e) Connon

SJ, Blechert S (2002) Bioorg Med Chem Lett 12:1873; (f) Connon SJ, Dunne AM, Blechert S (2002) Angew Chem Int Ed 41:3835; (g) Varray S, Lazaro R, Martinez J, Lamaty F (2003) Organometallics 22:2426; (h) Yao Q, Zhang Y (2003) Angew Chem Int Ed 42:3395
62. Kang B, Kim D, Do Y, Chang S (2003) Org Lett 5:3041
63. Denk K, Fridgen J, Herrmann WA (2002) Adv Synth Catal 344:666
64. Ghosh AK, Cappiello J, Shin D (1998) Tetrahedron Lett 39:4651
65. Langer P, Albrecht U (2002) Synlett 1841
66. Anand RV, Baktharaman S, Singh VK (2002) Tetrahedron Lett 43:5393
67. Cossy J, Bauer D, Bellosta V (2002) Synlett 715
68. Cossy J, Taillier C, Bellosta V (2002) Tetrahedron Lett 43:7263
69. Boiteau J-G, van de Weghe P, Eustache J (2001) Org Lett 3:2737
70. (a) Grigg R, Hodgson A, Morris J, Sridharan V (2003) Tetrahedron Lett 44:1023; (b) Grigg R, Martin W, Morris J, Sridharan V (2003) Tetrahedron Lett 44:4899
71. Choi T-L, Chatterjee AK, Grubbs RH (2001) Angew Chem Int Ed 40:1277
72. Choi TL, Lee CW, Chatterjee AK, Grubbs RH (2001) J Am Chem Soc 123:10417
73. Royer F, Vilain C, Elkaim L, Grimaud L (2003) Org Lett 5:2007
74. (a) Sturino CF, Wong JCY (1998) Tetrahedron Lett 39:9623; (b) Williams DR, Cortez GS, Bogen SL, Rojas CM (2000) Angew Chem Int Ed 39:4612; (c) Clark JS, Hamelin O (2000) Angew Chem Int Ed 39:372
75. (a) Postema MHD, Calimente D (1999) Tetrahedron Lett 40:4755; (b) Calimente D, Postema MHD (1999) J Org Chem 64:1770; (c) Postema MHD, Calimente D, Liu L, Behrmann TL (2000) J Org Chem 65:6061
76. (a) Allwein SP, Cox JM, Howard BE, Johnson HWB, Rainier JD (2002) Tetrahedron 58:1997; (b) Majumder U, Cox JM, Rainier JD (2003) Org Lett 5:913; (c) Cox JM, Rainier JD (2001) Org Lett 3:2919; (d) Rainier JD, Cox JM, Allwein SP (2001) Tetrahedron Lett 42:179; (e) Rainier JD, Allwein SP, Cox JM (2001) J Org Chem 66:1380; (f) Postema MHD, Piper JL, Liu L, Shen J, Faust M, Andreana P (2003) J Org Chem 68:4748; (g) Liu L, Postema MHD (2001) J Am Chem Soc 123:8602; (h) Holson EB, Roush WR (2002) Org Lett 4:3719; (i) van Otterlo WAL, Ngidi EL, Coyanis EM, de Koning CB (2003) Tetrahedron Lett 44:311; (j) van Otterlo WAL, Ngidi EL, de Koning CB (2003) Tetrahedron Lett 44:6483. For an example of enol ether–alkyne cross metathesis, see: (k) Giessert AJ, Snyder L, Markham J, Diver ST (2003) Org Lett 5:1793
77. (a) Sutton AE, Seigal BA, Finnegan DF, Snapper ML (2002) J Am Chem Soc 124:13390; (b) Schmidt B (2003) Eur J Org Chem 816
78. (a) Okada A, Ohshima T, Shibasaki M (2001) Tetrahedron Lett 42:8023; (b) Arisawa M, Theeraladanon C, Nishida A, Nakagawa M (2001) Tetrahedron Lett 42:8029; (c) Schramm MP, Reddy DS, Kozmin SA (2001) Angew Chem Int Ed 40:4274; (d) Aggarwal VK, Daly AM (2002) Chem Commun 2490; (e) Whitehead A, Moore JD, Hanson PR (2003) Tetrahedron Lett 44:4275
79. Arisawa M, Terada Y, Nakagawa M, Nishida A (2002) Angew Chem Int Ed 41:4732
80. (a) Saito N, Sato Y, Mori M (2002) Org Lett 4:803; (b) Huang J, Xiong H, Hsung RP, Rameshkumar C, Mulder JA, Grebe TP (2002) Org Lett 4:2417
81. Witulski B, Stengel T, Fernández-Hernández JM (2000) Chem Commun 1965
82. Ahmad I, Falck-Pedersen ML, Undheim K (2001) J Organomet Chem 625:160
83. Trost BM, Ball ZT (2001) J Am Chem Soc 123:12726
84. Schuman M, Gouverneur V (2002) Tetrahedron Lett 43:3513
85. (a) Denmark SE, Yang S (2001) Org Lett 3:1749; (b) Denmark SE, Yang SM (2002) J Am Chem Soc 124:2102; (c) Denmark SE, Yang S-M (2002) J Am Chem Soc 124:15196
86. (a) Pietraszuk C, Marciniec B, Fischer H (2000) Organometallics 19:913; (b) Pietraszuk C, Fischer H, Kujawa M, Marciniec B (2001) Tetrahedron Lett 42:1175; (c) Kujawa-Welten M,

Marciniec B (2002) J Mol Catal A Chem 190:79; (d) Kujawa-Welten M, Pietraszuk C, Marciniec B (2002) Organometallics 21:840

87. (a) Marciniec B, Pietraszuk C (1997) Organometallics 16:4320; (b) Marciniec B, Pietraszuk C, Kujawa M (1998) J Mol Catal A Chem 133:41; (c) Marciniec B, Kujawa M, Pietraszuk C (2000) New J Chem 24:671
88. (a) Blanco OM, Castedo L (1999) Synlett 557; (b) Liu B, Das SK, Roy R (2002) Org Lett 4:2723; (c) Goldberg SD, Grubbs RH (2002) Angew Chem Int Ed 41:807
89. (a) Østergaard N, Pederson BT, Skjærbæk N, Vedsø P, Begtrup M (2002) Synlett 1889; (b) Chao W, Weinreb SM (2003) Org Lett 5:2505; (c) Salim SS, Bellingham RK, Satcharoen V, Brown RCD (2003) Org Lett 5:3403
90. Armstrong SK, Christie BA (1996) Tetrahedron Lett 37:9373
91. Leconte M, Jourdan I, Pagano S, Lefebvre F, Basset J-M (1995) Chem Commun 857
92. Schuman M, Trevitt M, Redd A, Gouverneur V (2000) Angew Chem Int Ed 39:2491
93. Moore JD, Sprott KT, Hanson PR (2001) Synlett 605
94. Spagnol G, Heck M-P, Nolan SP, Mioskowski C (2002) Org Lett 4:1767
95. Smulik JA, Giessert AJ, Diver ST (2002) Tetrahedron Lett 43:209
96. Yao Q (2002) Org Lett 4:427
97. (a) Hanson PR, Probst DA, Robinson RE, Yau M (1999) Tetrahedron Lett 40:4761; (b) Brown RCD, Castro JL, Moriggi J-D (2000) Tetrahedron Lett 41:3681; (c) Long DD, Termin AP (2000) Tetrahedron Lett 41:6743; (d) Lane C, Snieckus V (2000) Synlett 1294; (e) Wanner J, Harned AM, Probst DA, Poon KWC, Klein TA, Snelgrove KA, Hanson PR (2002) Tetrahedron Lett 43:917
98. Karsch S, Schwab P, Metz P (2002) Synlett 2019
99. (a) Hanson PR, Stoinova DS (1998) Tetrahedron Lett 39:3939; (b) Hanson PR, Stoianova DS (1999) Tetrahedron Lett 40:3297; (c) Sprott KT, McReynolds MD, Hanson PR (2001) Synthesis 612; (d) Stoianova DS, Hanson PR (2001) Org Lett 3:3285; (e) Sprott KT, McReynolds MD, Hanson PR (2001) Org Lett 3:3939; (f) Whitehead A, Moore JD, Hanson PR (2003) Tetrahedron Lett 44:4275
100. Timmer MSM, Ovaa H, Filippov DV, van der Marel GA, van Boom JH (2001) Tetrahedron Lett 42:8231
101. (a) Lera M, Hayes CJ (2001) Org Lett 3:2765; (b) Chatterjee AK, Choi T-L, Grubbs RH (2001) Synlett 1034; (c) Chatterjee AK, Grubbs RH (2002) Angew Chem Int Ed 41:3171; (d) Demchuk OM, Pietrusiewicz KM, Michrowska A, Grela K (2003) Org Lett 5:3217
102. (a) Morrill C, Grubbs RH (2003) J Org Chem 68:6031; (b) Renaud J, Ouellet SG (1998) J Am Chem Soc 1998:7995
103. Yamamoto Y, Takahashi M, Miyaura N (2002) Synlett 128
104. (a) Micalizio GC, Schreiber SL (2002) Angew Chem Int Ed 41:152; (b) Micalizio GC, Schreiber SL (2002) Angew Chem Int Ed 41:3272; (c) Renaud J, Graf C-D, Oberer L (2000) Angew Chem Int Ed 39:3101; (d) Garcia PG, Hohn E, Pietruszka J (2003) J Organomet Chem 680:281
105. Kirkland TA, Grubbs RH (1997) J Org Chem 62:7310
106. (a) Mori M, Sakakibara N, Kinoshita A (1998) J Org Chem 63:6082. For a study dealing with the regioselectivity of enyne metathesis, see: (b) Kitamura T, Sato Y, Mori M (2002) Adv Synth Catal 344:678
107. Mentink G, van Maarseveen JH, Hiemstra H (2002) Org Lett 4:3497
108. Wagner J, Martin-Cabrejas LM, Grossmith CE, Papageorgiou C, Senia F, Wagner D, France J, Nolan SP (2000) J Org Chem 65:9255
109. (a) Garbaccio RM, Danishefsky SJ (2000) Org Lett 2:3127; (b) Garbaccio RM, Stachel SJ, Baeschlin DK, Danishefsky SJ (2001) J Am Chem Soc 123:10903; (c) Yamamoto K, Biswas K, Gaul C, Danishefsky SJ (2003) Tetrahedron Lett 44:3297

110. Paquette LA, Basu K, Eppich JC, Hofferberth JE (2002) Helv Chim Acta 85:3033
111. Lazarova TI, Binet SM, Vo NH, Chen JS, Phan LT, Or YS (2003) Org Lett 5:443
112. Ahmed M, Arnauld T, Barrett AGM, Braddock DC, Flack K, Procopiou PA (2000) Org Lett 2:551
113. Weck M, Mohr B, Sauvage J-P, Grubbs RH (1999) J Org Chem 64:5463
114. Horn CR, Martín-Alvarez JM, Gladysz JA (2002) Organometallics 21:5386
115. Ng PL, Lambert JN (1999) Synlett 1749
116. Dijkstra HP, Chuchuryukin A, Suijkerbuijk BMJM, van Klink GPM, Mills AM, Spek AL, van Koten G (2002) Adv Synth Catal 344:771
117. Ruwwe J, Martín-Alvarez JM, Horn CR, Bauer EB, Szafert S, Lis T, Hampel F, Cagle PC, Gladysz JA (2001) Chem Eur J 7:3931
118. Paley RS, Estroff LA, Gauguet J-M, Hunt DK, Newlin RC (2000) Org Lett 2:365
119. Maity BC, Swamy VM, Sarkar A (2001) Tetrahedron Lett 42:4373
120. (a) Seshadri H, Lovely CJ (2000) Org Lett 2:327; (b) Locke AJ, Jones C, Richards CJ (2001) J Organomet Chem 637–639:669; (c) Ogasawara M, Nagano T, Hayashi T (2002) J Am Chem Soc 124:9068; (d) Ogasawara M, Nagano T, Hayashi T (2003) Organometallics 22:1174
121. (a) Hüerländer D, Kleigrewe N, Kehr G, Erker G, Fröhlich R (2002) Eur J Inorg Chem 2633; (b) Sierra JC, Hüerländer D, Hill M, Kehr G, Erker G, Fröhlich R (2003) Chem Eur J 9:3618
122. Sierra MA, del Amo JC, Mancheno MJ, Gómez-Gallego M (2001) J Am Chem Soc 123:851
123. Sültemeyer J, Dötz KH, Hupfer H, Nieger M (2000) J Organomet Chem 606:26
124. Licandro E, Maiorana S, Vandoni B, Perdicchia D, Paravidino P, Baldoli C (2001) Synlett 757
125. (a) Green JR (2001) Synlett 353; (b) Burlison JA, Gray JM, Young DGJ (2001) Tetrahedron Lett 42:5363; (c) Young DGJ, Burlison JA, Peters U (2003) J Org Chem 68:3494; (d) Yang Z-Q, Danishefsky SJ (2003) J Am Chem Soc 125:9602; (e) Ono K, Nagata T, Nishida A (2003) Synlett 1207; (f) Rosillo M, Cararrubios L, Domínguez G, Pérez-Castells J (2003) Org Biomol Chem 1:1450

Topics Organomet Chem (2004) 13: 269–366
DOI 10.1007/978-3-540-40910-6

Diene, Enyne, and Diyne Metathesis in Natural Product Synthesis

Johann Mulzer (✉) · Elisabeth Öhler

Institut für Organische Chemie, Universität Wien, Währinger Strasse 38, 1090 Wien, Austria
Johann.mulzer@univie.ac.at

Abstract With the commercial availability of well-defined ruthenium metathesis catalysts which combine high stability and broad functional group compatibility, metathesis has firmly established itself in the toolbox of target-oriented chemists. RCM is now routinely integrated in the retrosynthetic planning of natural product syntheses. The availability of metathesis catalysts with different activity, and hence chemoselectivity, allows one increasingly to influence the regio- and stereoselectivity of metathesis events. With the advent of the highly active NHC-bearing catalysts, CM also began to emerge from the shadow of RCM as a novel and economical alternative for the formation of electron-deficient and highly substituted double bonds. This progress in olefin CM translates into an increasing number of natural product-directed fragment syntheses and has also been useful for convergent assembly of main fragments. An increasing number of uniquely short and atom-economical natural product syntheses feature sequences of several metathesis events, by combining ring-opening metathesis with RCM and/or CM with concomitant chirality transfer to transform one ring into a thermodynamically more stable one. Enyne metathesis, which can be performed by the same catalysts, produces synthetically useful 1,3-dienes that lend themselves to further structural elaboration through subsequent cycloadditions or metathesis cascades. However, although many general studies have demonstrated the preparative use of inter- and intramolecular enyne metathesis, natural product-directed applications are, if compared with diene metathesis, still relatively scarce. The remaining drawback of macrocyclic RCM is the lack of control over the stereochemistry of the newly formed double bond. A solution to this problem is provided by a sequence of ring-closing diyne metathesis and stereoselective partial hydrogenation. As diyne metathesis demands different catalyst systems, sequences of chemoselective olefin, alkene, and alkyne metathesis events have been used in the synthesis of increasingly complex natural products.

Keywords Alkenes · Alkynes · Carbene complexes · Ruthenium · Tandem reactions

Abbreviations

ACM	Alkyne cross metathesis
ADMET	Acyclic diene metathesis
ARCM	Asymmetric ring-closing metathesis
AROM	Asymmetric ring-opening metathesis
CM	Cross metathesis
CsA	Cyclosporin A
Cy	Cyclohexyl
DET	Diethyl tartrate
epoA (B, C...)	Epothilone A (B, C...)
Mes	2,4,6-Trimethylphenyl (mesityl)
NHC	*N*-Heterocyclic carbene
PMB	*p*-Methoxybenzyl
PMP	*p*-Methoxyphenyl
Pyr	Pyridine
RCAM	Ring-closing alkyne metathesis
RCM	Ring-closing metathesis
ROCM	Ring-opening cross metathesis
ROM	Ring-opening metathesis
ROMP	Ring-opening metathesis polymerization
RRM	Ring-rearrangement metathesis
TADA	Transannular Diels–Alder
TBS	*Tert*-butyldimethylsilyl
TES	Triethylsilyl

1 Introduction

Over the past few years, metathesis reactions catalyzed by well-defined alkylidene complexes that combine excellent activity with broad functional group tolerance have emerged as a powerful tool for carbon–carbon bond formation in natural product-directed organic chemistry. Many general reviews on metathesis reactions in organic synthesis [1] and catalyst development [2] have appeared during the past 5 years, including the most recent contributions concerning enyne metathesis [3], olefin cross metathesis [4], and alkyne metathesis in natural product synthesis [5]. A number of reviews have focused on special applications in organic synthesis, such as olefin metathesis in carbohydrate chemistry [6], RCM leading to medium-sized rings [7] or to heterocycles, alkaloids, and peptidomimetics [8], macrocyclizations leading to (*E*) double bonds [9], RCM in the synthesis of epothilones and polyether natural products [10], and in the total synthesis of laulimalide [11].

Depending on the types of unsaturated functional units involved in the metathesis process, the reactions can be classified into three major categories: diene, enyne, and diyne metathesis (Figs. 1–3). Another mode of classification

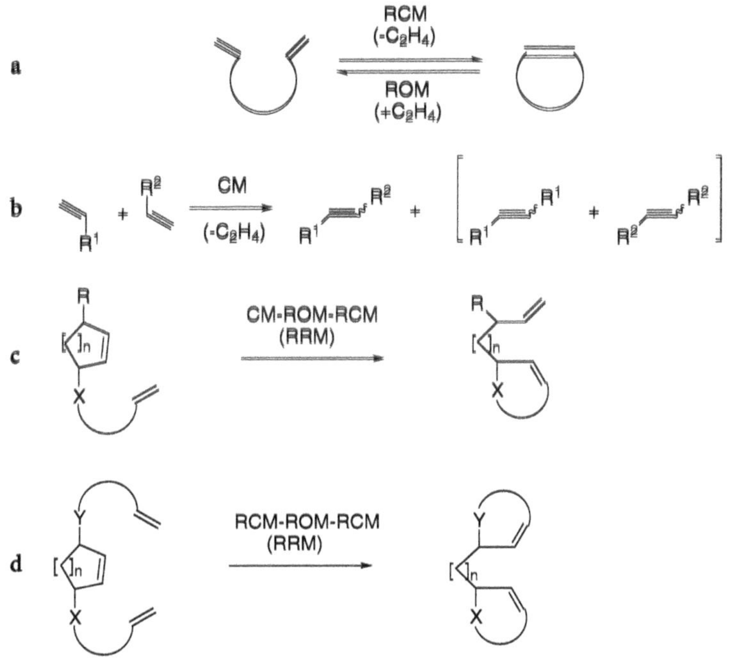

Fig. 1a–d Typical alkene metathesis reactions: ring-closing (RCM) and ring-opening (ROM) metathesis (**a**), diene cross metathesis (CM, **b**), ROM–RCM (**c**), and ROM–double RCM (**d**) sequences (ring-rearrangement reactions, RRM)

results from the structural changes during the metathesis reaction: ring closing (RCM), ring opening (ROM), cross metathesis (CM), and tandem reactions that combine two or more of these reaction types. Among these subclasses, ring-closing diene metathesis (Fig. 1a) has drawn by far the most attention in natural product synthesis, despite the fact that in the macrocyclic series the reaction often proceeds with low stereoselectivity. Olefin cross metathesis (Fig. 1b), which for decades has found numerous industrial uses is, due to problems with chemo- and stereoselectivity, not yet common in the area of natural products. However, with the advent of a second generation of highly potent and commercially available ruthenium catalysts (see below), CM reactions with an electron-deficient olefin as one of the reaction partners can be considered as a useful alternative for the formation of C=C double bonds. A growing number of interesting natural products have been prepared by combining sequential ROM–RCM (Fig. 1c) and ROM–double RCM (Fig. 1d) processes, also known as ring-rearrangement metathesis (RRM) reactions, which lead from a (strained) cycloalkene to one or two thermodynamically favored heterocycles.

While diene metathesis or diyne metathesis are driven by the loss of a (volatile) alkene or alkyne by-product, *enyne metathesis* (Fig. 2) cannot benefit from this contributing feature to the ΔS term of the reaction, since the event is entirely atom economic. Instead, the reaction is driven by the formation of conjugated dienes, which ensures that once these dienes have been formed, the process is no longer a reversible one. Enyne metathesis can also be considered as an alkylidene migration reaction, because the alkylidene unit migrates from the alkene part to one of the alkyne carbons. The mechanism of enyne metathesis is not well described, as two possible complexation sites (alkene or alkyne) exist for the ruthenium carbene, leading to different reaction pathways, and the situation is further complicated when the reaction is conducted under an atmosphere of ethylene. Despite its enormous potential to form mul-

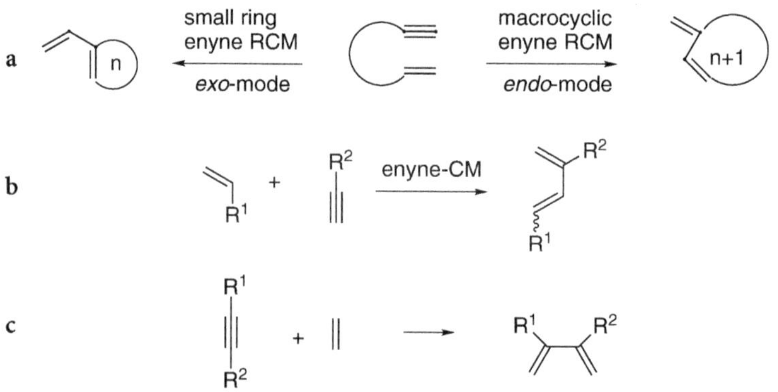

Fig. 2a–c Typical enyne metathesis reactions: ring-closing enyne metathesis (**a**); enyne cross metathesis (**b** and **c**)

tiple C–C bonds and polycyclic systems, enyne metathesis, which is catalyzed by the same catalysts, has rarely been exploited in the area of natural products. In particular, enyne CM between a terminal olefin and a terminal alkyne is hampered by the formation of isomers (Fig. 2b), albeit recent progress was made in this field when the reactions were performed in the presence of ethylene (Fig. 2c).

The regiochemical course of *enyne RCM* (ring closure by the *exo* or *endo* mode shown in Fig. 2a) is not only determined by the substitution pattern of the substrate, but also by the size of the ring to be formed and the choice of catalyst. Therefore, the reaction can lead either to vinylcycloalkenes (*exo* path) or to the enlarged ring system, where both alkyne carbons are integrated in the ring (*endo* path). Both types of the resulting 1,3-diene systems possess high potency for further manipulation. Since the first enyne RCM-based natural product synthesis in 1996 by Mori, the synthetic potential of ring-closing enyne and dienyne metathesis following the *exo* path has been highlighted by several natural product syntheses. Macrocyclic enyne RCM, which apparently follows the *endo* mode by integrating both alkyne carbons in the cycloalkene, seems to be an area of interest and deeper investigation [12] but is to date only represented by a single total synthesis.

Ring-closing alkyne metathesis (RCAM, Fig. 3a) and *alkyne cross metathesis* (ACM, Fig. 3b), which are mediated by tungsten and molybdenum catalysts (see below), seem to offer a simple solution to stereochemistry problems frequently encountered in the macrocyclic olefin series, as the resulting (cyclo)alkynes can be transformed to *E*- or *Z*-(cyclo)alkenes by appropriate semihydrogenation methods. Although the power of the method has been demonstrated in various natural product syntheses by Fürstner and coworkers, alkyne metathesis has remained in the shadow of alkene-based metathesis reactions and was – until now – not widely used by other target-orientated groups, certainly due to the lack of commercially available and stable catalyst systems.

The acceptance of a (new) catalytically mediated methodology by the target-directed synthetic community strongly depends on the availability, stability, and functional group tolerance of the respective catalysts. With the commercial availability of Grubbs' benzylidene ruthenium catalyst **A** [13] and Schrock's even more active, yet highly air- and moisture-sensitive molybdenum catalyst **B** [14]

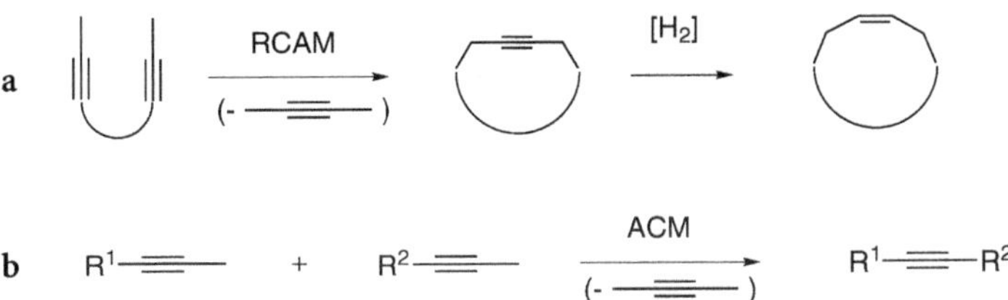

Fig. 3a,b Typical diyne metathesis reactions: ring-closing alkyne metathesis (RCAM, **a**); diyne cross metathesis (ACM, **b**)

(Fig. 4), RCM started rapidly to establish itself in the "toolbox" of target-orientated organic chemists as a method to generate the heterocyclic and macrocyclic motifs present in many natural products. The exchange of one PCy_3 ligand of the "classical" Grubbs' catalyst **A** by an *N*-heterocyclic carbene (NHC) ligand, reported in 1999 independently and almost simultaneously by three groups [15–17], led to a "second generation" of metathesis catalysts (**C** and **E**) with even superior reactivity and increased stability. The NHCs are particularly strong σ-donors but poor π-acceptor ligands with little tendency to dissociate from the metal center. The sterically demanding mesityl substituents on their N atoms are able to stabilize the catalytically relevant intermediates by electronic and steric means against decomposition. These properties translate into highly improved reactivity as well as higher stability which surpass those of the parent complex **A** and also, in many cases, those of molybdenum complex **B**. In 2000, the number of innovative ruthenium catalysts was expanded by catalyst **D** [18], now commonly designed as Hoveyda's catalyst. The phosphine-free catalyst **D** bears, in addition to the NHC ligand, a styrenyl ether that allows for the easy recovery of the catalyst by chromatography. Catalysts **C** and **D** are now also commercially available, and have set new standards in the rapidly expanding field of diene and enyne metathesis reactions, as they also allow the formation of tri- and in some cases tetra-substituted cycloalkenes and the cyclization of conformationally handicapped substrates, as well as stereoselective CM reactions with electron-deficient conjugated olefins. The less common indenylidene ruthenium complex **F** is a catalyst of the first generation. It is not commercially available, but is readily prepared from commercial products, and exhibited favorable selectivity in several natural product syntheses by Fürstner.

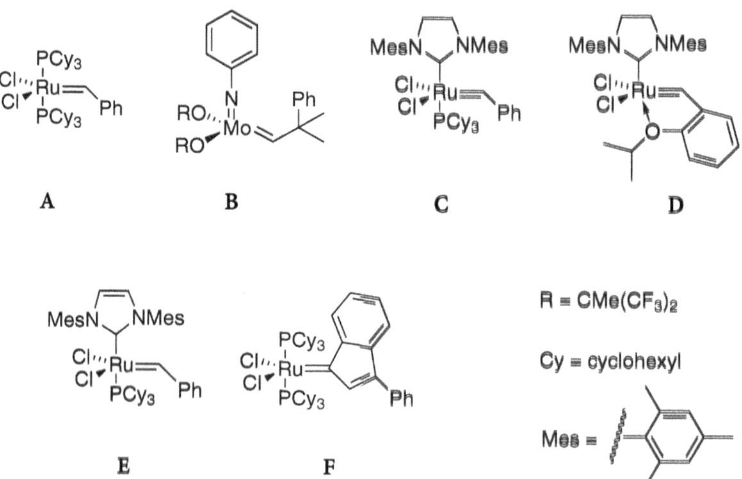

Fig. 4 Commonly used (**A–D**) and less commonly used (**E, F**) initiators for diene and enyne metathesis

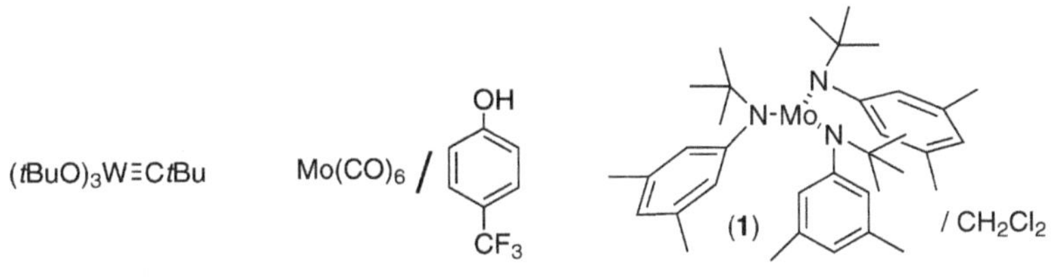

$(tBuO)_3W{\equiv}CtBu$ G

$Mo(CO)_6$ / OH, CF_3 H

N-Mo (1) / CH_2Cl_2 I

Fig. 5 Diyne metathesis initiators used in Fürstner's natural product syntheses

The alkyne metathesis catalysts used to date in natural product-directed syntheses (Fig. 5), fall into two categories [19]. Tungsten complex **G** belongs to the structurally well-defined Schrock-type alkylidyne transition metal complexes, that are thoroughly studied from the mechanistic point of view. Catalyst systems **H** and **I** belong to the other class of initiator systems, in which a structurally unknown catalyst is formed in situ from two reagents. The "user-friendly" combination of $Mo(CO)_6$ and a phenol additive (**H**) is restricted in its scope by the harsh reaction conditions required. Catalyst system **I**, formed in situ from the trisamido-molybdenum complex **1** and dichloromethane, features broad functional group tolerance and high reactivity under mild conditions. However, due to the extreme reactivity of the basic complex toward small molecules (including nitrogen) [20], its broad acceptance by the synthetic community will be limited.

Following the guidelines of typical metathesis reactions outlined in Figs. 1–3, the present review will concentrate – with only a few exceptions – on the most recent applications of metathesis reactions in the total synthesis of natural products.

2
Diene Metathesis

2.1
Ring-Closing Diene Metathesis (RCM)

Without question, the area of olefin metathesis that has expanded most dramatically in recent years is ring-closing diene metathesis. RCM is proving useful in total syntheses where it has been applied to targets ranging from relatively small molecules or cyclic fragments of natural products, to highly complex ones that contain multiple unsaturations. One of the major applications of RCM has been the synthesis of natural products with medium-sized (8–11-membered) carbo- and heterocyclic rings, as well as chemo- and stereoselective macrocyclizations of increasingly complex substrates such as diene-enes and even precursors with two terminal 1,3-diene units.

2.1.1 Formation of 5-, 6-, and 7-Membered Rings

In recent years, a wealth of information has accumulated on RCM reactions leading to 5-, 6-, and 7-membered carbocycles and heterocycles, so that it is impossible to refer to all the new, natural product-directed work. Therefore, we will concentrate here on a few selected examples that can illustrate (1) the progress made by the advent of the second-generation ruthenium catalysts C–E, (2) the use of RCM in concert with other innovative methodology, and (3) the use of RCM in total syntheses of newly discovered natural products which, due to an outstanding biological profile, have attracted specific interest by the synthetic community.

2.1.1.1 Carbocycles

Madindoline A (**7**) and B (*ent*-**8**) are potent inhibitors of interleukin 6. In a total synthesis [21] that also intended to determine the relative and absolute configurations of these novel antibiotics, the densely functionalized cyclopentene-1,3-dione ring of **7** and **8** was elaborated via RCM of diene-diol **2** (Scheme 1).

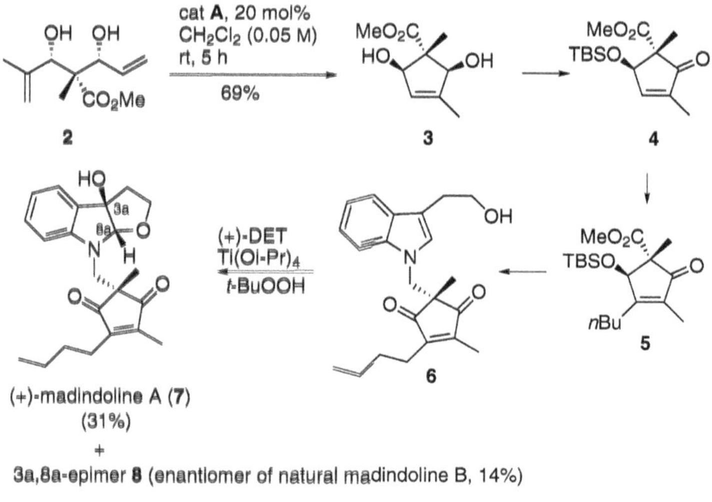

cat A, 20 mol%
CH2Cl2 (0.05 M)
rt, 5 h
69%
OH OH
CO2Me
2
MeO2C
HO
OH
3
MeO2C
TBSO
O
4
MeO2C
TBSO
O
nBu
5
OH
N
O
O
6
(+)-DET
Ti(Oi-Pr)4
t-BuOOH
HO
3a
8a
O
N
H
O
O
(+)-madindoline A (7)
(31%)
+
3a,8a-epimer 8 (enantiomer of natural madindoline B, 14%)

Scheme 1 RCM-based formation of a densely substituted cyclopentene in the first total synthesis of madindoline A (**7**) and *ent*-madindoline B (**8**) [21]

The densely functionalized cyclopentenyl core **11** of the potent antitumor antibiotic viridenomycin (**12**) was most recently prepared by treatment of enone **9** with second-generation Ru catalyst **C** (Scheme 2) [22]. This reaction proved to be very slow, requiring 3.5 days to give only incomplete formation of cyclization product **10** in 69% yield (86%, based on recovered **9**).

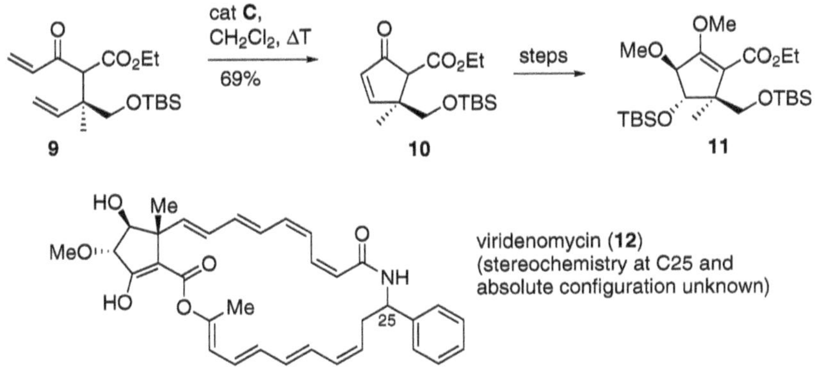

cat **C**, CH_2Cl_2, ΔT, 69%; steps

9 **10** **11**

viridenomycin (**12**) (stereochemistry at C25 and absolute configuration unknown)

Scheme 2 RCM-based formation of a highly substituted cyclopentenone in Trost's synthesis of the cyclopentenyl core of viridenomycin (**12**) [22]

A striking example of the power of NHC-bearing catalysts with sterically demanding substrates was disclosed by Chavez and Jacobsen [23], who presented a novel route to several iridoid natural products, exemplified by the enantio- and diastereoselective synthesis of boschnialactone (**17**) outlined in Scheme 3. Chiral aldehyde **13**, available from citronellal by Eschenmoser methylenation in a single step, reacted despite the presence of an isoprenyl moiety and a *gem*-disubstituted double bond, in the presence of second-generation catalyst **C** smoothly to form cyclopentene carboxaldehyde **14**. Aldehyde **14** in turn underwent, in the presence of tridentate (Schiff base) Cr(III) complex **15**, an efficient and highly selective inverse-electron-demand hetero-Diels–Alder reaction with ethyl vinyl ether to produce cycloadduct **16** in 85% yield. Compound **16** was then converted into boschnialactone (**17**) by hydrogenation and subsequent introduction of the carbonyl group.

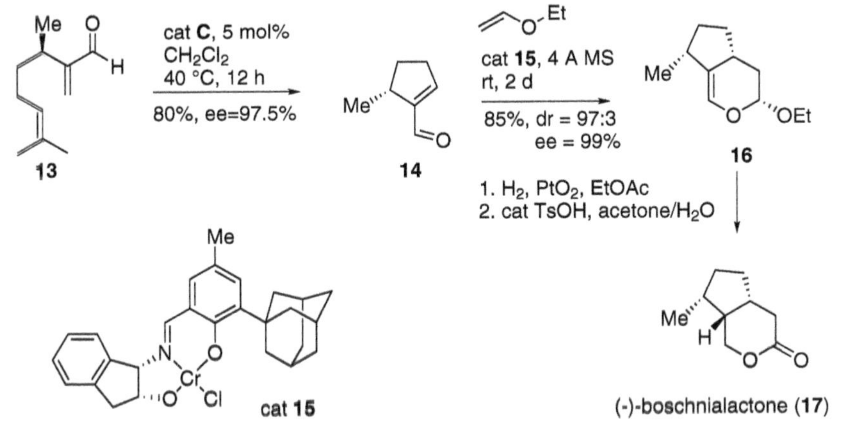

cat **C**, 5 mol%, CH_2Cl_2, 40 °C, 12 h, 80%, ee=97.5%

cat **15**, 4 A MS, rt, 2 d, 85%, dr = 97:3, ee = 99%

1. H_2, PtO_2, EtOAc
2. cat TsOH, acetone/H_2O

13 **14** **16** cat **15** (-)-boschnialactone (**17**)

Scheme 3 RCM of a diene with trisubstituted and *gem*-disubstituted double bond en route to iridoid natural products [23]

In a recent total synthesis of the novel neurotrophic agent merrilactone A (**22**, Scheme 4) by Inoue and Hirama [24], key intermediate **21** with the *cis*-bicyclo[3.3.0]octane framework embedded within the caged pentacycle **22** was elaborated from cyclobutane **18** by a sequence of RCM and immediate cleavage of the resulting bicyclic vicinal diol **19** to *meso*-diketone **20**. Cyclooctenedione **20** then underwent regioselective transannular aldol reaction at low temperature (LHMDS, THF, –100 °C) to produce a 3:1 mixture of isomers in 85% combined yield. The major isomer **21** with the required stereochemistry was then converted into the racemic natural compound (±)-**22** in 19 steps.

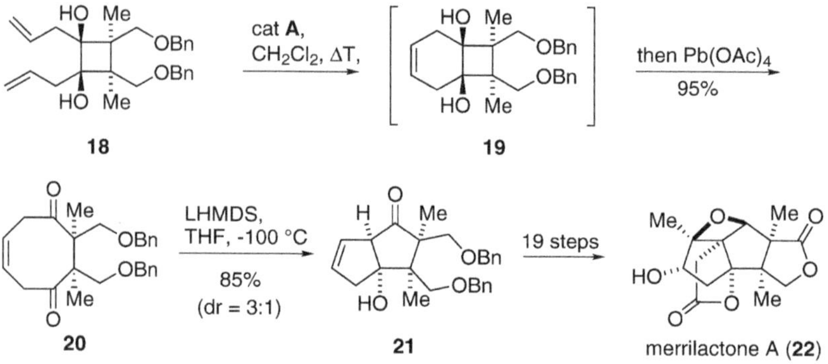

Scheme 4 A sequence of RCM–glycol cleavage–transannular aldol reaction in Inoue's total synthesis of merrilactone A (**22**) [24]

The widespread occurrence and biological significance of polyoxygenated carbocycles provided the impetus to apply RCM to sugar-derived dienes. Carbohydrate carbocyclization based on a sequence of Vasella reductive opening of iodo-substituted methyl glycosides [25], and RCM of the dienes available from the resulting unsaturated aldehydes, were used to prepare a series of natural compounds (Schemes 5–7).

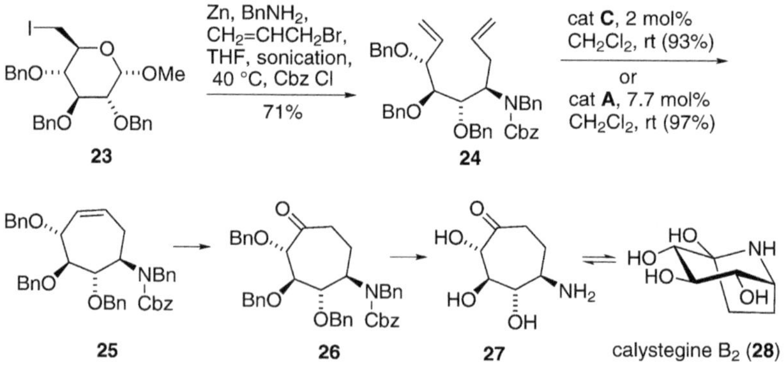

Scheme 5 Carbohydrate carbocyclization in the total synthesis of calystegine B_2 (**28**) [27,28]

Two groups reported independently the synthesis of the potent glucosidase inhibitor calystegine B_2 (**28**), a polyhydroxylated alkaloid with a nortropane ring system (Scheme 5). The RCM precursor **24** was prepared from the iodo-substituted methyl pyranoside **23** by using a zinc-mediated triple domino reaction (ultrasound-accelerated reductive fragmentation of **23** to generate the 5,6-unsaturated aldehyde, trapping of the aldehyde as the benzylimine, and zinc-mediated allylation of the latter) [26]. After Cbz protection, diene **24** was exposed to first-generation catalyst **A** [27] or to the second-generation catalyst **C** [28], to provide in both cases the desired cycloheptene **25** in good yield. The synthesis was then completed by regioselective introduction of the carbonyl group. The 5-amino-cycloheptanone **27** formed in the deprotection step finally cyclized to the bicyclic aminoketal structure of **28**.

(−)-Pentenomycin (**33**), a highly oxygenated cyclopentenoid with a quaternary chiral center (Scheme 6), was prepared by a similar reaction sequence [29]. The RCM precursor **31** was prepared in eight steps from D-mannose via iodo compound **29** and aldehyde **30** (1:1 diastereomeric mixture). RCM of **31** led to the epimeric cyclopentenols **32**.

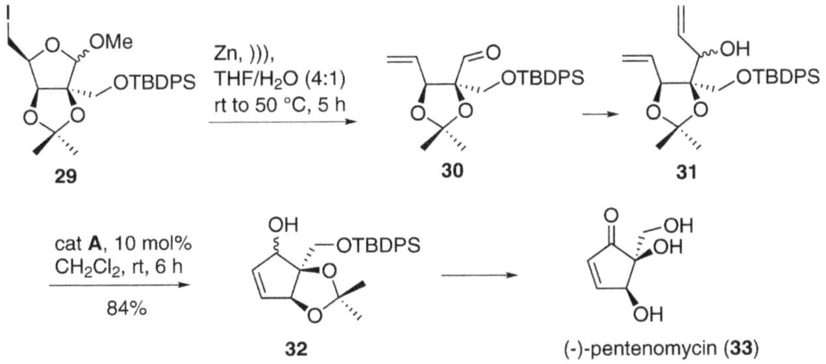

Scheme 6 Synthesis of pentenomycin (**33**) via RCM-based carbocyclization [29]

Recently, a formal total synthesis of the antitumor agent and glycogen synthase kinase-3β-inhibiting alkaloid (−)-agelastatin A (**38**) was disclosed by a British team (Scheme 7) [30]. The highly functionalized diene **35** was prepared from iodo compound **34** via Vasella-type reductive ring opening [25], followed by Julia–Kocienski methylenation of the resulting aldehyde. The ring closure to cyclopentene **36** in the presence of Hoveyda's highly active ruthenium catalyst **D** proceeded smoothly (benzene, ΔT, 14 h), notwithstanding the presence of the urethane and sulfonamido groups in **35**. (Due to difficulties in removing a tetrazole by-product formed in the Julia olefination, only the overall yield for the transformation to bicycle **37** was given).

Solanoeclepin A (**39**), a natural hatching agent of potato cyst nematodes, possesses a seven-membered ring in a complex pentacyclic framework. Hiemstra and coworkers achieved the synthesis of several analogs **42** containing the

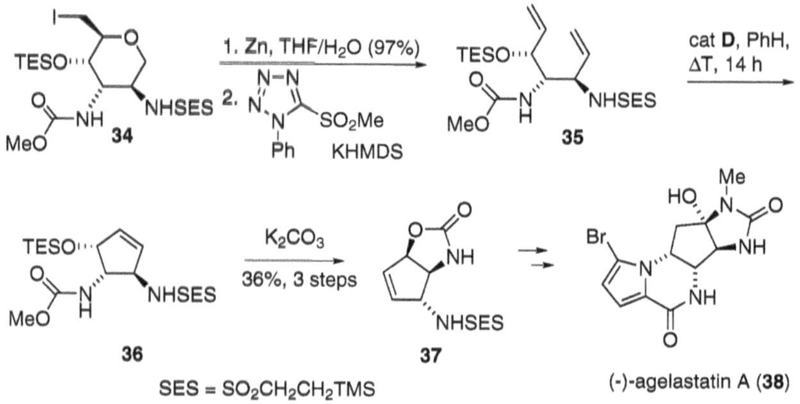

Scheme 7 Cyclopentene ring closure in the presence of urethane and sulfonamido groups in total synthesis of agelastatin (**38**) [30]

enantiopure tetracyclic left-hand substructure of **39** (Scheme 8) [31]. When the cyclization experiments on triene **40a** were performed with Grubbs' catalyst **A**, the cycloheptadiene-forming process to **41a** was very slow requiring a stoichiometric amount of the catalyst for completion. The use of the more reactive second-generation catalyst **E**, however, provided the tetracyclic diene **41a**

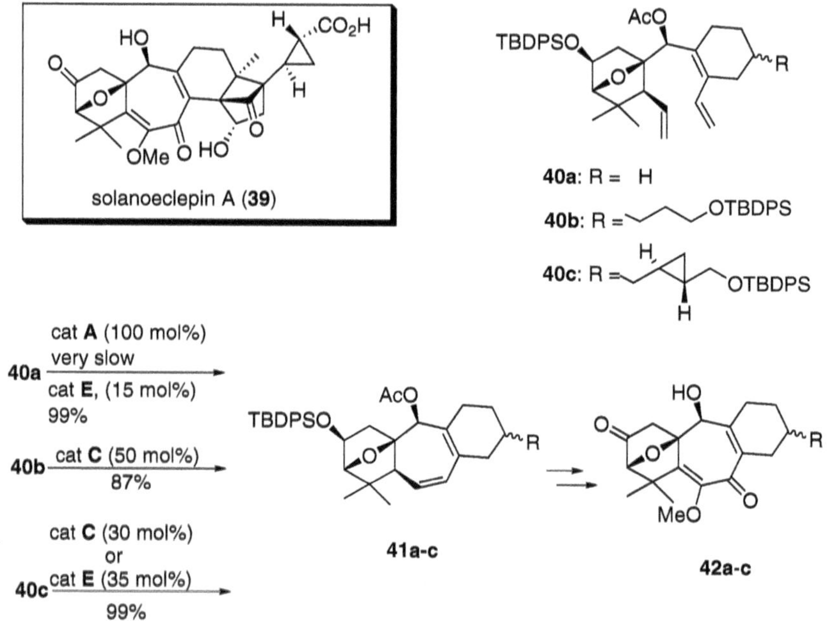

Scheme 8 Efficient formation of the conjugated cycloheptadiene core in tetracyclic compounds **42** during studies toward solanoeclepin A (**39**) [31]

with only 15 mol% of **E** after 16 h in refluxing toluene. The more elaborate precursors **40b** and **40c** reacted sluggishly, but the use of 0.35–0.50 equivalents of catalysts **C** or **E** led to cyclization products **41b** and **41c** in high yield.

2.1.1.2
Cyclic Ethers

5,6-Dihydro-2*H*-pyran-2-ones (α-pyrones **II**) and dihydropyrans (**VIII**) are present in a large number of biologically active natural products. Both types of compounds are now routinely prepared by RCM, either via path A or via path B in Fig. 6. Additionally, chiral lactones **II** can be used to induce stereospecificity to the neighboring carbons via substrate-controlled reactions, as illustrated by the transformation **II**→**IV** or **II**→**V** in the scheme. While the formation of pentenolides by RCM of acrylates (path A) mediated by Grubbs' first-generation catalyst **A** often proceeded sluggishly, needing $Ti(O\textit{i}Pr)_4$ as an additive in many cases, the ring closure occurs generally without problems in the presence of second-generation catalysts **C**, **D**, and **E**. Disubstituted dihydropyrans of the type **VIII** are prepared preferably via path B, by RCM of mixed acrolein acetals **VI**, rather than via the corresponding lactones **II**, as the former cyclize uneventfully with catalyst **A**.

The difference in reactivity is perfectly revealed in Metz's total synthesis of the molluscicidal furanosesquiterpene lactones ricciocarpin A (**50**) and B (**51**) (Scheme 9) [32]. Attempts to convert acrylate **43** to lactone **44** using Grubbs' catalyst **A** or Schrock's molybdenum catalyst **B** resulted in very low yields of the

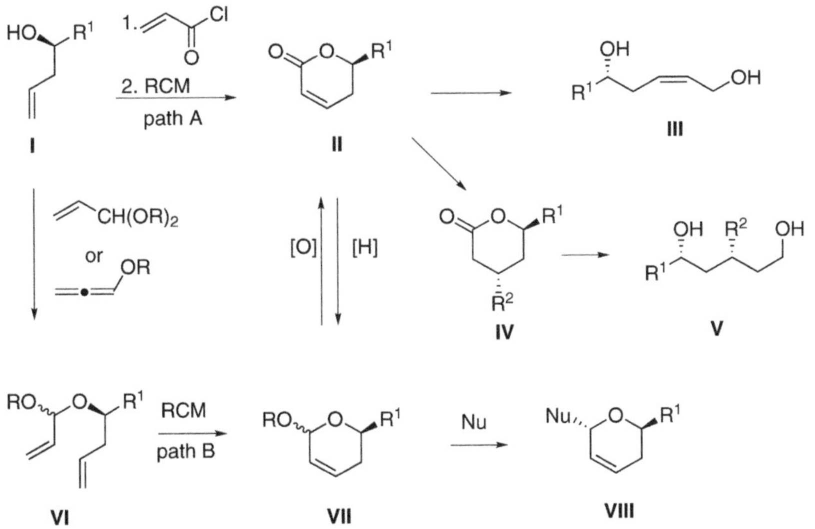

Fig. 6 RCM-based formation and synthetic potential of dihydropyrans **VIII** and α-pyrones **II**

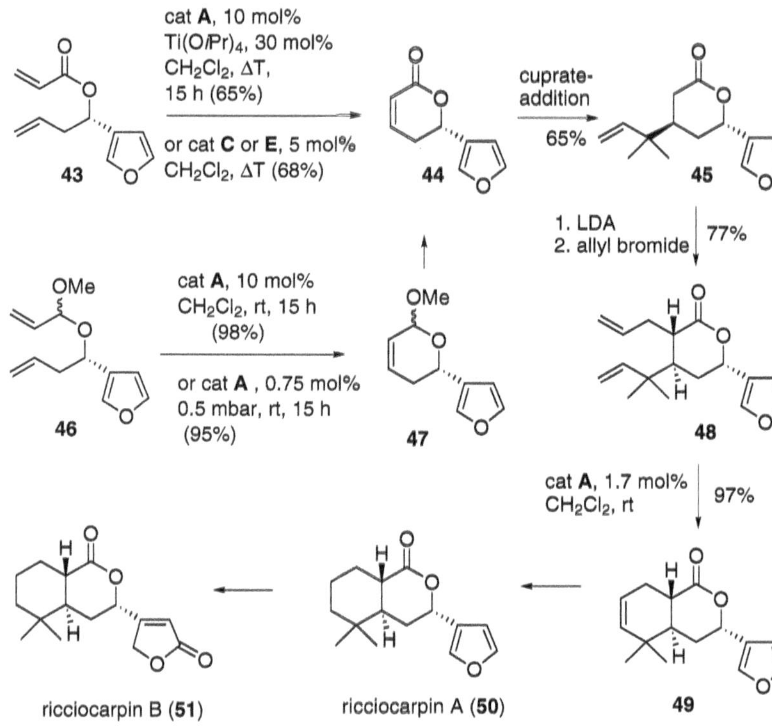

Scheme 9 Comparison of substrate reactivity and catalyst activity in total synthesis of ricciocarpin A (**50**) and B (**51**) [32]

desired product. With $Ti(OiPr)_4$ as an additive, which is suggested to reduce deactivation of catalytic intermediates by the Lewis basic carbonyl oxygen [33], the yield was improved to 65%. With second-generation catalysts **C** and **E** the catalyst loading could be reduced, and the reaction led – without additive – to comparable yields of **44**. RCM of mixed acetal **46** [34] proceeded smoothly with catalyst **A** at room temperature leading to dihydropyran **47** in almost quantitative yield. The best results, however, were obtained when the reaction was performed without any solvent using only 0.75 mol% of catalyst **A** at reduced pressure, which guaranteed the efficient removal of ethylene produced during metathesis. The conversion of lactone **44** to the natural compounds was continued by sequential *trans*-selective conjugate addition of a cuprate and α-allylation of intermediate **45**. The resulting diene **48** was subjected to another high-yielding RCM reaction to **49**, which finally was transformed into **50** and **51**.

The unique power of Hoveyda's recyclable ruthenium catalyst **D** in RCM with electron-deficient and sterically demanding substrates is illustrated in Honda's total synthesis of the simple marine lactone (–)-malyngolide (**54**), which contains a chiral quaternary carbon center (Scheme 10) [35]. Attempted RCM of diene **52** with 5 mol% of NHC catalyst **C** for 15 h produced the desired

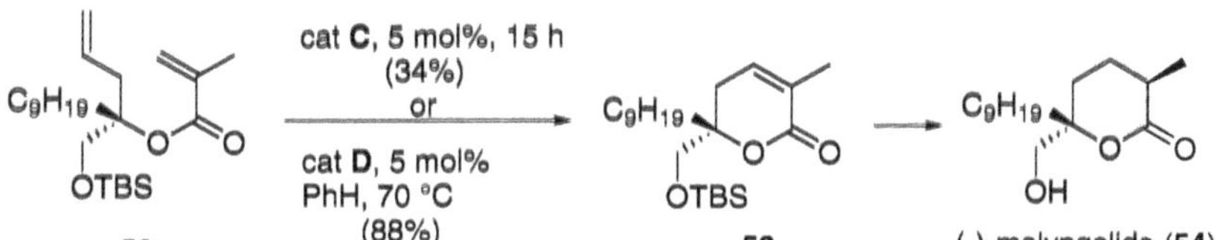

Scheme 10 The power of Hoveyda's catalyst **D** in total synthesis of malyngolide (**54**) [35]

product **53** only in 34% yield. When 5 mol% of catalyst **D** was used, the yield was improved to 88%.

RCM was also one of the key steps in many other total or partial syntheses of natural products with a δ-lactone moiety. The α-pyrone moiety of the potent antitumor agent (+)-fostriecin has been closed in Hatakeyama's total synthesis [36] and in Cossy's synthesis of an advanced intermediate [37]. Also the cytotoxic styryllactone (+)-goniodiol [38] and the plant metabolite (+)-boronolide [39] were prepared via ruthenium-catalyzed RCM. RCM-based synthesis of some lactones with the proposed structures of passifloricin A [40], and a recent total synthesis [41] have led to a correction of the structure of the natural compound. Syntheses of an advanced fragment of the microbial metabolite and dimeric polyketide SCH 351448 [42], of the saturated lactone moiety of the potent HMG-Co A reductase inhibitors compactin and mevinolin [43], and Ghosh's recent synthesis of the highly functionalized C1–C9 segment of the novel microtubule-stabilizing agent peloruside A [44] are examples of the additional introduction of stereocenters to the lactone after the RCM step.

The utility of RCM methodology for the synthesis of open-chain building blocks from α,β-unsaturated δ-lactones is exemplified by the partial syntheses of Cossy aimed for (+)-methynolide (the aglycon of the methymicin family of macrolide antibiotics) [45], and the anticancer agent discodermolide [46], as well as during a recent total synthesis of the highly cytotoxic marine natural depsipeptide apratoxin A by Forsyth and Chen [47].

The orally active antifungal agent ambruticin S (**59**), that exhibits activity against a variety of pathogenic fungi, has attracted intense synthetic interest. In two of the more recent total syntheses the 2,6-*cis*-disubstituted tetrahydropyran unit in **59** was prepared by RCM (Scheme 11). In Martin's synthesis [48], secondary alcohol **55** was used as the metathesis substrate, and the reaction led to ketone **56** in 60% yield after TPAP oxidation (substrate concentration, catalyst loading, and reaction time were not given). In Lee's work [49], the ring closure of diene **57** was effected under high dilution with catalyst **A**, leading to cyclization product **58** in 98% yield, when $Pb(OAc)_4$ was added to the reaction mixture before workup to remove traces of ruthenium and phosphine by-products derived from the catalyst [50].

The novel marine natural product laulimalide (**65**), a metabolite of various sponges, has received attention as a potential antitumor agent due to its "taxol-like" ability to stabilize microtubules. There has been considerable synthetic effort toward **65**, culminating within not more than 2 years in as many as ten

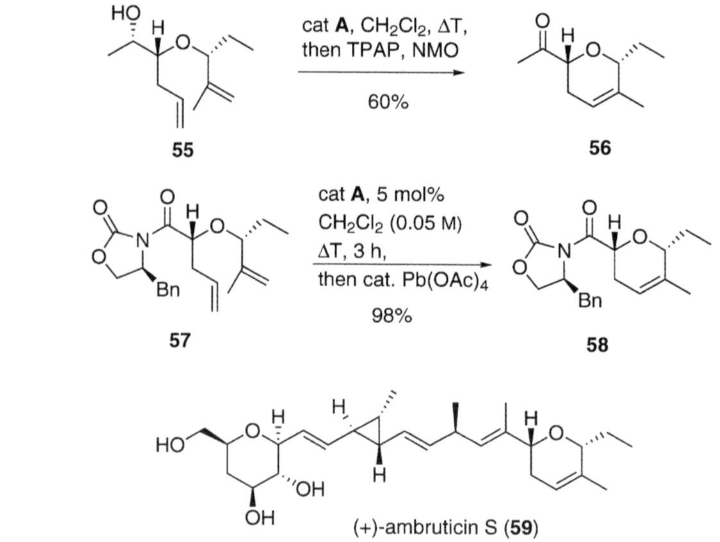

Scheme 11 RCM-based synthesis of dihydropyran fragments **56** [48] and **58** [49] in total synthesis of the antifungal agent ambruticin S (**59**)

total syntheses by seven groups and numerous fragment syntheses [51]. Both, the exocyclic and the inner 2,6-*trans*-disubstituted dihydropyran unit in **65** have been prepared by RCM [52], and it was shown that the ring closure can also be performed chemoselectively in the presence of additional double bonds leading to the advanced intermediates **60–64** depicted in Fig. 7. Intermediate **64**

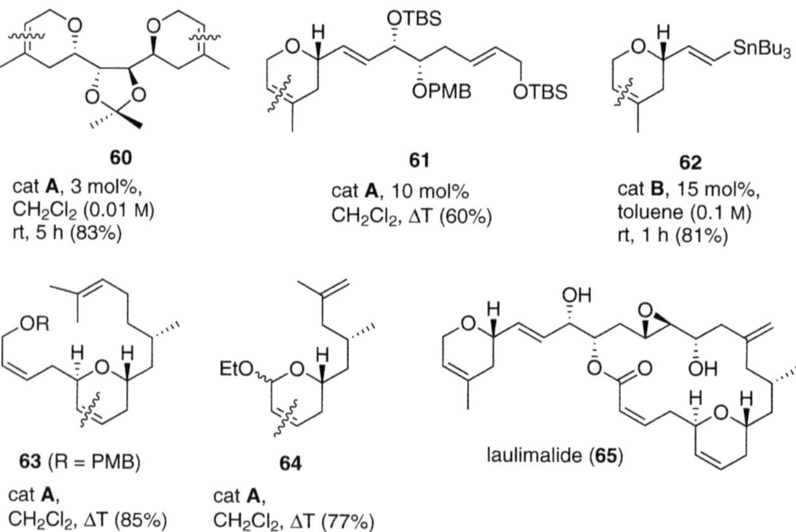

Fig. 7 Advanced dihydropyran fragments used in total syntheses of laulimalide (**65**)

was prepared by twodirectional RCM under high dilution from the corresponding D-mannose-derived tetraene, and served as an efficient precursor for the volatile dihydropyran carboxaldehyde [53]. The RCM reactions leading to dihydropyrans **60, 61** [54], **63** [55], and **64** [56] were all performed with first-generation Grubbs' catalyst **A**, while the tributylstannyl-substituted dihydropyran **62** [57] was prepared with Schrock's molybdenum catalyst **B**.

The first total synthesis of the marine dolabellane diterpene (+)-4,5-deoxyneodolabelline (**70**) was accomplished by D. R. Williams et al. [58]. The *trans*-disubstituted dihydropyran moiety in key intermediate **69** was efficiently prepared from mixed acetal **66** by RCM with second-generation catalyst **C** and subsequent Lewis acid-catalyzed allylation of ethyl glycosides **67** with allylsilane **68** (Scheme 12) [59].

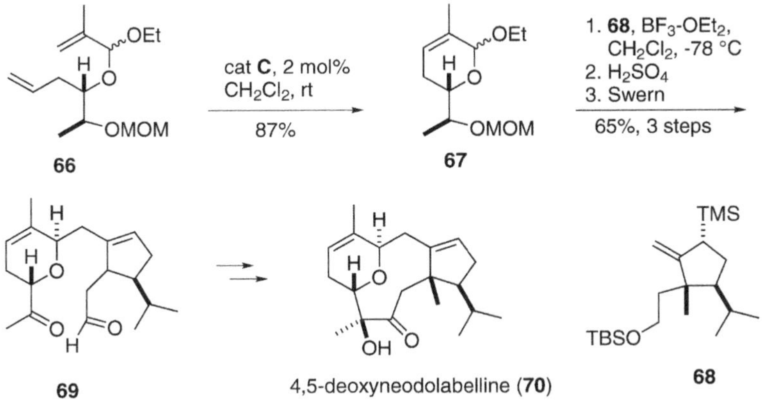

Scheme 12 Stereoselective synthesis of main fragment **69** by sequential RCM and allylation in total synthesis of 4,5-deoxyneodolabelline (**70**) [58]

The formation of a highly complex 2,5-disubstituted dihydropyran by RCM was one of the key steps in Snapper's total synthesis of the cytotoxic marine natural product (+)-cacospongionolide B (**74**) (Scheme 13) [60]. Despite the use of second-generation catalyst **C**, RCM of triene **71** proceeded regioselectively to produce only the dihydropyran ring, leading to compound **72** in 91% yield. The first total synthesis of **74** was then completed in four steps, by selective reduction of the *exo* methylene group, followed by introduction of the methylene group at C4 and photooxygenation of the furan ring in intermediate **73**.

In the first convergent total synthesis of the marine neurotoxin hemibrevetoxin B (**77**) [61], dienetriol **75** was used as an RCM substrate for elaboration of the seven-membered A ring in **77** (Scheme 14). While first-generation catalyst **A** was ineffective in this case, the ring closure occurred smoothly with second-generation catalyst **C**, providing the tetracyclic intermediate **76** in high yield.

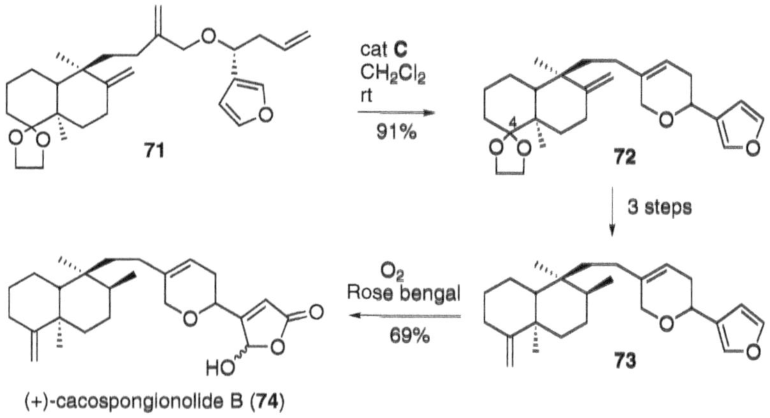

Scheme 13 RCM-based formation of the advanced dihydropyran fragment **72** in the first total synthesis of cacospongionolide B (**74**) [60]

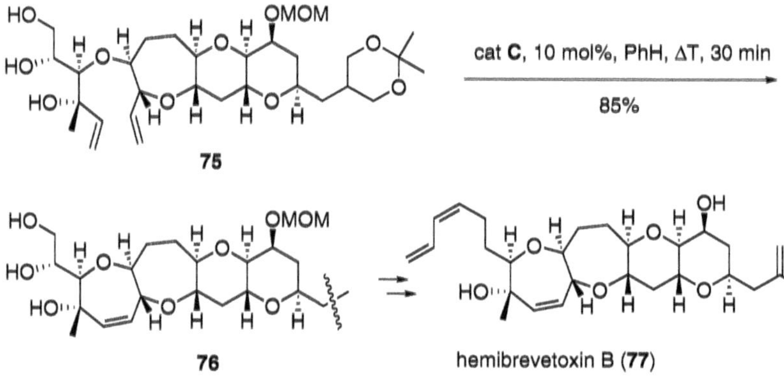

Scheme 14 Efficient formation of the seven-membered A ring from unprotected triol **75** in Holton's total synthesis of hemibrevetoxin B (**77**) [61]

RCM was also used in Yamamoto's total synthesis of the marine neurotoxin gambierol (**81**) [62], to close the central seven-membered E ring, thereby completing the octacyclic polyether core **80** (Scheme 15). Following previously developed methodology [63], metathesis precursor **79** was produced as the major epimer, by boron trifluoride etherate-mediated intramolecular allylation of α-chloroacetoxy ether **78**. Subsequent treatment of **79** with catalyst **C** produced the octacyclic ether **80** in 88% yield.

Following a previously developed strategy in the group for the synthesis of medium-sized rings containing a 1,3-*cis,cis*-diene unit [64], Denmark disclosed the first total synthesis of the nine-membered cyclic ether (+)-brasilenyne (**85**), through a sequence of RCM with formation of a six-membered cyclic silyl ether followed by silicon-assisted intramolecular cross coupling (Scheme 16) [65]. When RCM precursor **82** was subjected to Schrock's catalyst **B**, compound **83**

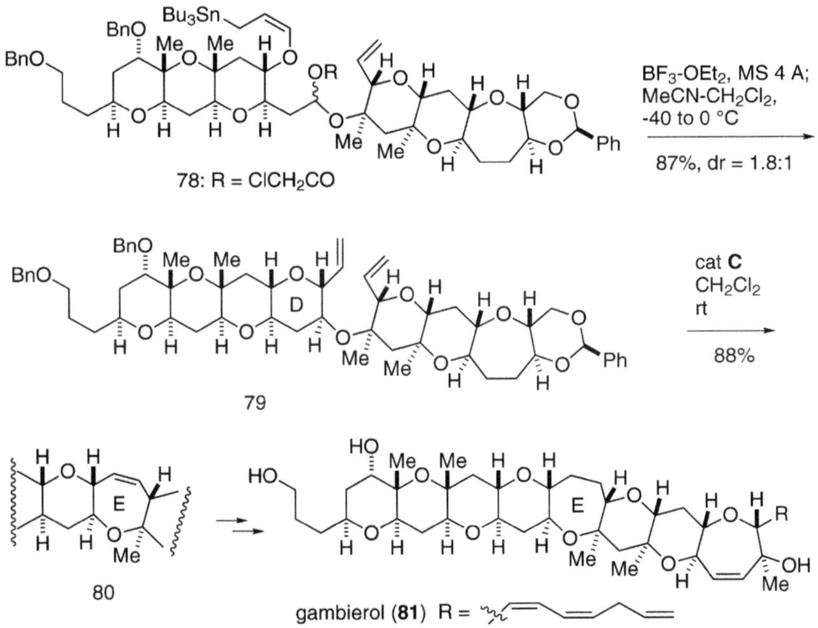

Scheme 15 Formation of the central rings D and E by sequential intramolecular allylation and RCM in Yamamoto's total synthesis of the marine neurotoxin gambierol (**81**) [62]

with the silicon-based temporary linker was formed in 92% yield. The intramolecular cross coupling leading to the nine-membered ring was then carried out with $[allylPdCl]_2$ as the catalyst and TBAF as the activator, which led to intermediate **84** in 61% yield. (For the total synthesis of various other medium-sized ring ethers of marine origin by direct RCM methods, see Schemes 31–35).

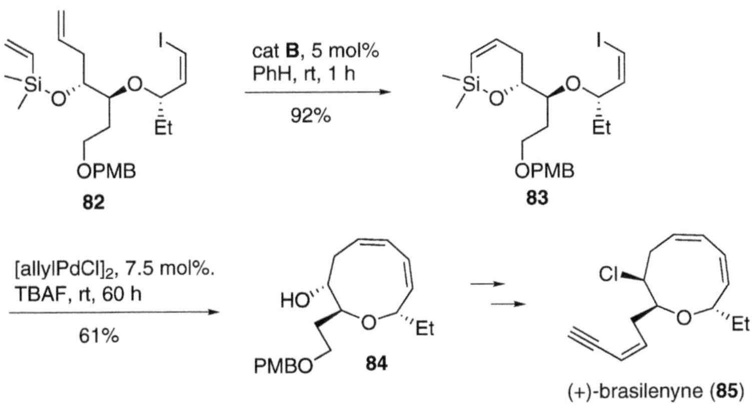

Scheme 16 Construction of a nine-membered cyclic ether with a (*Z*,*Z*)-1,3-diene unit by sequential RCM and silicon-assisted intramolecular cross coupling in Denmark's synthesis of brasilenyne (**85**) [65]

2.1.1.3
Alkaloids

The marine natural product dynosin A (**92**) is a new member of the aeruginosin family and a novel inhibitor of thrombin and Factor VIIa. In Hanessian's total synthesis of **92** [66], both the dihydroxyoctahydroindole **88** and the Δ3 pyrroline moiety **91** were prepared by RCM-based routes (Scheme 17).

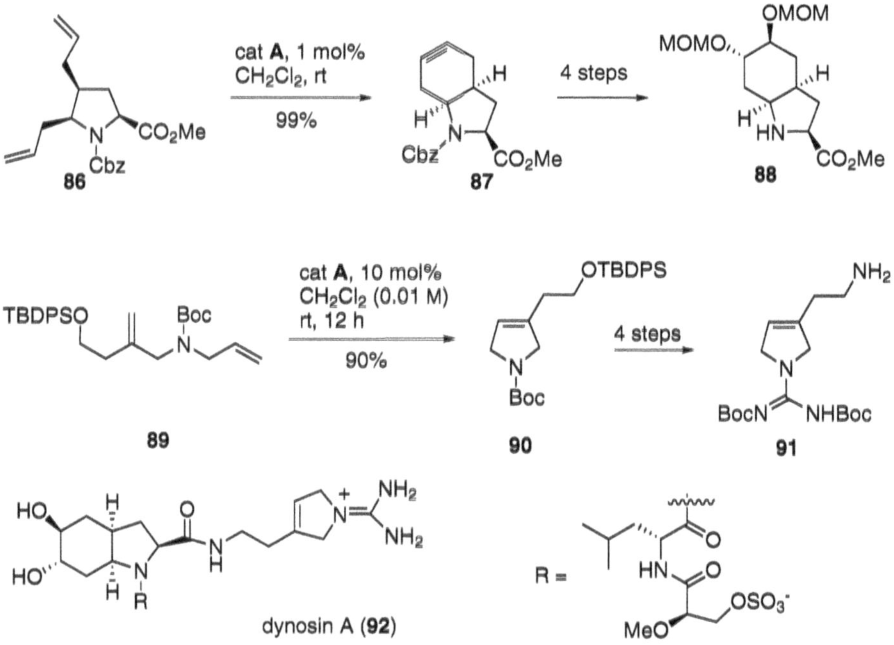

Scheme 17 RCM-based synthesis of octahydroindole **87** and pyrroline **90** in the first total synthesis of dynosin A (**92**) [66]

The phenanthroindolizidine alkaloid (–)-antofine (**95**) exhibits high cytotoxicity to drug-sensitive and multidrug-resistant cancer cells by arresting the G2/M phase of the cell cycle. In the first asymmetric total synthesis of (–)-**95**, the late-stage construction of pyrrolidine **94** for the final Pictet–Spengler cyclomethylenation to **95** was performed by RCM and subsequent hydrogenation (Scheme 18) [67].

A concise total synthesis of the indole alkaloid dihydrocorynantheol (**101**) (Scheme 19), that features two RCM steps and a zirconocene-catalyzed carbomagnesation [68], is a further example of Martin's interest in applying RCM as a key reaction for the construction of alkaloid frameworks [69]. The first RCM step was applied to bis-allyl amide **96**. The resulting intermediate **97** was directly subjected to carbomagnesation and subsequent elimination to deliver **98** in 71% yield from **96**. Amide **98** was then transformed into acrylamide **99** in

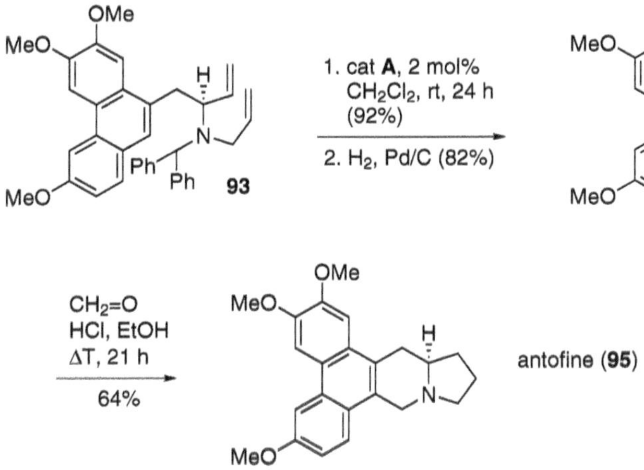

OMe MeO H N Ph Ph MeO 93 1. cat A, 2 mol% CH_2Cl_2, rt, 24 h (92%) 2. H_2, Pd/C (82%) OMe MeO H HN MeO 94

CH_2=O HCl, EtOH ΔT, 21 h 64% OMe MeO H N MeO antofine (95)

Scheme 18 Synthesis of the phenanthroindolizine alkaloid antofine (95) by a sequence of RCM, hydrogenation, and Pictet–Spengler cyclomethylenation [67]

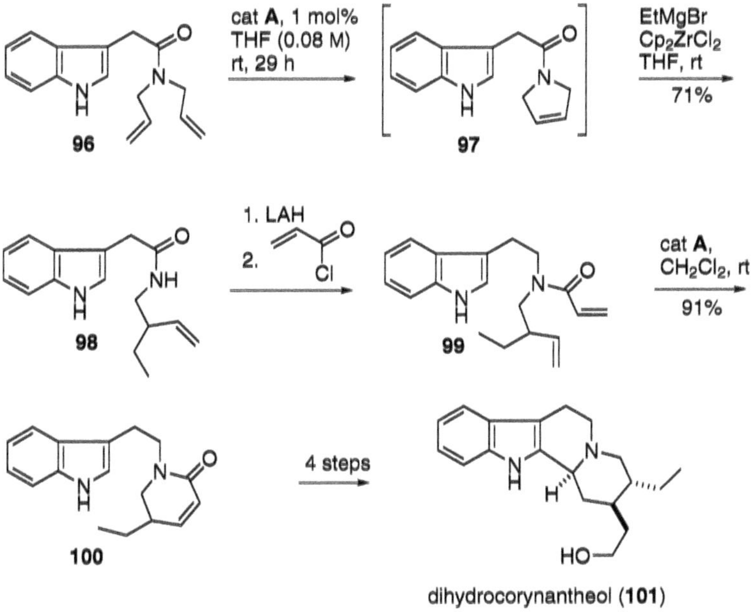

O N N H 96 cat A, 1 mol% THF (0.08 M) rt, 29 h O N N H 97 EtMgBr Cp_2ZrCl_2 THF, rt 71%

O NH N H 98 1. LAH 2. O Cl N O N H 99 cat A, CH_2Cl_2, rt 91%

N O N H 100 4 steps N N H H HO dihydrocorynantheol (101)

Scheme 19 Construction of key lactam 100 via two RCM steps and a zirconocene-mediated carbomagnesation in Martin's total synthesis of dihydrocorynantheol (101) [68]

two steps. RCM of **99** furnished lactam **100** in 91% yield, which was converted into racemic **101** in four additional steps.

Also, a novel RCM-based approach to the 6-aza[3.2.1]bicyclooct-3-ene **103**, and hence a formal total synthesis of the antitumor antibiotic (–)-peduncularine (**104**) (Scheme 20), was recently disclosed by Martin's group [70]. Initial ex-

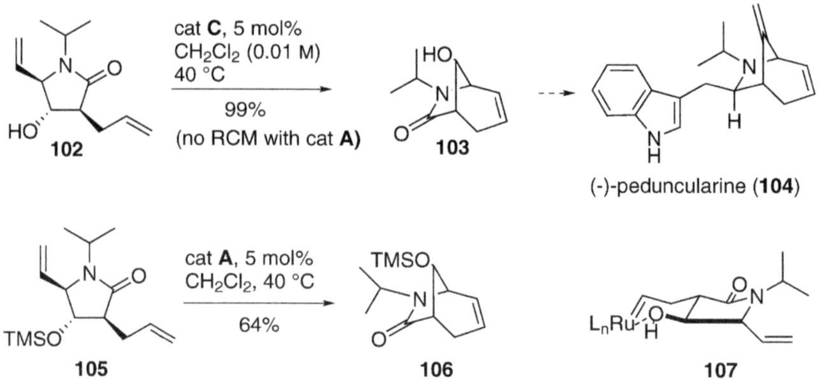

Scheme 20 RCM-based construction of key intermediate **103** in a formal total synthesis of the antitumor antibiotic peduncularine (**104**) [70]

periments to effect the RCM of alcohol **102** with first-generation catalyst **A** were not successful (possibly by formation of unreactive intermediate **107**), while TMS ether **105** led to cyclization product **106** in 64% yield. When alcohol **106** was treated with second-generation catalyst **C**, the ring closure proceeded smoothly leading to **103** in nearly quantitative yield.

A further example of the rapid progress in Ru-catalyzed metathesis reactions is Wipf's total synthesis of (–)-tuberostemonine (**113**) [71]. This complex polycycle belongs to the family of *Stemona* alkaloids, which cover a broad range of biological activities including applications in Eastern folk medicine against pulmonary tuberculosis and bronchitis. The first total synthesis of **113** (Scheme 21) highlights the threefold use of Ru catalysts, first by an azepine ring-closing step and, in the endgame of the synthesis, by a Ru-catalyzed allyl to 1-propenyl isomerization/Ru-catalyzed cross metathesis (CM) sequence leading to a propenyl–vinyl interchange. When key intermediate **108** was exposed to 2–5% of NHC catalyst **C**, tricyclic azepine **109** was smoothly formed in high yield. After stereoselective elaboration of the complete pentacyclic skeleton, the allyl substituent in the advanced intermediate **110** was isomerized using a modification of a method developed by Roy et al. for allyl ethers [72]. Thus, heating a solution of **110** in toluene in the presence of catalyst **C**, allyl tritylamine, and diisopropylethylamine led to the 1-propenyl-substituted isomer **111** in high yield. Subsequent CM of **111** with ethylene in the presence of Hoveyda's catalyst **D** and TsOH gave access to the desired vinyl group in **112**, which was hydrogenated to provide (–)-**113**.

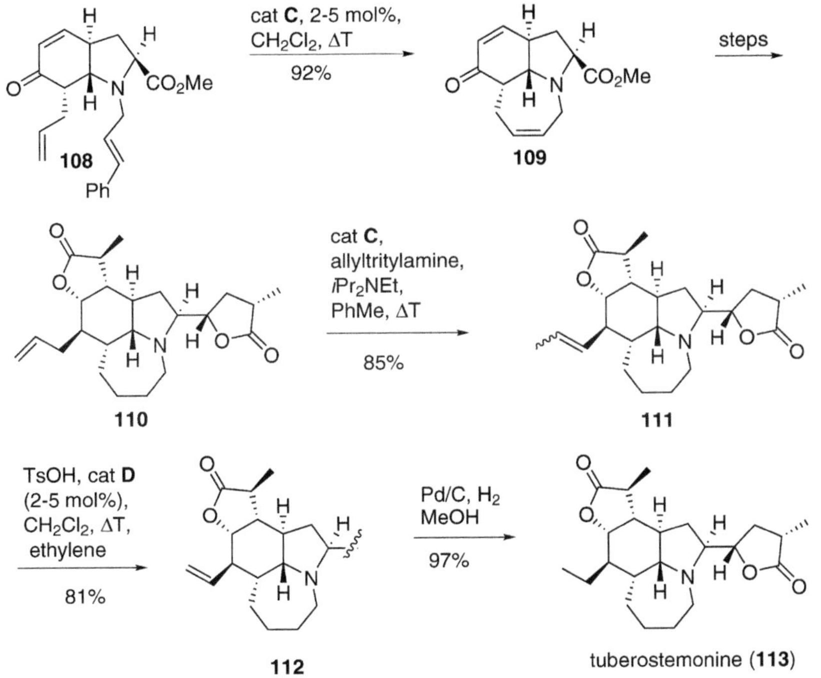

Scheme 21 Threefold use of ruthenium catalysts in the first total synthesis of *Stemona* alkaloid tuberostemonine (**113**) [71]

2.1.2
Formation of Medium-Sized Rings

Because of enthalpic (increasing strain in the transition state) and entropic influence (probability of the chain ends meeting), medium-sized rings are the most difficult to prepare. Additionally, the formation of medium-sized rings by RCM may pose considerable challenges as, due to the inherent ring strain, 8–11-membered cycloalkenes are prone to the reverse process, that is, to ROM or ring-opening metathesis polymerization (ROMP) sequences. One approach to circumvent this problem is to incorporate control elements (cyclic conformational constraints by preexisting rings or acyclic constraints by the substitution pattern of the cyclization precursor) that force the cyclization substrate to adopt a conformation suitable for ring closure. These constraints will facilitate RCM and stabilize the product against the competing ROMP pathway. While up to early 2000 only a limited number of successful RCM reactions for the synthesis of natural products with medium-sized rings were reported [73], the number has rapidly increased during the last 3 years [74]. Most importantly, we will see that in some cases the stereochemical outcome of the reactions could also be mediated by the choice of the catalyst, which is deemed to reflect kinetic versus thermodynamic control of the cyclization reaction.

2.1.2.1
Carbocycles

The importance of conformational restriction for the ring-closing reaction is nicely demonstrated during Paquette's concise total synthesis of natural (+)-asteriscanolide (**116**) [75], whose framework consists of a rather uncommon bicyclo[6.3.0]octane ring system bridged by a butyrolactone fragment (Scheme 22). Despite the presence of a conjugated diene unit and a *gem*-disubstituted double bond in precursor triene **114**, the cyclooctene ring in **115** was formed in high yield (93%, based on recovered starting material) when a total of 30 mol% of Ru catalyst **A** was sequentially added within 48 h to a boiling solution of **114** in dichloromethane.

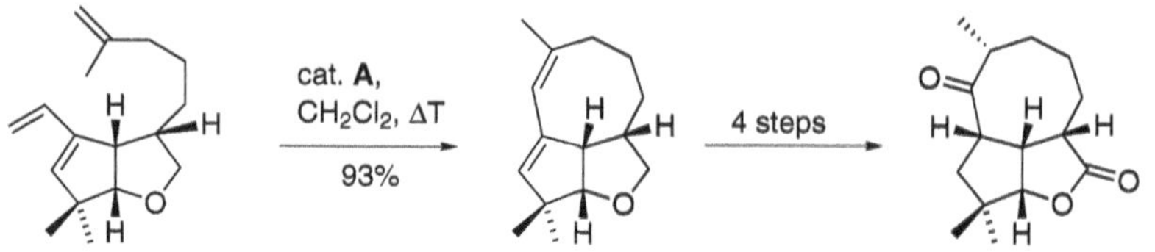

cat. **A**, CH_2Cl_2, ΔT, 93%; 4 steps

114 **115** (+)-asteriscanolide (**116**)

Scheme 22 Efficient cyclooctene-forming diene-ene RCM, facilitated by cyclic constraint in Paquette's total synthesis of asteriscanolide (**116**) [75]

An illustrative example of the potency of the second-generation Ru catalyst **C** is found in Paquette's highly efficient total synthesis of the natural products teubrevin G (**122**) and teubrevin H (**123**), which feature a cyclooctane core fused and spiroannulated to smaller oxygen-containing rings [76]. In the retrosynthetic analysis, the viability of an RCM step for annulation of a cyclooctenone ring to the furan played a central role.

Despite the presence of a conformational constraint by the furan ring in the cyclization substrate, only poor results were obtained when Grubbs' first-generation Ru catalyst **A** was examined to effect the ring closure of TES ether **117b** and ketone **117c** (Scheme 23). Using very high catalyst loading and reaction times up to 1 week in boiling dichloromethane produced the desired RCM products **120** and **121** only in low yield (53 and 35%, respectively). In the case of allylic alcohol **117a**, catalyst **A** provided no cyclization product (**119**) at all, leading instead to ethyl ketone **118a** (68%) and methyl ketone **118b** (20%) as the sole reaction products. When the cyclization was performed in the presence of 10 mol% of Ru catalyst **C**, the RCM reaction of allylic alcohol **117a** and vinyl ketone **117c** proceeded smoothly within several hours to furnish alcohol **119** and ketone **121** in high yield. Cyclooctenone **121** was then successfully converted to **122** and **123**.

En route to a planned total synthesis of the phytotoxic natural compound cornexistin (**128**), Stephen Clark recently reported the first example of the

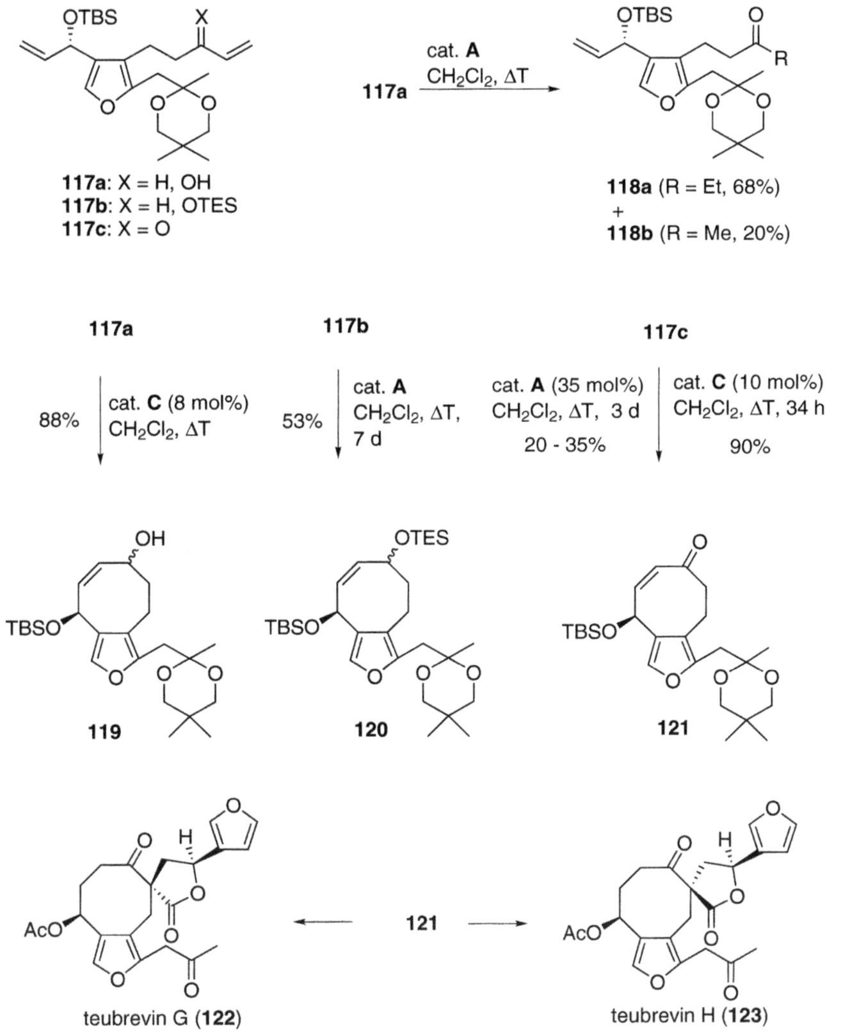

Scheme 23 Improved formation of cyclooctene ring with second-generation catalyst **C** in Paquette's total synthesis of teubrevins G (**122**) and H (**123**) [76]

direct construction of a nine-membered carbocycle, using a novel sequence of Pd-catalyzed fragment coupling followed by RCM (Scheme 24) [77]. With Grubbs' second-generation catalyst **C** in toluene at 80 °C, the RCM of precursor **124** (1:1 mixture of diastereomers) was complete within 3 h leading to isomers **125** in 61% yield. When the cyclization was performed with catalyst **A**, the reaction was only complete after 3 days. However, when the RCM reaction was attempted with model compound **126** containing a *gem*-disubstituted double bond in conjugation to the furan ring, both catalysts failed to provide the ring closure to **127**.

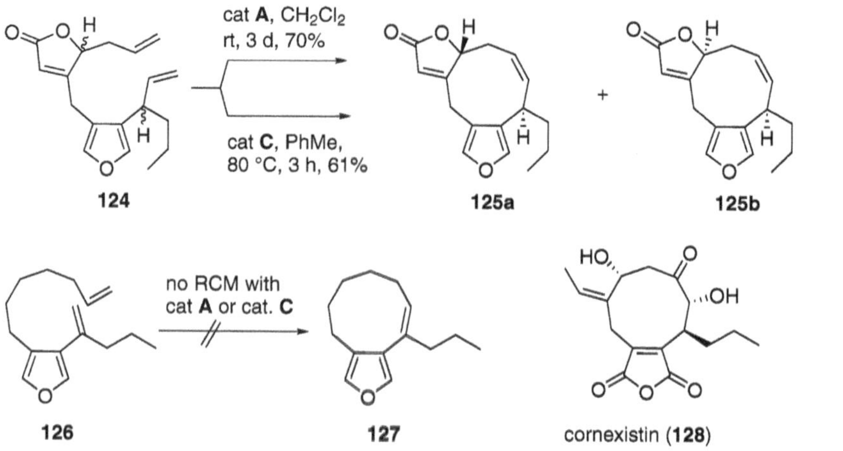

cat **A**, CH_2Cl_2 rt, 3 d, 70%

cat **C**, PhMe, 80 °C, 3 h, 61%

124 **125a** + **125b**

no RCM with cat **A** or cat. **C**

126 **127** cornexistin (**128**)

Scheme 24 First example of RCM-based construction of a nine-membered carbocycle by altering the key disconnection in cornexistin (**128**)-directed work [77]

2.1.2.2 Alkaloids

In 1999, a total synthesis of ircinal A (**129**) and hence a formal synthesis of the potent antitumor agent manzamine A (**130**) was disclosed by the team of Stephen Martin (Scheme 25) [78]. Two RCM reactions were exploited to elaborate sequentially the requisite 13- and 8-membered rings. When triene **131** (0.005 M in dichloromethane) was exposed to Ru catalyst **A** (13 mol%), a facile and regioselective RCM reaction occurred to furnish a mixture (*Z*/*E*=ca. 8:1) of geometric isomers from which the major isomer **132** was isolated in 67% yield. In contrast to a previous observation [79], protonation of the tertiary amine prior to the metathesis reaction was not necessary in this case. Hydrolytic removal of the cyclic carbamate in **132** followed by acylation led to the precursor **133** for the second RCM reaction. However, the formation of the eight-membered lactam was problematic, leading to the desired reaction product **134** in only 26% yield despite the use of as much as 1.1 equivalents of catalyst **A**. In the subsequent full account from 2002, additional details concerning the RCM steps were revealed. In initial experiments, it had been shown that model compound **135** underwent smooth ring closure with Schrock's molybdenum catalyst **B** to provide the tetracyclic product **138**. It was also attempted to effect double RCM to construct the pentacyclic skeleton in a single operation. However, compound **137** reacted only to form tetracycle **140** with a 15-membered ring. Also, compound **136** underwent rapid ring closure in the presence of catalyst **A**, leading to **139** as a mixture (ca. 1:1) of *E*/*Z* isomers. The inability to effect double RCM made it necessary to elaborate the 8- and 13-membered rings in a serial fashion.

Nakadomarin A ((–)-**145**) is a marine natural product with a unique hexacyclic structure (Scheme 26). Recently, the first total synthesis of its enantiomer

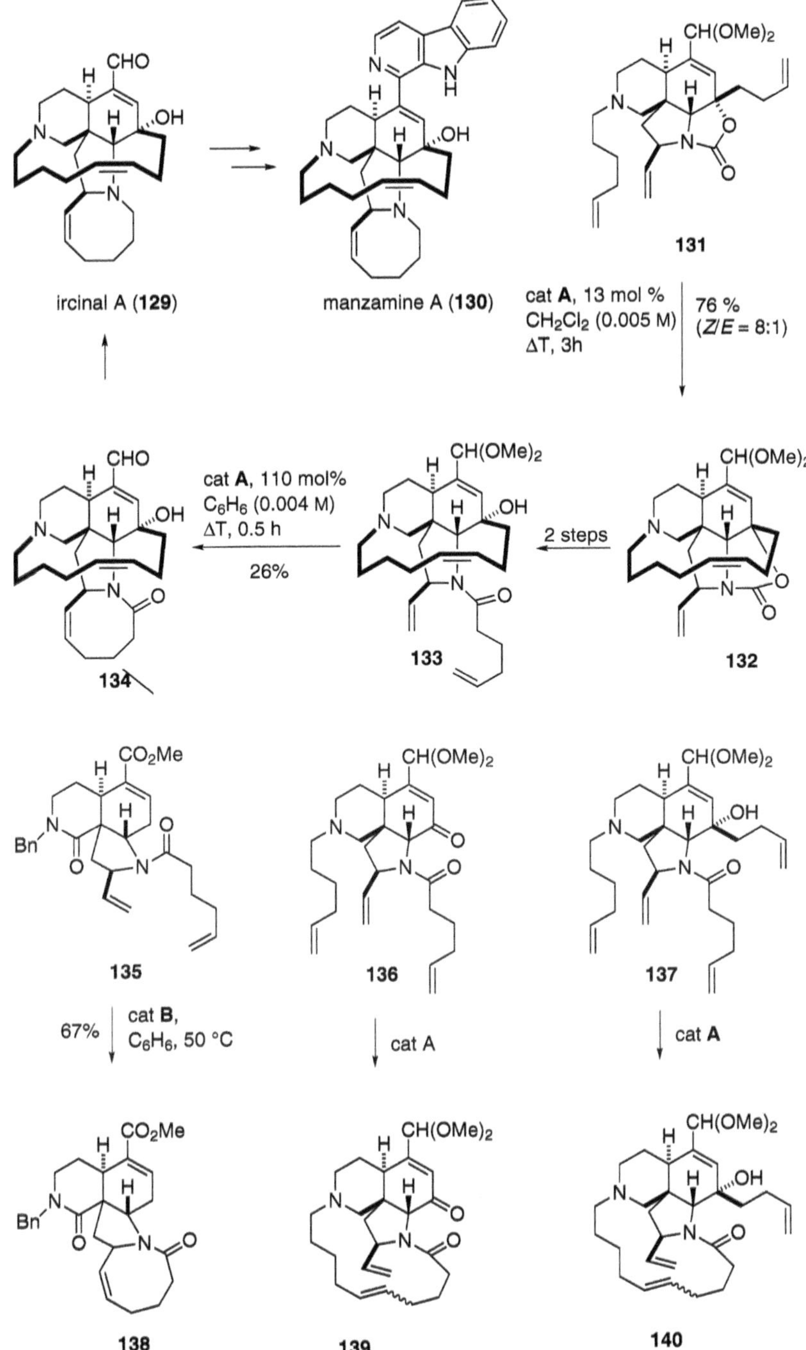

Scheme 25 Sequential formation of 13- and 8-membered azacycles in Martin's total synthesis of ircinal A (**129**) and related manzamine alkaloids [78]

(+)-**145** was achieved, which also features two sequential RCM reactions to form the 8- and 15-membered azacycles [80]. However, compared with the above synthesis of ircinal A, the order of ring-closing steps was reversed. When diene **141** (0.002 M in dichloromethane) was exposed to Grubbs' catalyst **C**, a facile RCM reaction ensued leading within 1.5 h to azocine lactam **142** in 70% yield. Note that when **141** was exposed to catalyst **A**, **142** was obtained in only 15% yield after 48 h with recovery of **141** (36%), underlining again the high potential of the second-generation Ru catalysts. Pentacyclic compound **142** was then elaborated to diene **143** in five steps. The second RCM reaction to close the 15-membered lactam was performed with Ru catalyst **A** and delivered a mixture ($Z/E \approx 2{:}3$) of isomers, from which the desired minor isomer (*Z*)-**144** was separated in only 26% yield. Reductive removal of both carbonyl groups in bis-lactam (*Z*)-**144** finally led to (+)-**145**.

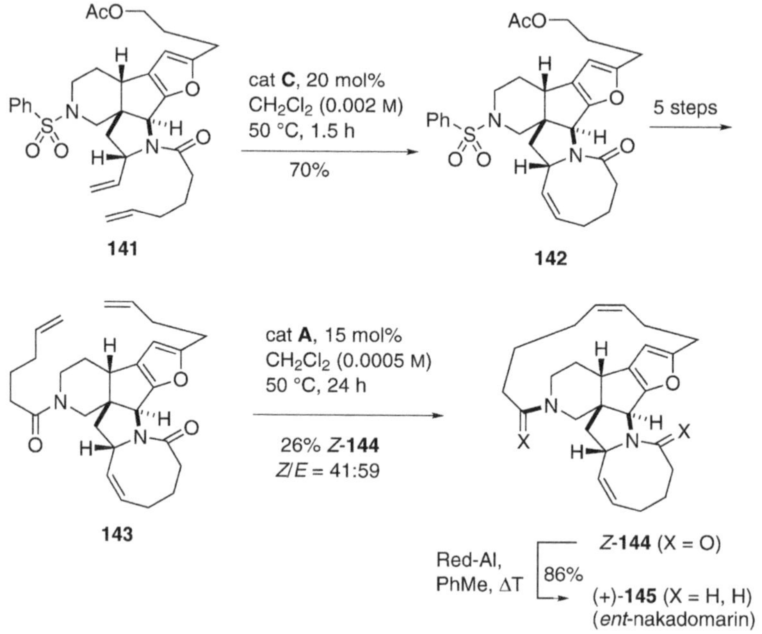

AcO
H
Ph S N O O
H H N O
141
cat **C**, 20 mol%
CH_2Cl_2 (0.002 M)
50 °C, 1.5 h
70%
AcO
H
Ph S N O O
H H N O
142
5 steps
H
N O
H H N O
143
cat **A**, 15 mol%
CH_2Cl_2 (0.0005 M)
50 °C, 24 h
26% *Z*-**144**
Z/*E* = 41:59
H
N X
H H N X
Red-Al, PhMe, ΔT
86%
Z-**144** (X = O)
(+)-**145** (X = H, H)
(*ent*-nakadomarin)

Scheme 26 Sequential formation of 8- and 15-membered azacycles in total synthesis of *ent*-nakadomarin A (**145**) [80]

2.1.2.3
Lactones

An example of a surprisingly facile and stereoselective formation of an eight-membered lactone from an acyclic precursor diene ester was observed during the total synthesis of the antitumor agent octalactin A (**148**) (Scheme 27) [81]. The dense substitution pattern in cyclization substrate **146** presumably imposes

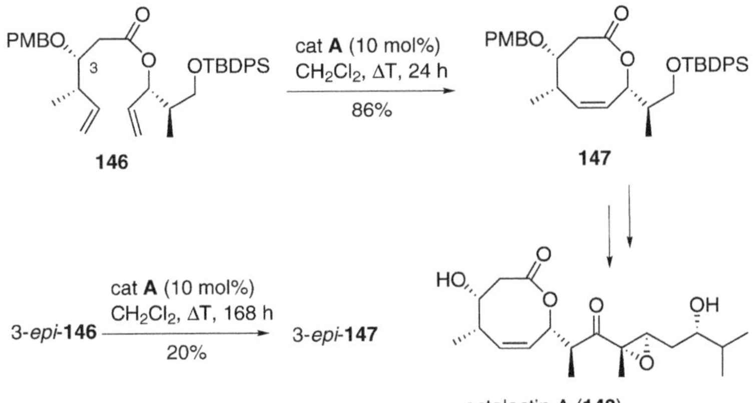

Scheme 27 Influence of remote substrate substituents on RCM efficiency, observed during total synthesis of octalactin A (**148**) [81]

conformational constraints in a way that leads to a conformation favorably disposed for the ring closure. Thus, exposure of **146** to 10–20 mol% of catalyst A afforded cyclization product **147** within 24 h in 86% yield. In contrast, the diene ester with epimeric PMB ether group (3-*epi*-**146**) underwent ring closure under analogous conditions only with difficulty, leading to lactone 3-*epi*-**147** in 20% yield after 7 days in boiling dichloromethane.

Halicholactone (**151**), a marine metabolite with lipoxygenase inhibitory activity, belongs to the family of oxylipins which all contain a lactone moiety substituted by a *trans*-disubstituted cyclopropane subunit. Stereoselective RCM for the formation of the nine-membered lactone core in **151** was the penultimate step (**149**→**150**) in an asymmetric total synthesis of **151** by a Japanese group (Scheme 28) [82]. After extensive experimentation, it was found that reaction of **149** with catalyst **A** under high dilution (0.1 mM in boiling dichloromethane) in the presence of a catalytic amount of $Ti(O\textit{i}Pr)_4$ gave rise to the desired (*Z*)-isomer **150** in 72% yield along with the corresponding dimer (11%). When the reaction was performed under more than 1.0 mM concentration, monomer **150** and the dimer were formed in almost equal amounts (each

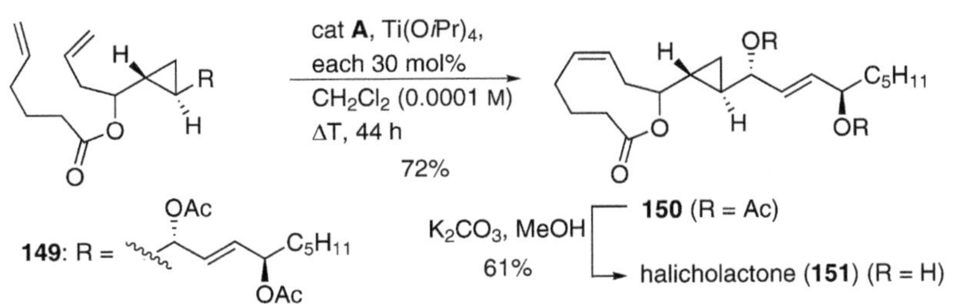

Scheme 28 (*Z*)-selective RCM-based macrocyclization in the penultimate step in the total synthesis of halicholactone (**151**) [82]

20–30%). Note that the (*E*)-isomer of **150** was not detected under any reaction conditions. The total synthesis of **151** was then completed by methanolysis of the two acetyl groups.

The marine natural product ascidiatrienolide (**157**) is a strong inhibitor of phospholipase A2. Compound **157** and the closely related didemnilactones **158–160** feature a common hydroxy-substituted (*anti* to the ring oxygen) (*Z*)-nonenolide core. Lactone **156**, that constituted the key intermediate in a previous total synthesis of **157** and can also be elaborated to lactones **158–160**, has been the subject of an interesting study by Fürstner's group [83] that revealed once more the very subtle and cooperative influence of different parameters on the stereochemical course of metathesis reactions. Thus, it was shown that the *E*/*Z* ratio obtained in an RCM step is not only dependent on the relative configuration of the cyclization substrate, but also on the chosen catalyst (Scheme 29). When applied to the *anti*-configured diene ester **152**, both ruthenium indenylidene complex **F** and second-generation catalyst **E** induced the preferential (*E*/*Z*≈8:1) formation of the undesired lactone (*E*)-**153** in compara-

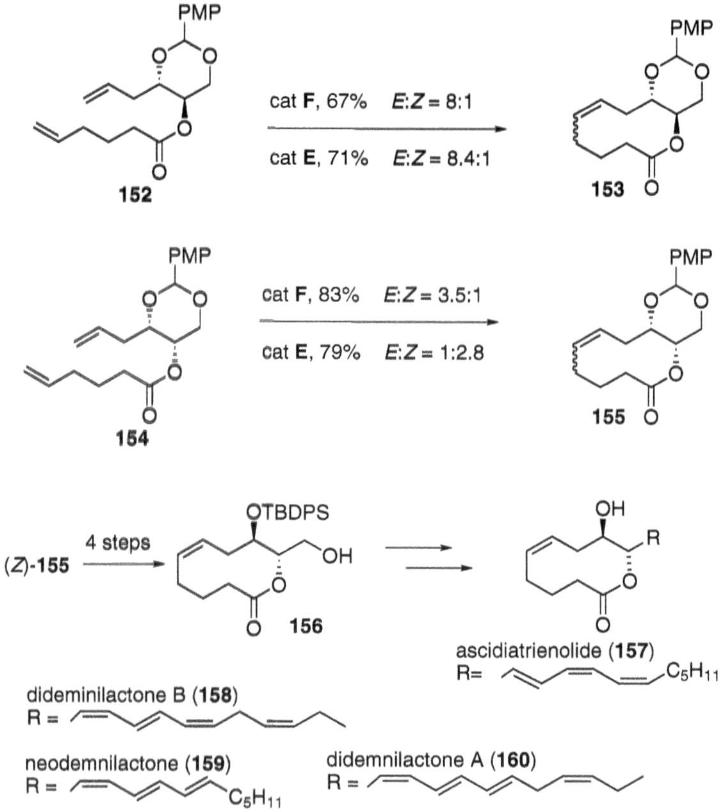

Scheme 29 Effect of substrate substituents and catalyst activity on RCM stereochemistry, observed during the total synthesis of ascidiatrienolide (**157**) [83]

ble yield, but opposite results were obtained with the *syn* analog **154**. Specifically, indenylidene catalyst **F** still favored (3.5:1) the (*E*)-isomer of **155**, while NHC catalyst **E** favored the formation of the required nonenolide (*Z*)-**155** (*Z*:*E*=2.8:1), which was converted to target **156** in four steps.

In related contributions, Fürstner disclosed a concise RCM-based approach to a family of potent herbicidal ten-membered lactones with an (*E*) double bond, which led to the first total syntheses of herbarium I (**163**) [84a] and II (**164**) [84b], and also allowed the stereostructure of pinolidoxin (**165**) to be established (Scheme 30) [84b]. Again, the stereochemical outcome of the ring-closing step could be controlled by the choice of the catalyst, which is deemed to reflect kinetic versus thermodynamic control. En route to herbarium I (**163**), cyclization precursor **161** containing an isopropylidene protecting group, which should align the olefinic side chain in a "cyclization-friendly" conformation, was prepared in six steps from protected D-ribonolactone. Semiempirical calculations carried out for both possible cyclization products derived from **161** indicated that isomer (*Z*)-**162** is about 3.5 kcal mol^{-1} more stable than (*E*)-**162**. That means that only under kinetic control would it be possible to obtain the desired (*E*)-isomer, and that highly active catalysts known to favor the retro reaction, and hence leading to equilibration, would be counterproductive. The results obtained with indenylidene catalyst **F** and with the second-generation NHC catalyst **E** were fully consistent with the above predictions: catalyst **F** exhibited activity similar to Grubbs' benzylidene catalyst **A** and produced mainly (7.7:1) the less stable and desired (*E*)-**162** (the *E*/*Z* ratio did not evolve with time), while catalyst **E** led exclusively to the thermodynamically more stable (*Z*)-isomer. It seems that complex **E** and congeners, due to their higher overall activity, are able to isomerize the cycloalkenes formed during the course of the reaction and

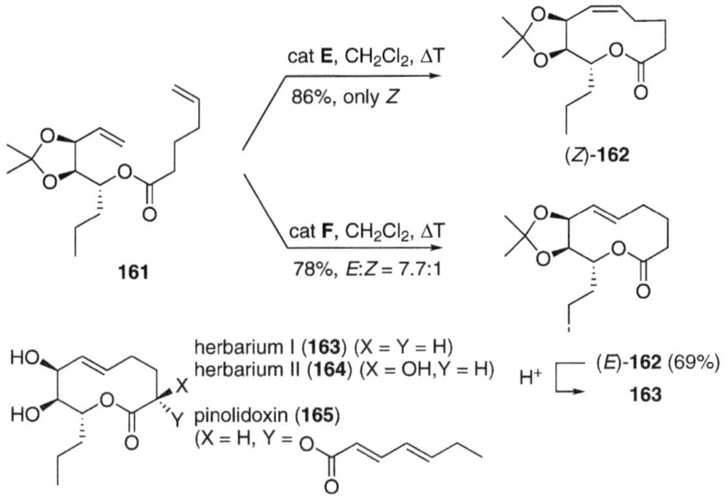

cat **E**, CH_2Cl_2, ΔT
86%, only *Z*
(*Z*)-**162**
161
cat **F**, CH_2Cl_2, ΔT
78%, *E*:*Z* = 7.7:1
herbarium I (**163**) (X = Y = H)
herbarium II (**164**) (X = OH, Y = H)
pinolidoxin (**165**)
(X = H, Y = O...)
H+
(*E*)-**162** (69%)
163

Scheme 30 Kinetically and thermodynamically controlled RCM in Fürstner's total synthesis of herbarium I (**163**) [84]

hence enrich the mixture in the thermodynamically favored product. Further support for this interpretation was provided by a control experiment showing that pure (*E*)-**162** was slowly isomerized to (*Z*)-**162** in the presence of catalyst **E** when the reaction was performed under an atmosphere of ethylene.

2.1.2.4
Cyclic Ethers

A range of topographically unique structures with seven- to nine-membered ether rings are produced by marine organisms. Several total syntheses of natural monocyclic eight-membered ring ethers (oxocenes) and the less common homologous oxonenes, produced by *laurencia* red algae, were reported by Crimmins' team by merging asymmetric aldol addition (or alkylation) of glycolates with an RCM reaction. Thereby it was demonstrated that medium-sized cyclic ethers are readily available without cyclic conformational constraint by exploiting the acyclic bias of the gauche effect of substituents on the carbons flanking the ether linkage.

A year after the total synthesis of (+)-laurencin [85], Crimmins disclosed the total syntheses of (+)-prelaureatin (**169**) and (+)-laurallene (**168**) by applying a similar strategy (Scheme 31) [86]. The critical RCM reaction was undertaken

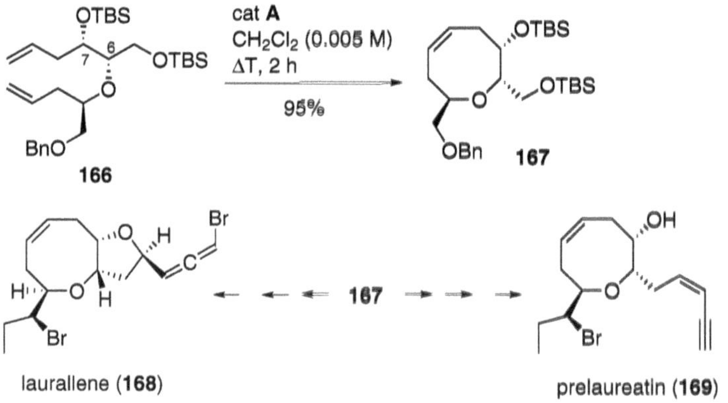

Scheme 31 Efficient RCM to Δ4-oxocene **167** in Crimmins' total syntheses of the marine natural products laurallene (**168**) and prelaureatin (**169**) [86]

with precursor **166**, anticipating that the gauche effect of the C6 and C7 oxygens would accelerate the ring closure. Exposure of **166** (0.005 M in dichloromethane) to Grubbs' catalyst **A** proceeded smoothly to provide the key Δ^4-oxocene **167** in 95% yield with no detectable dimerization.

In subsequent reports [87], the principle of asymmetric glycolate alkylation/RCM sequence was applied to the first total synthesis of isolaurallene (**172**), that contains a densely functionalized Δ^5-oxonene core (Scheme 32). Anticipating that the gearing effect created by two synergistic gauche effects at C6–C7 and

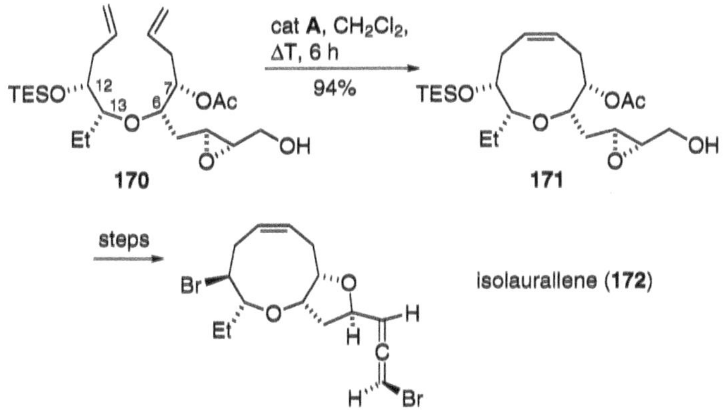

Scheme 32 RCM-based synthesis of the Δ5-oxonene **171** in Crimmins' total synthesis of isolaurallene (**172**) [87]

C12–C13 would facilitate the ring closure, monocyclic diene **170** was chosen as the metathesis substrate. Indeed, exposure of **170** to catalyst **A** provided cyclization product **171** in 94% yield within 6 h, without the aid of a cyclic conformational constraint.

To date, the most recent of Crimmins' contributions in this series are the total syntheses of the nine-membered cyclic ether obtusenyne (**177**) [88] and the oxepene rogioloxepane A (**180**) [89], which both feature a *trans*-orientation of the substituents flanking the ether linkage (Scheme 33). Three different RCM precursors (**173a–c**) were investigated during the synthesis of **177**. Attempts to form the nine-membered ring from the bromo-substituted triene **173a** resulted in loss of the vinyl halide by regioselective formation of cyclohexene derivative **174** in 80% yield. Triene **173b** with a trisubstituted double bond provided a 3:1 mixture of oxonene **175b** and cyclohexene **174**. Finally, conversion of **173b** to epoxy-diene **173c** followed by treatment with catalyst **A** effected rapid closure to **175c** that was converted to **177** in 13 steps.

In the total synthesis of rogioloxepane A (**180**), oxazolidinone **178a** was primarily examined as the metathesis substrate. However, the subsequent removal of the auxiliary with sodium borohydride proceeded with low yield due to concomitant hydrogenation of the oxepene by remaining traces of the ruthenium catalyst. Therefore the order of steps was reversed and the RCM step performed with primary alcohol **178b**, which additionally could bias the diene conformation by a hydrogen bond with the ether oxygen. Treatment of **178b** with catalyst **A**, followed by DMSO workup to remove traces of catalyst-derived materials [50b], then led to key intermediate **179** in excellent yield.

A highlight in the application of RCM methodology in natural product synthesis is Hirama's total synthesis of ciguatoxin CTX3C (**183**) [90], including the more recent improved protective group strategy, as depicted in Scheme 34 [90b]. The structure of **183** spans more than 3 nm and is characterized by 12 six- to nine-membered *trans*-fused cyclic ethers and a spiroannulated terminal tetra-

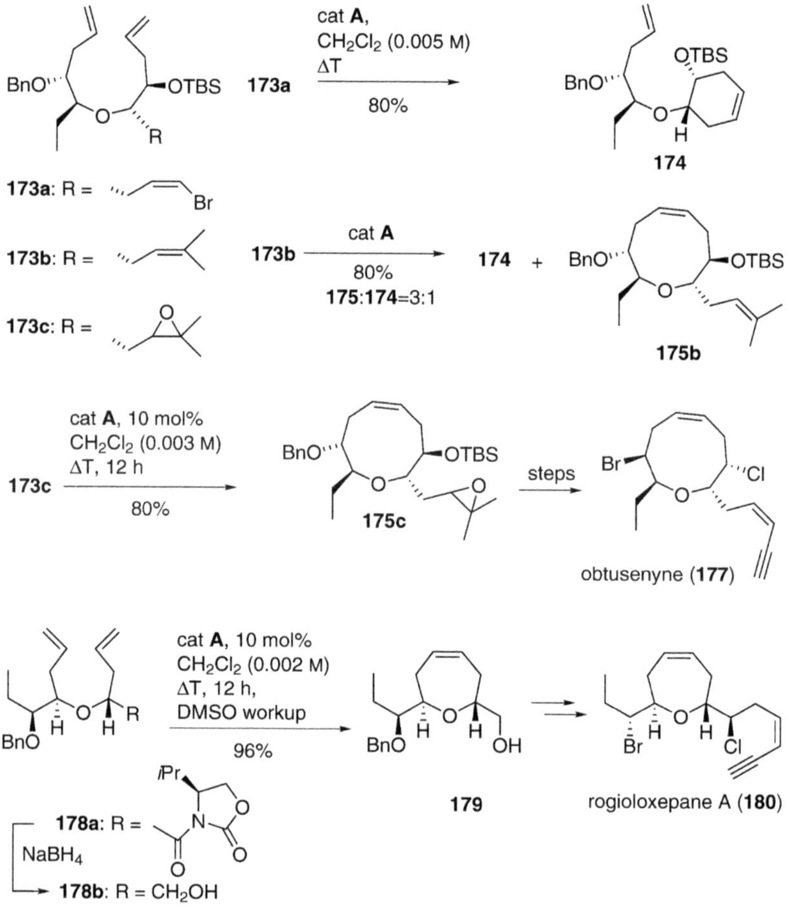

Scheme 33 RCM-based total synthesis of the marine cyclic ethers obtusenyne (**177**) [88] and rogioloxepane (**180**) [89] by Crimmins and coworkers

hydrofuran ring. Causative toxins such as **183** are produced by the marine dinoflagellate *Gambierdiscus toxicus* and accumulate in fish of many species through the food chain. In the penultimate step of the improved total synthesis, pentaene **181**, that is only missing the central nine-membered ring, was exposed to catalyst **A** in boiling dichloromethane to provide 2-naphthylmethyl (NAP)-protected CTX3C (**182**) chemoselectively in 90% yield and to set all rings in place. The three NAP groups in **182** (the deprotection of the corresponding tris-benzyl ether in the original synthesis proceeded with low yield) were then removed with DDQ to furnish the natural compound **183** in 63% yield.

Intramolecular allylation of α-chloroacetoxy ether **185** followed by RCM (Scheme 35) was used by Yamamoto and coworkers to construct the eight-membered cyclic ether in the F–K ring segment **186** of the marine neurotoxin brevetoxin B (**184**) [91].

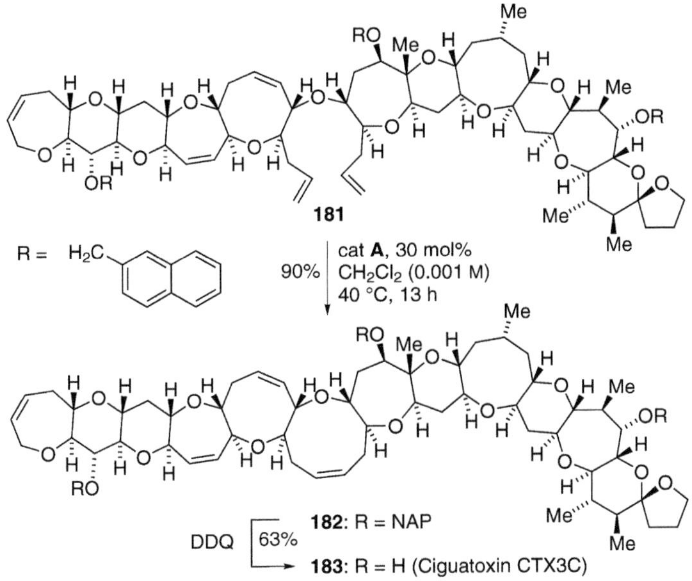

Scheme 34 Formation of the central nine-membered ring and completion of the carbon skeleton in Hirama's improved total synthesis of the marine neurotoxin ciguatoxin CTX3C (**183**) [90b]

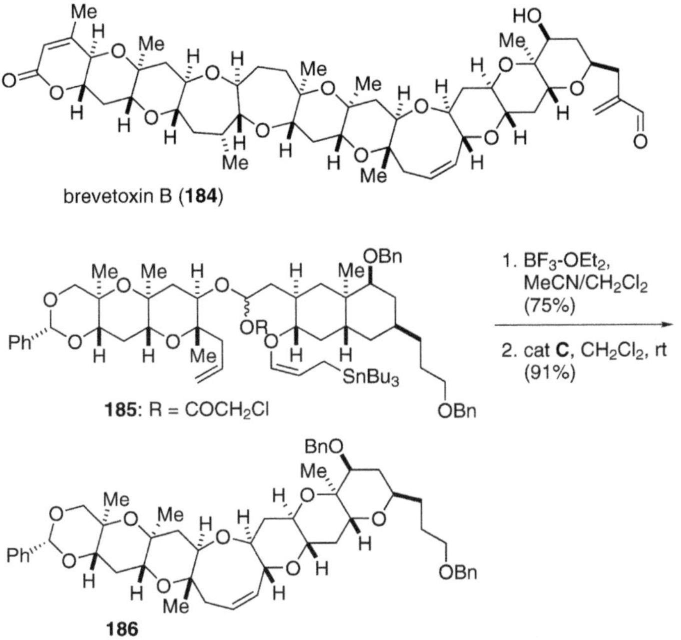

Scheme 35 Sequential formation of rings I and H by intramolecular allylation/RCM in Yamamoto's synthesis of the F–K ring segment **186** of brevetoxin B (**184**) [91]

2.1.3
Formation of Macrocycles

RCM-based formation of – unstrained – macrocycles is, due to the concomitant loss of a volatile alkene, mainly entropically driven and therefore a high yielding process. However, there exists still a lack of prediction for the configuration of the newly formed double bond of cycloalkenes with more than ten ring atoms. The products formed are frequently obtained as *E*/*Z* mixtures with the (*E*)-isomer dominating in most of the recorded cases. This obvious drawback in target-oriented synthesis was already evident from the early and most prominent RCM-based epothilone syntheses [92], which suffered from very low stereoselectivity in the formation of the required (*Z*)-12,13 double bond. The following examples of RCM-based syntheses of macrocyclic natural products will reveal that the success and/or the stereochemical outcome of macrocyclic RCM is highly sensitive to steric or electronic substituent effects in the precursor diene, and can also depend on the choice of the catalyst, as well as on the solvent and the reaction temperature applied in the metathesis process. Additionally we will see that, for the formation of strained products, large enthalpic barriers can be overcome by altering the shape of the metathesis substrate through the introduction of additional conformational constraints.

2.1.3.1
Macrolides

Three RCM-based syntheses of the 18-membered α,β-unsaturated macrolide aspicilin (**189**), all performed with Grubbs' first-generation catalyst **A** and differently protected precursor trienes **187a–d** (Scheme 36) [93], illustrate the importance of substituent effects on the regio- and stereochemistry of the metathesis reaction, albeit in this case the stereochemical outcome of the ring-closing step is inconsequential. Hatakeyama's isopropylidene-protected precursor **187a** led exclusively to macrolactone **188a** with (*Z*) configuration at the newly formed double bond. In contrast, Banwell's first precursor **187c** reacted regioselectively with formation of the undesired cyclohexene **190c**, while the open-chain precursor **187b** furnished a 3:1 mixture of macrolides in favor of the (*E*)-isomer. Partial cyclohexene formation was also observed by Ley, who isolated from the cyclic metathesis substrate **187d** a mixture of macrolide **188d** (1.5:1 *Z*/*E*-mixture) and cyclohexene **190d**.

Migrastatin (**192**) (Scheme 37) is a novel macrolide natural product that displays an inhibitory effect on the migration of human tumor cells. After an RCM-based synthesis of the 14-membered macrolide core of **192** [94], Danishefsky also achieved the first total synthesis of the natural compound [95], using the fully functionalized tetraene **191** as the metathesis precursor. Under the conditions shown in Scheme 37, the ring-closing step proceeded (*E*)-selectively with exclusive participation of the two terminal double bonds in **191**, delivering only the (*E*,*E*,*Z*)-trienyl arrangement present in **192**.

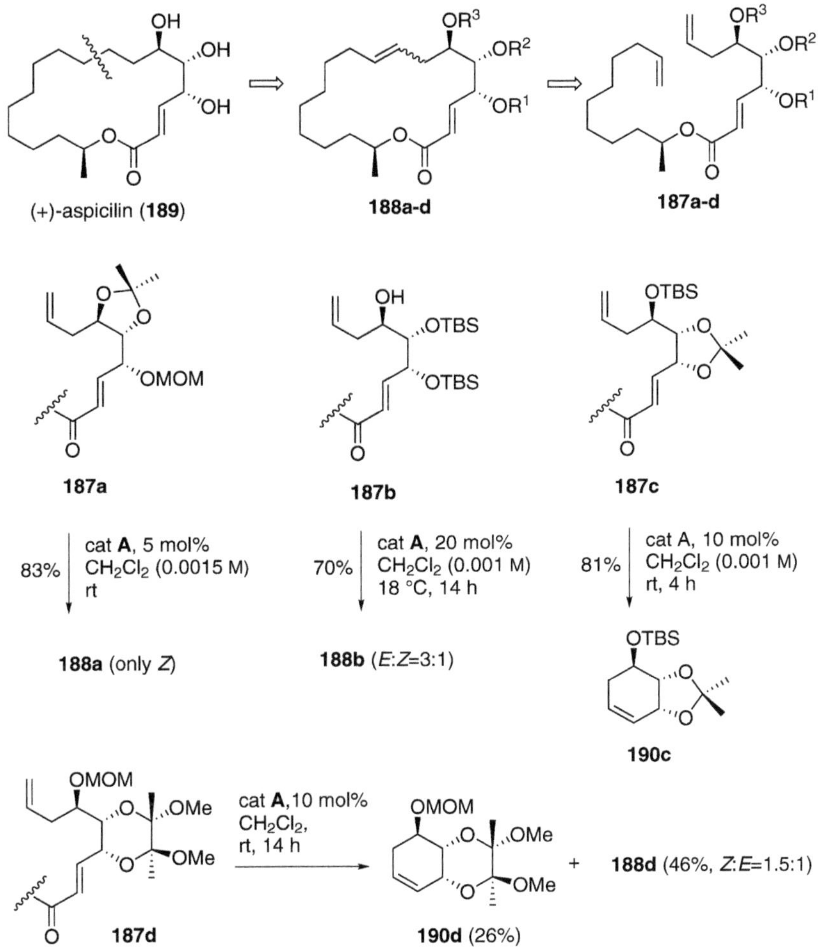

(+)-aspicilin (**189**) ⇒ **188a-d** ⇒ **187a-d**

187a — cat **A**, 5 mol%, CH_2Cl_2 (0.0015 M), rt, 83% → **188a** (only *Z*)

187b — cat **A**, 20 mol%, CH_2Cl_2 (0.001 M), 18 °C, 14 h, 70% → **188b** (*E*:*Z*=3:1)

187c — cat A, 10 mol%, CH_2Cl_2 (0.001 M), rt, 4 h, 81% → **190c**

187d — cat **A**,10 mol%, CH_2Cl_2, rt, 14 h → **190d** (26%) + **188d** (46%, *Z*:*E*=1.5:1)

Scheme 36 RCM-based synthesis of aspicilin (**189**): effect of substrate substitution on regio- and stereochemistry [93]

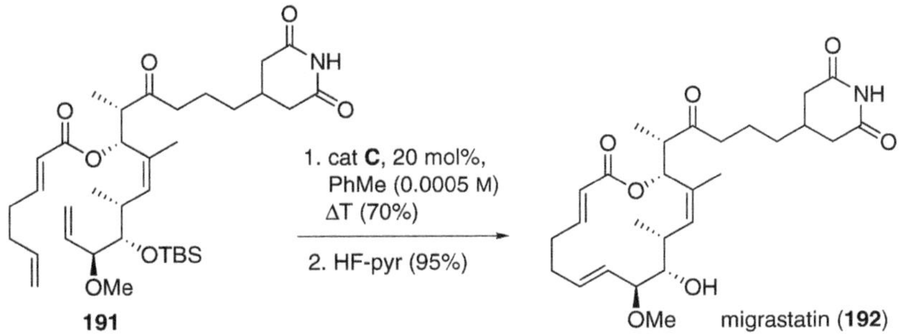

191 — 1. cat **C**, 20 mol%, PhMe (0.0005 M), ΔT (70%); 2. HF-pyr (95%) → migrastatin (**192**)

Scheme 37 Regioselective RCM of tetraene **191** in Danishefsky's total synthesis of migrastatin (**192**) [95]

Three total syntheses of the highly unsaturated macrolactone **195**, featuring the structure proposed for the cytotoxic marine natural product amphidinolide A (Scheme 38) were disclosed in 2002 [96], which all confirmed that the structure of **195** proposed for the natural product needs to be revised. In Maleczka's synthesis [96a], the highly unsaturated 20-membered ring of **195** was formed by a late-stage RCM reaction. Given the array of olefinic functionality in metathesis substrate **193**, the authors used the less active first-generation catalyst **A** in their first attempt, which should guarantee regioselectivity, but this catalyst only truncated the allylic alcohol in **193** leading to the corresponding methyl ketone [97]. With second-generation catalyst **C**, the ring closure occurred, but 0.5 equivalents of the catalyst were necessary to provide regio- and (*E*)-stereoselectively macrolide **194** in low yield.

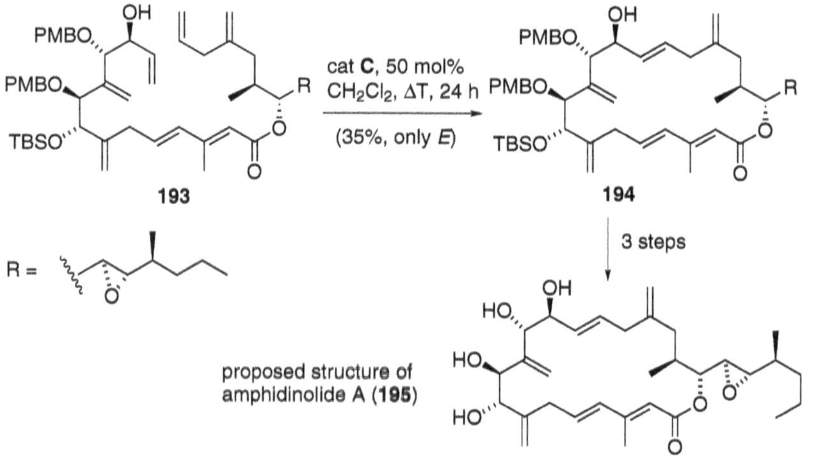

Scheme 38 Regioselective RCM of heptaene **193** in Maleczka's synthesis of the structure **195** proposed for amphidinolide A [96a]

Several members of the structurally quite diverse amphidinolide T family containing a saturated 19-membered lactone core were recently synthesized by Fürstner and coworkers [98]. Amphidinolide T4 (**199**, Scheme 39) [98a] and also T1 and T5 (not shown in the scheme) [98b] were prepared from compound **196** as a common intermediate. The macrocyclic ring was efficiently formed by a high-yielding RCM of diene **196** in the presence of second-generation catalyst **E** bearing an imidazol-2-ylidene ligand. The efficiency of the RCM transformation was mainly attributed to the conformational bias introduced by the *syn–syn* configured stereotriad at C12–C14 in **196**. The resulting cycloalkenes obtained in 86% yield as an inconsequential isomeric mixture (*E*:*Z*=6:1) were hydrogenated to **197**. After methylenation to **198**, the synthesis of **199** was completed by three additional steps. En route to amphidinolide T3 (12-*epi*-**199**) [98b] via RCM of 12-*epi*-**196**, it turned out that the efficiency and stereochemical outcome of the ring closure was distinctly affected by the configurational

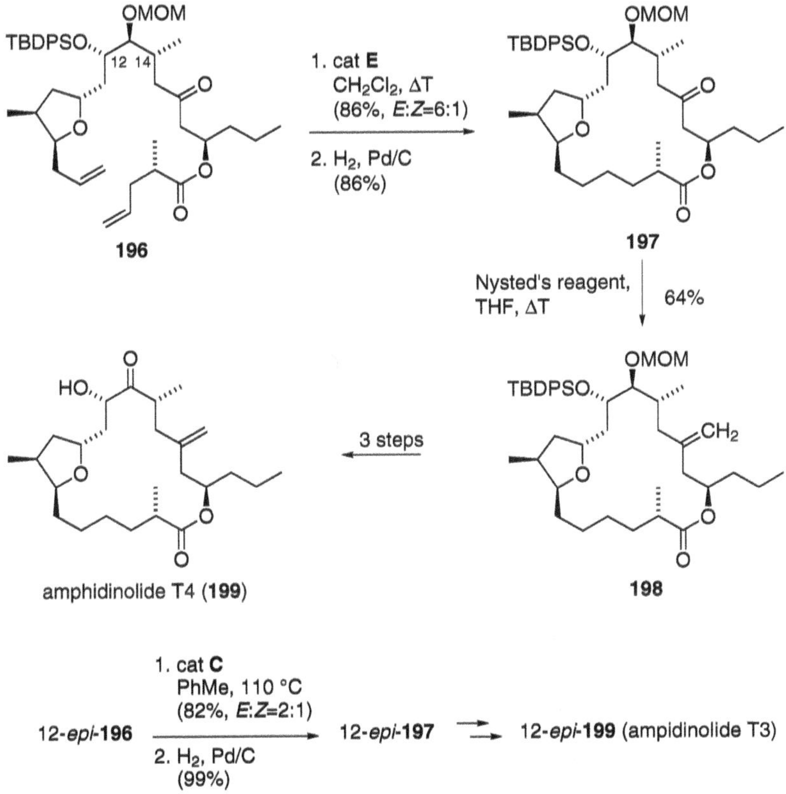

Scheme 39 Influence of a remote substituent on efficiency and stereochemistry of the RCM step in Fürstner's total synthesis of amphidinolide T4 (**199**) [98a] and amphidinolide T3 (12-*epi*-**199**) [98b]

change (*anti–syn* stereotriad at C12–C14) at the seemingly remote stereocenter C12. Good conversion could only be attained in the presence of catalyst C (bearing a saturated NHC ligand), and by exchanging the solvent from dichloromethane to toluene with concomitant increase of the reaction temperature to 110 °C.

Recently, Hoye described an RCM-based total synthesis of the 20-membered marine macrolide dactylolide (**202**) and its subsequent conversion to the natural carbinolamide zampanolide (**203**) (Scheme 40), which feature a common highly unsaturated macrolide core, bridging a *cis*-2,6-disubstituted 4-methylene tetrahydropyran unit [99]. When the polyunsaturated acyclic lactone **200** (1:1 epimeric mixture around the TBS-protected carbinol center) was in situ protected with bis-trimethylsilylacetamide (BSA) and then treated with catalyst C in benzene at 60 °C, each diastereomer smoothly cyclized to the corresponding cycloalkene with exclusive (*E*) geometry at the newly formed double bond, demonstrating that configurational change at this position had (in this special case) no influence on the results.

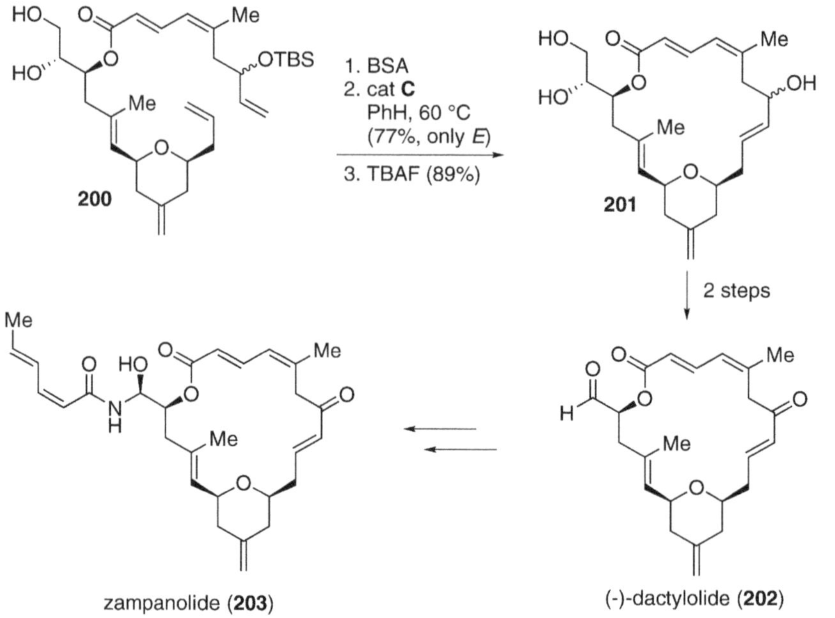

1. BSA
2. cat **C**
PhH, 60 °C
(77%, only *E*)
3. TBAF (89%)

2 steps

zampanolide (**203**) (-)-dactylolide (**202**)

Scheme 40 (*E*)-Selective macrocyclization in the total synthesis of the marine macrolides dactylolide (**202**) and zampanolide (**203**) [99]

An example of the efficient formation of an electron-deficient double bond by RCM was disclosed by a Japanese group in a novel total synthesis of the macrosphelides A (**209**) and B (**208**) (Scheme 41) [100]. When the PMB-protected compound **204** was examined as a metathesis substrate, the ring closure did not proceed at all in dichloromethane using catalysts **A** or **C**. When the reaction was carried out using equimolar amounts of catalyst **C** in refluxing 1,2-dichloroethane, the cyclized product **205** was obtained in 65% yield after 5 days. On the other hand, the free allylic alcohol **206** reacted smoothly at room temperature leading to the desired macrocycle **207** in improved yield.

Also the novel antifungal antibiotic (–)-PF1163B (**211**), isolated from *Streptomyces* sp., which features a 13-membered macrocycle incorporating both a lactone and a lactam unit, was synthesized by an RCM route (Scheme 42) [101]. While only poor results were obtained by treatment of diene **210** (containing 8% of an unidentified epimer) with catalyst **A**, the use of NHC catalyst **C** led, under the conditions outlined in the scheme, to the corresponding cyclization product in 60% yield along with 10% of a diastereomer resulting from epimerization in a previous step.

The salicylihalamides A (**215a**) [102] and B (**215b**) are the first members of a growing class of secondary marine metabolites with a 12-membered benzolactone core incorporating salicylic acid in conjunction with a dienylenamide side chain (Scheme 43). Salicylihalamide A (**215a**) was reported to be a unique and highly differential cytotoxin and a potent inhibitor of the mammalian vac-

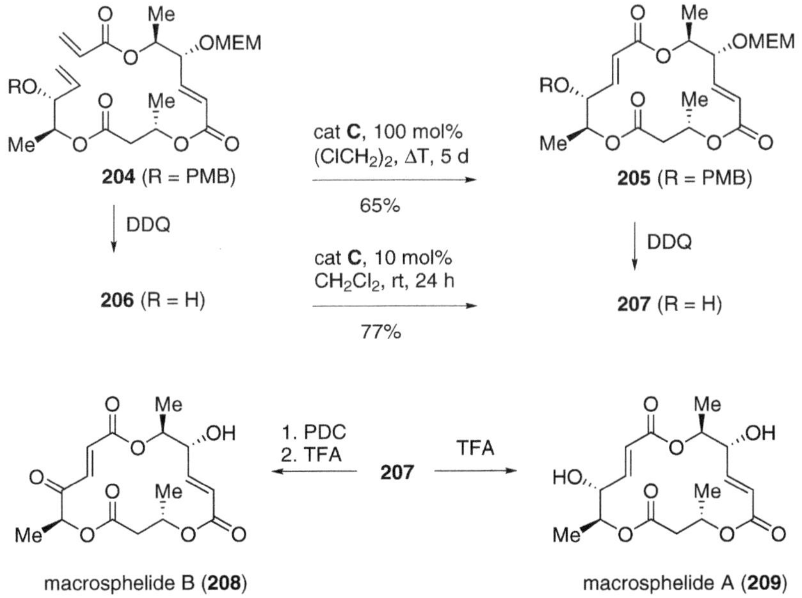

Scheme 41 (*E*)-Selective RCM of acrylate **204** in the total syntheses of the macrosphelides A (**209**) and B (**208**) [100]

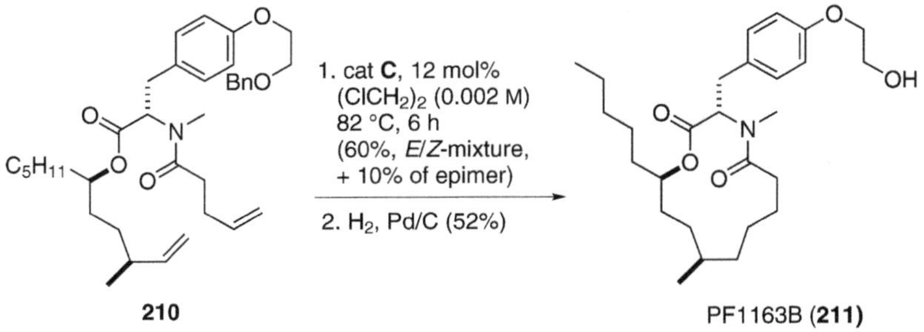

Scheme 42 Macrocyclization by RCM in the total synthesis of the antifungal antibiotic PF1163B (**211**) [101]

uolar (H^+)-ATPase. To date, there exist several total syntheses of **215a** that rely on an (*E*)-selective RCM of dienes **212** or **213** to construct the benzolactone core **214** [103].

The results obtained with the various metathesis substrates depicted in Scheme 44 demonstrate the lack of a stereopredictive model for the RCM-based formation of macrocycles, not only by the strong influence that may be exhibited by *remote* substituents, but also by the fact that the use of more reactive second-generation catalysts may be unfavorable for the stereochemical outcome of the reaction. Dienes **212a–f** illustrate the influence of the substitution pattern. All reactions were performed with Grubbs' first-generation catalyst **A**

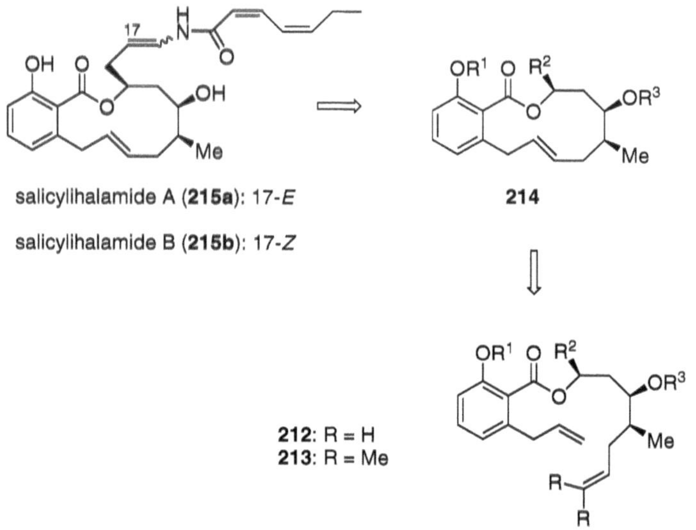

salicylihalamide A (**215a**): 17-*E*

salicylihalamide B (**215b**): 17-*Z*

214

212: R = H
213: R = Me

Scheme 43 Structure and retrosynthetic analysis of the salicylate macrolides salicylihalamide A (**215a**) and B (**215b**) by various groups [103]

in dichloromethane, but the isomeric ratio varied from the favorable *E*/*Z*=10:1 obtained by Smith [103d] with **212a**, to the "mainly *Z*" observed by Snider [103a] with **212e** featuring a remote free phenolic hydroxy group, while no ring closure occurred with Snider's bicyclic model **212f.** The RCM precursors **213a–d** used by Fürstner's group [103c] differ from compounds of the type **212** mainly by the *gem*-disubstitution at one of the olefinic moieties, so that the ring-closing step had – in this case – to be conducted with a more reactive ruthenium catalyst of the second generation. The macrocyclizations with compounds **213a–d** were all performed with catalyst **E** in toluene at 80 °C, and again it turned out that the stereochemical outcome was strongly dependent on the phenolic protective group, ranging from "only *Z*" for the unprotected phenol **213a,** 1.5:1 in favor of the (*Z*)-isomer for the corresponding silyl ether **213b,** to a 2:1 ratio in favor of the required (*E*)-isomer for both the methyl and the MOM ether derivatives **213c,d.** Finally, a detailed study of the metathesis step conducted with dienes **212g** and **212h** in De Brabander's full account [103e] brought partial light to this confusing situation, identifying the high (*E*) stereoselection obtained with catalyst **A** at room temperature as a result of a kinetically favored process. On the other hand, with second-generation catalyst **C** (or **E**), an equilibrium is quickly reached, so that the identical isomeric ratios (*E*:*Z*≈2:1) obtained with Fürstner's precursors **103c,d** and Brabander's substrates **101g,h** reflect a thermodynamic distribution, where secondary metathesis isomerization can compete at the timescale of the experiment. (It should be pointed out, however, that the pronounced influence of a remote phenolic OH group, which favors the undesired (*Z*) stereochemistry with catalysts **A** and **E**, still remains unclear).

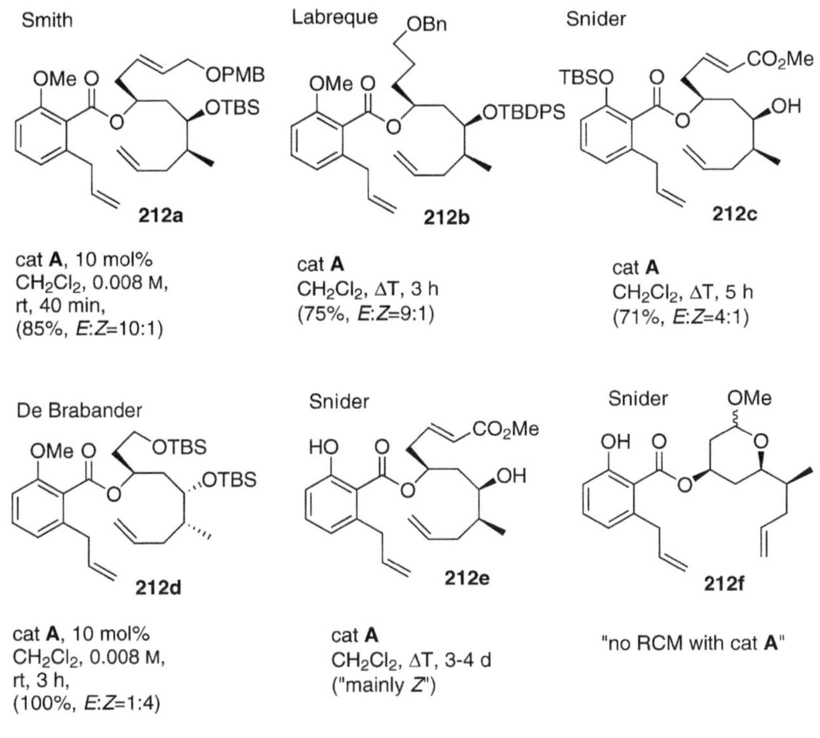

Fürstner (all reactions performed with cat **E**)

OR O, OPMB, OMOM, O, Me, Me, Me

213a-d

213a (R=H) cat **E**, 10 mol% PhMe, 80 °C, 20 h (69%, only *Z*)

213b (R=TBS) cat **E**, 5 mol% PhMe, 80 °C, 1 h (91%, *E*:*Z*=1:1.5)

213c (R=Me) cat **E**, 5 mol% PhMe, 80 °C, 1.5 h (93%, *E*:*Z*=1.9:1)

213d (R=MOM) cat **E**,10 mol% PhMe, 80 °C, 3 h (91%, *E*:*Z*=2.1:1)

De Brabander (reactions performed with cat **A** or cat **C**, 10 mol%, CH_2Cl_2, rt, all combined yields > 93%)

OR O, OPMB, OMOM, O, Me

212g,h

	cat **A**	cat **C**
212g (R=Me)	1.3 h (*E*:*Z*=9:1)	1.3 h (*E*:*Z*=1.9:1)
	18.3 h (*E*:*Z*=4:1)	18.3 h (*E*:*Z*=2:1)
212h (R=MOM)	1.3 h (*E*:*Z*=9:1)	1.3 h (*E*:*Z*=1.8:1)
	18.3 h (*E*:*Z*=5:1)	18.3 h (*E*:*Z*=2.1:1)

Scheme 44 Influence of remote substituents in RCM precursors **212** and **213** and of catalyst activity on stereochemistry in salicylihalamide synthesis [103]

A recent example of RCM-based macrocyclization with a highly complex metathesis substrate is the formation of macrolide **216** from the corresponding diene precursor in Fürstner's total synthesis of the resin glycoside woodrosin I (**218**) (Scheme 45) [104]. The site of ring closure was chosen far away from potential donor sites in the oligosaccharide scaffold, so that the formation of unreactive metal chelate complexes was avoided. Accordingly, a virtually quantitative formation of **216** (*E*:*Z*=9:1) was observed on treatment with catalysts A or F. Subsequent exposure of **216** to glycosyl donor **217** led not only to the introduction of the missing rhamnose unit, but also to concomitant rearrangement of the *ortho* ester into the desired *β*-glycoside. The synthesis of **218** was then completed in two steps.

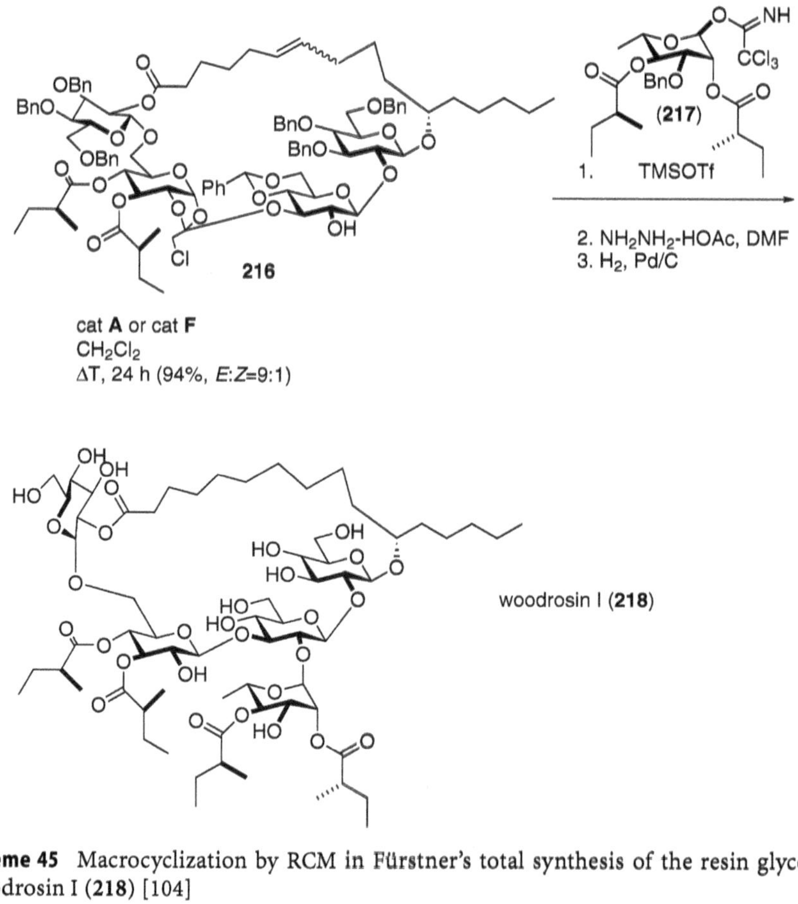

216

cat **A** or cat **F**
CH_2Cl_2
ΔT, 24 h (94%, *E*:*Z*=9:1)

(**217**)

1. TMSOTf
2. NH_2NH_2-HOAc, DMF
3. H_2, Pd/C

woodrosin I (**218**)

Scheme 45 Macrocyclization by RCM in Fürstner's total synthesis of the resin glycoside woodrosin I (**218**) [104]

2.1.3.2
RCM of Diene-Ene Systems

An increasing number of natural product syntheses feature the RCM of diene-ene systems to produce macrocyclic dienes. However in some cases, divergences in regioselectivity of catalyst attack were observed depending on the structural features of the metathesis substrate and also on the catalyst used to promote the metathesis event [105]. The first example of regioselective diene-ene metathesis was contributed by a Novartis group in 1999, during synthesis of simplified macrolide analogs of the immunosuppressant sanglifehrin (Scheme 46) [106].

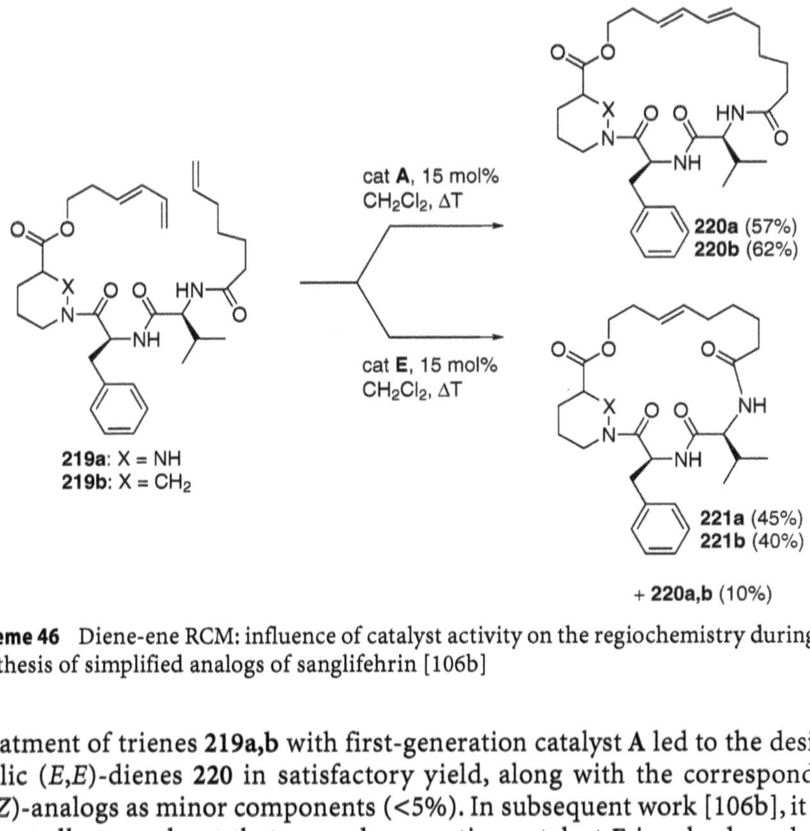

Scheme 46 Diene-ene RCM: influence of catalyst activity on the regiochemistry during the synthesis of simplified analogs of sanglifehrin [106b]

Treatment of trienes **219a,b** with first-generation catalyst **A** led to the desired cyclic (*E*,*E*)-dienes **220** in satisfactory yield, along with the corresponding (*E*,*Z*)-analogs as minor components (<5%). In subsequent work [106b], it unexpectedly turned out that second-generation catalyst **E** involved predominantly the more substituted internal double bond in precursors **219**, leading to the ring-contracted cyclic monoenes **221** in moderate yield, while the desired cyclodienes **220** were detected only as minor components.

Another example of macrocyclic RCM with a diene-ene was disclosed in 2000 by Meyers and coworkers in the first total synthesis of griseoviridin (**223**) [107]. Griseoviridin is a highly complex member of the family of streptogramin antibiotics, featuring a 23-membered unsaturated bis-lactam core incorporat-

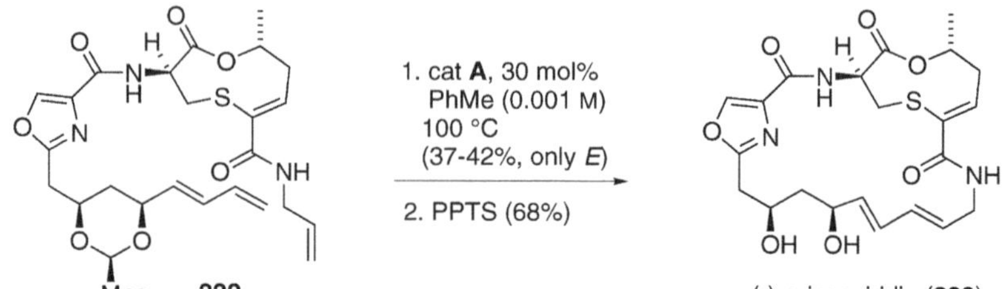

222

1. cat **A**, 30 mol%
PhMe (0.001 M)
100 °C
(37-42%, only *E*)
2. PPTS (68%)

(-)-griseoviridin (**223**)

Scheme 47 Macrocyclization via diene-ene RCM in the first total synthesis of griseoviridin (**223**) [107]

ing an oxazole and a nine-membered lactone with an ene–thio linkage. The macrocyclic ring of **223** was (*E*)-selectively elaborated in the penultimate step, by exposing allylamine **222** to catalyst **A** (Scheme 47).

Hsp90 is a molecular chaperon required for the refolding of proteins in cells exposed to environmental stress. It contains an ATP-binding pocket in its amino terminus. Several natural products, for example radicicol (**230**) (Scheme 48), bind to this pocket and inhibit its chaperon function, which is mirrored in enhanced proteosomal degradation of Hsp90 client proteins, so that compounds like **230** are of interest as novel anticancer agents.

Danishefsky's total synthesis of **230** and its chlorine-free precursor monocillin I (**229**) [108] features a novel RCM reaction with a substrate (**224**) that, in addition to a dithiane protective group, contains a vinyl epoxide and a diene moiety at both the termini involved in the metathesis process (Scheme 48). Reaction of **224** with catalyst **A** furnished only traces of the desired product. Application of catalyst **C** gave the desired 14-membered benzolactone **226** with (*Z*) configuration at the newly formed double bond, which was deprotected to **229** and finally chlorinated to **230**. Later on, with the aim of improving the unfavorable pharmacokinetics of **230**, a similar RCM-based route was examined to obtain the cyclopropa- analog **228** [109]. Under the reaction conditions applied to **224**, cyclopropa-derivative **225** furnished the desired cyclization product **227** in only 20% yield together with substantial amounts of dimers. Carrying out RCM in refluxing toluene at higher dilution afforded an improved yield of the monomeric macrocycle when the reaction was quenched after a few minutes. Runs with prolonged reaction times resulted in the formation of more dimer, indicating that the monomer might eventually revert to the thermodynamically more favored dimers.

In a more recent and improved approach to cyclopropa-radicicol (**228**) [110], also outlined in Scheme 48, the synthesis was achieved via ynolide **231** which was transformed to the stable cobalt complex **232**. RCM of **232** mediated by catalyst **C** led to cyclization product **233** as a 2:1 mixture of isomers in 57% yield. Oxidative removal of cobalt from this mixture followed by cycloaddition of the resulting cycloalkyne **234** with the cyclic diene **235** led to the benzofused macrolactone **236**, which was converted to cyclopropa-radicicol (**228**).

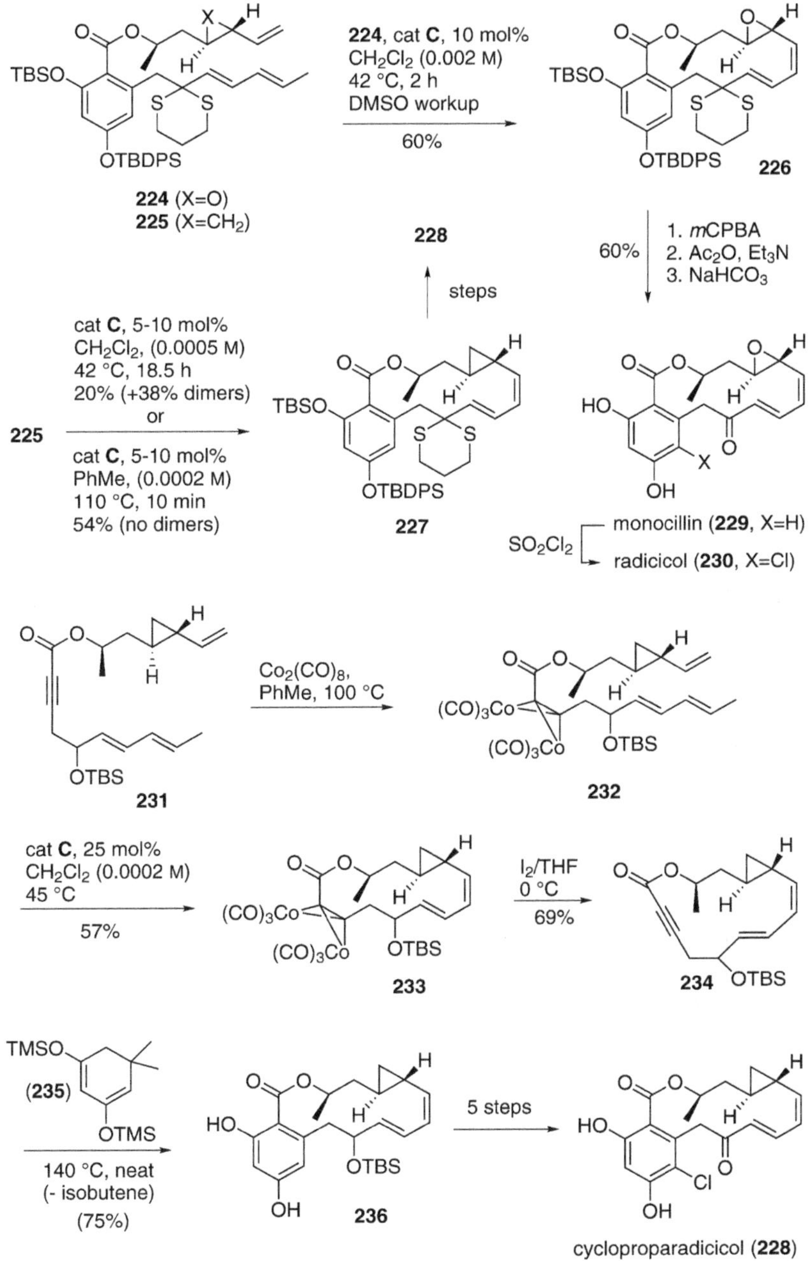

Scheme 48 Diene-ene RCM in Danishefsky's total synthesis of radicicol (**230**) [108] and its cyclopropa-analog **228** [109, 110]

The epothilones **237** and **238** are 16-membered macrolides isolated from *myxobacteria* (Fig. 8). These compounds possess a taxol-like mode of action, epoB (**238b**) being the most active compound, and function through the stabilization of cellular microtubules, exhibiting cytotoxicity even in taxol-resistant cell lines. The emergence of epothilones as promising anticancer drug candidates has led to a worldwide effort to synthesize new analogs with a view to identifying and developing later-generation derivatives for clinical evaluation [111]. Previous attempts at applying RCM to epothilone syntheses have been repeatedly plagued by complete lack of stereocontrol in the generation of the desired (*Z*)-12,13-olefin geometry [112].

An alternative RCM-based *bond connection between C10 and C11* in the epothilone series was used in Danishefsky's total synthesis of epo490 (**240d**), a naturally occurring, recently discovered cometabolite, that differs from epoD (**237d**) by the presence of an additional (*E*)-10,11 double bond [113]. This alternative macrocyclization also proved to be a viable and novel route to **237d** (Scheme 49). Initially, the metathesis step was performed with differently protected precursor diene-enes **239a–c** using catalyst **C** in refluxing dichloromethane. It turned out that triene **239a** led to a mixture of two compounds in a 2.3:1 ratio with a total yield of 50% (no reaction at all was observed with ruthenium catalyst **A**, while molybdenum catalyst **B** led to decomposition of **239a**). The major component of the mixture was the desired RCM product **240a**, while the 14-membered by-product **241a** arose from extrusion of a propene unit from attack at the internal olefin [114]. When the cyclization of **239a** was performed in refluxing toluene for a few minutes, the yield of **240a** was distinctly improved, while the amount of the by-product decreased. (A similar beneficial effect by performing ene-diene RCM in toluene was also observed for analogous compounds with only slight structural variations). Performing the ring closure as the last synthetic step with unprotected diol **239d** led directly to epo490 (**240d**) in 64% yield; as both the C3 and C7 alcohols in **139d** are β to carbonyl groups, it is assumed that intramolecular H bonding contributes a higher degree of favorable rigidity to the cyclization precursor. Finally, selective diimide reduction of **240d** led to epoD (**237d**), a current clinical candidate in the epothilone series [115].

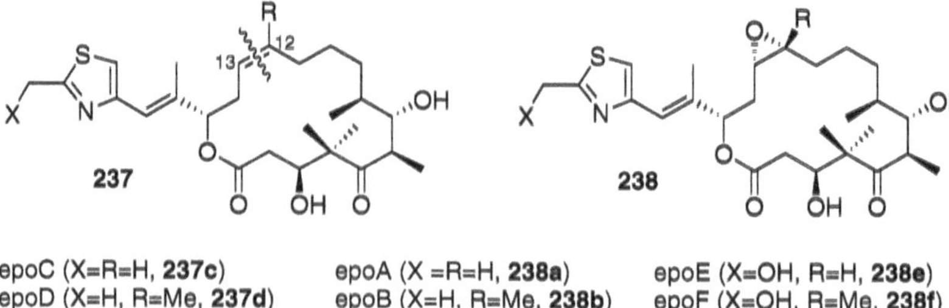

Fig. 8 Structure of epothilones A–F

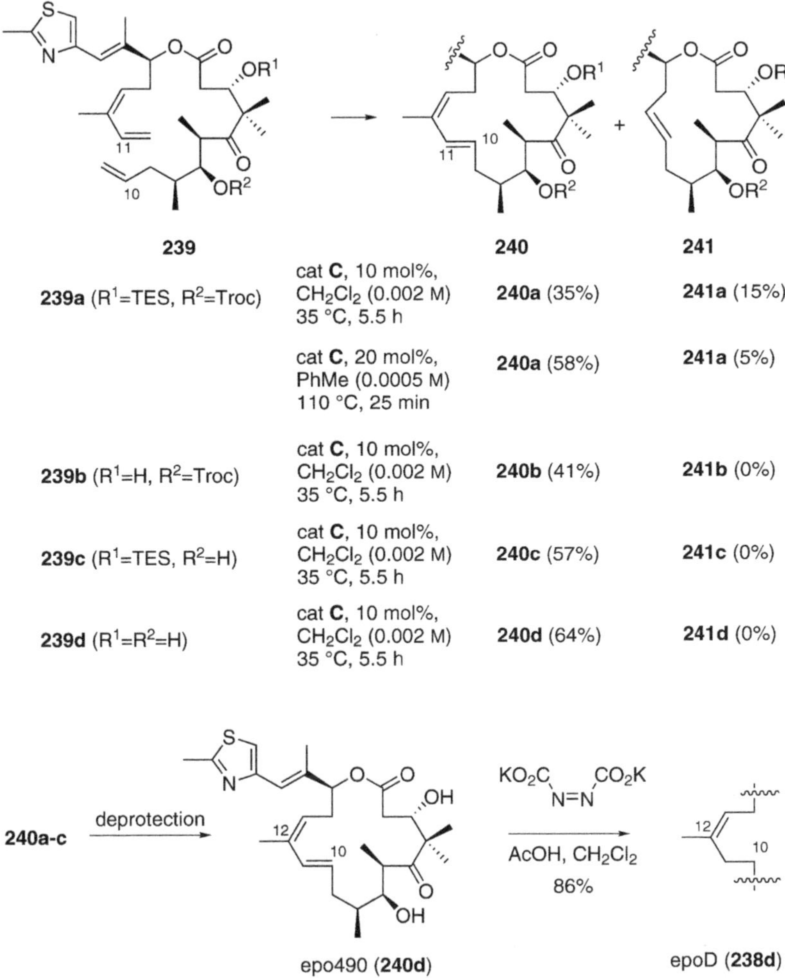

Scheme 49 Total syntheses of epo490 (**240d**) and epoD (**237d**) by diene-ene RCM between C10 and C11: favorable influence of solvent toluene or unprotected hydroxy groups in metathesis substrates **239** [113]

The synthesis of an epothilone model system via an alternative *C9–C10 disconnection* was first examined by Danishefsky in 1997. However, extension of this C9–C10 strategy to a fully functionalized epothilone intermediate was not successful, demonstrating the limitations of RCM with the early catalysts **A** and **B** [116]. In 2002, Sinha and Sun disclosed the stereoselective total syntheses of epoA (**238a**) and epoB (**238b**) by the RCM of epoxy compounds **242** in the presence of catalyst **C** (Scheme 50) [117]. The reaction furnished an inconsequential mixture of isomers **243** (*E*/*Z*≈1:1) in high yield. Subsequent selective hydrogenation of the newly formed double bond followed by deprotection led to epothilones A and B.

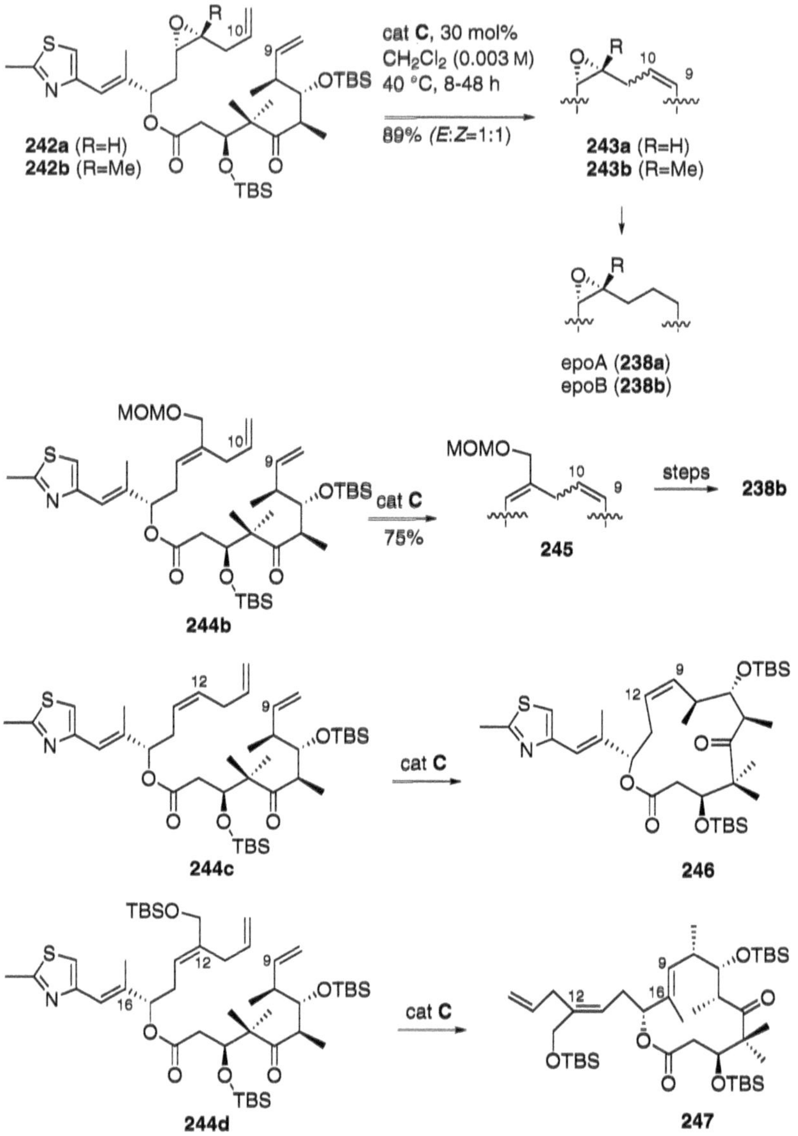

Scheme 50 Epothilone synthesis via RCM between C9 and C10: dependence of chemoselectivity on the size of the C12 substituent in metathesis substrates **244** [117]

Alternatively, diene-ene **244b** was also efficiently cyclized in the presence of catalyst **C** to produce macrolides **245b** (*E*/*Z* mixture at the newly formed double bond) in 75% yield. Global deprotection of **245b**, followed by a sequence of selective hydrogenation at C9–C10, Sharpless asymmetric epoxidation, and deoxygenation of the primary hydroxy group provided an alternative route to epoB (**238b**). In contrast analog **244c** with an unsubstituted 1,4-diene moiety changed the course of the metathesis reaction, leading to the 13-membered macrocycle **246** instead, while tetraene **244d** with a bulky $TBSOCH_2$ group at C12 reacted with the thiazole-substituted double bond [118].

A similar approach, outlined in Scheme 51, was disclosed recently by Danishefsky using methyl ketones **248a** and **248b** as the RCM substrates [119]. Treatment of diene-enes **248** in refluxing toluene with catalyst **C** for a few minutes afforded exclusively (*E*)-isomers **249** in high yield. The thiazole moiety was then installed (*E*)-selectively by olefination with tributylphosphonium salt **250**. Subsequent deprotection of the olefination product obtained from **249a** led to (*E*)-9,10-dehydro-epoB (**251a**), which was *not identical* to a previously reported compound presumed to be the same entity [120]. Moreover, the novel compound **251a** proved to exhibit highly promising in vitro and in vivo potencies, as well as encouraging pharmacokinetic properties. Site-selective diimide reduction of **251a** led to epoD (**237d**). Note that the 12-CF_3 analog **251b** was recognized to feature even more favorable therapeutic activities, and alternative routes to the key fragments leading to metathesis substrate **248b** have been developed [119b]. (For an approach to epoA by ring-closing alkyne metathesis, see Scheme 90; for a sequence of ROM–CM to obtain epothilone analogs, see Scheme 81).

The first example of successful *diene–diene RCM* to construct a macrocyclic conjugated triene was disclosed by Wang and Porco in the first total synthesis of oximidine II (**258**) [121]. Oximidine II belongs to the family of salicylate en-

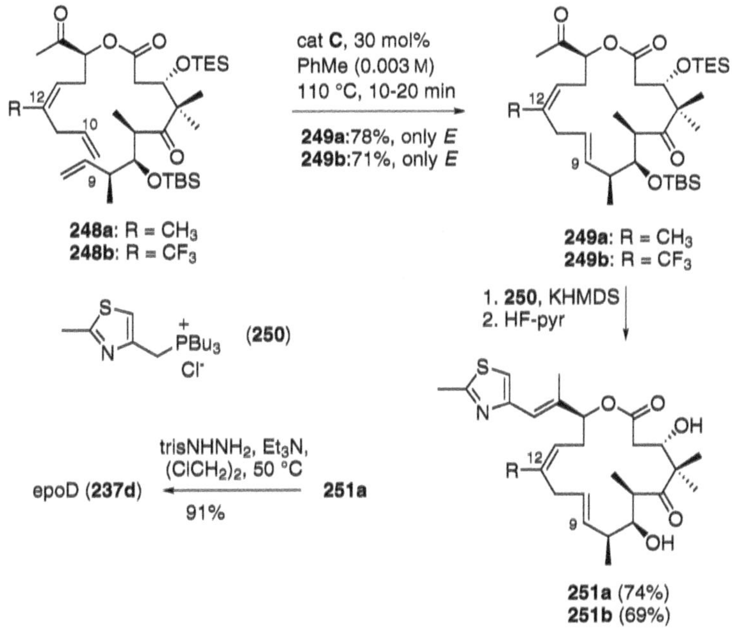

Scheme 51 Synthesis of novel (*E*)-9,10-dehydro analogs **251** of epoD, and a novel route to epoD (**237d**) via RCM between C9 and C10 [119]

amide macrolides. It contains an (*E,Z,Z*)-conjugated triene unit in a 12-membered macrolactone core (Scheme 52). In the first attempts, tetraenes **252a–d** were chosen as metathesis substrates. Treatment with catalyst **A** afforded only products resulting from reaction of the *trans*-diene moiety, giving no further conversion, as exemplified by the formation of ruthenium complex **253** from **252b**. When treated with catalyst **C**, substrates **252a** and **252b** again reacted with the *trans*-diene, while the constrained substrates **252c** and **252d** afforded oligomeric products within 20 min, and only traces of undesired ten-membered product(s) were detected by HPLC–MS analysis. In an effort to initiate the ring closure at the *cis*-diene site, attempts were then undertaken with substrates **254** bearing an additional methyl substituent at the *trans*-diene moiety. Phenol **254b** afforded only oligomeric products, while the silyl ether **254a** eventually furnished the oximidine core **255** under the conditions outlined in the scheme. Cyclization product **255** with the required (*E,Z,Z*) geometry was selectively obtained in 39% yield, together with 49% of the precursor tetraene and oligomeric products. Extended reaction time resulted in decomposition of both starting material and product. The synthesis of **258** was then completed by conversion of **255** to (*Z*)-vinyl iodide **256** and copper-mediated amidation with **257** to construct the enamide side chain.

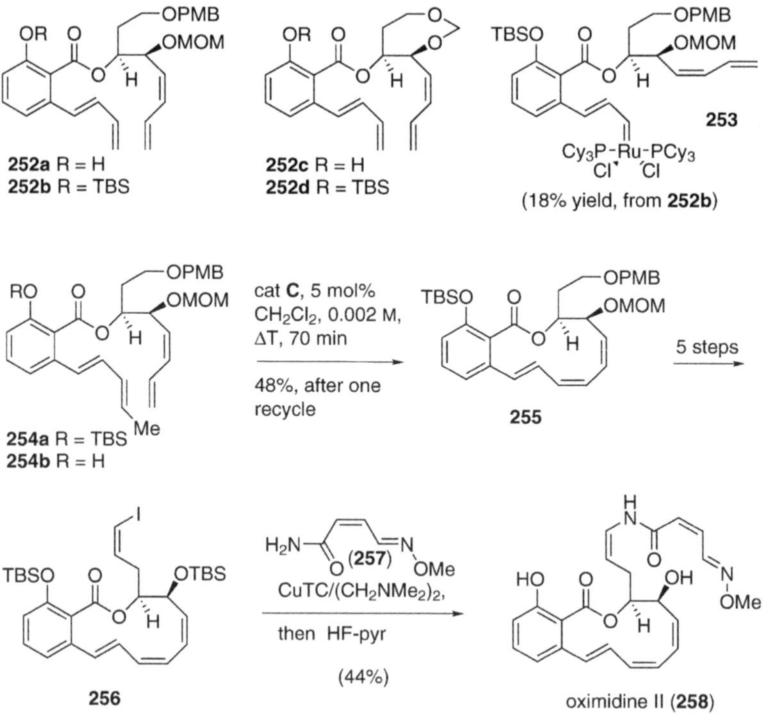

Scheme 52 Stereo- and regioselective diene–diene RCM of tetraene **254a** in the first total synthesis of oximidine II (**258**) [121]

2.1.3.3
Bridged Compounds

The possibility of distinguishing between medium-sized rings and macrocycles, as well as between terminal and internal (disubstituted) double bonds, and of performing "tandem catalysis" events (RCM/ADMET–hydrogenation) with a single ruthenium complex as the catalytically competent precursor, was nicely demonstrated in Fürstner's total synthesis of the natural *meta*-pyridinophane (+)-muscopyridine (**264**) (Scheme 53) [122]. Sequential coupling of pyridine derivative **259** with two different Grignard reagents mediated by iron complex **260** led to an inseparable 4:1 mixture of disubstituted pyridines **261** and **262**, with compound **262** formed in the first coupling step. As ruthenium-based metathesis catalysts are poisened by amines, pyridines **261** and **262** were pro-

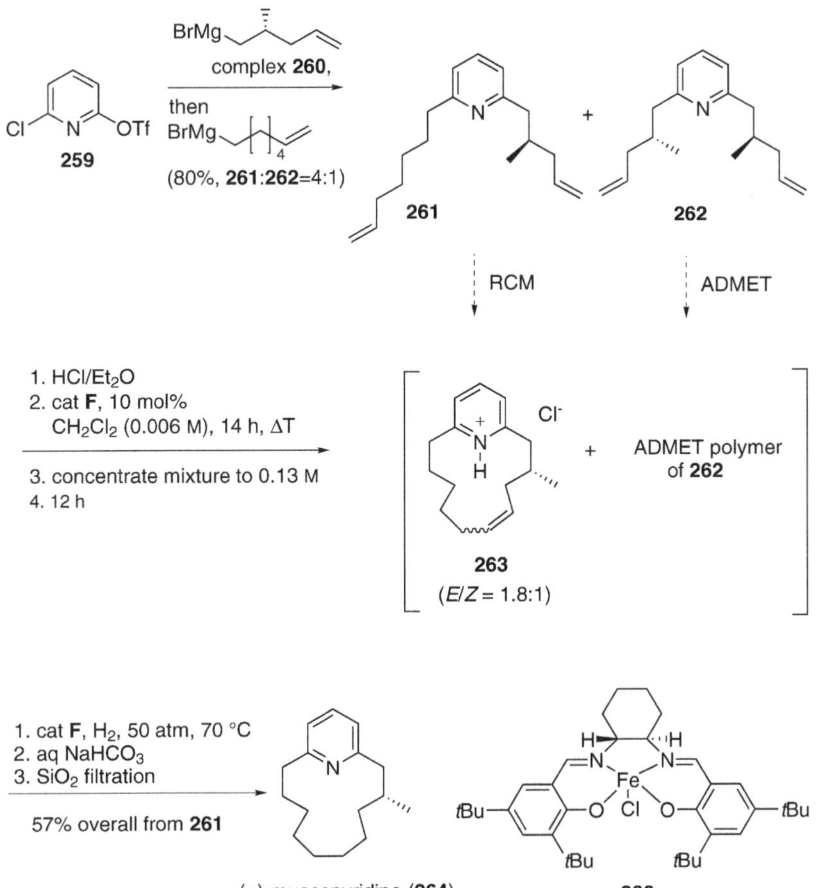

Scheme 53 Fürstner's RCM-based synthesis of muscopyridine (**264**) with integrated "self-clearance" by proper choice of catalyst activity [122]

tonated and the resulting hydrochlorides subjected to RCM under high dilution conditions (0.006 M) in the presence of Fürstner's catalyst **F** [123], which features a similar activity to Grubbs' first-generation catalyst **A**. Under these conditions, only the kinetically and thermodynamically favored 13-membered ring **263** derived from diene **261** was formed. On concentration of the reaction mixture (to 0.13 M), the residual hydrochloride of **262** was forced to polymerize by acyclic diene metathesis (ADMET), while hydrochloride **263**, now containing a less reactive (disubstituted) double bond, persisted. Notably, this would not be the case using a more reactive second-generation ruthenium catalyst. Further postponing workup, the crude mixture containing monomer **263**, the ADMET polymer, and the still-intact catalyst **F** was transferred into an autoclave and stirred under hydrogen for 14 h. Thereby, carbene complex **F** was converted into an active Ru hydride species, acting as an effective hydrogenation catalyst, so that after this procedure target **264** could be isolated in 57% overall yield from the starting mixture of dienes **261** and **262** by passing the mixture through silica.

The marine natural product (+)-chatancin (**272**), a platelet factor antagonist with several interesting biological activities, features a *cis-anti-cis*-dodecahydrophenanthrene framework possessing seven stereogenic centers (Scheme 54). In a recent attempt by Deslongchamps to prepare **272** by one of two proposed pathways involving transannular Diels–Alder (TADA) reaction [124], furanophane **269** was projected as a key intermediate to generate tetracycle **270** with stereofacial and diastereocontrol. Subsequent hydride-shift-mediated oxygen transposition should then generate **272**. The furanophane **269** in turn, featuring a trisubstituted double bond with (*E*) configuration (necessary for the success of the TADA reaction), was to be generated by RCM.

Diene **265**, substituted by a bulky silyl ether to prevent cycloaddition before the metathesis process, produced in the presence of catalyst **C** the undesired furanophane **266** with a (*Z*) double bond as the sole reaction product in high yield. The same compound was obtained with Schrock's molybdenum catalyst **B**, while first-generation catalyst **A** led even under very high dilution only to an isomeric mixture of dimerized products. The (*Z*)-configured furanophane **266** after desilylation did not, in accordance with earlier observations, produce any TADA product. On the other hand, dienone **267** furnished the desired macrocycle (*E*)-**268**, though as minor component in a 2:1 isomeric mixture with (*Z*)-**268**. Alcohol **269** derived from *E*-**268** then underwent the projected TADA reaction selectively to produce cycloadduct **270** (70% conversion) in a reversible process after 3 days. The final Lewis acid-mediated conversion to **272** however did not occur, delivering anhydrochatancin **271** instead.

Roseophilin (**273**), a deeply red-colored pentacyclic compound isolated from the culture broth of *Streptomyces griseoviridis*, is a novel antitumor antibiotic. Compound **273** possesses a topologically unique pentacyclic skeleton, consisting of a 13-membered macrocycle incorporated in an *ansa*-bridged azafulvene, which in turn is linked to a conjugated heterocyclic ring system. The absolute stereochemistry of roseophilin, as depicted in Fig. 9, was unknown until the first total synthesis published by Tius and Harrington in 2001 [125]. All syn-

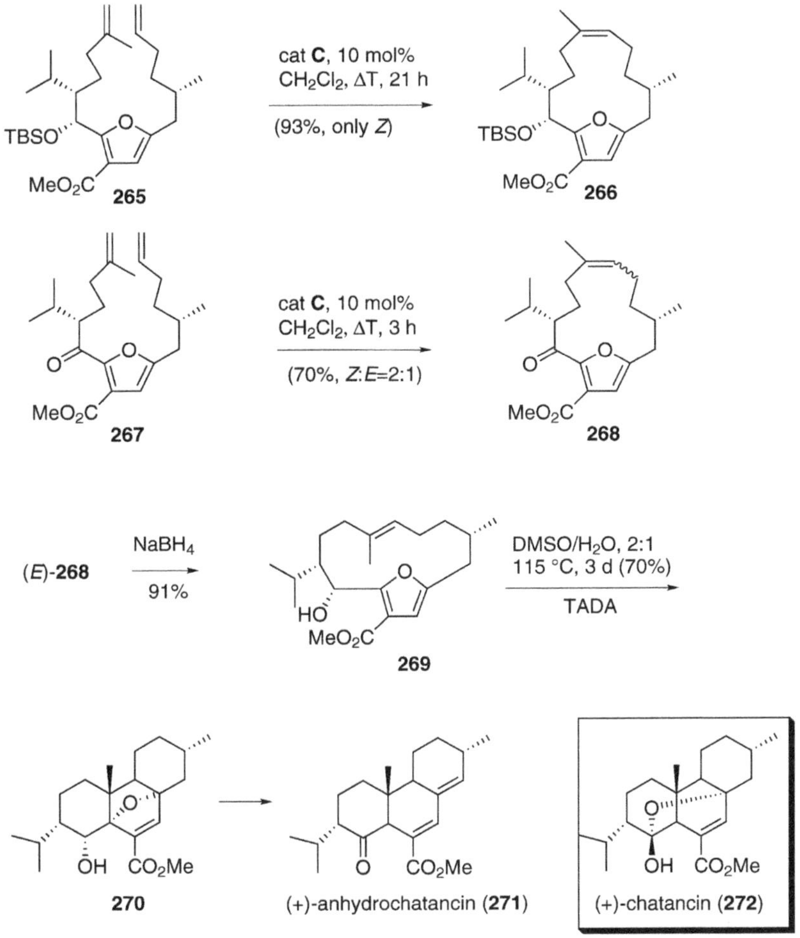

Scheme 54 Sequential RCM and TADA reactions in Deslongchamps' biomimetic synthesis of anhydrochatancin **271** [124]

thetic approaches toward **273** known to date rely on tricyclic ketone **274** as one of two main fragments [126]. Various approaches to **274** (*ent*-**274** or *rac*-**274**) were performed via an RCM step to form the 13-membered macrocycle.

The respective metathesis substrates used by the different groups, as well as the reaction conditions used in the RCM step, are presented in Fig. 9. In the approach pursued by Fuchs [127], the racemic dienes **275a–e** were investigated. The unsubstituted compound **275a** and the diastereomeric alcohols **275b,c** did not cyclize in the presence of catalyst **A**. From the bulky silyl ethers derived from alcohols **275b,c**, only one (**275e**) underwent cyclization. Evidently, in this special case, the bulky TIPS ether helped to orient the olefinic side chains into a favorable conformation. In the approach of Hiemstra and coworkers leading to *ent*-**274** [128], the phenylsulfonyl-substituted diene **276** proved to be a very

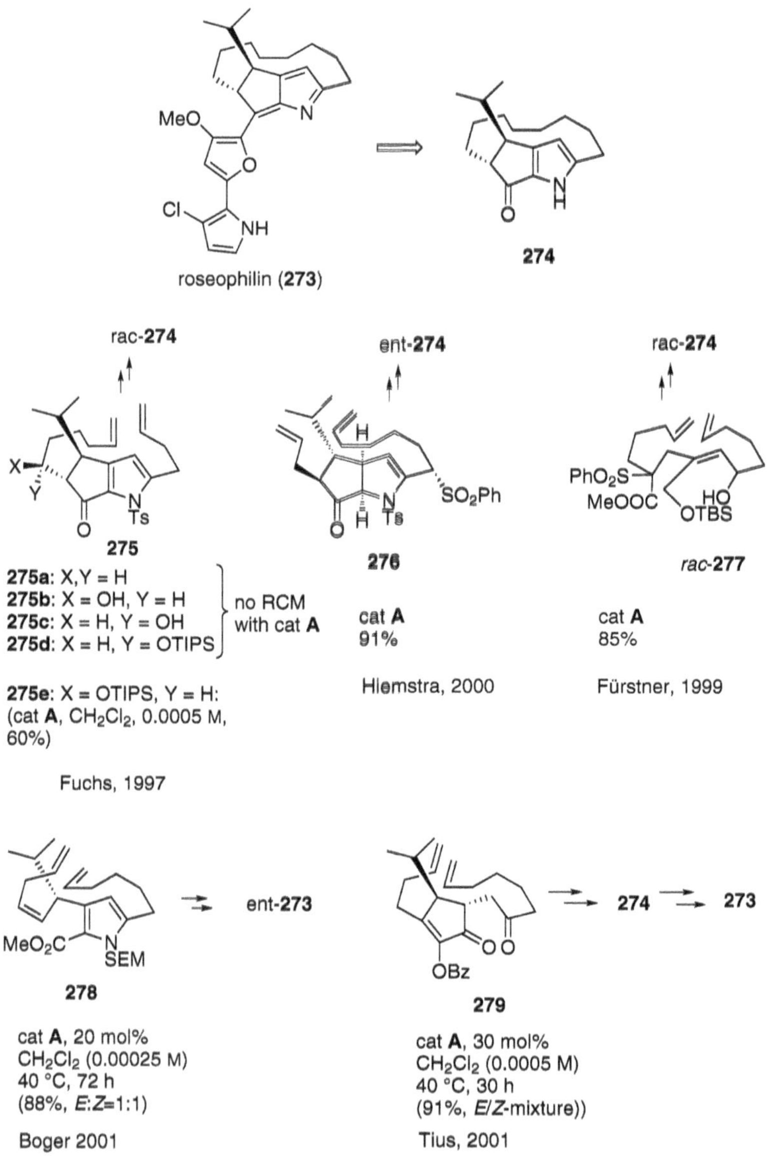

MeO
Cl
NH
N
O
roseophilin (**273**)

274

rac-**274**

275

275a: X,Y = H
275b: X = OH, Y = H
275c: X = H, Y = OH
275d: X = H, Y = OTIPS
no RCM with cat **A**

275e: X = OTIPS, Y = H: (cat **A**, CH_2Cl_2, 0.0005 M, 60%)

Fuchs, 1997

ent-**274**

SO_2Ph

276

cat **A**
91%

Hiemstra, 2000

rac-**274**

PhO_2S
MeOOC
HO
OTBS

rac-**277**

cat **A**
85%

Fürstner, 1999

MeO_2C
SEM

278

ent-**273**

cat **A**, 20 mol%
CH_2Cl_2 (0.00025 M)
40 °C, 72 h
(88%, *E*:*Z*=1:1)

Boger 2001

OBz

279

274 **273**

cat **A**, 30 mol%
CH_2Cl_2 (0.0005 M)
40 °C, 30 h
(91%, *E*/*Z*-mixture))

Tius, 2001

Fig. 9 Various RCM substrates used in total syntheses of roseophilin (**273**)

efficient metathesis substrate, providing the desired macrocycle (mixture of *E*/*Z*-isomers) in 91% yield. The efficiency of this reaction was ascribed to both the conformational restriction induced by the phenylsulfonyl group and the concave shape of the *cis*-fused bicyclic system present in **276**, which cooperatively bring the reacting double bonds in close proximity. In the case of Fürstner's acyclic metathesis substrate *rac*-**277** [129], no additional conformational assistance was necessary, and treatment with catalyst **A** led to the corresponding macrocycles in high yield. In Boger's total synthesis of *ent*-**273** [130], the macrocycle was closed efficiently by treatment of the monocyclic triene **278** with catalyst **A**. The formation of the *ansa*-macrocycle prior to formation of the cyclopentanone avoids, to a large extent, the strain to be overcome in compounds **275**. The cyclopentanone ring was subsequently introduced by a 5-*exo-trig* radical-alkene cyclization of the acyl selenide derived from the ester group. Also in Tius' first total synthesis of enantiomerically pure **273** [125], a monocyclic diene (**279**) was used to produce the macrocycle efficiently. After selective hydrogenation of the newly formed double bond, the missing pyrrole ring was formed by involving the 1,4-dicarbonyl moiety in a Paal–Knorr reaction.

The marine alkaloid sarain A (**285**) features an exceptionally challenging pentacyclic architecture (Fig. 10). To date, **285** has not succumbed to a total synthesis. Two groups, however, have completed the tricyclic core of **285** and have annulated the western 13-membered ring using quite similar RCM approaches [131]. The results obtained with different metathesis substrates and catalysts are outlined in Fig. 10. RCM of Weinreb's dienes **280**, **281**, and **282** [131a], that differ by the site of ring closure, were mediated by first-generation catalyst **A**. Dienes **280** and **282** furnished comparable results leading to the corresponding cyclization products in moderate yield together with substantial amounts of *cyclic dimers*. RCM of diene **281**, however, proceeded very sluggishly leading to an inseparable mixture of the desired macrocycle along with a *linear dimer* in poor overall yield, suggesting that the allylic side chain was positioned too close to the tricyclic core to participate efficiently in the metathesis event. Four years later, when the strategy was adapted by Cha and coworkers [131b], RCM of dienes **283** and **284** was performed with catalyst **C**. In contrast to the uncomplicated ring closure of the *N*-PMB-protected derivative **283** (71% yield within 5 h), it was surprising that diene **284**, bearing a more elaborate alkyl chain instead, produced the macrocycle in distinctly lower yield (42%) along with a dimer. Silylation of **284** prior to RCM gave a reliable higher overall yield.

In the organization of RCM to strained products, it is essential to preorganize the substrate into a conformation that favors cyclization. In the above roseophilin case, the stereochemical outcome of the ring-closing step was inconsequential for the successful formation of a saturated macrocyclic ring. In Deslongchamps' synthesis of a strained furanophane, attempts to obtain the required (*E*) stereochemistry were only partially successful by alcohol to carbonyl interconversion in the RCM precursor. A highlight among RCM-based natural product syntheses, that pushes the limits of olefin metathesis as a means to construct highly strained and complex targets with total stereocontrol, is found in Nicolaou's first total synthesis of two members of the coleophomone family, namely coleophomone B (**287**) and C (**286**) [132]. These compounds differ only in the geometry of the double bond in the macrocyclic *ansa* bridge (Scheme 55). In addition to an interesting biological profile, the coleophomones feature a strained and rigid framework with a sensitive tricarbonyl system tethered to an 11-membered macrocycle, whose strain is derived from a fused aryl ring and

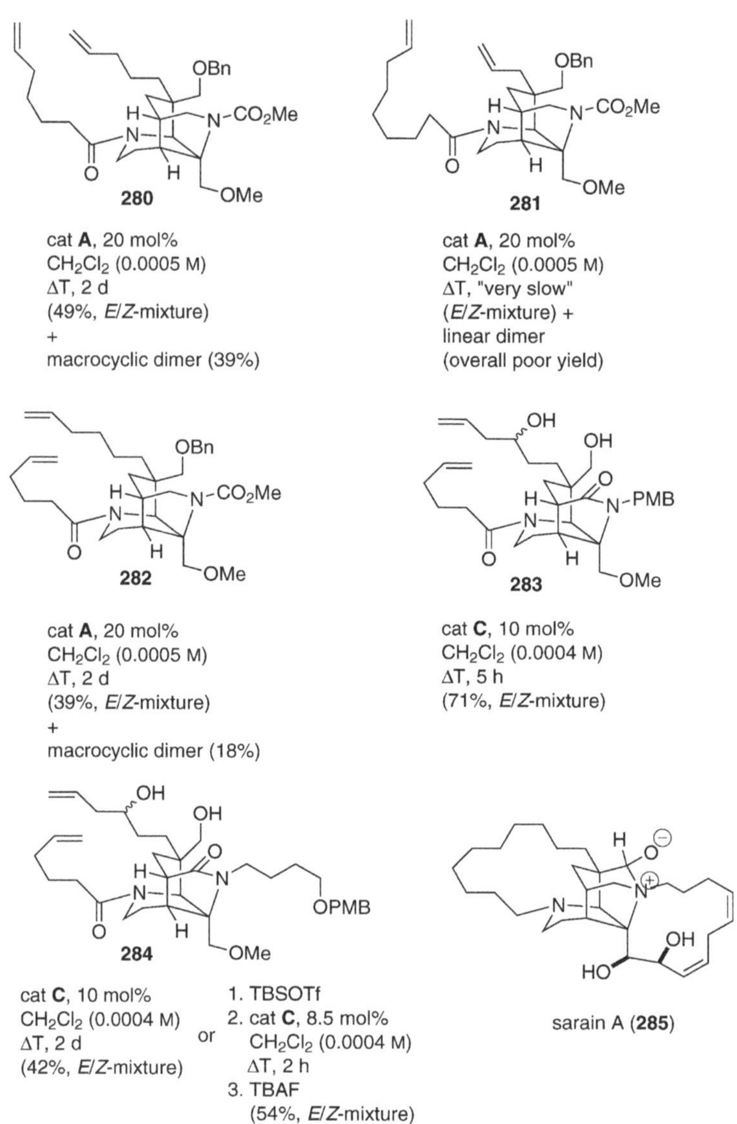

Fig. 10 Various RCM substrates used in synthetic work directed to the marine alkaloid sarain (**285**)

an internal cyclohexadienone. During the exploration of the crucial metathesis step, various dienes **288** and enol ether **291** synthesized from **288a** were investigated. In the case of the simplest substrate **288a** bearing a monoalkylated cyclohexadione moiety, first-generation catalyst **A** failed to induce ring closure. With second-generation catalyst **C**, the tricycle **289** with a (*Z*) double bond was formed in 60% yield. Also with diene **288b**, bearing a di- and a trisubstituted

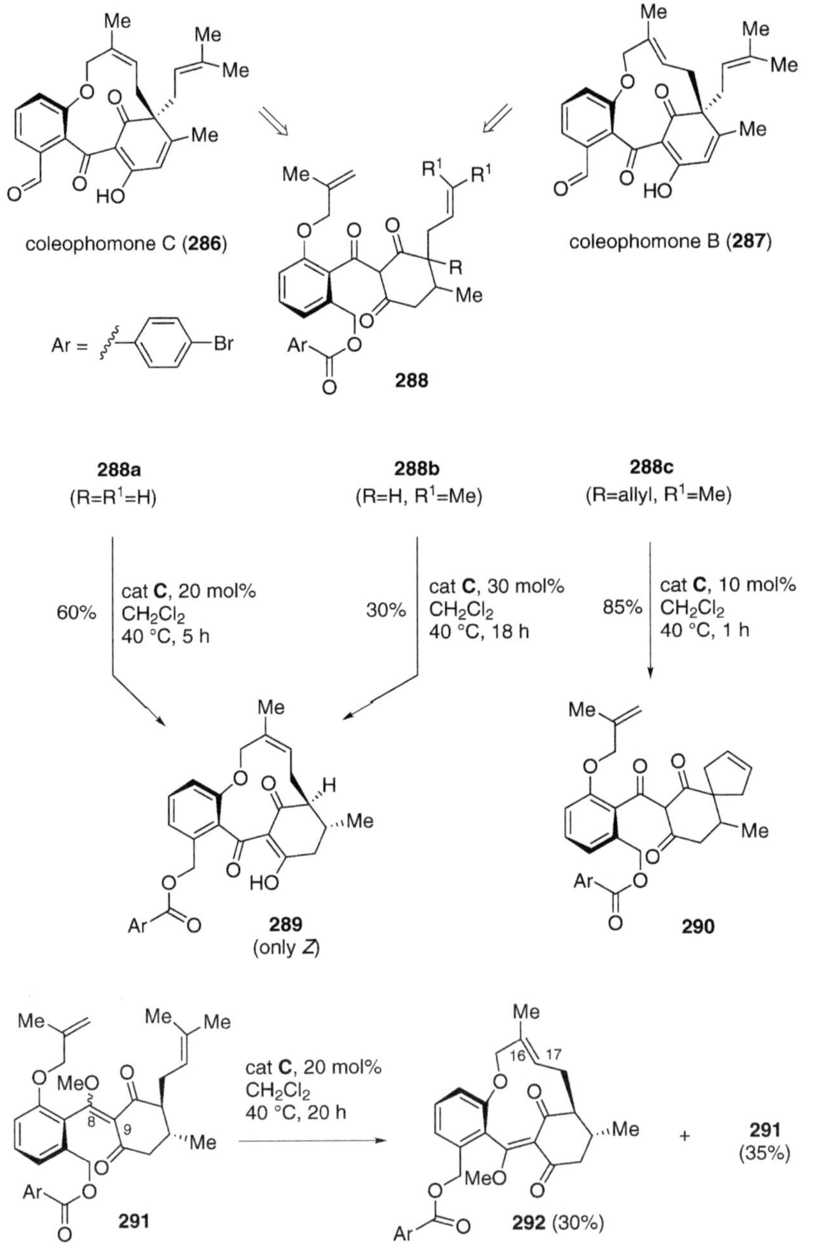

coleophomone C (**286**)

coleophomone B (**287**)

Ar =

288

288a (R=R[1]=H)

288b (R=H, R[1]=Me)

288c (R=allyl, R[1]=Me)

60% cat **C**, 20 mol% CH_2Cl_2 40 °C, 5 h

30% cat **C**, 30 mol% CH_2Cl_2 40 °C, 18 h

85% cat **C**, 10 mol% CH_2Cl_2 40 °C, 1 h

289 (only *Z*)

290

291

cat **C**, 20 mol% CH_2Cl_2 40 °C, 20 h

292 (30%) + **291** (35%)

Scheme 55 First retrosynthetic analysis and RCM substrates investigated during synthetic work directed to coleophomone B (**287**) and C (**286**) [132]

double bond, the exclusive formation of **289** was observed, albeit in reduced yield (30%). However, the simplest dialkylated cyclohexadione derivative, triene **288c** (bearing an allyl and a prenyl substituent), did produce rapidly within 1 h only spirocyclopentene **290** in 85% yield. Additional and unexpected information was gained by the RCM reaction of enol ether(s) **291** (1:1 mixture of $\Delta^{8,9}$-isomers, each of which consists of a 1:1 pair of atropisomers). When this mixture was subjected to the usual metathesis conditions, a single macrocycle

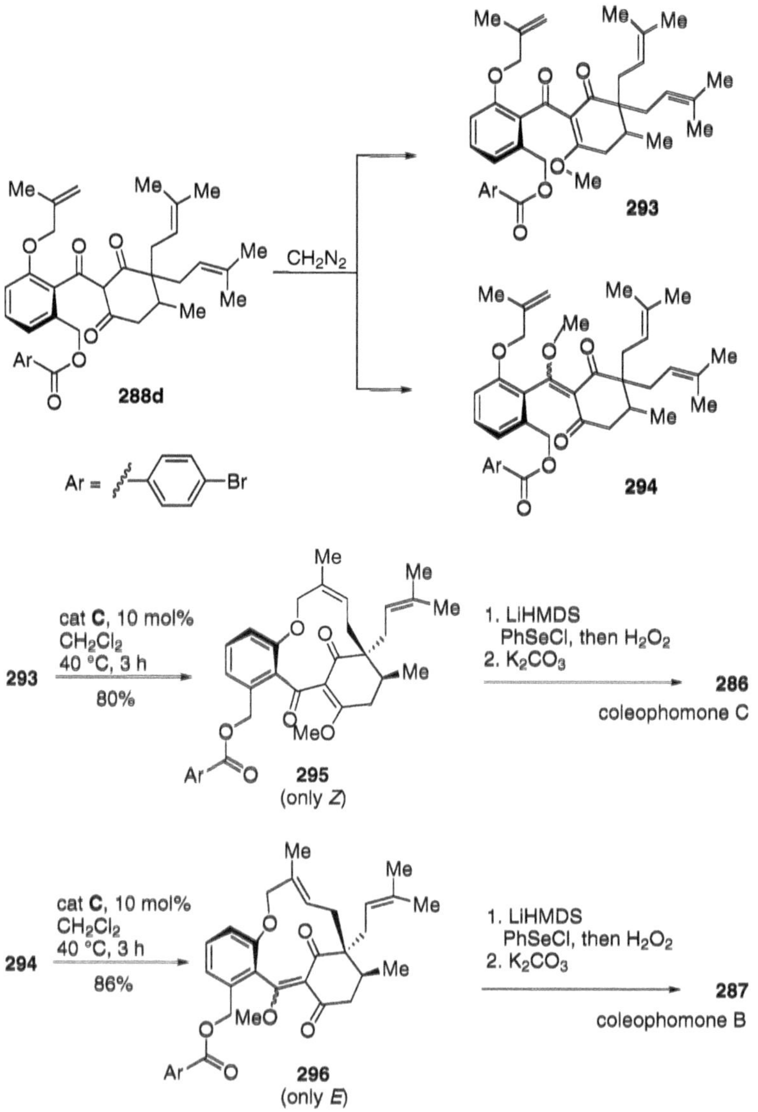

Me Me Me Me Me O O O Me O Ar O O 288d
Ar = Br
CH_2N_2
293
294
293 cat **C**, 10 mol% CH_2Cl_2 40 °C, 3 h 80% → 295 (only *Z*) → 1. LiHMDS PhSeCl, then H_2O_2 2. K_2CO_3 → 286 coleophomone C
294 cat **C**, 10 mol% CH_2Cl_2 40 °C, 3 h 86% → 296 (only *E*) → 1. LiHMDS PhSeCl, then H_2O_2 2. K_2CO_3 → 287 coleophomone B

Scheme 56 Final solution of the coleophomone problem: stereoselective macrocyclization by RCM of enol ether derivatives **293** and **294** [132]

(**292**) with (*E*) configuration at both the $\Delta^{8,9}$ and the newly formed $\Delta^{16,17}$ double bond, was isolated in 30% yield, together with a considerable amount of the starting material which was found to be enriched in the (*Z*)-isomer around the enol ether ($\Delta^{8,9}$) double bond.

The final solution of the coleophomone problem is outlined in Scheme 56. The fully substituted diprenylated compound **288d**, itself a very poor RCM substrate, was treated with diazomethane, which led to a separable mixture of regioisomeric enol ethers **293** and **294**, the latter being a ca. 1.3:1 mixture of geometrical isomers ($\Delta^{8,9}$). Treatment of **293** with catalyst **C** led within 3 h exclusively to the (*Z*)-configured macrocycle **295** in 80% yield. Regioisomer **294** in turn furnished under the same conditions the (*E*)-configured macrocycles **296** (ca. 1:1 mixture of isomers at $\Delta^{8,9}$). Remarkably, in both cases only the prenyl group in the *cis* position to the vicinal C12 methyl group participated in the ring-closing step. Thus, two different coleophomone frameworks were obtained stereospecifically from a single precursor (**288d**). Conversion of compounds **295** and **296** to coleophomone C (**286**) and B (**287**), respectively, was accomplished in both series by introducing the missing $\Delta^{11,12}$ double bond and global deprotection.

2.2 Olefin Cross Metathesis (CM)

Olefin CM can be formally described as the mutual intermolecular exchange of alkylidene (or carbene) fragments between two olefins promoted by metal-carbene complexes [133]. For decades CM has found numerous industrial uses, but it is not yet in such widespread use in natural product synthesis as the more entropically favorable RCM reaction. This is largely due to inherent difficulties of controlling selectivities. Minimization of unproductive self-coupled alkenes and maximization of the crossed product is one of the crucial issues to be optimized in CM chemistry, as well as stereocontrol on the newly generated double bond. It is only in the last few years that, with the development of a second generation of highly active and stable ruthenium catalysts bearing *N*-heterocyclic carbene (NHC) ligands, the synthetic community has begun to accept CM reactions as a useful alternative in natural product synthesis. In the presence of NHC catalysts **C** and **D**, the range of substrates amenable to CM has been also expanded to trisubstituted and electron-deficient conjugated olefins. The latter are known to be poor substrates for homodimerization and allow the (*E*)-selective introduction of functionality to an α-olefin. CM with these substrates can therefore be considered as a formal vinylic C–H activation or a formal allylic oxidation [134]. The impressive developments in olefin CM, including a number of applications in the syntheses of biologically important molecules and natural products, have been the subject of an excellent review by Connon and Blechert [4]. In a recent and comprehensive article, Grubbs and coworkers have developed an important general model for the prediction of product selectivity and stereoselectivity in CM [135], by categorizing the olefins into four different types according to their relative abilities to undergo homodimerization and the susceptibility of their homodimers toward secondary metathesis reactions. By means of many examples it was demonstrated that by employing a metathesis catalyst with the appropriate activity, selective CM reactions are possible with a variety of electron-rich, electron-deficient, and sterically demanding olefins. In the following, we will concentrate only on a few recent examples that highlight the growing use of CM in natural product syntheses.

An "early" example of nonstereoselective CM used in the synthesis of biologically interesting trisubstituted γ-lactones is the late-stage introduction of a large saturated alkyl chain by a CM–hydrogenation sequence in Reiser's synthesis of (+)-roccellaric acid (**299**, Scheme 57) [136]. The required tridecyl side chain was elaborated by exposing the allyl-substituted intermediate **297** and 1-dodecene (1.5 equiv) to Ru catalyst **A**. No self-condensation of **297** was observed, the homodimer of dodecene being the only by-product. CM product **298** was obtained in 64% yield as a 3.5:1 mixture of (*E*/*Z*)-isomers, which was hydrogenated and converted to **299**.

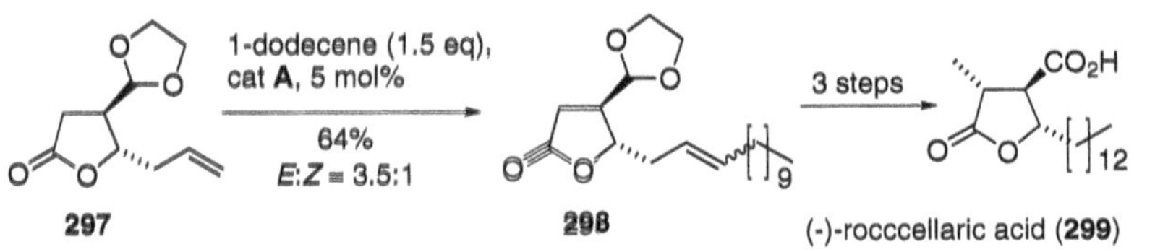

Scheme 57 CM in the total synthesis of roccellaric acid (**299**) [136]

The CM reaction between 2-methyl-2-butene (a *gem*-disubstituted olefin that served in this case also as solvent) and the allylated compound **300**, possessing the bicyclo[3.3.1]nonane core of the potential Alzheimer therapeutic garsubellin A (**302**) [137], underlines the increased activity of the second-generation ruthenium catalysts (Scheme 58). In the presence of 10 mol% of NHC catalyst **C**, the prenylated compound **301** was formed after only 2 h in 88% yield.

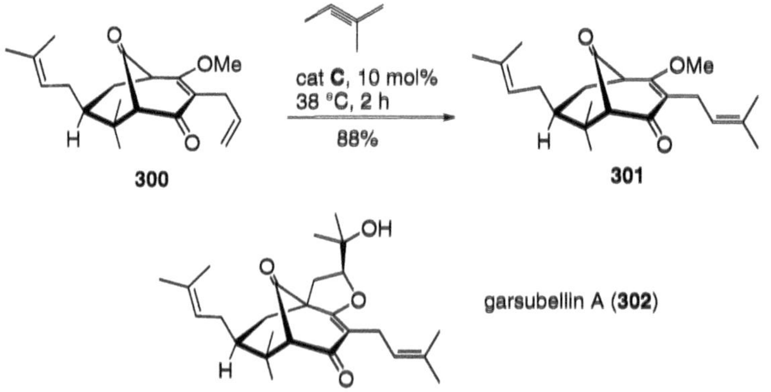

Scheme 58 Allyl to prenyl interconversion by the use of a trisubstituted olefin in CM, in garsubellin A (**302**)-directed work [137]

In an article dealing with applications of olefin CM to a series of commercial products [138], solvent-free CM between (*E*)-3-hexene (produced by homocoupling of 1-butene) and 11-eicosenyl acetate **303** (produced from jojoba oil) was used to produce acetate **304** (Scheme 59), which is – as a natural 82:18 (*E*/*Z*) mixture – the pheromone of omnivorous leafroller, and serves as an environment-friendly pest controlling agent. The CM reaction was performed without solvent at 5 °C with a 4:1 mixture of (*E*)-3-hexene and **303**, in the presence of only 0.2 mol% catalyst **C**, and furnished after 20 h coupling product **304** (*E*:*Z*=83:17) in 50% yield.

(4 equiv)
OAc
303
cat **C**, 0.2 mol%
5 °C, 16 h
84% conversion
OAc
E:*Z* = 83:17 50%
Omnivorous Leafroller pheromone (**304**)

Scheme 59 Synthesis of omnivorous leafroller pheromone **304** via CM [138]

The high activity of the second-generation Ru catalysts also allows the use of conjugated electron-deficient olefins as efficient CM partners with low tendency to self-dimerization. Several syntheses of biologically interesting natural compounds [139], or advanced intermediates therefrom [140], have been disclosed by the group of Cossy using an iterative sequence of stereoselective allyl titanations and CM reactions of the resulting homoallylic alcohols [141]. En route to the natural compound (+)-strictifolione (**308**), metathesis precursor **305** (Scheme 60) was prepared from 3-phenylpropionaldehyde via two sequential allylations. CM with acrolein (3 equiv) in the presence of Hoveyda's recyclable catalyst **D** furnished enal **306** with complete (*E*) stereocontrol in 70% yield. Aldehyde **306** was then elaborated in four steps to the natural compound, the dihydropyrone ring being closed in the penultimate step by RCM of acrylate **307** with catalyst **A** [139b].

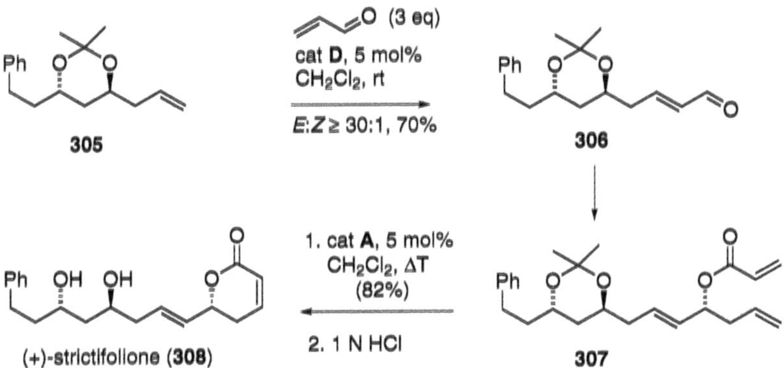

Scheme 60 (*E*)-Selective CM with electron-deficient alkene in Cossy's total synthesis of strictifolione (**308**) [139b]

An (*E*)-selective CM reaction with an acrylate (Scheme 61) was applied by Smith and O'Doherty in the enantioselective synthesis of three natural products with cyclooxygenase inhibitory activity (cryptocarya triacetate (**312**), cryptocaryolone (**313**), and cryptocaryolone diacetate (**314**)) [142]. CM reaction of homoallylic alcohol **309** with ethyl acrylate mediated by catalyst **C** led (*E*)-selectively to δ-hydroxy enoate **310** in near quantitative yield. Subsequent Evans acetal-forming reaction of **310**, which required the *trans* double bond in **310** to prevent lactonization, led to key intermediate **311** that was converted to **312–314**.

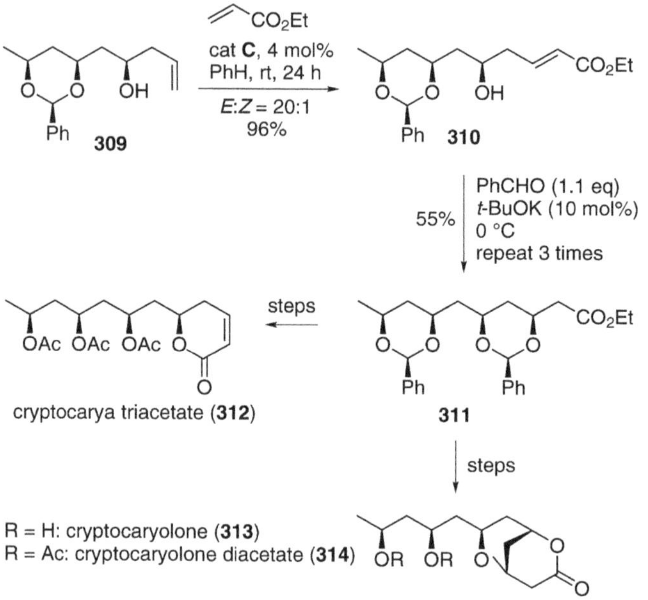

Scheme 61 (*E*)-Selective CM with ethyl acrylate in total syntheses of *Cryptocaria* natural products **312–314** [142]

In Ghosh's enantioselective total synthesis of the cytotoxic marine macrolide (+)-amphidinolide T1 (**318**) [143], the C1–C10 fragment **317** was constructed by CM of subunits **315** and **316** (Scheme 62). The reaction mediated by catalyst **C** (5 mol%) afforded in the first cycle an inconsequential 1:1 mixture of (*E*/*Z*)-isomeric CM products **317** in 60% yield, along with the homodimers of **315** and **316**. The self-coupling products were separated by chromatography and exposed to a second metathesis reaction to provide olefins **317** in additional 36% yield [144].

The fungal metabolite (+)-brefeldin A (**325**) displays potent antitumor, antifungal, antiviral, antimitotic, and immunosuppressive activities. Recently, Romo and Wang described a highly concise total synthesis of **325** by a combined β-lactone–CM approach (Scheme 63), that again underlines the high tolerance of sensitive functionality exhibited by the second-generation Ru catalysts [145].

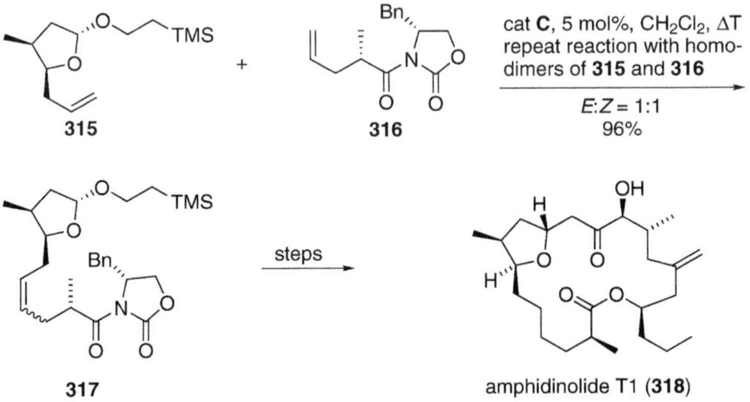

Scheme 62 Efficient coupling of fragments **315** and **316** via CM in Ghosh's total synthesis of the cytotoxic marine macrolide amphidinolide T1 (**318**) [143]

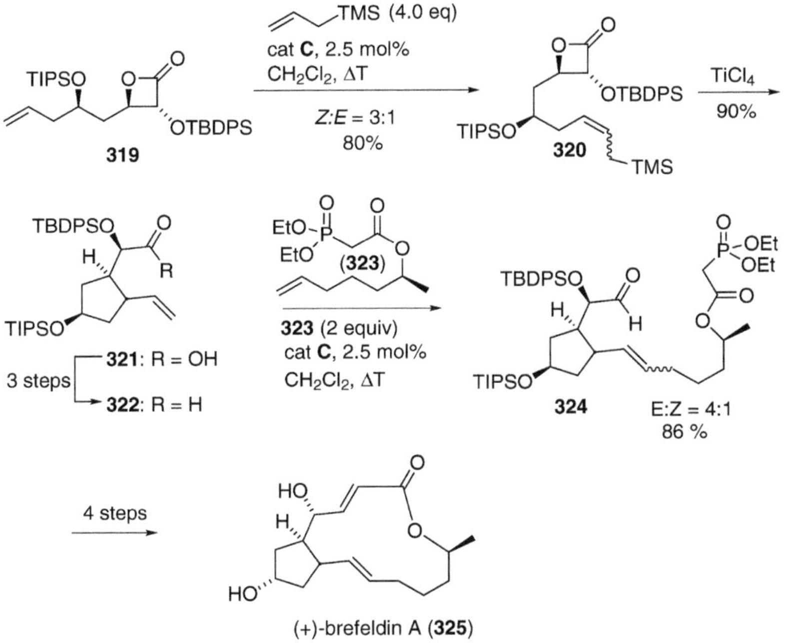

Scheme 63 Twofold use of CM in the total synthesis of brefeldin A (**325**) [145]

An allylsilane-generating CM using catalyst **C** between the sensitive β-lactone **319** and allyltrimethylsilane served to introduce the allylsilane moiety in intermediate **320** as an inconsequential mixture (ca. 3:1) of (*E*/*Z*)-isomers in 80% yield. Cyclization of β-lactone **320** with $TiCl_4$ smoothly delivered cyclopentane **321** with inversion at the β-carbon. Acid **321** was converted to key aldehyde **322** in three steps. The convergent fragment coupling was performed by a uniquely complex CM reaction between epimerizable aldehyde **322** and phosphonoacetate **323**, again in the presence of catalyst **C**. Homodimerization of phosphonate **323** was found to be competitive with the CM process. However, it was shown that this dimer could also be used in a CM reaction with **322**. The metathesis product **324**, thus obtained in 86% yield as an isomeric mixture (*E*:*Z*=4:1), was then subjected to (*E*)-selective Horner–Wadsworth–Emmons cyclization. After deprotection, both hydroxy groups were inverted by the Mitsunobu protocol leading to **325**.

Application of vinyl boronate CM in epothilone chemistry, leading to epo490 (**240d**, a naturally occurring minor cometabolite, cf Scheme 49) and to novel 11-hydroxy and 11-fluoro analogs of epoD (Scheme 64), was reported by Dani-

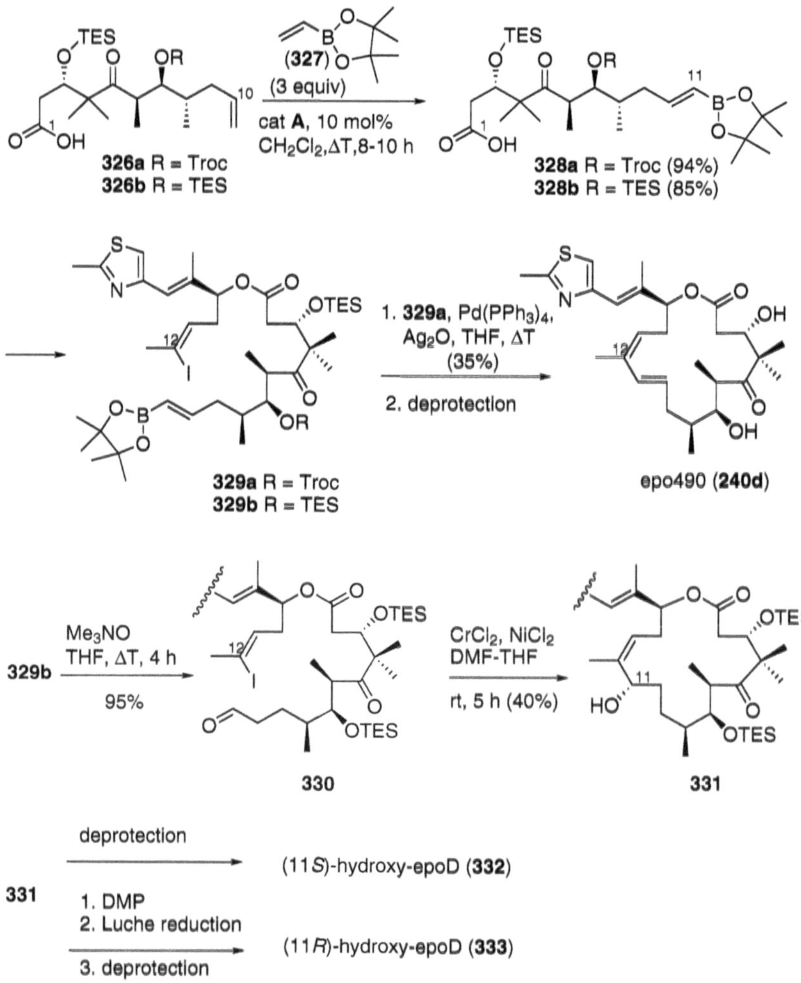

326a R = Troc
326b R = TES
(**327**) (3 equiv)
cat **A**, 10 mol%
CH_2Cl_2, ΔT, 8-10 h
328a R = Troc (94%)
328b R = TES (85%)

1. **329a**, $Pd(PPh_3)_4$, Ag_2O, THF, ΔT (35%)
2. deprotection
329a R = Troc
329b R = TES
epo490 (**240d**)

329b Me_3NO, THF, ΔT, 4 h, 95% → **330**
$CrCl_2$, $NiCl_2$, DMF-THF, rt, 5 h (40%) → **331**

331 deprotection → (11*S*)-hydroxy-epoD (**332**)
331 1. DMP 2. Luche reduction 3. deprotection → (11*R*)-hydroxy-epoD (**333**)

Scheme 64 CM with vinyl boronate **327** in novel syntheses of epo490 (**240d**) and 11-hydroxy analogs **332** and **333** of epoD [146]

shefsky and coworkers [146]. Treatment of acids **326** representing the C1–C11 moiety of epothilones, with an excess (3 equiv) of vinyl-pinacol boronate ester **327** in the presence of catalyst **A**, provided the corresponding CM products **328** almost exclusively as (*E*)-isomers in high yield. Esterification of compounds **328** provided seco compounds **329**. Suzuki macrocyclization of **329a** (R=Troc) under the conditions shown in the scheme led, after stepwise deprotection, to epo490 (**240d**) in low yield. (For alternative routes to **240d** by RCM macrocyclization, see Scheme 49). Through a novel alternative for the conventional hydroboration–oxidation method, analog **329b** (R=TES) was converted into aldehyde **330** by oxidation with trimethylamine *N*-oxide. Nozaki–Kishi macrocyclization of **330** led to protected (11*S*)-hydroxy-epoD (**331**) as the only diastereomer. Compound **331** was then converted to both possible 11-hydroxyepothilones and also to the pair of 11-fluoro analogs.

The important immunosuppressant cyclosporin A (CsA, **334**) (Scheme 65), known also by its trade name Sandimmune, is widely used to prevent organ rejection in transplant patients. CsA is also effective in the treatment of asthma patients, but in this case its chronic use is limited by the nephrotoxicity, caused by inhibition of calcineurin. CsA is a cyclic undecapeptide with an unusual unsaturated amino acid (MeBmt), which offers the possibility of structural modifications by CM. With the aim of investigating **334** as a metathesis substrate, and thus to gain new affinity reagents useful for detecting novel cyclophilins from cellular extracts, Diver and coworkers [147] treated **334** with several terminal olefins **335** in the presence of NHC catalyst **C** (benzylidene catalyst **A** was completely ineffective). CM products **336a–c** were obtained in good yields as a mixture of geometric isomers. The CM with olefin **335c** is notable because of the direct installation of an active ester onto an unprotected polypeptide. Metathesis product **336c** was then coupled with a sepharose resin **337** in three steps, as shown in Scheme 65, and the resulting resin-bound cyclosporin **338** could be used for the detection of novel cyclophilins from cell lysates.

A different task was pursued by the CM of CsA with various maleates **339** [148]. The CM demanded in this case the highly active Hoveyda catalyst **D**, that exhibits potency not reached by the phosphine-containing catalysts **C** and E. Under the conditions given in Scheme 65, metathesis with maleates **339** led (*E*)-selectively to the α,β-unsaturated ester derivatives **340** in high yield. Compounds **340** still demonstrated activity comparable to that of CsA and are thus potential "soft drugs" via esterase-mediated biotransformation to the corresponding inactive carboxylic acids **341**.

An interesting example of regioselective CM with ethylene as a tool in natural product *degradation* was recently disclosed by Hawaiian authors [149]. Thus, CM using catalyst **C** and ethylene gas was used to degrade the plant polyacetylene oxylipin (+)-falcarindiol (**342**) with uncertain stereochemistry at C3. As the reaction provided a *meso* product (**343**) in 81% yield by regioselective attack at the aliphatic side chain, the natural compound **342**, isolated from a Hawaiian endemic plant, had the 3*R*,8*S* configuration shown in Scheme 66.

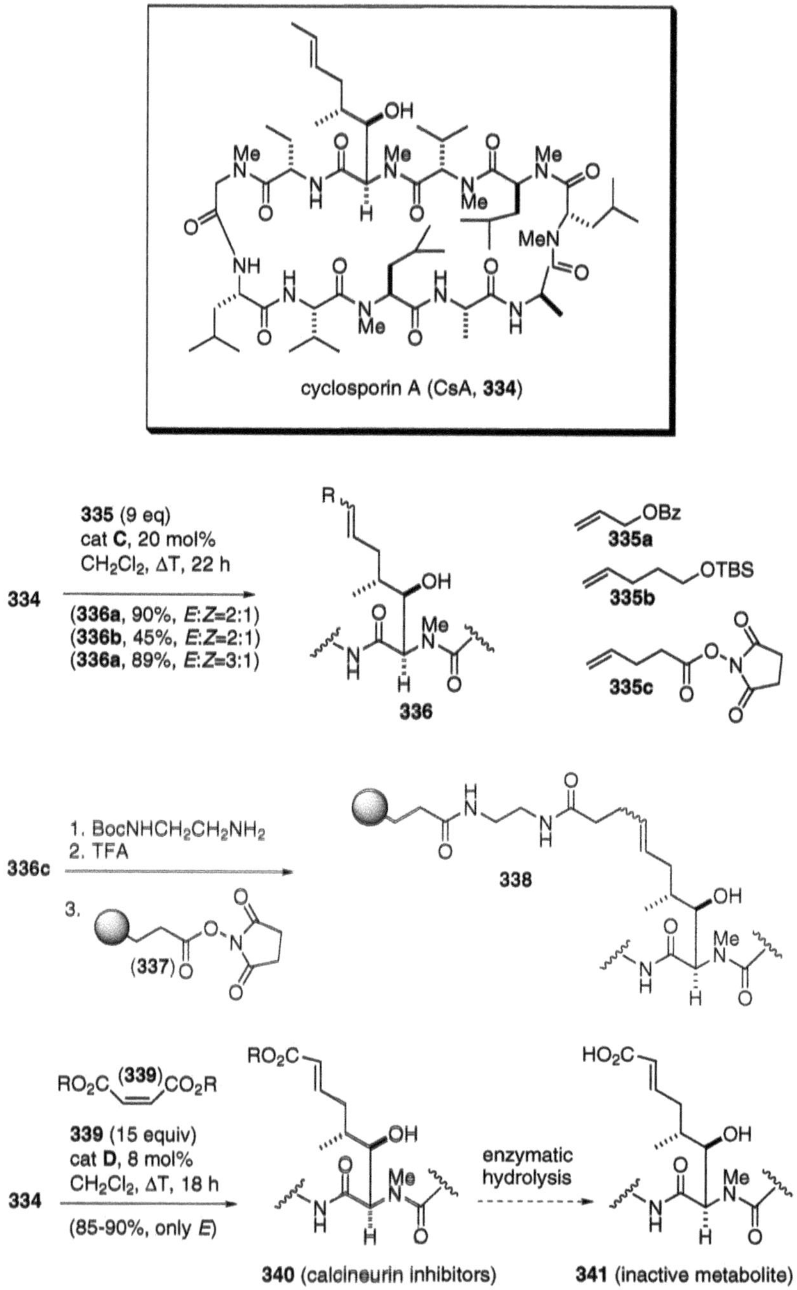

Scheme 65 Various modifications of cyclosporin A (**334**) via CM [147, 148]

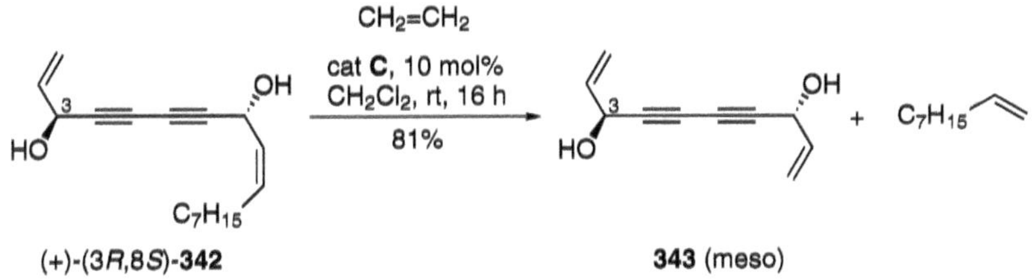

Scheme 66 Olefin CM in natural product degradation: configurational assignment of falcarindiol (**342**) [149]

Recently, a microscale CM/exciton chirality protocol for the determination of absolute configurations of natural products with allylic alcohol or amine moieties was developed [150]. In Scheme 67, the method is exemplified on prostaglandin A_1 ethyl ester (PGA_1, **344**). Because **344** exhibits an intense Cotton effect due to the twisted enone chromophore and moreover is unstable under basic acylation conditions, the usual allylic benzoate method for determination of its absolute configuration is not applicable. When **344** was treated with excess styrene in the presence of 10 mol% of second-generation catalyst **C**, the resulting (*E*)-styrenoids **345** and **346a** (the latter in the form of its *p*-phenylbenzoate **346b**) were easily amenable to conventional configurational assignment.

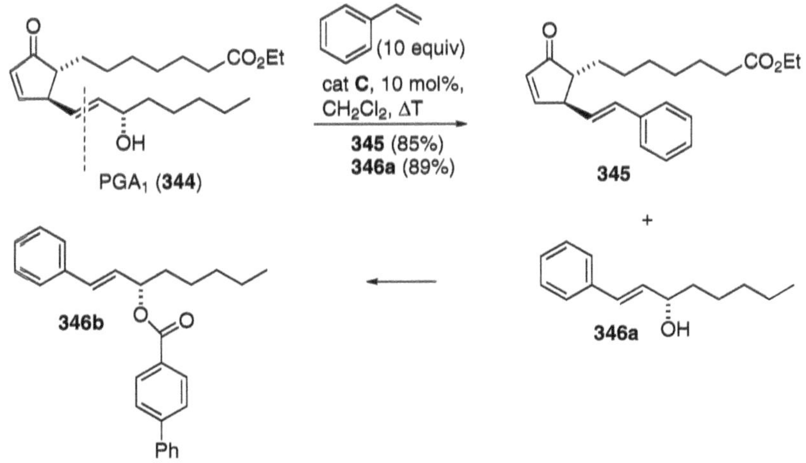

Scheme 67 Olefin CM with styrene for configurational assignment of natural products [150]

The reversible nature of cross metathesis is of synthetic importance because, by the use of a sufficiently active metathesis catalyst, it generally ensures the preferential formation of the most thermodynamically stable product. This results in the transformation of terminal olefins into internal ones, and we have seen that undesired self-metathesis products can be recycled by exposing them to a second CM process.

Exploiting the reversible nature of CM, Smith and coworkers reported a highly impressive total synthesis of the naturally occurring cylindrophanes A (**348a**) and F (**348f**) [151]. Initially, the [7,7]-paracyclophane skeleton of (–)-**348f** was elaborated by "conventional" RCM macrocyclization, which led from diene **347** stereoselectively to the cyclization product with an (*E*) double bond in 88% yield (Scheme 68). In a second-generation strategy, however, a remark-

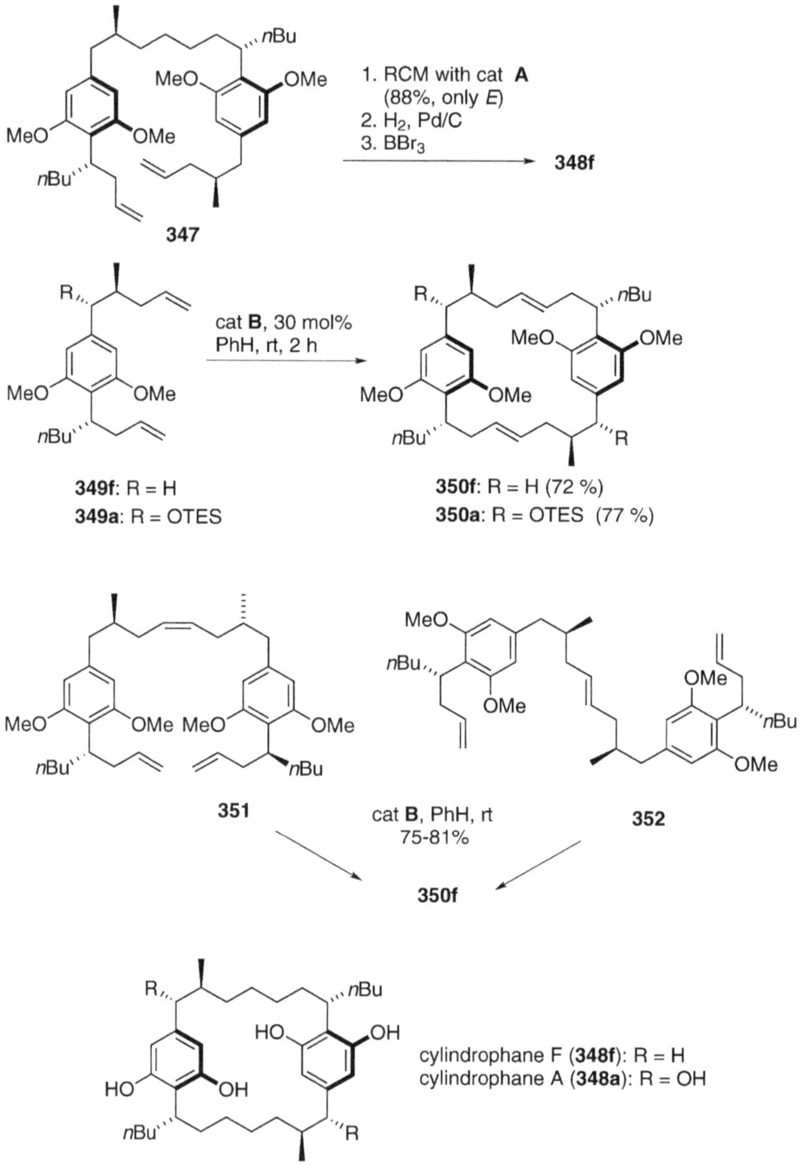

1. RCM with cat **A** (88%, only *E*)
2. H_2, Pd/C
3. BBr_3

347 → **348f**

cat **B**, 30 mol%
PhH, rt, 2 h

349f: R = H
349a: R = OTES

350f: R = H (72 %)
350a: R = OTES (77 %)

351, **352**: cat **B**, PhH, rt, 75-81% → **350f**

cylindrophane F (**348f**): R = H
cylindrophane A (**348a**): R = OH

Scheme 68 Synthesis of cylindrophanes A (**348a**) and F (**348f**) via reversible olefin CM [151]

able CM dimerization cascade was described which served to assemble **348a** and **348f** from dienyl monomers **349a** and **f**. Ruthenium catalysts **A** and **C** and Schrock's Mo catalyst **B** were investigated under different conditions, and it turned out that catalyst **B** was the most effective, leading exclusively to cyclodimers **350**. In both cases, only the head-to-tail (*E*,*E*)-isomer was formed, and no "head-to-head" dimers or dimers with a (*Z*) double bond were detected, suggesting that compounds **350** are the most stable out of all other possible isomers. This was not only corroborated by MM2 force field calculations, but also by the outcome of metathesis reactions performed with trienes **351** and **352**, which both gave in a self-editing process only cyclodimer **350f** with the [7,7]-paracyclophane skeleton in good yield, despite their disposition to form [8,6]-cyclophanes.

2.3
Metathesis on Solid Support

Chemistry on solid support has gained tremendous importance during the last few years, mainly driven by the needs of the pharmaceutical sciences. Due to the robust and tolerable nature of the available catalysts, metathesis was soon recognized as a useful technique in this context. Three conceptually different, RCM-based strategies are outlined in Fig. 11. In the approach delineated in Fig. 11a, a polymer-bound diene **353** is subjected to RCM. The desired product **354** is formed with concomitant traceless release from the resin. This strategy is very favorable, since only compounds with the correct functionality will be liberated, while unwanted by-products remain attached to the polymer. However, as the catalyst is captured in this process by the matrix (**355**), a higher catalyst loading will be required, or "ancillary" alkenes have to be added to liberate the catalyst.

With polymer-bound diene **356**, two different strategies are possible. Following path A, RCM results in the formation of a (volatile) alkene **357a** and a

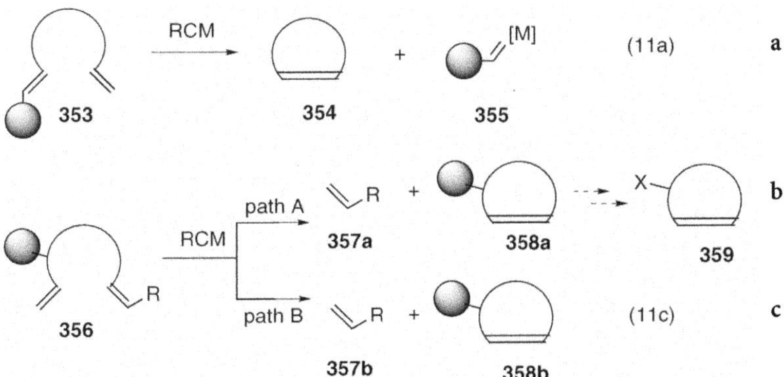

Fig. 11 Variations of RCM with polymer-bound substrates

cyclic product **358a**, which remains attached to the polymer support. This product can undergo further manipulation, with cleavage from the resin at a later stage (**358a**→**359**, Fig. 11b). Alternatively, RCM of diene **356** can also be used for the traceless release of a polymer-bound cycloalkene (**358b**), with concomitant formation of a terminal alkene (**357b**) as the desired reaction product (Fig. 11c).

The feasibility of multistep natural product total synthesis via solid-phase methodology, and its application to combinatorial chemistry, was first demonstrated by Nicolaou and coworkers in epothilone synthesis and in the generation of an epothilone library [152]. The traceless release of TBS-protected epoC **361** by RCM of resin-bound precursor **360** (Scheme 69) is an early and most prominent example for the strategy outlined in Fig. 11a.

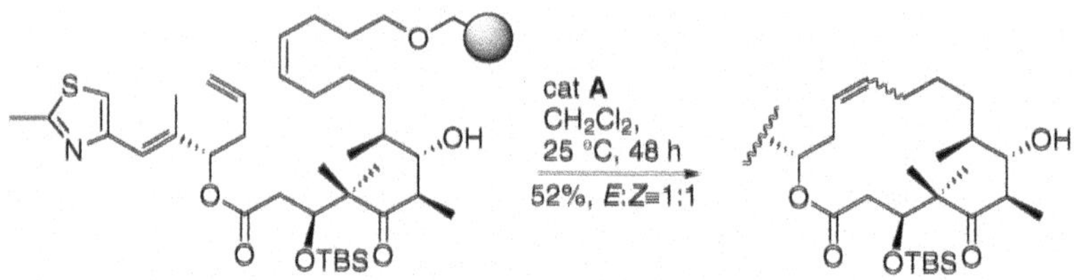

cat **A**
CH_2Cl_2,
25 °C, 48 h
52%, *E*:*Z*=1:1

360 361

Scheme 69 Traceless release of epoC derivative **361** from solid support by RCM [152]

An illustrative example of an alternative strategy (cf Fig. 11c) involving the use of a novel traceless linker is found in the multistep synthesis of 6-*epi*-dysidiolide (**363**) and several dysidiolide-derived phosphatase inhibitors by Waldmann and coworkers [153], outlined in Scheme 70. During the synthesis, the growing skeleton of **363** remained attached to a robust dienic linker. After completion of intermediate **362**, the terminal olefin in **363** was liberated from the solid support by the final metathesis process with concomitant formation of a polymer-bound cyclopentene **364**. Notably, during the synthesis it turned out that polymer-bound intermediate **365a**, in contrast to soluble benzoate **365b**, produced diene **367** only in low yield. After introduction of an additional linker (cf intermediate **366**), diene **367** was released in distinctly improved yield by RCM.

A short and efficient synthetic approach to hydroxy-substituted (*E*)-stilbenoids, as exemplified by the natural compound resveratrol (**371b**) via solid-phase CM, was reported by a Korean group (Scheme 71) [154]. When two different stilbenes were allowed to couple by catalyst **C**, all three kinds of possible stilbenes were obtained as an inseparable mixture. Anchoring 4-vinylphenol to Merrifield resin, followed by exposing the supported styrenyl ether **368** and diacetoxy styrene **369** (10 equiv) to the catalyst, inhibited self-metathesis of the supported substrate. Sequential separation of the homodimer formed from **369** by washing and subsequent cleavage of the resin **370** with acid provided (*E*)-stilbene **371a** with complete stereocontrol in 61% yield.

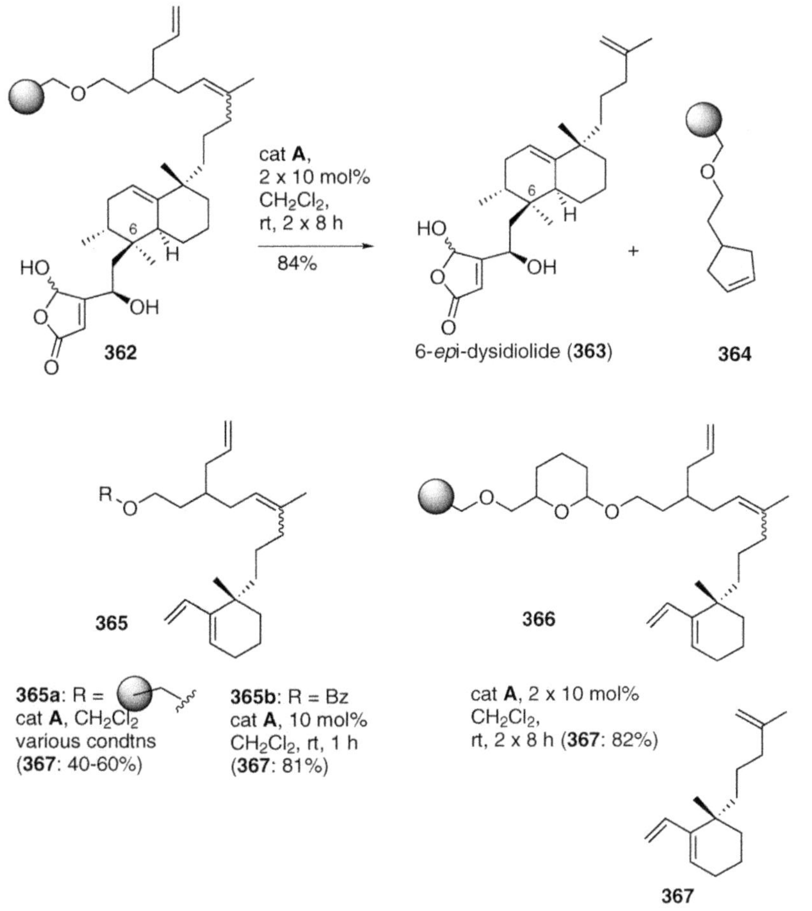

cat A,
2 x 10 mol%
CH2Cl2,
rt, 2 x 8 h
84%
HO
OH
H
6
O
362
6-epi-dysidiolide (363)
364
R
365
366
365a: R =
cat A, CH2Cl2
various condtns
(367: 40-60%)
365b: R = Bz
cat A, 10 mol%
CH2Cl2, rt, 1 h
(367: 81%)
cat A, 2 x 10 mol%
CH2Cl2,
rt, 2 x 8 h (367: 82%)
367

Scheme 70 Traceless removal of polymer **364** by RCM in the synthesis of 6-*epi*-dysidiolide (**363**) [153]

OAc
(369)
(10 eq)
OAc
cat C, 20 mol%,
PhH, ΔT, 12 h
TFA, CH2Cl2
61%
OH
RO
OR
AcO
OAc
368
370
371a : R = Ac
371b : R = H

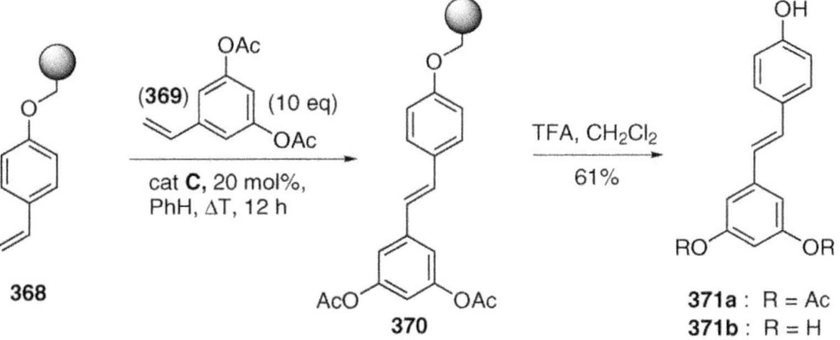

Scheme 71 Improved synthesis of (*E*)-stilbenoids by olefin CM on a solid phase [154]

2.4 Domino Metathesis Reactions

2.4.1 Ring-Rearrangement Reactions

An alternative access to complex heterocyclic structures is the Ru- or Mo-catalyzed ring-rearrangement metathesis (RRM), in which a strained carbocyclic alkene is transformed into a heterocyclic product by an intramolecular ring-opening/ring-closing or double ring-closing domino metathesis. Due to the reversibility of the processes involved, the amount of rearrangement product depends on thermodynamic effects, e.g., ring strain and substitution pattern of the starting cycloalkene. A particularly attractive aspect of these transformations is the catalytic transfer of stereochemical information from readily available carbocyclic olefins to one or two newly formed heterocyclic rings (cf Fig. 1c and d). This methodology, initially investigated by Grubbs [155], was extensively applied in natural product synthesis by Blechert et al.

Four members of the tetraponerine family (the major constituents of the contact poison of the New Guinean ant *Tetraponera* sp.) were prepared by RRM methods [156]. The key step leading to tetraponerine T7 (**374**) from the readily available cyclopentene precursor **372** is shown in Scheme 72. When compound **372** was exposed to catalyst **A** in the presence of ethylene, the desired ROM–RCM sequence proceeded smoothly to furnish heterocycle **373** with complete conversion, whereas the corresponding di-nosyl (2-nitrophenylsulfonyl)-protected analog of **372** led only to a 1:2 equilibrium mixture of starting material and RRM product.

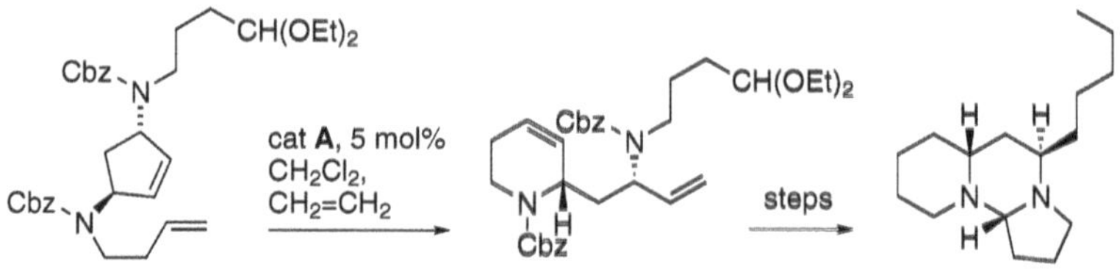

Scheme 72 Synthesis of tetraponerine T7 (**374**) via RRM of cyclopentene **372** [156]

The same principle of sequential cyclopentene-opening RCM resulting in the formation of a dihydropyrrole ring was the key step in Blechert's novel approach to the polyhydroxylated indolizine alkaloid (–)-swainsonine (**378**) via RRM of **375** (Scheme 73) [157].

An early example of cyclopentene-opening/double RCM leading to bis-dihydropyran **380** (the C22–C34 segment of the potent antitumor agent halichondrin A) was disclosed by Burke et al. (Scheme 74) [158]. In this case, the ROM–RCM sequence was performed with catalyst **B**, leading from cyclopentene **379** to **380** in 71% yield. When metathesis precursor **379** was exposed to catalyst **A**, only one

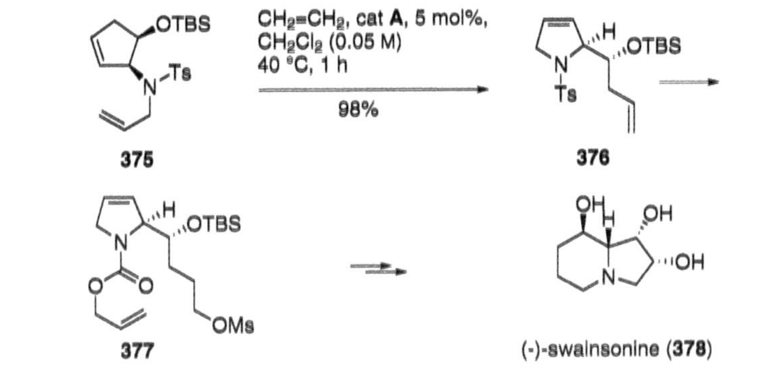

Scheme 73 Synthesis of swainsonine (378) via RRM of cyclopentene 375 [157]

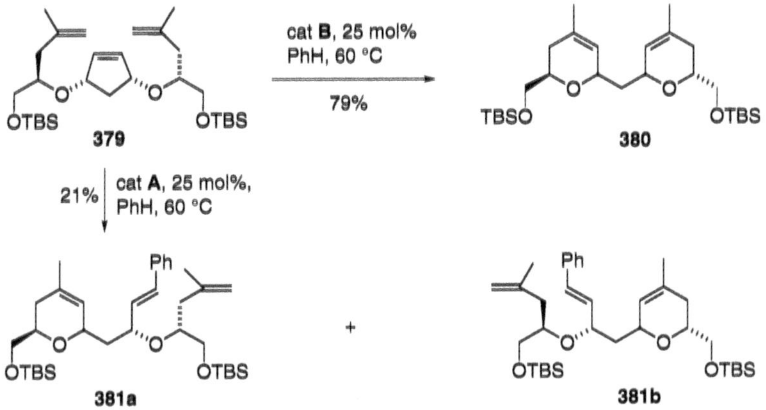

Scheme 74 Synthesis of bis-dihydropyran fragment 380 of halichondrin by tandem ROM–double RCM of cyclopentene 379 [158]

dihydropyran ring was formed and the reaction led to a mixture of the isomeric compounds **381a** and **381b** in low yield.

RRM of enantiopure cyclopentene **382**, induced by commercially available catalyst C, was the key step in Blechert's total synthesis of the bis-piperidine alkaloid (+)-astrophylline (**384**) [159]. Exposure of metathesis precursor **382** to only 1 mol% C provided within 2 h bicycle **383** in 82% yield (Scheme 75).

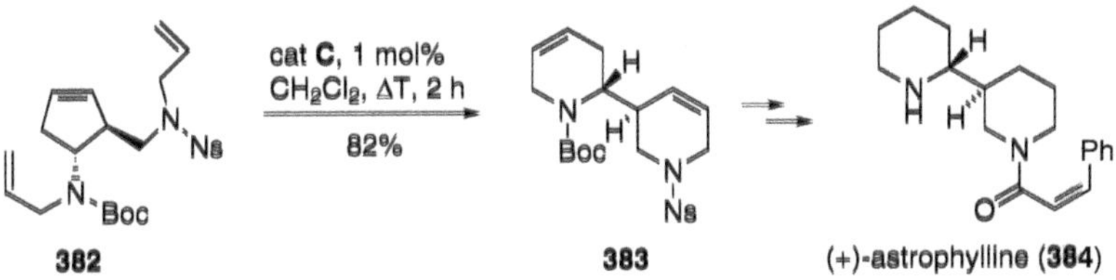

Scheme 75 Synthesis of astrophylline (384) via RRM of cyclopentene 382 [159]

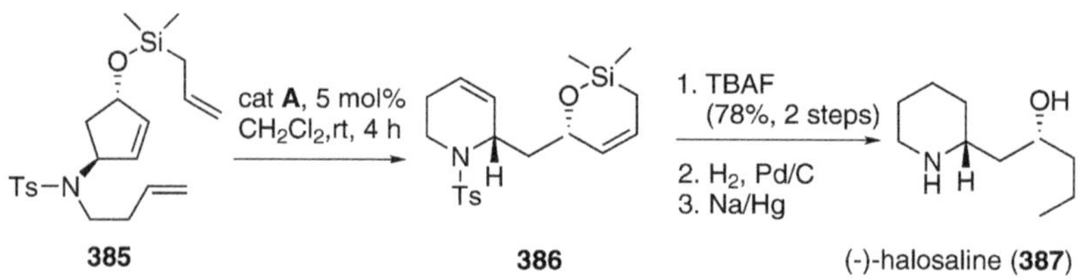

cat **A**, 5 mol%
CH_2Cl_2, rt, 4 h

1. TBAF (78%, 2 steps)
2. H_2, Pd/C
3. Na/Hg

385 **386** (-)-halosaline (**387**)

Scheme 76 Total synthesis of halosaline (**387**) via RRM of cyclopentene **385** [160]

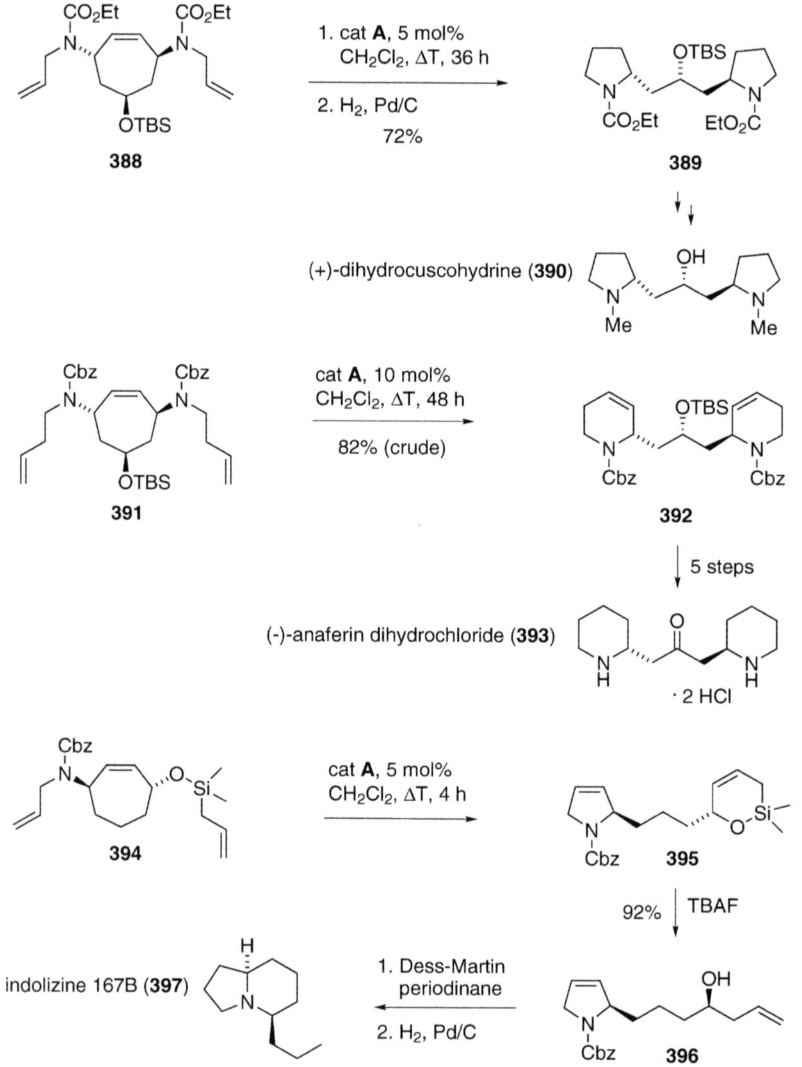

1. cat **A**, 5 mol%
CH_2Cl_2, ΔT, 36 h
2. H_2, Pd/C
72%

388 **389**

(+)-dihydrocuscohydrine (**390**)

cat **A**, 10 mol%
CH_2Cl_2, ΔT, 48 h
82% (crude)

391 **392**

5 steps

(-)-anaferin dihydrochloride (**393**) · 2 HCl

cat **A**, 5 mol%
CH_2Cl_2, ΔT, 4 h

394 **395**

92% TBAF

indolizine 167B (**397**)

1. Dess-Martin periodinane
2. H_2, Pd/C

396

Scheme 77 RRM reactions of enantiopure cycloheptenes leading to dihydrocuscohydrine (**390**) [161], anaferin dihydrochloride (**393**) [162], and indolizine 167B (**397**) [163]

Blechert's synthesis of the piperidine alkaloid (–)-halosaline (**387**) by Ru-catalyzed RRM is outlined in Scheme 76 [160]. In the presence of 5 mol% of catalyst **A**, the ring rearrangement of metathesis precursor **385** proceeded cleanly with formation of both heterocyclic rings in **386.** In situ deprotection of the cyclic silyl ether in **386,** followed by selective reduction and removal of the tosyl group led to **387.**

The utility of strained disubstituted cycloheptenes in alkaloid syntheses is highlighted by Blechert's total syntheses of the bis-pyrrolidine alkaloid (+)-dihydrocuscohydrine (**390**) [161], the bis-piperidine alkaloid (–)-anaferin (in the form of its dihydrochloride **393**) [162], and indolizine 167B (**397**) [163] (Scheme 77).

Recently, a novel type of ROM–RCM-based tandem reaction was discovered by Lazarova et al. [164]. When various natural 16-membered macrolide antibiotics with a 1,3-diene unit in the marocyclic core (e.g., josamycin (**398**)) were exposed to catalyst **E** (20 mol%) in the presence of 1-hexene (2 equiv), a ROM–RCM sequence occurred with excision of ethylene and ring contraction to 14-membered-ring lactones **400** (Scheme 78) [165]. The reaction did not occur with catalyst **A**, and demanded – without additives – a stoichiometric amount of catalyst **E**. 1-Hexene, which was added instead of ethylene to initiate and propagate the catalytic cycle by generating the highly active $L_nRu{=}CH_2$ species, could also be replaced by titanium isopropoxide, which is known to destabilize catalyst-deactivating chelates between the catalyst and hydrogen bond acceptors in the metathesis substrate [33].

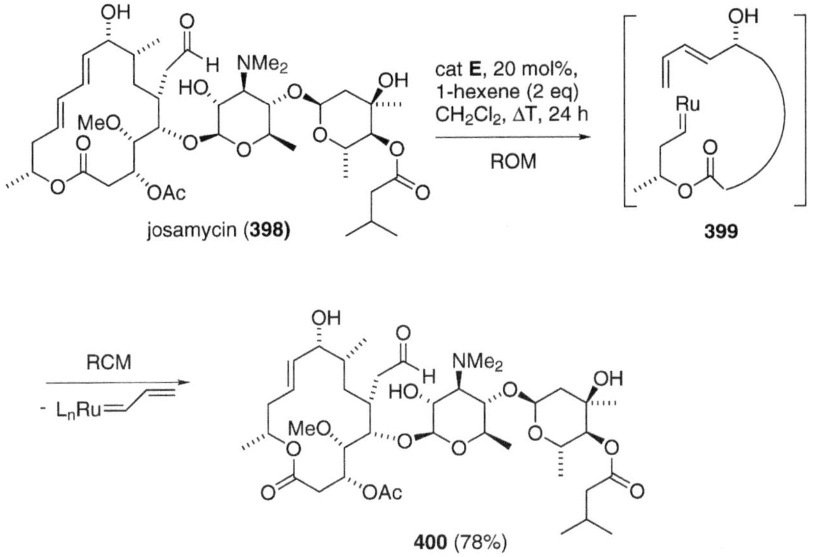

Scheme 78 Novel ring-contraction metathesis of macrolide antibiotics with a 1,3-diene moiety [164]

2.4.2 Ring-Opening Cross Metathesis (ROCM)

An important variation of CM methodology is ring-opening cross metathesis (ROCM) where one of the reacting substrates is a (strained) cycloalkene. ROCM reactions of highly strained cyclobutenes have been extensively studied by Snapper and coworkers and provided firm precedent for successful conversion to 1,5-dienes through the use of ethylene as the second component [166]. These investigations resulted in a novel approach to asteriscanolide (**116**, cf Scheme 22) by a novel sequence of intramolecular cyclobutadiene cycloaddition–cyclobutene ROCM–Cope rearrangement (Scheme 79) [167]. The highly functionalized cyclobutene **402** with a trisubstituted double bond was obtained by heating iron tricarbonyl-protected cyclobutadiene **401** with trimethylamine *N*-oxide in acetone. Cycloadduct **402** was the precursor for the ROCM reaction with ethylene. When a solution of **402** in benzene was exposed to catalyst **C** (5 mol%, 50 °C, 10 h) under an ethylene atmosphere, followed by refluxing (10 h), cyclooctadiene **404** was directly produced in 74% yield. Evidently, the initial ROCM product **403** underwent the Cope rearrangement to **404** under such mild reaction conditions. Completion of the formal synthesis of **116** was accomplished by allylic oxidation of **404** to lactone **405**, which was an intermediate in a previous synthesis of **116** [168, 169].

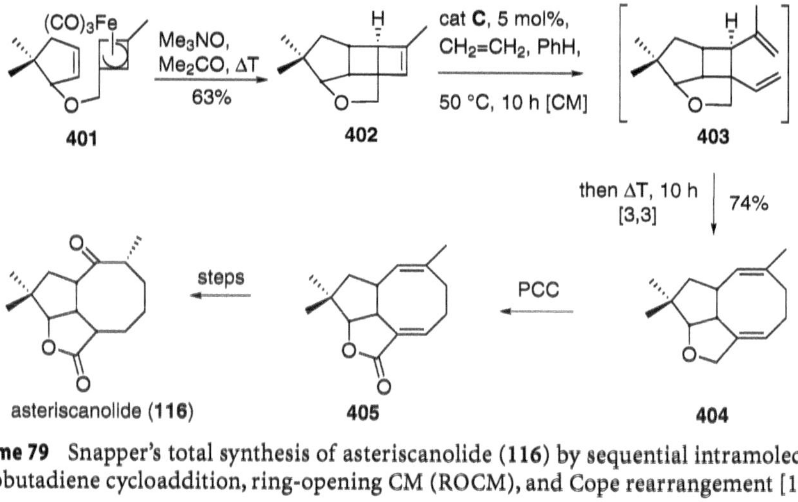

Scheme 79 Snapper's total synthesis of asteriscanolide (**116**) by sequential intramolecular cyclobutadiene cycloaddition, ring-opening CM (ROCM), and Cope rearrangement [167]

A cyclobutene ROCM sequence was also used in a synthesis of racemic sporochnol (**410**), a naturally occurring feeding deterrent toward herbivorous fish (Scheme 80) [170]. Exposing cyclobutene **406** (0.01 M in boiling 1,2-dichloroethane) in the presence of ethylene to second-generation catalyst **C** (8 mol%) led to 1,5-diene **407** in 73% yield, along with 9% of the homodimer derived from **407** by involving the less hindered double bond. Site-selective hy-

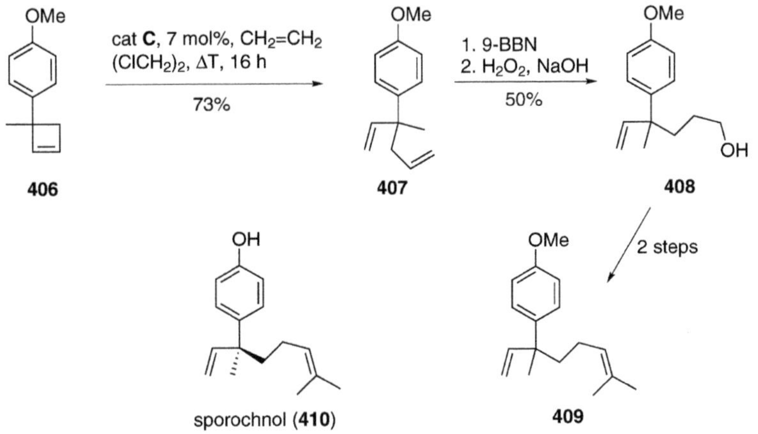

OMe
cat **C**, 7 mol%, $CH_2=CH_2$
$(ClCH_2)_2$, ΔT, 16 h
73%
OMe
1. 9-BBN
2. H_2O_2, NaOH
50%
OMe
OH
406
407
408
OH
OMe
2 steps
sporochnol (**410**)
409

Scheme 80 Use of ROCM of cyclobutene **406** in synthetic work directed to sporochnol (**410**) [170]

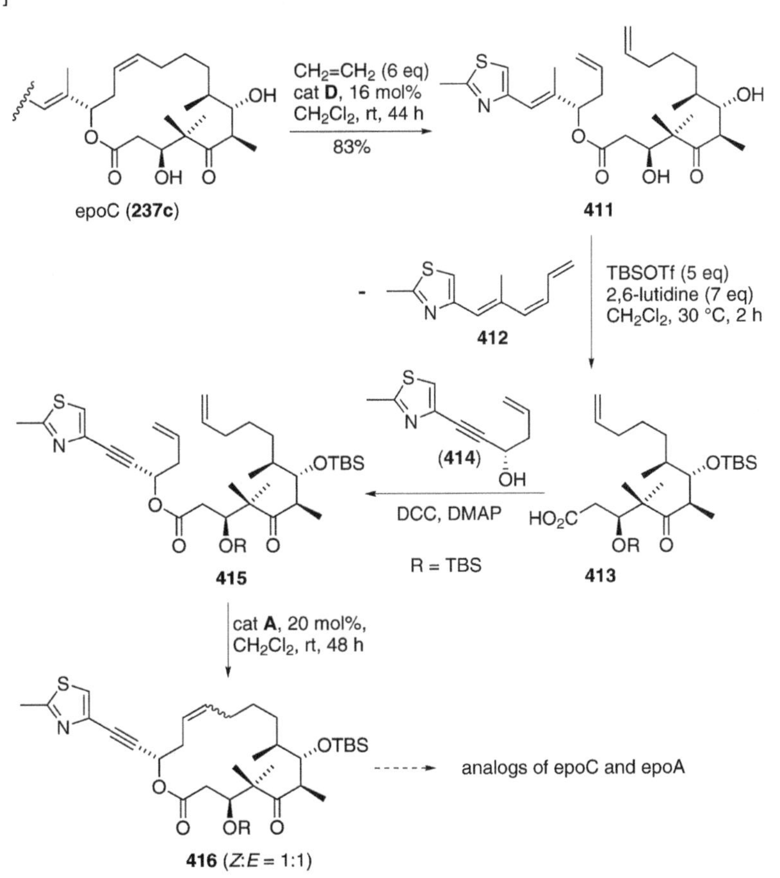

OH
$CH_2=CH_2$ (6 eq)
cat **D**, 16 mol%
CH_2Cl_2, rt, 44 h
83%
S
N
OH
O
O OH O
epoC (**237c**)
411
S
N
412
TBSOTf (5 eq)
2,6-lutidine (7 eq)
CH_2Cl_2, 30 °C, 2 h
S
N
OTBS
(**414**)
OH
OTBS
DCC, DMAP
HO_2C
O OR O
R = TBS
OR O
415
413
cat **A**, 20 mol%,
CH_2Cl_2, rt, 48 h
OTBS
analogs of epoC and epoA
O OR O
416 (Z:E = 1:1)

Scheme 81 ROCM performed on epoC (**237c**), an unstrained natural macrocycle: synthesis of epoC analogs with modified side chain [171]

droboration of **407**, followed by alcohol to aldehyde interconversion and a Wittig reaction, led to the racemic methyl ether analog **409** of **410**.

The first example of a ROM–CM sequence performed on an *unstrained macrocyclic* natural compound was recently presented by Höfle et al. [171]. ROCM of epoC (**237c**, cf Fig. 8), produced by fermentation with excess ethylene in the presence of catalyst **D** (16 mol%, added in two portions during 44 h), led to seco compound **411** in high yield (Scheme 81). Silylation of **411** with an excess of reagents not only protected the hydroxy groups but also cleaved the ester by elimination of triene **412** to furnish carboxylic acid **413** in 80% yield. Known building block **413** was then used for the synthesis of the novel 16,17-alkyne analogs **416** of epoC (**237c**) and epoA (**238a**), through esterification with **414** followed by (nonstereoselective) RCM with catalyst **A**.

3 Enyne Cross Metathesis and Ring-Closing Enyne Metathesis

In contrast to diene metathesis, enyne metathesis (cf Fig. 2), which is catalyzed by the same catalysts, has not been developed as much, although the reaction produces synthetically useful 1,3-dienes [172]. Until recently, *intermolecular enyne metathesis* reactions were thought of as unselective with regard to both (*E*/*Z*) and chemoselectivity [173]. Competing CM homodimerization of the alkene, alkyne metathesis, and polymerization hampered the development of the intermolecular variant as a tool in natural product synthesis. Thus, it seems that to date, Mori's synthesis of the natural HIV-1 reverse transcriptase inhibitors anolignan A (**419**) and B (**422**), which both feature a 2,3-dibenzyl-substituted 1,3-butadiene skeleton [174], is a single example of the successful use of enyne CM in natural product synthesis (Scheme 82). In previous work in the group, it was shown that Ru-catalyzed CM between mono- or disubstituted alkynes and ethylene led to the introduction of a methylene group to both alkyne carbons providing substituted 1,3-butadienes (cf Fig. 2c) in good yield [175]. Applying this method, alkynes **417** and **420** (both compounds bear rate-accelerating acetoxy groups at each propargylic position) were exposed to catalyst **C** (10 mol%) under an atmosphere of ethylene, to furnish the desired 1,3-dienes **418** and **421** in high yield. The synthesis of **419** and **422** was completed by sequential hydrogenolytic removal of the propargylic acetoxy groups and deprotection.

More research efforts have focused on the *ring-closing enyne metathesis*, which usually [176] provides conjugated vinyl cycloalkenes (cf Fig. 2a, *exo* mode) useful for further manipulation, but also allows tandem metathesis processes for the formation of polycyclic compounds.

Clark and coworkers utilized enyne RCM for constructing the AB ring fragment of the manzamine alkaloids (Scheme 83) [177]. Exposing metathesis precursor **423** and ethylene gas to catalyst **A** provided bicycle **424** in near quantitative yield. Regioselective hydroboration of the vinyl group in **424**, followed

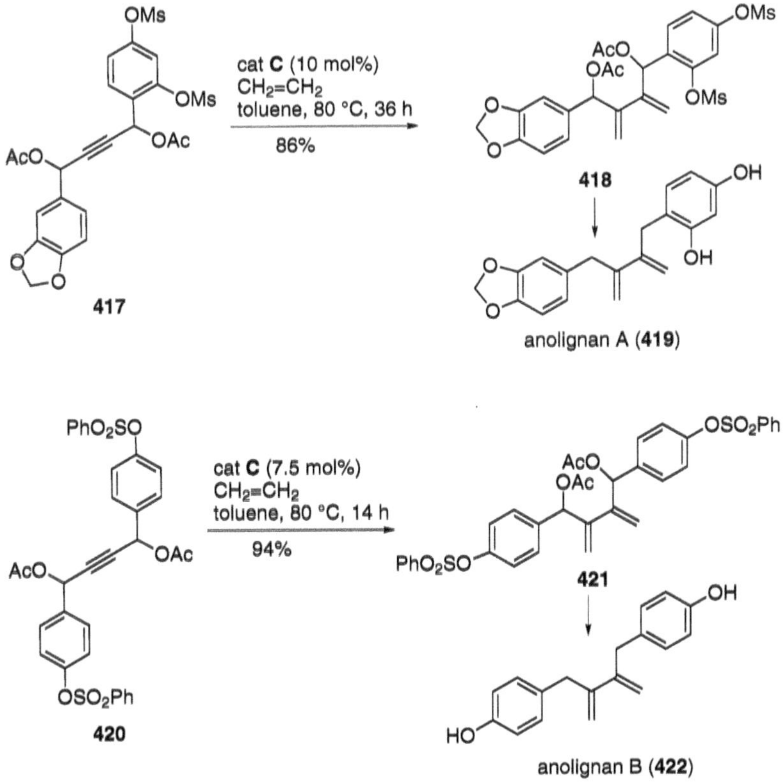

Scheme 82 Synthesis of anolignans A (**419**) and B (**422**) via enyne CM [174]

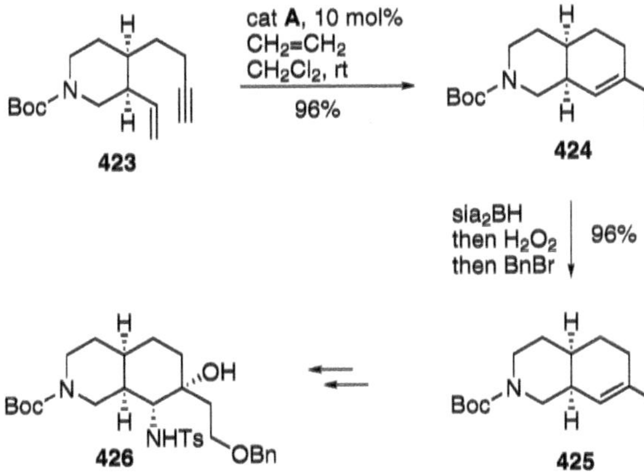

Scheme 83 Synthesis of the manzamine AB ring segment **426** via enyne RCM [177]

by O-benzylation to **425** and Sharpless aminohydroxylation led to fragment **426**, possessing most of the functionality required for further elaboration to the manzamine skeleton [178].

Clark's group also reported on ring-closing enyne metathesis for the preparation of six- and seven-membered cyclic enol ethers **428** (n=1,2) as potential building blocks for the synthesis of marine polyether natural compounds such as brevetoxins and ciguatoxins. Metathesis products **428** were obtained from ene-ynes **427** in 72–98% yield when the NHC-bearing catalyst **C** was used (Scheme 84) [179].

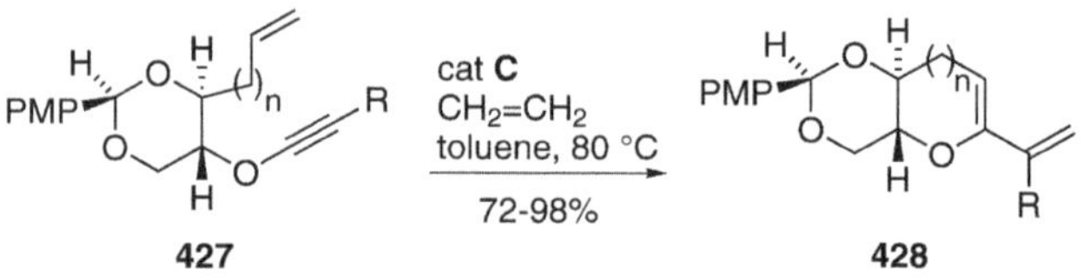

427: R = H, Me, SiMe$_3$, CH$_2$OAc, CH$_2$OTBDPS; n = 1,2 — cat **C**, CH$_2$=CH$_2$, toluene, 80 °C, 72-98% → 428

Scheme 84 Synthesis of cyclic enol ethers **428** by enyne RCM [179]

The first examples of *macrocyclization* by enyne RCM were used in Shair's impressive biomimetic total synthesis of the cytotoxic marine natural product longithorone A (**429**) [180]. This unique compound features an unusual heptacyclic structure which, in addition to the stereogenic centers in rings A–E, is also chiral by atropisomerism arising from hindered rotation of quinone ring G through macrocycle F (Scheme 85). It was assumed that biosynthesis of **429** could occur via an intermolecular Diels–Alder reaction between [12]paracy-

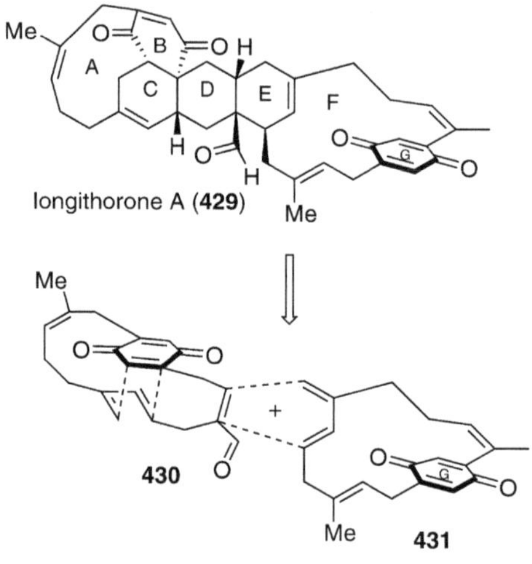

longithorone A (**429**) ⇒ **430** + **431**

Scheme 85 Biomimetic retrosynthetic analysis of longithorone A (**429**)

clophanes **430** and **431** to form ring E, and a transannular Diels–Alder reaction across **430** to simultaneously assemble rings A, C, and D [181].

According to this hypothesis, Shair's synthesis began with the construction of the protected versions **438** and **439** of key fragments **430** and **431** as single atropisomers, by using for the first time enyne macrocyclization to generate the required 1,3-diene units present in both key fragments (Scheme 86). This plan

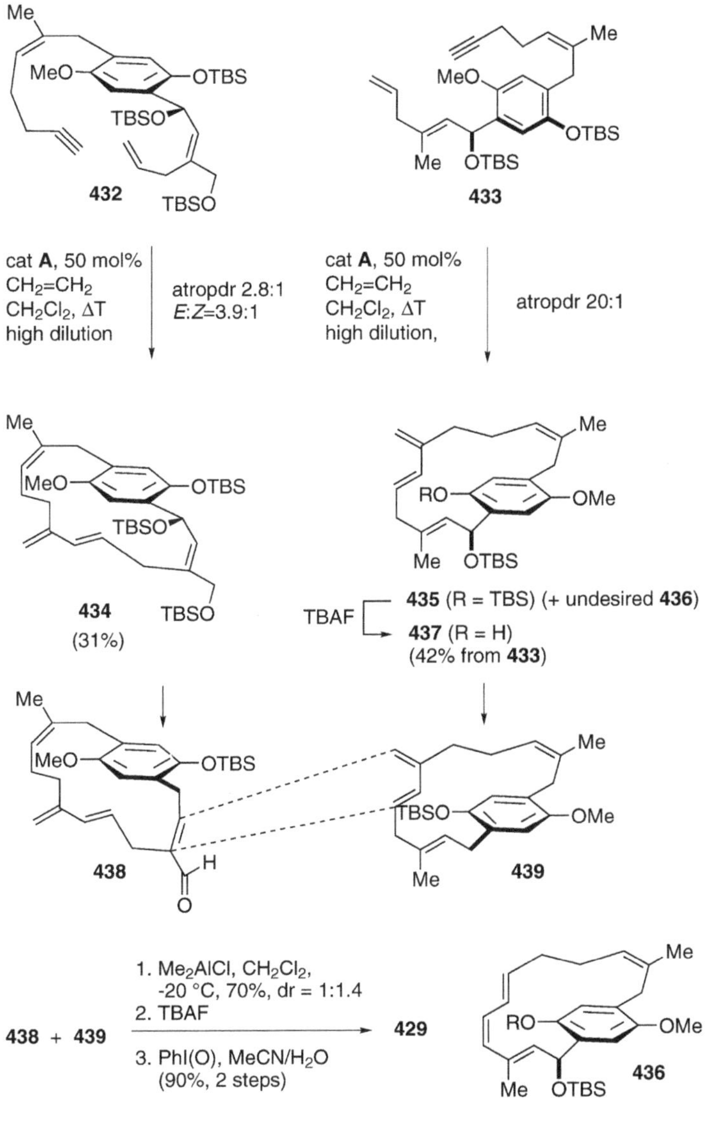

Scheme 86 Regioselective enyne RCM for the synthesis of 1,3-disubstituted macrocyclic 1,3-dienes **433** and **434**, main fragments in Shair's total synthesis of longithorone A [180]

implicated that macrocyclization of metathesis substrates **432** and **433** would resemble *intermolecular* enyne metathesis and generate *1,3-disubstituted dienes* (via the *endo*-type cyclization, cf Fig. 2a), since the resulting [12]paracyclophanes would be less strained than the [11]paracyclophanes resulting from 1,2-disubstituted diene formation (cf *exo*-mode in Fig. 2a). The bulky benzylic silyl ethers in metathesis precursors **432** and **433** were used to gear the aromatic rings during the metathesis process in order to control the atropisomerism and enforce atropdiastereoselection during the ring closure. Exposure of **433** to catalyst **A** and ethylene at high dilution in dichloromethane at 40 °C did indeed afford the desired [12]paracyclophane **435** with ≥20:1 atropdiastereoselectivity. However, **435** was obtained as an inseparable 2.2:1 mixture with the undesired paracyclophane **436** that had lost a molecule of propene during cyclization and could only be separated after selective deprotection of **435** to **437**. An analogous *endo*-type enyne macrocyclization was performed by exposing the more complex substrate **432** to the same conditions. However, this reaction resulted in a 2.8:1 mixture of atropdiastereomers and in a 3.9:1 (*E*/*Z*) ratio of double bond isomers, favoring [12]paracyclophane **434**. Compounds **437** and **434** were then transformed in a few steps (including reductive removal of the benzylic silyl ethers that had served their purpose as control elements) into precursors **438** and **439** for the intermolecular Diels–Alder reaction. During oxidation of the cycloadduct generated from **438** and **439** to the corresponding bis-quinone, the transannular cycloaddition occurred, leading directly to longithorone A (**429**).

Guanacastepene A (**444**) is a novel tricyclic diterpene with fused five-, seven-, and six-membered rings. The possibility of constructing polycyclic compounds via *tandem RCM of dienynes* was used in Hanna's synthesis of a highly functionalized tricyclic system **443** related to **444**. Under the conditions outlined in Scheme 87, trienyne **440** provided the desired tricycle **442** in a single step, as a result of sequential enyne RCM followed by RCM of intermediate **441**. Compound **442** was then further functionalized to **443** [182].

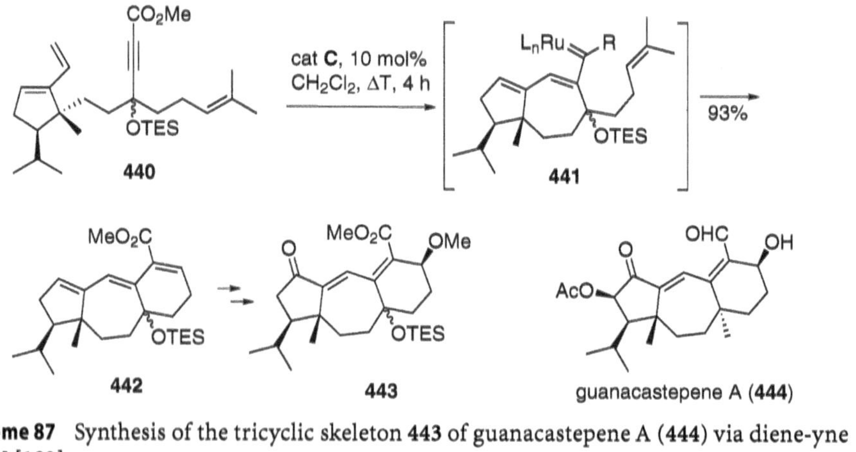

Scheme 87 Synthesis of the tricyclic skeleton **443** of guanacastepene A (**444**) via diene-yne RCM [182]

RCM of a dienyne was also a key step in Mori's recent total synthesis of the alkaloid erythrocarine (**447**) [183]. The tetracyclic framework of **447** was elaborated in the penultimate step, by exposing the hydrochloride of metathesis precursor **445** (1:1 diastereomeric mixture at the carbinol center) to first-generation catalyst **A**. The tandem process occurred smoothly within 18 h at room temperature leading to tetracycles **446** (1:1 mixture) in quantitative yield. Deprotection of the α-acetoxy isomer **446a** led to **447** (Scheme 88).

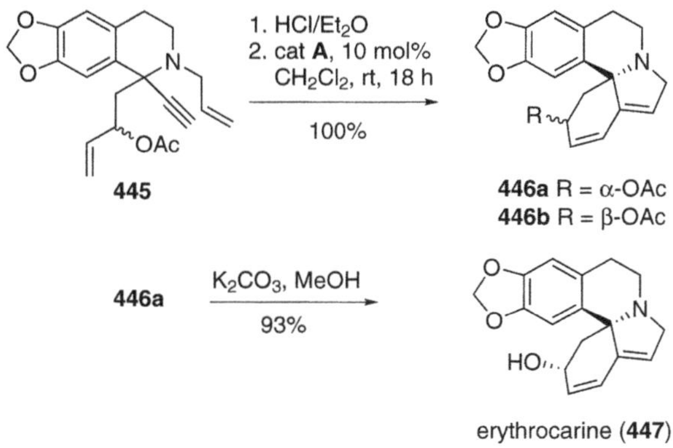

Scheme 88 Total synthesis of erythrocarine (**447**) via RCM of diene-yne **445** [183]

4
Ring-Closing Alkyne Metathesis (RCAM) and Alkyne Cross Metathesis (ACM)

An obvious drawback in RCM-based synthesis of unsaturated macrocyclic natural compounds is the lack of control over the newly formed double bond. The products formed are usually obtained as mixture of (*E*/*Z*)-isomers with the (*E*)-isomer dominating in most cases. The best solution for this problem might be a sequence of RCAM followed by (*E*)- or (*Z*)-selective partial reduction. Until now, alkyne metathesis has remained in the shadow of alkene-based metathesis reactions. One of the reasons may be the lack of commercially available catalysts for this type of reaction. When alkyne metathesis as a new synthetic tool was reviewed in early 1999 [184], there existed only a single report disclosed by Fürstner's laboratory [185] on the RCAM-based conversion of functionalized diynes to triple-bonded 12- to 28-membered macrocycles with the concomitant expulsion of 2-butyne (cf Fig. 3a). These reactions were catalyzed by Schrock's tungsten-carbyne complex **G**. Since then, Fürstner and coworkers have achieved a series of natural product syntheses, which seem to establish RCAM followed by partial reduction to (*Z*)- or (*E*)-cycloalkenes as a useful macrocyclization alternative to RCM. As work up to early 2000, including the development of alternative alkyne metathesis catalysts, is competently covered in Fürstner's excellent review [2a], we will concentrate here only on the most recent natural product syntheses, which were all achieved by Fürstner's team.

RCAM of diyne **448** catalyzed by the molybdenum-based system **I** followed by Lindlar reduction of the resulting cycloalkyne was the key step in the first total synthesis of the complex glycoconjugate and 26-membered macrolide sophorolipid lactone (**449**) [186] (Scheme 89), that together with the corresponding seco acid constitutes the major component of extracellular biosurfactants produced by the yeast *candida bombicola*. Applying the not rigorously defined catalyst system **I** (prepared in situ from Mo[N(*t*-Bu)(Ar)]$_3$ (**1**, R= 3,5-dimethylphenyl) and CH_2Cl_2 in toluene at 80 °C) to diyne **448** led smoothly to the desired macrocycle in 78% yield. Neither the PMB ethers nor the glycosidic linkages were damaged by the Lewis acidic metal center of the catalyst. Notably, RCM of a (differently protected) terminal diene mediated by various ruthenium catalysts of the first generation previously led to a mixture (*E*:*Z*≈3:1) of isomers.

In a subsequent report [187b], the high functional group tolerance of catalyst system **I** in RCAM reactions was demonstrated by the formation of an impressive number of nonnatural cycloalkynes with ring sizes varying from 12-membered to very large systems, and it was also shown that double bonds (isolated and conjugated) present in the cyclization substrate remained intact. Limits were encountered only with substrates containing acidic protons, including the protons of secondary amides. But it was also remembered that compounds of the basic type Mo[N(*t*-Bu)(Ar)]$_3$ are extremely reactive, being able to activate even molecular nitrogen at or below room temperature [20]. Therefore N_2 must not be used as a protecting atmosphere for any reactions involving these reagents. Application of the above catalyst system **I** culminated

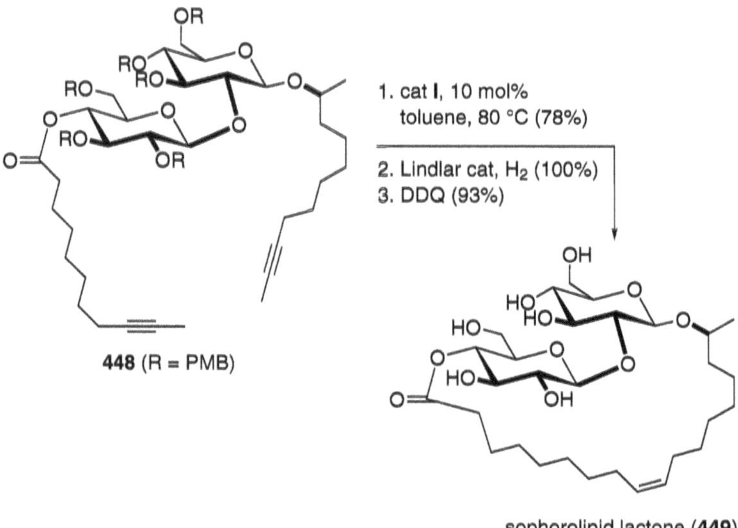

Scheme 89 Total synthesis of sophorolipid lactone **449** by sequential RCAM and (*Z*)-selective partial hydrogenation [186]

in a novel macrocyclization strategy for epoC (**237c**) by RCAM of the polyfunctional diyne **450** (Scheme 90) [187]. When **450** was exposed to catalyst system **I** (10 mol%, toluene, 80 °C, 8 h) the ring closure proceeded smoothly leading to cycloalkyne **451** in 80% yield. Neither the basic nitrogen nor the sulfur of the thiazole moiety interfered with the catalyst. No racemization at the chiral center *α* to the carbonyl group was encountered, and the protecting silyl ethers as well as the double bonds remained intact. The total synthesis of epoC (**237c**) was easily completed by Lindlar reduction of **451** followed by deprotection.

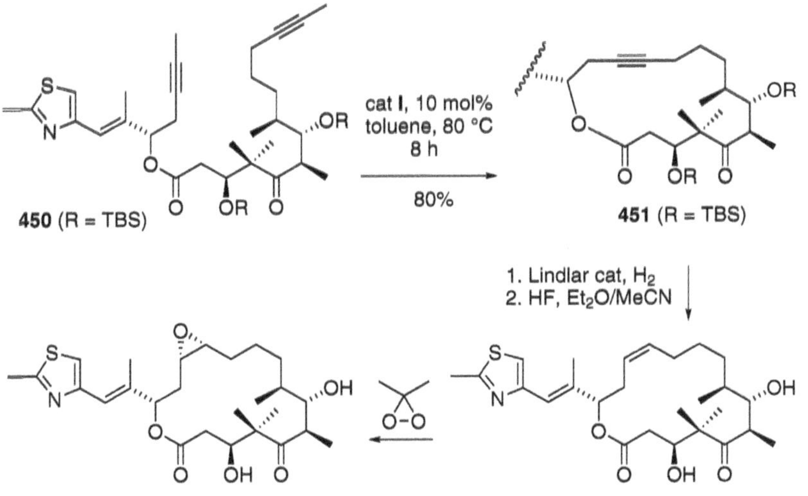

450 (R = TBS)

cat **I**, 10 mol%
toluene, 80 °C
8 h
80%

451 (R = TBS)

1. Lindlar cat, H_2
2. HF, Et_2O/MeCN

epoA (**238a**)

epoC (**237c**)

Scheme 90 Fürstner's total synthesis of epoC (**237c**) via sequential RCAM of diyne **450** and (*Z*)-selective partial hydrogenation [187]

The "user-friendly" catalyst **H**, prepared in situ from $Mo(CO)_6$ and *p*-trifluoromethylphenol, and also the well-defined tungsten complex **G** were used in the first total syntheses of the naturally occurring cyclophane derivatives **454** and **455** belonging to the turriane family [188] (Scheme 91). Exposing metathesis substrates **452** and **457** to tungsten complex **G** (10 mol%, toluene 80 °C, 16 h) led to cyclization products **453** and **456** in 64 and 61% yield, respectively. By the use of the less reactive, but more conveniently available, catalyst system **H** (10 mol%, chlorobenzene, 135 °C, 4 and 6 h, respectively), the yields were increased to 83 and 76%, and when the latter reactions were assisted by microwave heating, the RCAM proceeded within 5 min, leading to **453** and **456** in 69 and 71% yield, respectively.

A sequence of RCAM followed by transannular cycloaromatization in Fürstner's total synthesis of the natural 11-membered macrolide (+)-citreofuran (**461**) nicely demonstrates that RCAM has a broader scope than just the prepa-

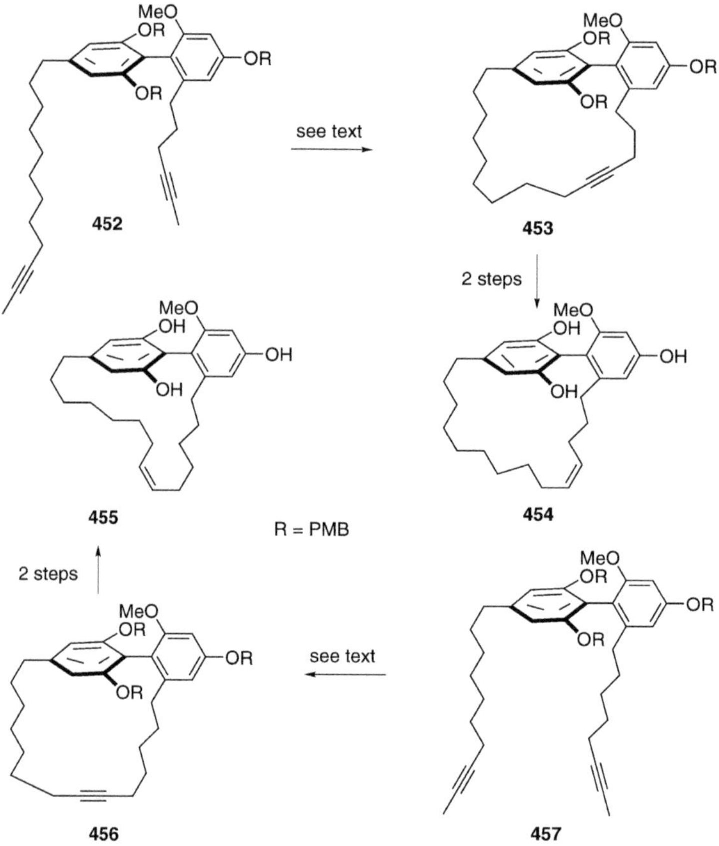

Scheme 91 RCAM-based synthesis of turrianes **454** and **455** by Fürstner and coworkers [188]

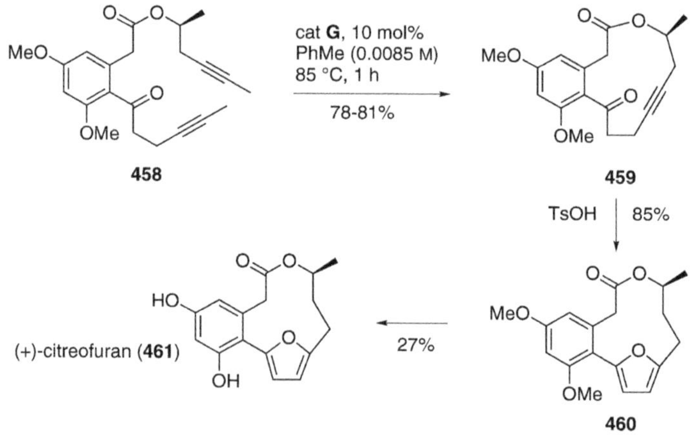

Scheme 92 Total synthesis of citreofuran (**461**) via RCAM of diyne **458** and subsequent transannular cycloaromatization [189]

ration of stereo-defined olefins [189] (Scheme 92). Exposure of metathesis precursor **458** to Schrock's tungsten complex **G** (10 mol%, toluene, 85 °C, 1 h) led to cyclization product **459** in 78% yield, provided the reaction mixture was devoid of any trace impurities, which underlines the high sensitivity of this catalyst. Subsequent treatment of **459** with TsOH smoothly generated the furan ring in **460** and completed the skeleton of the natural compound. The final removal of the methyl ethers from **460**, however, was very sluggish leading to **461** only in unsatisfactory yield.

A particularly flexible and novel entry into prostaglandins and analogs, either by RCAM (**463**→**464**) or by the intermolecular variant ACM (**462**+**466** →**467**) from a common intermediate (**462**), is outlined in Scheme 93 [190]. Prostaglandin E_2-1,15-lactone (**465**), an ichthyotoxic compound produced by a marine nudibranch for defense purposes, was produced in Fürstner's laboratory along the RCAM-based sequence **462**→**463**→**464**→**465**. Alternatively, the

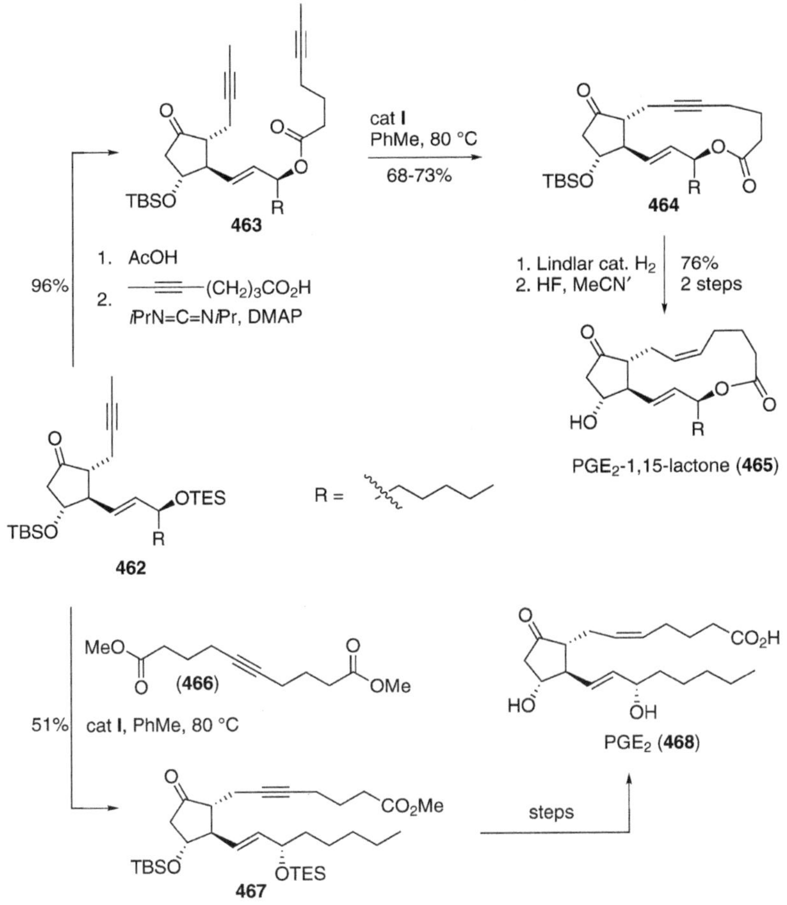

Scheme 93 Prostaglandin synthesis based on RCAM or ACM [190]

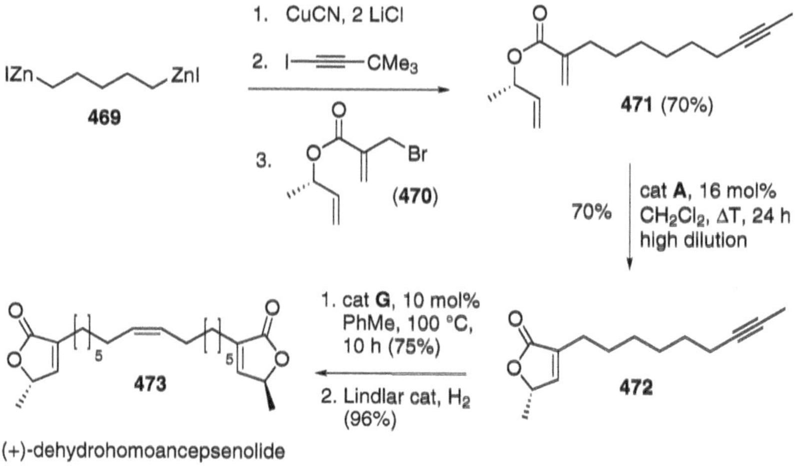

Scheme 94 Total synthesis of the natural compound dehydrohomoancepsenolide (**473**) through sequential application of chemoselective ruthenium-catalyzed RCM and tungsten-catalyzed alkyne homodimerization [191]

parent prostaglandin **468** was prepared via the *inter*molecular ACM mode, by exposing alkynes **462** and **466** to the same catalyst system (**I**). The conversion **462→467→468**, which in the metathesis step proceeded without homodimerization of key fragment **462**, represents the first example of an ACM-based natural product synthesis.

An elegant combination of sequential ruthenium-catalyzed RCM and tungsten-catalyzed homodimerization of an alkyne (both types of metathesis reactions being totally selective with respect to the π-systems involved) is found in Fürstner's total synthesis of the marine metabolite dehydrohomoancepsenolide (**473**) [191] (Scheme 94). Copper-mediated "three-component coupling" of the bimetallic species **469** with 1-iodo-1-propine and the chiral methacrylate **470** led to the precursor **471** for the RCM reaction. The ring closure of dienyne **471** to butenolide **472** proceeded readily and chemoselectively in 70% yield when catalyst **A** was used and when the reaction was performed under high dilution. The more powerful NHC-bearing second-generation catalysts, however, turned out to be too reactive in this case as they did not distinguish between the alkyne and the alkene moieties in the metathesis substrate **471**. The dimerization of alkyne **472** with tungsten complex **G** (10 mol%, toluene, 100 °C, 10 h) provided with selective involvement of the triple bond the C2-symmetric alkyne that was partially hydrogenated to furnish **473**.

5 Conclusions and Outlook

The last few years have witnessed an exponential growth in the application of ruthenium-catalyzed metathesis reactions in target-oriented synthesis. The development of highly active metathesis catalysts, that are commercially available and combine high functional group tolerance with "user-friendly" low sensitivity to moisture and air, has rendered metathesis a mature tool for the rapid construction of small-, medium-, and large-ring carbo- and heterocycles. Consequently, the logic of modern retrosynthetic planning is strongly affected by metathesis, since this transformation can now be applied to increasingly complex targets, as exemplified by metathesis-based total syntheses of polyether marine toxins, as well as by regio- and stereoselective macrocyclizations of diene-enes in the epothilone series.

Olefin cross metathesis starts to compete with traditional C=C bond-forming reactions such as the Wittig reaction and its modifications, as illustrated by the increasing use of electron-deficient conjugated alkenes for the (*E*)-selective construction of enals and enoates.

The use of metathesis cascades applied in various ring-rearrangement reactions allowed for a uniquely short access to various heterocyclic natural compounds, while diene-yne metathesis led to the formation of complex polycyclic structures. Also, tandem sequences combining a metathesis event with other reactions in the current synthetic repertoire, such as [3.3]-sigmatropic rearrangement, Pd-catalyzed alkene coupling, or Diels–Alder reaction, have been used as key steps in total syntheses of highly complex natural products. Particularly attractive tandem processes occur when two or more sequential reactions are mediated by the same catalytic precursor. The ability of ruthenium alkylidenes to function directly, or by simple modifications also as precatalysts for nonmetathetic processes (radical additions, olefin and carbonyl hydrogenations, hydrogen transfer reactions, olefin isomerizations) [192], broadens their synthetic utility toward efficient catalytic tandem sequences that combine metathesis events with one or more nonmetathesis reactions. To date, this strategy has led to highly efficient syntheses of relatively simple natural products [122, 193] and will certainly be utilized for more complex targets in future work.

Thus far, chemists have been able to influence the stereoselectivity of macrocyclic RCM through steric and electronic substrate features or by the choice of a catalyst with appropriate activity, but there still exists a lack of prediction over the stereochemistry of macrocyclic RCM. One of the most important extensions of the original metathesis reaction for the synthesis of stereochemically defined (cyclo)alkenes is alkyne metathesis, followed by selective partial hydrogenation.

An area in which catalytic olefin metathesis could have a significant impact on future natural product-directed work would be the desymmetrization of achiral molecules through asymmetric RCM (ARCM) or asymmetric ROM

(AROM)-RCM- and -CM sequences initiated by chiral molybdenum-based catalysts [194] or, more recently, also by ruthenium-based [195] catalysts.

Ongoing research efforts will lead to the arrival of even more efficient and selective metathesis catalysts with specifically tailored properties [196]. Due to the synergistic relationship between catalyst design and subsequent application in advanced synthesis [197], this progress will further expand the scope of metathesis and its popularity amongst the synthetic community.

References

1. (a) Armstrong SK (1998) J Chem Soc Perkin Trans I 371; (b) Grubbs RH, Chang S (1998) Tetrahedron 54:4413; (c) Ivin KJ (1998) J Mol Catal A 133:1; (d) Randall ML, Snapper ML (1998) J Mol Catal A 133:29; (e) Wright DL (1999) Curr Org Chem 3:211
2. (a) Fürstner A (2000) Angew Chem Int Ed 39:3012; (b) Trnka TM, Grubbs RH (2001) Acc Chem Res 34:18; (c) Schrock RR, Hoveyda AH (2003) Angew Chem Int Ed 42: 4592
3. Storm Poulsen C, Madsen R (2003) Synthesis 1
4. Connon SJ, Blechert S (2003) Angew Chem Int Ed 42:1900
5. Lindel T (2003) Alkyne metathesis in natural product synthesis. In: Schmalz HG, Wirth T (eds) Organic synthesis highlights, vol V. Wiley-VCH, Weinheim, p 27
6. (a) Roy R, Das SK (2000) Chem Commun 519; (b) Jørgensen M, Hadwiger P, Madsen R, Stütz AE, Wrodnigg TM (2000) Curr Org Chem 4:565
7. Maier ME (2000) Angew Chem Int Ed 39:2073
8. Phillips AJ, Abell AD (1999) Aldrichim Acta 32:75
9. Prunet J (2003) Angew Chem Int Ed 42:2826
10. Nicolaou KC, King NP, He Y (1998) Top Organomet Chem 1:73
11. Mulzer J, Öhler E, Enev V, Hanbauer M (2002) Adv Synth Catal 344:573
12. Hansen EC, Lee D (2003) J Am Chem Soc 125:9582
13. Schwab B, Grubbs RH, Ziller JW (1996) J Am Chem Soc 118:100
14. Schrock RR, Murdzek JS, Bazan GZ, Robins J, DiMare M, O'Regan M (1990) J Am Chem Soc 112:3875
15. Huang J, Stevens ED, Nolan SP, Peterson JL (1999) J Am Chem Soc 121:2674
16. Scholl M, Trnka TM, Morgan JP, Grubbs RH (1999) Tetrahedron Lett 40:2247
17. Ackermann L, Fürstner A, Weskamp T, Kohl FJ, Herrmann WA (1999) Tetrahedron Lett 40:4787
18. (a) Garber SB, Kingsbury JS, Gray BL, Hoveyda AH (2000) J Am Chem Soc 122:8168; (b) Gessler S, Randl S, Blechert S (2000) Tetrahedron Lett 41:9973
19. For details, see: Ref. [2]
20. For a review on the preparation and stoichiometric reactions of complex **1** with small molecules, see: Cummins CC (1998) Chem Commun 1777
21. Sunazuka T, Hirose T, Shirahata T, Harigaya Y, Hayashi M, Komiyama K, Omura S, Smith AB III (2000) J Am Chem Soc 122:2122
22. Trost BM, Jiang C (2003) Org Lett 5:1563
23. Chavez DE, Jacobsen EN (2003) Org Lett 5:2563
24. Inoue M, Sato T, Hirama M (2003) J Am Chem Soc 125:10772
25. Bernet B, Vasella A (1979) Helv Chim Acta 62:1990
26. For leading references, and the application of this method to ring-closing enyne metathesis, see: Storm Poulsen C, Madsen R (2002) J Org Chem 67:4441

27. Boyer F-D, Hanna I (2001) Tetrahedron Lett 42:1275
28. Skaanderup PR, Madsen R (2001) Chem Commun 1106
29. Ramana GV, Rao BV (2003) Tetrahedron Lett 44:5103
30. Hale KJ, Domostoj MM, Tocher DA, Irving E, Scheinmann F (2003) Org Lett 5:2927
31. (a) Benningshof JCJ, Blaauw RH, van Ginkel AE, Rutjes FPJT, Fraanje J, Goubitz K, Schenk H, Hiemstra H (2000) Chem Commun 1465; (b) Benningshof JCJ, Ijsselstijn M, Wallner SR, Koster AL, Blaauw RH, van Ginkel AE, Brière J-F, van Maarseveen JH, Rutjes FPJT, Hiemstra H (2002) J Chem Soc Perkin Trans I 1701
32. Held C, Fröhlich R, Metz P (2002) Adv Synth Catal 344:720
33. Fürstner A, Langemann K (1997) J Am Chem Soc 119:9130
34. For the previous preparation of a mixed acrolein acetal and its use in an RCM reaction during callystatin A total synthesis, see: Crimmins MT, King BW (1998) J Am Chem Soc 120:9084
35. Mizutani H, Watanabe M, Honda T (2002) Tetrahedron 58:8929
36. Esumi T, Okamoto N, Hatakeyama S (2002) Chem Commun 3042
37. Cossy J, Pradaux F, BouzBouz S (2001) Org Lett 3:2233
38. (a) Banwell MG, Coster MJ, Edwards AJ, Karunaratne OP, Smith JA, Welling LL, Willis AC (2003) Aust J Chem 56:585; (b) Ramachandran PV, Chandra JS, Reddy MVR (2002) J Org Chem 67:7547
39. Murga J, Falomir E, Carda M, Marco JA (2002) Tetrahedron Asymmetry 13:2317
40. BouzBouz S, Cossy J (2003) Tetrahedron Lett 44:4471
41. Murga J, Garcia-Fortanet J, Carda M, Marco JA (2003) Tetrahedron Lett 44:7909
42. Bhattacharjee A, Soltani O, De Brabander JK (2002) Org Lett 4:481
43. Ghosh AK, Lei H (2000) J Org Chem 65:4779
44. Ghosh AK, Kim J-H (2003) Tetrahedron Lett 44:3967
45. Cossy J, Bauer D, Bellosta V (2002) Tetrahedron 58:5909
46. BouzBouz S, Cossy J (2003) Org Lett 5:3029
47. Chen J, Forsyth CJ (2003) J Am Chem Soc 125:8734
48. Kirkland TA, Colucci J, Geraci LS, Marx MA, Schneider M, Kaelin DE Jr, Martin SF (2001) J Am Chem Soc 123:12432
49. Lee E, Choi SJ, Kim H, Han HO, Kim YK, Min SJ, Son SH, Lim SM, Jang WS (2002) Angew Chem Int Ed 41:176
50. Paquette LA, Schloss JD, Efremov I, Fabris F, Gallou F, Méndez-Andino J, Yang J (2000) Org Lett 2:1259. For other methods to remove ruthenium by-products generated during metathesis reactions, see: (a) Maynard HD, Grubbs RH (1999) Tetrahedron Lett 40:4137; (b) Ahn YM, Yang KL, Georg GI (2001) Org Lett 3:1411; (c) Cho JH, Kim BM (2003) Org Lett 5:531
51. The complete laulimalide-directed work has been reviewed: Mulzer J, Öhler E (2003) Chem Rev 103:3753
52. For a review, see Ref. [11]
53. (a) Ahmed A, Öhler E, Mulzer J (2001) Synthesis 2007; (b) Mulzer J, Öhler E (2001) Angew Chem Int Ed 40:3842
54. Williams DR, Mi L, Mullins RJ, Stites RE (2002) Tetrahedron Lett 43:4841
55. Crimmins MT, Stanton MG, Allwein SP (2002) J Am Chem Soc 124:5958
56. Pitts MR, Mulzer J (2002) Tetrahedron Lett 43:8471
57. Nelson SG, Cheung WS, Kassick AJ, Hilfiker MA (2002) J Am Chem Soc 124:13654
58. Williams DR, Heidebrecht RW Jr (2003) J Am Chem Soc 125:1843
59. For related methodology in laulimalide synthesis, see Ref. [11]
60. Cheung AK, Snapper ML (2002) J Am Chem Soc 124:11584
61. Zakarian A, Batch A, Holton RA (2003) J Am Chem Soc 125:7822

62. (a) Kadota I, Takamura H, Sato K, Ohno A, Matsuda K, Yamamoto Y (2003) J Am Chem Soc 125:46; (b) Kadota I, Takamura H, Sato K, Ohno A, Matsuda K, Satake M, Yamamoto Y (2003) J Am Chem Soc 125:11893; (c) For the synthesis of the A–F ring segment of yessotoxin and adriatoxin by analogous methodology, see: Kadota I, Ueno H, Yamamoto Y (2003) Tetrahedron Lett 44:8935
63. Kadota I, Ohno A, Matsuda K, Yamamoto Y (2002) J Am Chem Soc 124:3562
64. Denmark SE, Yang S-M (2002) J Am Chem Soc 124:2102
65. Denmark SE, Yang S-M (2002) J Am Chem Soc 124:15196
66. Hanessian S, Margarita R, Hall A, Johnstone S, Tremblay M, Parlanti L (2002) J Am Chem Soc 124:13342
67. Kim S, Lee J, Lee T, Park H-G, Kim D (2003) Org Lett 5:2703
68. Deiters A, Martin SF (2002) Org Lett 4:3243
69. For additional examples, see Ref. [68]
70. Washburn DG, Heidebrecht RW Jr, Martin SF (2003) Org Lett 5:3523
71. Wipf P, Rector SR, Takahashi H (2002) J Am Chem Soc 124:14848
72. Hu Y-J, Dominique R, Das KS, Roy R (2000) Can J Chem 78:838
73. For a review, see Ref. [7]
74. For a collection of references up to 2002, see: Fürstner A, Radkowski K, Wirtz C, Goddard R, Lehmann CW, Mynott R (2002) J Am Chem Soc 124:7061
75. Paquette LA, Tae J, Arrington MP, Sadoun AH (2000) J Am Chem Soc 122:2742
76. (a) Efremov I, Paquette LA (2000) J Am Chem Soc 122:9324; (b) Paquette LA, Efremov I (2001) J Am Chem Soc 123:4492
77. Clark JS, Marlin F, Nay B, Wilson C (2003) Org Lett 5:89
78. (a) Martin SF, Humphrey JM, Ali A, Hillier MC (1999) J Am Chem Soc 121:866; (b) Humphrey JM, Liao Y, Ali A, Rein T, Wong Y-L, Chen H-J, Courtney AK, Martin SF (2002) J Am Chem Soc 124:8584
79. Fu GC, Nguyen ST, Grubbs RH (1993) J Am Chem Soc 115:9856
80. Nagata T, Nakagawa M, Nishida A (2003) J Am Chem Soc 125:7484
81. Buszek KR, Sato N, Jeong Y (2002) Tetrahedron Lett 43:181
82. Baba Y, Saha G, Nakao S, Iwata C, Tanaka T, Ibuka T, Ohishi H, Takemoto Y (2001) J Org Chem 66:81
83. Fürstner A, Schlede M (2002) Adv Synth Catal 344:657
84. (a) Fürstner A, Radkowski K (2001) Chem Commun 671; (b) Fürstner A, Radkowski K, Wirtz C, Goddard R, Lehmann CW, Mynott R (2002) J Am Chem Soc 124:7061
85. (a) Crimmins MT, Choy AL (1999) J Am Chem Soc 121:5653; (b) Crimmins MT, Emmitte KA (1999) Org Lett 1:2029
86. Crimmins MT, Tabet EA (2000) J Am Chem Soc 122:5473
87. (a) Crimmins MT, Emmitte KA (2001) J Am Chem Soc 123:1533; (b) Crimmins MT, Emmitte KA, Choy AL (2002) Tetrahedron 58:1817
88. Crimmins MT, Powell MT (2003) J Am Chem Soc 125:7592
89. Crimmins MT, DeBaillie AC (2003) Org Lett 5:3009
90. (a) Hirama M, Oishi T, Uehara H, Inoue M, Maruyama M, Oguri H, Satake M (2001) Science 294:1904; (b) Inoue M, Uehara H, Maruyama M, Hirama M (2002) Org Lett 4:4551. For the use of RCM in the synthesis of ciguatoxin fragments, see: (c) Maruyama M, Inoue M, Oishi T, Oguri H, Ogasawara Y, Shindo Y, Hirama M (2002) Tetrahedron 58:1835
91. Kadota I, Nishina N, Nishii H, Kikuchi S, Yamamoto Y (2003) Tetrahedron Lett 44:7929
92. For a review, see Ref. [10]
93. (a) Nishioka T, Iwabuchi Y, Irie H, Hatakeyama S (1998) Tetrahedron Lett 39:5597; (b) Banwell MG, McRae KJ (2000) Org Lett 2:3583; (c) Dixon DJ, Foster AC, Ley SV (2000) Org Lett 2:123

94. Gaul C, Danishefsky SJ (2002) Tetrahedron Lett 43:9039
95. Gaul C, Njardarson JT, Danishefsky SJ (2003) J Am Chem Soc 125:6042
96. (a) Maleczka RE Jr, Terrell LR, Geng F, Ward JS III (2002) Org Lett 4:2841; (b) Trost BM, Chisholm JD, Wrobleski ST, Jung M (2002) J Am Chem Soc 124:12420; (c) Lam HW, Pattenden G (2002) Angew Chem Int Ed 41:508
97. Hoye TR, Zhao H (1999) Org Lett 1:1123
98. (a) Fürstner A, Aïssa C, Riveiros R, Ragot J (2002) Angew Chem Int Ed 41:4763; (b) Aïssa C, Riveiros R, Ragot J, Fürstner A (2003) J Am Chem Soc 125:15512
99. Hoye TR, Hu M (2003) J Am Chem Soc 125:9576
100. Matsuya Y, Kawaguchi T, Nemoto H (2003) Org Lett 5:2939
101. Bouazza F, Renoux B, Bachmann C, Gesson J-P (2003) Org Lett 5:4049
102. The absolute configuration of **215a**, initially assigned through Mosher ester experiments, was later revised by De Brabander: (a) Wu Y, Esser L, De Brabander JK (2000) Angew Chem Int Ed 39:4308; (b) Wu Y, Seguil OR, De Brabander JK (2000) Org Lett 2:4241
103. (a) Snider BB, Song F (2001) Org Lett 3:1817; (b) Labrecque D, Charron S, Rej R, Blais C, Lamothe S (2001) Tetrahedron Lett 42:2645; (c) Fürstner A, Dierkes T, Thiel OR, Blanda G (2001) Chem Eur J 7:5286; (d) Smith AB III, Zheng J (2002) Tetrahedron 58:6455; (e) Wu Y, Liao X, Wang R, Xie X-S, De Brabander JK (2002) J Am Chem Soc 124:3245; (f) Yang KL, Haack T, Blackman B, Diederich WE, Roy S, Pusuluri S, Georg GI (2003) Org Lett 5:4007; (g) For a series of additional RCM substrates, see: Yang KL, Blackman B, Diederich W, Flaherty PT, Mossman CJ, Roy S, Ahn YM, Georg GI (2003) J Org Chem ASAP; (h) For a review on the chemistry and biology of salicylihalamide A and related compounds, see: Yet L (2003) Chem Rev 103:4283
104. Fürstner A, Jeanjean F, Razon P, Wirtz C, Mynott R (2003) Chem Eur J 9:320
105. For a systematic study, see: Paquette LA, Basu K, Eppich JC, Hofferberth JE (2002) Helv Chim Acta 85:3033
106. (a) Cabrejas LMM, Rohrbach S, Wagner D, Kallen J, Zenke G, Wagner J (1999) Angew Chem Int Ed 38:2443; (b) Wagner J, Cabrejas LMM, Grossmith CE, Papageorgiou C, Senia F, Wagner D, France J, Nolan SP (2000) J Org Chem 65:9255; (c) Sedrani R, Kallen J, Cabrejas LMM, Papageorgiou CD, Senia F, Rohrbach S, Wagner D, Thai B, Eme A-MJ, France J, Oberer L, Rihs G, Zenke G, Wagner J (2003) J Am Chem Soc 125:3849
107. Dvorak CA, Schmitz WD, Poon DJ, Pryde DC, Lawson JP, Amos RA, Meyers AI (2000) Angew Chem Int Ed 39:1664
108. Garbaccio RM, Stachel SJ, Baeschlin DK, Danishefsky SJ (2001) J Am Chem Soc 123:10903
109. Yamamoto K, Biswas K, Gaul C, Danishefsky SJ (2003) Tetrahedron Lett 44:3297
110. Yang ZQ, Danishefsky SJ (2003) J Am Chem Soc 125:9602
111. For reviews, see: (a) Mulzer J (2000) Monatsh Chem 131:205; (b) Nicolaou KC, Ritzén A, Namoto K (2001) Chem Commun1523; (c) For more recent leading references on epothilone analogs, see: Nicolaou KC, Sasmal PK, Rassias G, Reddy MV, Altmann K-H, Wartmann M, O'Brate A, Giannakakou P (2003) Angew Chem Int Ed 42:3515
112. For a review on RCM in early epothilone syntheses, see Ref. [10]
113. Biswas K, Lin H, Njardarson JT, Chappell MD, Chou T-C, Guan Y, Tong WP, He L, Horwitz SB, Danishefsky SJ (2002) J Am Chem Soc 124:9825
114. For a high-yielding ring contraction in the series of macrolide antibiotics containing a 1,3-diene unit, see Scheme 78
115. For the synthesis of ring-enlarged epothilone analogs by the same principle, see: (a) Rivkin A, Njardarson JT, Biswas K, Chou T-C, Danishefsky SJ (2002) J Org Chem 67:7737; (b) Rivkin A, Biswas K, Chou T-C, Danishefsky SJ (2002) Org Lett 4:4081
116. Meng D, Bertinato B, Balog A, Su D-S, Kamenecka T, Sorensen EJ, Danishefsky SJ (1997) J Am Chem Soc 119:10073

117. Sun J, Sinha SC (2002) Angew Chem Int Ed 41:1381
118. In the Supporting Information provided by Sinha et al., the formation of an unalike *E/Z* mixture of cycloheptadienes arising from **244d** was presented
119. (a) Rivkin A, Yoshimura F, Gabarda AE, Chou T-C, Dong H, Tong WP, Danishefsky SJ (2003) J Am Chem Soc 125:2899; (b) Chou T-C, Dong H, Rivkin A, Yoshimura F, Gabarda AE, Cho YS, Tong WP, Danishefsky SJ (2003) Angew Chem Int Ed 42:4762
120. White JD, Carter RG, Sundermann KF, Wartmann M (2001) J Am Chem Soc 123:5407; corrigendum (2003) J Am Chem Soc 125:3190
121. Wang X, Porco JA Jr (2003) J Am Chem Soc 125:6040
122. Fürstner A, Leitner A (2003) Angew Chem Int Ed 42:308
123. For the synthesis and a collection of previous applications of catalyst **F** in natural product syntheses, see Ref. [122]
124. (a) Toró A, Deslongchamps P (2003) J Org Chem 68:6847; (b) Recently, the total synthesis of **272** has been achieved via the alternative biomimetic approach, by TADA of an in situ generated macrocyclic pyranophane pseudobase: Soucy P, L'Heureux A, Toró A, Deslongchamps P (2003) J Org Chem 68:9983
125. (a) Harrington PE, Tius MA (1999) Org Lett 1:649; (b) Harrington PE, Tius MA (2001) J Am Chem Soc 123:8509
126. For an excellent review on the chemistry and biology of **273** and the related prodigiosin alkaloids, see: Fürstner A (2003) Angew Chem Int Ed 42:3582
127. Kim SH, Figueroa I, Fuchs PL (1997) Tetrahedron Lett 38:2601
128. Bamford SJ, Luker T, Speckamp WN, Hiemstra H (2000) Org Lett 2:1157
129. Fürstner A, Gartner T, Weintritt H (1999) J Org Chem 64:2361
130. Boger DL, Hong J (2001) J Am Chem Soc 123:8515
131. (a) Irie O, Samizu K, Henry JR, Weinreb SM (1999) J Org Chem 64:587; (b) Sung MJ, Lee HI, Lee HB, Cha JK (2003) J Org Chem 68:2205
132. Nicolaou KC, Vassilikogiannakis G, Montagnon T (2002) Angew Chem Int Ed 41:3276
133. For a recent review, see Ref. [4]
134. Chatterjee AK, Grubbs RH (2002) Angew Chem Int Ed 41:3172
135. Chatterjee AK, Choi T-L, Sanders DP, Grubbs RH (2003) J Am Chem Soc 125:11360
136. Böhm C, Reiser O (2001) Org Lett 3:1315
137. Spessard SJ, Stoltz BM (2002) Org Lett 4:1943
138. Pederson RL, Fellows IM, Ung TA, Ishihara H, Hajela SP (2002) Adv Synth Catal 344:728
139. (a) Cossy J, Willis C, Bellosta V, BouzBouz S (2002) J Org Chem 67:1982; (b) BouzBouz S, Cossy J (2003) Org Lett 5:1995
140. BouzBouz S, Cossy J (2001) Org Lett 3:1451
141. For a review, see: Cossy J, BouzBouz S, Pradaux F, Willis C, Bellosta V (2002) Synlett 1595
142. Smith CM, O'Doherty GA (2003) Org Lett 5:1959
143. Ghosh AK, Liu C (2003) J Am Chem Soc 125:2374
144. For a discussion on the beneficial effect of using homodimers of one CM substrate in cross metathesis reactions, see: Blackwell HE, O'Leary DJ, Chatterjee AK, Washenfelder RA, Bussmann DA, Grubbs RH (2000) J Am Chem Soc 122:58
145. Wang Y, Romo D (2002) Org Lett 4:3231
146. Njardarson JT, Biswas K, Danishefsky SJ (2002) Chem Commun 2759
147. Smulik JA, Diver ST, Pan F, Liu JO (2002) Org Lett 4:2051
148. Lazarova T, Chen JS, Hamann B, Kang JM, Homuth-Trombino D, Han F, Hoffmann E, McClure C, Eckstein J, Or YS (2003) J Med Chem 46:674
149. Ratnayake AS, Hemscheidt T (2002) Org Lett 4:4667
150. Tanaka K, Nakanishi K, Berova N (2003) J Am Chem Soc 125:10802

151. (a) Smith AB III, Kozmin SA, Adams CM, Paone DV (2000) J Am Chem Soc 122:4984; (b) Smith AB III, Adams CM, Kozmin SA (2001) J Am Chem Soc 123:990; (c) Smith AB III, Adams CM, Kozmin SA, Paone DV (2001) J Am Chem Soc 123:5925
152. (a) Nicolaou KC, Winssinger N, Pastor J, Ninkovic S, Sarabia F, He Y, Vourloumis D, Yang Z, Li T, Giannakakou P, Hamel E (1997) Nature 387:268; (b) Nicolaou KC, Vourloumis D, Li T, Pastor J, Winssinger N, He Y, Ninkovic S, Sarabia F, Vallberg H, Roschangar F, King NP, Finlay MRV, Giannakakou P, Verdier-Pinard P, Hamel E (1997) Angew Chem Int Ed 36:2097
153. (a) Brohm D, Metzger S, Bhargava A, Müller O, Lieb F, Waldmann H (2002) Angew Chem Int Ed 41:307; (b) Brohm D, Philippe N, Metzger S, Bhargava A, Müller O, Lieb F, Waldmann H (2002) J Am Chem Soc 124:13171
154. Chang S, Na Y, Shin HJ, Choi E, Jeong LS (2002) Tetrahedron Lett 43:7445
155. Zuercher WJ, Hashimoto M, Grubbs RH (1996) J Am Chem Soc 118:6634
156. Stragies R, Blechert S (2000) J Am Chem Soc 122:9584
157. Buschmann N, Rückert A, Blechert S (2002) J Org Chem 67:4325
158. Burke SD, Quinn KJ, Chen VJ (1998) J Org Chem 63:8626
159. Schaudt M, Blechert S (2003) J Org Chem 68:2913
160. Stragies R, Blechert S (1999) Tetrahedron 55:8179
161. Stapper C, Blechert S (2002) J Org Chem 67:6456
162. Blechert S, Stapper C (2002) Eur J Org Chem 2855
163. Zaminer J, Stapper C, Blechert S (2002) Tetrahedron Lett 43:6739
164. Lazarova TI, Binet SM, Vo NH, Chen JS, Phan LT, Or YS (2003) Org Lett 5:443
165. For the formation of undesired ring-contracted (by)products during macrocyclic diene-ene RCM with second-generation catalysts, see Sect. 1.3.3
166. (a) Snapper ML, Tallarico JA, Randall ML (1997) J Am Chem Soc 119:1478; (b) For an extension of this methodology, see: White BH, Snapper ML (2003) J Am Chem Soc 125:14901
167. Limanto J, Snapper ML (2000) J Am Chem Soc 122:8071
168. Wender PA, Ihle NC, Correia CRD (1988) J Am Chem Soc 110:5904
169. For an RCM-based synthesis of asteriscanolide (**116**), see Scheme 22
170. Bassindale MJ, Hamley P, Harrity JPA (2001) Tetrahedron Lett 42:9055
171. Karama U, Höfle G (2003) Eur J Org Chem 1042
172. For reviews on enyne metathesis, see: (a) Mori M (1998) Top Organomet Chem 1:133; (b) Ref. [3]
173. For progress in selectivity by the use of second-generation Ru catalysts and/or the presence of ethylene, see: (a) Lee H-Y, Kim BG, Snapper ML (2003) Org Lett 5:1855, and references therein; (b) Giessert AJ, Brazis NJ, Diver ST (2003) Org Lett 5:3819
174. Mori M, Tonogaki K, Nishiguchi N (2002) J Org Chem 67:224
175. Tonogaki K, Mori M (2002) Tetrahedron Lett 43:2235, and references therein
176. For recent examples of small-ring enyne RCM following the *endo* mode, see: (a) Kitamura T, Sato Y, Mori M (2002) Adv Synth Catal 344:678; (b) Dolhem F, Lièvre C, Demailly G (2003) Eur J Org Chem 2336
177. Clark JS, Townsend RJ, Blake AJ, Teat SJ, Johns A (2001) Tetrahedron Lett 42:3235
178. For a total synthesis of ircinal A and related manzamine alkaloids via RCM, see Scheme 25
179. Clark JS, Elustondo F, Trevitt GP, Boyall D, Robertson J, Blake AJ, Wilson C, Stammen B (2002) Tetrahedron 58:1973
180. Layton ME, Morales CA, Shair MD (2002) J Am Chem Soc 124:773. For a more recent investigation of macrocyclic enyne metathesis, see Ref. [12]
181. Fu X, Hossain MB, van der Helm D, Schmitz FJ (1994) J Am Chem Soc 116:12125
182. Boyer F-D, Hanna I (2002) Tetrahedron Lett 43:7469

183. Shimizu K, Takimoto M, Mori M (2003) Org Lett 5:2323
184. Bunz UHF, Kloppenburg L (1999) Angew Chem Int Ed 38:478
185. Fürstner A, Seidel G (1998) Angew Chem Int Ed 37:1734
186. Fürstner A, Radkowski K, Grabowski J, Wirtz C, Mynott R (2000) J Org Chem 65:8758
187. (a) Fürstner A, Mathes C, Grela K (2001) Chem Commun 1057; (b) Fürstner A, Mathes C, Lehmann CW (2001) Chem Eur J 7:5299
188. Fürstner A, Stelzer F, Rumbo A, Krause H (2002) Chem Eur J 8:1856
189. Fürstner A, Castanet A-C, Radkowski K, Lehmann CW (2003) J Org Chem 68:1521
190. Fürstner A, Grela K, Mathes C, Lehmann CW (2000) J Am Chem Soc 122:11799
191. Fürstner A, Dierkes T (2000) Org Lett 2:2463
192. For recent reviews, see: (a) Alcaide B, Almendros P (2003) Chem Eur J 9:1258; (b) Schmidt B (2003) Angew Chem Int Ed 42:4996
193. Louie J, Bielawski CW, Grubbs RH (2001) J Am Chem Soc 123:11312
194. For a chiral molybdenum-based catalyst available in situ from commercial components, see: (a) Aeilts SL, Cefalo DR, Bonitatebus PJ, Houser JH, Hoveyda AH, Schrock RR (2001) Angew Chem Int Ed 40:1452; (b) For the first enantiomerically pure solid-supported Mo catalyst, see: Hultzsch KC, Jernelius JA, Hoveyda AH, Schrock RR (2002) Angew Chem Int Ed 41:589; (c) For a chiral Mo catalyst, allowing RCM to small- and medium-ring cyclic amines, see: Dolman SJ, Sattely ES, Hoveyda AH, Schrock RR (2002) J Am Chem Soc 124:6991; (d) For a novel adamantyl imido-molybdenum complex with advanced selectivity profiles, see: Tsang WCP, Jernelius JA, Cortez GA, Weatherhead GS, Schrock RR, Hoveyda AH (2003) J Am Chem Soc 125:2591
195. (a) Van Veldhuizen JJ, Garber SB, Kingsbury JS, Hoveyda AH (2002) J Am Chem Soc 124:4954; corrigendum (2003) J Am Chem Soc 125:12666; (b) Van Veldhuizen JJ, Gillingham DG, Garber SB, Kataoka O, Hoveyda AH (2003) J Am Chem Soc 125:12502, and references therein
196. (a) For a ruthenium catalyst with two pyridine ligands, see: Love JA, Morgan JP, Trnka TM, Grubbs RH (2002) Angew Chem Int Ed 41:4035; (b) For a phenyl-substituted analog of catalyst **D**, see: Wakamatsu H, Blechert S (2002) Angew Chem Int Ed 41:2403; (c) For a nitro-substituted analog of catalyst **D**, see: Grela K, Harutyunyan S, Michrowska A (2002) Angew Chem Int Ed 41:4038; (d) For a novel recyclable ROMP-based ruthenium catalyst, see: Connon SJ, Dunne AM, Blechert S (2002) Angew Chem Int Ed 41:3835; (e) For a novel ionic liquid-supported Ru carbene complex, see: Audic N, Clavier H, Mauduit M, Guillemin J-C (2003) J Am Chem Soc 125:9248; (f) For novel triaryl phosphine-based ruthenium catalysts with distinctly increased activity in RCM, see: Love JA, Sanford MS, Day MW, Grubbs RH (2003) J Am Chem Soc 125:10103; (g) For a novel in situ arene indenylidene ruthenium species, see: Castarlenas R, Dixneuf PH (2003) Angew Chem Int Ed 42:4524; (h) For a review dealing with the development of molybdenum- and tungsten-based metathesis catalysts, see: Ref. [2c]
197. For the use of a novel phosphine-free ruthenium catalyst with *m*-bromopyridine ligands [196a] in the CM-based release of azide-protected carbohydrates from a solid support, see: Kanemitsu T, Seeberger PH (2003) Org Lett 5:4541

Author Index Volumes 1–13

Volumes 9 and 12 are already in planning and are announced and will be published when the manuscripts are submitted to the publisher.
The volume numbers are printed in italics.

Subject Index

GPSR Compliance
The European Union's (EU) General Product Safety Regulation (GPSR) is a set of rules that requires consumer products to be safe and our obligations to ensure this.

If you have any concerns about our products, you can contact us on

ProductSafety@springernature.com

In case Publisher is established outside the EU, the EU authorized representative is:

Springer Nature Customer Service Center GmbH
Europaplatz 3
69115 Heidelberg, Germany

www.ingramcontent.com/pod-product-compliance
Ingram Content Group UK Ltd.
Pitfield, Milton Keynes, MK11 3LW, UK
UKHW021858190726
13853UKWH00003B/1317

* 9 7 8 3 6 6 2 1 4 5 4 7 0 *